Włodzisław Duch and Jacek Mańdziuk (Eds.)

Challenges for Computational Intelligence

Studies in Computational Intelligence, Volume 63

Editor-in-chief
Prof. Janusz Kacprzyk
Systems Research Institute
Polish Academy of Sciences
ul. Newelska 6
01-447 Warsaw
Poland
E-mail: kacprzyk@ibspan.waw.pl

Further volumes of this series
can be found on our homepage:
springer.com

Vol. 41. Mukesh Khare, S.M. Shiva Nagendra (Eds.)
Artificial Neural Networks in Vehicular Pollution Modelling, 2007
ISBN 978-3-540-37417-6

Vol. 42. Bernd J. Krämer, Wolfgang A. Halang (Eds.)
Contributions to Ubiquitous Computing, 2007
ISBN 978-3-540-44909-6

Vol. 43. Fabrice Guillet, Howard J. Hamilton (Eds.)
Quality Measures in Data Mining, 2007
ISBN 978-3-540-44911-9

Vol. 44. Nadia Nedjah, Luiza de Macedo Mourelle, Mario Neto Borges, Nival Nunes de Almeida (Eds.)
Intelligent Educational Machines, 2007
ISBN 978-3-540-44920-1

Vol. 45. Vladimir G. Ivancevic, Tijana T. Ivancevic
Neuro-Fuzzy Associative Machinery for Comprehensive Brain and Cognition Modeling, 2007
ISBN 978-3-540-47463-0

Vol. 46. Valentina Zharkova, Lakhmi C. Jain
Artificial Intelligence in Recognition and Classification of Astrophysical and Medical Images, 2007
ISBN 978-3-540-47511-8

Vol. 47. S. Sumathi, S. Esakkirajan
Fundamentals of Relational Database Management Systems, 2007
ISBN 978-3-540-48397-7

Vol. 48. H. Yoshida (Ed.)
Advanced Computational Intelligence Paradigms in Healthcare, 2007
ISBN 978-3-540-47523-1

Vol. 49. Keshav P. Dahal, Kay Chen Tan, Peter I. Cowling (Eds.)
Evolutionary Scheduling, 2007
ISBN 978-3-540-48582-7

Vol. 50. Nadia Nedjah, Leandro dos Santos Coelho, Luiza de Macedo Mourelle (Eds.)
Mobile Robots: The Evolutionary Approach, 2007
ISBN 978-3-540-49719-6

Vol. 51. Shengxiang Yang, Yew Soon Ong, Yaochu Jin Honda (Eds.)
Evolutionary Computation in Dynamic and Uncertain Environment, 2007
ISBN 978-3-540-49772-1

Vol. 52. Abraham Kandel, Horst Bunke, Mark Last (Eds.)
Applied Graph Theory in Computer Vision and Pattern Recognition, 2007
ISBN 978-3-540-68019-2

Vol. 53. Huajin Tang, Kay Chen Tan, Zhang Yi
Neural Networks: Computational Models and Applications, 2007
ISBN 978-3-540-69225-6

Vol. 54. Fernando G. Lobo, Cláudio F. Lima and Zbigniew Michalewicz (Eds.)
Parameter Setting in Evolutionary Algorithms, 2007
ISBN 978-3-540-69431-1

Vol. 55. Xianyi Zeng, Yi Li, Da Ruan and Ludovic Koehl (Eds.)
Computational Textile, 2007
ISBN 978-3-540-70656-4

Vol. 56. Akira Namatame, Satoshi Kurihara and Hideyuki Nakashima (Eds.)
Emergent Intelligence of Networked Agents, 2007
ISBN 978-3-540-71073-8

Vol. 57. Nadia Nedjah, Ajith Abraham and Luiza de Macedo Mourella (Eds.)
Computational Intelligence in Information Assurance and Security, 2007
ISBN 978-3-540-71077-6

Vol. 58. Jeng-Shyang Pan, Hsiang-Cheh Huang, Lakhmi C. Jain and Wai-Chi Fang (Eds.)
Intelligent Multimedia Data Hiding, 2007
ISBN 978-3-540-71168-1

Vol. 59. Andrzej P. Wierzbicki and Yoshiteru Nakamori (Eds.)
Creative Enivornments, 2007
ISBN 978-3-540-71466-8

Vol. 60. Vladimir G. Ivancevic and Tijana T. Ivacevic
Computational Mind: A Complex Dynamics Perspective, 2007
ISBN 978-3-540-71465-1

Vol. 61. Jacques Teller, John R. Lee and Catherine Roussey (Eds.)
Ontologies for Urban Development, 2007
ISBN 978-3-540-71975-5

Vol. 62. Lakhmi C. Jain, Raymond A. Tedman and Debra K. Tedman (Eds.)
Evolution of Teaching and Learning Paradigms in Intelligent Environment, 2007
ISBN 978-3-540-71973-1

Vol. 63. Włodzisław Duch and Jacek Mańdziuk (Eds.)
Challenges for Computational Intelligence, 2007
ISBN 978-3-540-71983-0

Włodzisław Duch
Jacek Mańdziuk
(Eds.)

Challenges for Computational Intelligence

With 119 Figures, 19 in colour and 6 Tables

Springer

Prof. Włodzisław Duch
Department of Informatics
Nicolaus Copernicus University
Ul. Grudziadzka 5
87-100 Torun
Poland
E-mail: wduch@is.umk.pl
and
Division of Computer Science
School of Computer Engineering
Nanyang Technological University
Singapore 639798
E-mail: ASWDuch@ntu.edu.sg

Prof. Jacek Mańdziuk
Faculty of Mathematics
and Information Science
Warsaw University of Technology
Plac Politechniki 1
00-661 Warsaw
Poland
E-mail: mandziuk@mini.pw.edu.pl

Library of Congress Control Number: 2007925677

ISSN print edition: 1860-949X
ISSN electronic edition: 1860-9503
ISBN 978-3-540-71983-0 Springer Berlin Heidelberg New York

Cover design: deblik, Berlin
Typesetting by SPi using a Springer LaTeX macro package
Printed on acid-free paper SPIN: 12047871 89/SPi 5 4 3 2 1 0

Preface

Włodzisław Duch[1,2] and Jacek Mańdziuk[3]

[1] Department of Informatics, Nicolaus Copernicus University, Toruń, Poland
[2] School of Computer Engineering, Nanyang Technological University, Singapore
Google:Duch
[3] Faculty of Mathematics and Information Science, Warsaw University of Technology, Warsaw, Poland
mandziuk@mini.pw.edu.pl

In the year 1900 at the International Congress of Mathematicians in Paris David Hilbert delivered what is now considered the most important talk ever given in the history of mathematics. In this talk Hilbert outlined his philosophy of mathematics and proposed 23 major problems worth working at in future. Some of these problems were in fact more like programs for research than problems to be solved. 100 years later the impact of this talk is still strong: some problems have been solved, new problems have been added, but the direction once set – identify the most important problems and focus on them – is still important.

As the year 2000 was approaching we started to wonder if something like that could be done for the new field of Computational Intelligence? Can we define a series of challenging problems that will give it a sense of direction, especially to our younger colleagues? Obviously the situation of a new, rapidly growing field, does not resemble that of the ancient queen of science, and no-one has such a deep insight into its problems as David Hilbert had at his time, but without setting up clear goals and yardsticks to measure progress on the way, without having a clear sense of direction many efforts will be wasted. Some period of rather chaotic exploration of new mathematical techniques developed in neural, fuzzy and evolutionary algorithms was necessary, leading to many new directions and sub-branches of Computational Intelligence. Good mathematical foundations have been gradually introduced in the last decade. However, some of the problems CI experts attempted to solve as well as some of the methods used were of the same type as pattern recognition, operation research or some branches of statistics were working on 40 years earlier. For example, introduction of basic results from approximation theory has led to the development of basis set expansion techniques and Gaussian classifiers, and the old ideas of wide margins and kernels developed into the support vector machines. Although these ideas were known for decades they have been greatly developed on the theoretical, as well as on the practical algorithm and software fronts.

New CI techniques are frequently better at solving the optimization, approximation, classification, clusterization or other pattern recognition problems, but is Computational Intelligence just an improved version of pattern recognition? How do we define CI as a branch of science? How do we measure progress in CI?

The first idea that one of us (WD) wanted to pursue in order to find an answer to some of these questions was to organize a millenium special issue of the IEEE Transactions on Neural Networks (TNN) devoted to challenges in Computational Intelligence. For many reasons this project has been delayed and eventually, at the suggestion of Jacek Zurada, the head editor of TNN journal at that time, turned into a book project. Unfortunately numerous duties did not allow Jacek to participate personally in realization of this project but we are deeply grateful for his initial help and encouragement. Our hope was that this book will provide clear directions and the much-needed focus on the most important and challenging research issues, illuminate some of the topics involved, start a discussion about the best CI strategies to solve complex problems, and show a roadmap how to achieve its ambitious goals. Obviously such a book could not be written by a single author, it had to be written by top experts in different branches of CI expressing their views on this subject.

In the call for contributions we wrote:

> Computational Intelligence is used as a name to cover many existing branches of science. Artificial neural networks, fuzzy systems, evolutionary computation and hybrid systems based on the above three disciplines form a core of CI. Such disciplines as probabilistic reasoning, molecular (DNA) computing, computational immunology, rough sets or some areas of machine learning may also be regarded as subfields of the Computational Intelligence. CI covers all branches of science and engineering that are concerned with understanding and solving problems for which effective computational algorithms do not exist. Thus it overlaps with some areas of Artificial Intelligence, and a good part of pattern recognition, image analysis and operations research.
> In the last few years the annual volume of CI-related papers has been visibly increasing. Several ambitious theoretical and application driven projects have been formulated. Counting only the number of published papers and ongoing research topics one can conclude that there is an undisputed progress in the field. On the other hand besides the sheer numbers of published papers and ongoing research projects several fundamental questions concerning the future of Computational Intelligence arise. What is the ultimate goal of Computational Intelligence, and what are the short-term and the long-term challenges to the field? What is it trying to achieve? Is the field really developing? Do we

actually observe any progress in the field, or is it mainly the increase of the number of publications that can be observed?
We believe that without setting up clear goals and yardsticks to measure progress on the way, without having a clear sense of direction many efforts will end up nowhere, going in circles and solving the same type of pattern recognition problems. Relevant topics for invited book chapters include the following subjects:

- defining the short-term and/or long-term challenges for Computational Intelligence, or any of its subfields (e.g. advanced human-computer interaction systems, the brain-like problem solving methods, efficient scene analysis algorithms, implementation of very complex modular network architectures), and the ultimate goal(s) of CI;
- describing recent developments in most ambitious, ongoing research projects in the CI area, with particular attention to the challenging, unsolved problems;
- discussion on creative artificial agents and societies of agents, their relation to neural computing, evolving connectionist learning paradigms, biologically plausible adaptive intelligent controllers, efficient and automatic large-scale problem decomposition and similar issues;
- discussion on potential, intrinsic research limitations in the field (e.g. the curse of dimensionality problem, or complexity and scalability issues in the brain modeling and in the real-life problem domains);
- discussion on cross-fertilization between the subfields of CI, neurosciences and cognitive sciences;
- plausibility of and alternatives to the CI-based methods for solving various research problems, both theoretical and practical ones.

The book should provide clear directions and the much-needed focus on the most important and challenging research issues, illuminate some of the topics involved, start a discussion about the best CI strategies to solve complex problems, and show a roadmap how to achieve ambitious goals. The attempt to address some of the topics listed above should be especially helpful for the young researchers entering the area of CI.

The work on the book has not been easy, as the prospective authors showed a strong tendency to write about their current projects or making the state-of-the-art reviews. We have not always been completely successful in enforcing the focus on grand challenges and "forward thinking"; the judgment is left to the readers. So finally here it is, 17 chapters on many aspects of CI written by 16 authors. The first chapter tries to define what exactly is Computational Intelligence, how is it related to other branches of science, what are the grand challenges, what the field could become and where is it going. In the second chapter "New Millennium AI and the Convergence of History" Jürgen Schmidhuber writes about recurrent networks and universal problem solvers, the great dream of Artificial Intelligence, speculating about the increased pace of future developments in computing.

In the next 6 chapters challenges for CI resulting from attempts to model cognitive and neurocognitive processes are presented. "The Challenges of Building Computational Cognitive Architectures" chapter by Ron Sun is focused on issues and challenges in developing computer algorithms to simulate cognitive architectures that embody generic description of the thinking processes based on perceptions. In the fourth chapter, "Programming a Parallel Computer: The Ersatz Brain Project" James Anderson and his colleagues speculate about the basic design, provide examples of "programming" and suggest how intermediate level structures could arise in a sparsely connected massively parallel, brain like computers using sparse data representations. Next John G. Taylor in "The Brain as a Hierarchical Adaptive Control System" considers various components of information processing in the brain, choosing attention, memory and reward as key elements, discussing how to achieve cognitive faculties. Soo-Young Lee proposes "Artificial Brain and OfficeMate based on Brain Information Processing Mechanism", capable of conducting essential human functions such as vision, auditory, inference, emergent behavior and proactive learning from interactions with humans. Stan Gielen's chapter "Natural Intelligence and Artificial Intelligence: Bridging the Gap between Neurons and Neuro-Imaging to Understand Intelligent Behavior" addresses some aspects of the hard problems in neuroscience regarding consciousness, storage and retrieval of information, the evolution of cooperative behavior and relation of these questions to major problems in Computational Intelligence. The final chapter in this part, written by DeLiang Wang, is devoted to the "Computational Scene Analysis", analysis and understanding of the visual and auditory perceptual input.

The next three chapters present a broad view of other inspirations that are important for the foundations of Computational Intelligence. First, the chapter "Brain-, Gene-, and Quantum Inspired Computational Intelligence: Challenges and Opportunities", by Nikola Kasabov discusses general principles at different levels of information processing, starting from the brain and going down to the genetic and quantum level, proposing various combinations, such as the neurogenetic, quantum spiking neural network and quantum neuro-genetic models. Robert Duin and Elżbieta Pękalska contributed "The Science of Pattern Recognition. Achievements and Perspectives", writing about challenges facing pattern recognition and promising directions to overcome them. In the "Towards Comprehensive Foundations of Computational Intelligence" chapter Włodzisław Duch presents several proposals for CI foundations: computing and cognition as compression, meta-learning as search in the space of data models, (dis)similarity based methods providing a framework for such meta-learning, quite general approach based on compositions of transformations, and learning from partial observations as a natural extension towards reasoning based on perceptions.

The remaining five chapters are focused on theoretical issues and specific sub-areas of CI. Witold Pedrycz writes about "Knowledge-Based Clustering in Computational Intelligence" governed by the domain knowledge articulated through proximity-based knowledge hints supplied through an interaction with experts, and other such forms of clustering. Věra Kůrková addresses the problem of "Generalization in Learning from Examples" exploring the relation of learning from data to inverse problems, regularization and reproducing kernel Hilbert spaces, and relating that to broader philosophical issues in learning. Lei Xu presents another theoretical chapter, "Trends on Regularization and Model Selection in Statistical Learning: A Perspective from Bayesian Ying Yang Learning", integrating regularization and model selection and providing a general learning procedure based on solid foundations. Jacek Mańdziuk writes about "Computational Intelligence in Mind Games", discussing challenging issues and open questions in the area of intelligent game playing with special focus on implementation of typical for human players concepts of intuition, abstraction, creativity, game-independent learning and autonomous knowledge discovery in game playing agents. Xindi Cai and Donald Wunsch focus on "Computer Go: A Grand Challenge to AI", providing a survey of methods used in computer Go and offering a basic overview for future study, including their own hybrid evolutionary computation algorithm. The final chapter of the book, "Noisy Chaotic Neural Networks for Combinatorial Optimization", written by Lipo Wang and Haixiang Shi, addresses the problem of using neural network techniques to solving combinatorial optimization problems and shows some practical applications of this approach.

We hope that the readers, especially the younger ones, will find in these chapters many new ideas, helpful directions for their own research and challenges that will help them to focus on unsolved problems and move the whole field of Computational Intelligence forward.

We would like to thank Mr Marcin Jaruszewicz, a Ph.D. student of JM, for his assistance in formatting.

Editors:

Włodzisław Duch and Jacek Mańdziuk

Singapore, Warsaw, December 2006

Contents

What Is Computational Intelligence and Where Is It Going?

Włodzisław Duch

Department of Informatics, Nicolaus Copernicus University, Grudziądzka 5, Toruń, Poland, and School of Computer Engineering, Nanyang Technological University, Singapore

Summary. What is Computational Intelligence (CI) and what are its relations with Artificial Intelligence (AI)? A brief survey of the scope of CI journals and books with "computational intelligence" in their title shows that at present it is an umbrella for three core technologies (neural, fuzzy and evolutionary), their applications, and selected fashionable pattern recognition methods. At present CI has no comprehensive foundations and is more a bag of tricks than a solid branch of science. The change of focus from methods to challenging problems is advocated, with CI defined as a part of computer and engineering sciences devoted to solution of non-algoritmizable problems. In this view AI is a part of CI focused on problems related to higher cognitive functions, while the rest of the CI community works on problems related to perception and control, or lower cognitive functions. Grand challenges on both sides of this spectrum are addressed.

1 Introduction

What exactly is Computational intelligence (CI)? How is it related to other branches of computer science, such as artificial intelligence (AI), classification, cognitive informatics, connectionism, data mining, graphical methods, intelligent agents and intelligent systems, knowledge discovery in data (KDD), machine intelligence, machine learning, natural computing, parallel distributed processing, pattern recognition, probabilistic methods, soft computing, multivariate statistics, optimization and operation research? This is a very confusing issue, hotly debated, but with no consensus in sight. Computational intelligence became a new buzzword that means different things to different people.

Branches of science are not defined, but slowly develop in the process of sharing and clustering of common interests. Even well-established sciences have problems with clear definition: for example, one of the best definition of physics is "physics is what physicist do". Herbert Simon, one of the AI

Włodzisław Duch: *What Is Computational Intelligence and Where Is It Going?*, Studies in Computational Intelligence (SCI) **63**, 1–13 (2007)
www.springerlink.com

fathers, answered the question "What is artificial intelligence?" writing "We define it in terms of the tasks that are done" [1]. Computational Intelligence experts focus on problems that are difficult to solve using artificial systems, but are solved by humans and some animals, problems requiring intelligence. Specific interests also focus on methods and tools that are applicable to this type of problems. Starting with seminal papers, special sessions, growing into separate conferences and specialized journals, different branches of CI evolve in many directions, frequently quite far from original roots and inspirations. New communities are formed and need to establish their identity by defining borders distinguishing them from other scientific communities.

Artificial Intelligence (AI) was the first large scientific community, established already in the mid 1950s, working on problems that require intelligence to be solved. Its evolution has been summarized in the 25th anniversary issue of the *AI Magazine* by Mackworth [2]: "In AI's youth, we worked hard to establish our paradigm by vigorously attacking and excluding apparent pretenders to the throne of intelligence, pretenders such as pattern recognition, behaviorism, neural networks, and even probability theory. Now that we are established, such ideological purity is no longer a concern. We are more catholic, focusing on problems, not on hammers. Given that we do have a comprehensive toolbox, issues of architecture and integration emerge as central."

IEEE Computational Intelligence Society defines its subjects of interest as neural networks, fuzzy systems and evolutionary computation, including swarm intelligence. The approach taken by the journals and by the book authors is to treat computational intelligence as an umbrella under which more and more methods are slowly added. A good definition of the field is therefore impossible, because different people include or exclude different methods under the same CI heading. Chess programs based on heuristic search are already in the superhuman computational intelligence category, but they do not belong to CI defined in such a way. In the early days of CI some experts tried to explicitly exclude problems requiring reasoning. Take for example this definition: "A system is computationally intelligent when it: deals only with numerical (low level) data, has a pattern recognition component, and does not use knowledge in the AI sense" [3].

As in the case of Artificial Intelligence the need to create strong identity by emphasizing specific methods defining Computational Intelligence as a field should be replaced by focus on problems to be solved, rather than hammers. Below some remarks on the current state of CI are made, based on analysis of journals and books with "computational intelligence" in their title. Then a new definition of CI is proposed and some remarks are made on what should the computational intelligence field really be in future. Finally grand challenges to computational intelligence are discussed.

2 CI Journals

The name "Computational Intelligence" has been used for over 20 years, although only recently it has gained a widespread popularity and somewhat different flavor. There are already at least 10 journals with "Computational Intelligence" in the title and the number of journals with "intelligent" or "intelligence" in the title is far greater.

The quarterly journal *Computational Intelligence. An International Journal* (Blackwell Publishing, since 1984) is the oldest among CI journals. It is focused on typical artificial intelligence problems, related to higher cognition: logic, reasoning, planning, complex agents, language understanding, rule-based machine learning and reinforcement learning. In the description of the journal it is clearly stated that: "This leading international journal promotes and stimulates research in the field of artificial intelligence (AI). ... The journal is designed to meet the needs of a wide range of AI workers in academic and industrial research." The main focus areas include AI applications in entertainment, software engineering, computational linguistics, web intelligence, business, finance, commerce and economics. Unfortunately this journal makes an impression that computational intelligence is just another name for artificial intelligence.

The *Journal of Computational Intelligence in Finance* (Finance & Technology Publishing, since 1993) was focused on the financial applications of CI predictive methods, but seems to have vanished by now.

The *International Journal of Computational Intelligence and Organizations* (Lawrence Erlbaum Associates, since 1996) is a quarterly journal focusing on theories, methods and applications of computational intelligence in organizations. This journal "publishes original, high-quality articles dealing with the design, development, implementation and management of neural networks, genetic algorithms, fuzzy logic, uncertain reasoning techniques, and related machine learning methods as they apply to organizations. Application of alternative techniques and comparisons to other artificial intelligence models, nonparametric statistics, and decision trees are encouraged. The emphasis is on how computational intelligence is being applied to decision making and problem solving in organizations." Note that this journal (unfortunately conspicuously absent in the Internet) encourages comparisons of results with alternative techniques that may be used to solve the same problem.

The *Journal of Advanced Computational Intelligence and Intelligent Informatics* (Fuji Technology Press, since 1997) is published bimonthly. This journal focuses on "the synergetic integration of neural networks, fuzzy logic and evolutionary computation", and building intelligent systems for industrial applications. Except for the standard fuzzy, neural and evolutionary computation triad, "hybrid systems, adaptation and learning systems, distributed intelligent systems, network systems, multi-media, human interface, biologically inspired evolutionary systems, artificial life, chaos, fractal, wavelet analysis, scientific applications and industrial applications" are also mentioned.

The *International Journal of Computational Intelligence and Applications* (World Scientific, since 2001) is "dedicated to the theory and applications of computational intelligence (artificial neural networks, fuzzy systems, evolutionary computation and hybrid systems). The main goal of this journal is to provide the scientific community and industry with a vehicle whereby ideas using two or more conventional and computational intelligence based techniques could be discussed." Areas include neural, fuzzy and evolutionary computation, pattern recognition, hybrid intelligent systems, symbolic machine learning, statistical models, image/audio/video compression and retrieval, encouraging "new ideas, combining two or more areas, such as neuro-fuzzy, neuro-symbolic, neuro-evolutionary, neuro-symbolic, neuro-pattern recognition, fuzzy-evolutionary, evolutionary-symbolic, fuzzy-evolutionary, evolutionary-symbolic, fuzzy-symbolic, etc."

The *International Journal of Computational Intelligence* (World Enformatika Society, since 2004) is a quarterly open access journal with a double-blind international review system. It is "focusing on theories, methods and applications in computational intelligence". There is no explanation what is meant by CI, just a statement that it deals "with any area of computational intelligence research". So far most papers in this journal are on various applications using a mixture of neural, fuzzy, and bio-inspired optimization methods.

The *International Journal of Computational Intelligence Research* (Research India Publications, since 2005) is a free online journal. In description of its aims the connection with biology is stressed: "Computational intelligence is a well-established paradigm, where new theories with a sound biological understanding have been evolving. The current experimental systems have many of the characteristics of biological computers and are beginning to be built to perform a variety of tasks that are difficult or impossible to do with conventional computers." CI is considered to be heterogeneous field involving "such technologies as neurocomputing, fuzzy systems, probabilistic reasoning, artificial life, evolutionary algorithms, multi-agent systems etc." All of these of course performed using conventional computers.

The *International Journal of Computational Intelligence Theory and Practice* (Serials Publications, since 2006) "aims at publishing papers addressing theories, methods and applications in artificial neural networks, fuzzy systems, evolutionary computation, intelligent agents, hybrid systems and other areas of artificial intelligence". No links to papers are provided, and no papers on classical AI have been published so far.

The *Journal of Computational Intelligence in Bioinformatics* (Research India Publications, 2006) covers "artificial intelligence and computational intelligence theory and their applications in bioinformatics". This journal tries to cover all "advances in computational molecular/structural biology, encompassing areas such as computing in biomedicine and genomics, computational proteomics and systems biology, and metabolic pathway engineering". The topics covered include many CI methods.

The *IEEE Computational Intelligence Magazine* (published by the IEEE Computational Intelligence Society, since 2006) covers "applications oriented developments, successful industrial implementations, design tools, technology reviews, computational intelligence education, and applied research". It also provides an overview of interesting CI topics in special issues.

The *Computational Intelligence and Neuroscience* (Hindawi Publishing, since 2007) is a new open access journal for "the interdisciplinary field of neural computing, neural engineering and artificial intelligence, where neuroscientists, cognitive scientists, engineers, psychologists, physicists, computer scientists, and artificial intelligence investigators among others can publish their work in one periodical that bridges the gap between neuroscience, artificial intelligence and engineering." This journal has a definite profile. "Artificial" probably means here "computational", and in most journal descriptions these words are treated as synonyms.

In the last year five new "computational intelligence" journals have been established. Unfortunately they all seem to be oriented towards methods rather than grand challenging problems to be solved. Some journals add fashionable topics like wavelet analysis, chaos, fractals, other go in the direction of AI, mentioning agents and reasoning as the main topics. The oldest CI journal happens to be a good old-fashioned AI in disguise.

For historical reasons these journals accept as valid CI topics selected statistical and mathematical techniques, such as Bayesian networks, probabilistic reasoning, rough sets and rough logic, basis set expansion methods for function approximation, support vector machines, kernel methods, or statistical natural language processing methods, while many other methods, including various statistical and logical approaches to clusterization, classification, approximation, first and higher-order logic in reasoning, numerical optimizations techniques, approximation theory, or search techniques used to solve the same type of problems as the "valid CI methods", are beyond their scope. One can predict with confidence that many other journals called "Computational intelligence in xxx" will appear in near future. Thus analysis of the topics covered by CI journals does not allow for clear understanding of what CI is or should be.

3 CI Books

Perhaps books with "Computational Intelligence" will define the field better than journal descriptions. So far there are only a few textbooks with this title. The oldest one, *Computational Intelligence – A Logical Approach* [4], is a typical symbolic AI book focusing on logic and reasoning. The authors acknowledge that "Artificial intelligence is the established name for the field we have defined as computational intelligence", but think that "the term 'artificial intelligence' is a source of much confusion" and therefore propose to change the name, creating even greater confusion. In the first chapter they write: "Computational intelligence is the study of the design of intelligent agents. [...] The central scientific goal of computational intelligence is to understand the principles that make intelligent behavior possible, in natural or artificial systems". This could make their view of CI rather broad, because there are many approaches to analyze and model such systems. Unfortunately they focus only on reasoning as computation, and logic as the basis for reasoning, forgetting that symbols have first to be derived from real perceptions, and therefore pattern analysis cannot be avoided.

In the book *Computational Intelligence for Decision Support* similar definition is given: "Computational intelligence is the field of studying how to build intelligent agents" [5]. This obviously does not include most of what is discussed by CI community, presented at conferences, and published in CI journals. People with AI background evidently tend to see CI through the perspective of intelligent agents.

The book *Computational Intelligence: An Introduction* [6] defines CI as "the study of adaptive mechanisms to enable or facilitate intelligent behavior in complex and changing environments. As such, computational intelligence combines artificial neural networks, evolutionary computing, swarm intelligence and fuzzy systems". These are the main topics covered in the book, leaving aside many other CI topics.

Finally, the book *Computational Intelligence: Principles, Techniques and Applications* [7] contains a whole chapter in which the author tries to come up with a definition of CI by adding more and more methods to the core set that includes fuzzy, neural and evolutionary computations. This book covers also possibilistic reasoning, belief calculus, fuzzy Petri nets, and various combinations of these methods. It is perhaps the most ambitious attempt to define computational intelligence, discussing many exotic approaches, but still it falls short of covering all major tracks of any large conference on computational intelligence, for example it completely ignores kernel methods and basis set expansion networks.

Springer *Studies in Computational Intelligence* series has published already many books covering various aspects of CI. IEEE Computational Intelligence Society sponsors a book series on computational intelligence that is published by IEEE Press/Wiley. Several books apply CI techniques to specific areas, for example *Computational Intelligence in Design and Manufacturing* [8], *Computational Intelligence in Software Quality Assurance* [9], *Computational Intelligence in Control Engineering* [10], *Computational Intelligence in Economics and Finance* [11] and many others. They all tend to see computational intelligence as "a consortium of data-driven methodologies which includes fuzzy logic, artificial neural networks, genetic algorithms, probabilistic belief networks and machine learning" [11]

The Artificial Intelligence Portal in Wikipedia defines Computational intelligence (CI) as "a branch of the study of artificial intelligence. Computational intelligence research aims to use learning, adaptive, or evolutionary computation to create programs that are, in some sense, intelligent. Computational intelligence research either explicitly rejects statistical methods (as is the case with fuzzy systems), or tacitly ignores statistics (as is the case with most neural network research). In contrast, machine learning research rejects non-statistical approaches to learning, adaptivity, and optimization." According to this view CI is a part of AI focused on learning, but ignoring statistical methods. If CI is what computational intelligence experts do, this view is obviously false: kernel methods and Bayesian approaches are statistical learning techniques that are very much part of their bag of tools. AI experts have initially explicitly excluded all pattern recognition methods that CI community is very much interested in, so CI cannot be a part of classical AI.

4 What Should Computational Intelligence Really Be?

For many CI experts biological inspirations are very important, but even if biology is extended to include all neural, psychological, and evolutionary inspirations this will only cover the main themes (neural, fuzzy and evolutionary) that the CI community works on. The whole Bayesian foundations of learning, probabilistic and possibilistic reasoning, other alternative approaches to handle uncertainty, kernel methods, information geometry and geometrical learning approaches, search algorithms and many other methods have little or no biological connections. Some neural methods are obviously more neural then others, with basis set function expansion methods having more to do with approximation theory than neurobiological inspirations. CI experts have a tendency to use only their own tools, for example only evolutionary algorithms for optimization, although there are many other methods for optimization with well-proven convergence properties. In real world applications we certainly should not restrict our toolbox to selected methods, but search for the best tools to solve the problem. If a straightforward numerical solution is possible the problem is not interesting to the CI community, but if intelligence is required to solve it all useful methods should be considered.

Physics studies nature and cannot be defined by its experimental or theoretical tools; the same is true for other branches of science. Computer science studies computable processes and information processing systems. What does computational intelligence study? CI studies problems for which there are no effective algorithms, either because it is not possible to formulate them or because they are NP-hard and thus not effective in real life applications. This is quite broad definition: **computational intelligence is a branch of science studying problems for which there are no effective computational algorithms**. Biological organisms solve such problems every day: extracting meaning from perception, understanding language, solving ill-defined computational vision problems thanks to evolutionary adaptation of the brain to the environment, surviving in a hostile environment. These problems require intelligence to solve but they may also be approached in different ways. Defining computational intelligence by the problems that the field studies has the advantage of removing the need to restrict the types of methods used for solution. Different fields obviously overlap with each other, and thus some problems will be of interest mainly for CI experts, while other problems will be of interests to experts in other fields. For example, optimization problems tackled by evolutionary, swarm, ant and other algorithms are of interest to operational research community. What problems are typical for computational intelligence?

A good part of CI research is concerned with low-level cognitive functions: perception, object recognition, signal analysis, discovery of structures in data, simple associations and control. Methods developed for this type of problems include supervised and unsupervised learning by adaptive systems, and they encompass not only neural, fuzzy and evolutionary approaches but also probabilistic and statistical approaches, such as Bayesian networks or kernel methods. These methods are used to solve the same type of problems in various fields such as pattern recognition, signal processing, classification and regression, data mining. Higher level cognitive functions are required to solve non-algorithmizable problems involving systematic thinking, reasoning, complex representation of knowledge, episodic memory, planning, understanding of symbolic knowledge. These problems are at present solved in a best way by AI community using methods based on search, symbolic knowledge representation, reasoning with frame-based expert systems, machine learning in symbolic domains, logics and linguistic methods. There is little overlap between problems solved using low and high-level mental functions, although they belong to the same broader category of non-algorithmizable problems.

From this point of view AI is a part of CI focusing on problems that require higher cognition and at present are easier to solve using symbolic knowledge representation. It is possible that other CI methods will also find applications to these problems in future. The main overlap areas between low and high-level cognitive functions are in sequence learning, reinforcement and associative learning, and distributed multi-agent systems. All tasks that require reasoning based on perceptions, such as robotics, automatic car driving, autonomous systems require methods for solving both low and high-level cognitive problems and thus are a natural meeting ground for AI experts with the rest of the CI community.

The idea that all intelligence comes from symbol manipulation has been perhaps misunderstood by AI community. Newell and Simon who originated this idea [12, 13] wrote about physical symbols, not about symbolic variables. Physical symbols are better represented as multi-dimensional patterns representing states of various brain areas. Symbolic models of brain processes certainly do not offer accurate approximation for vision, control or any other problem that is described by continuous rather then symbolic variables. Approximations to brain processes should be done at a proper level to obtain similar functions. Symbolic dynamics [14] and extraction of finite state atomata from recurrent networks [15] may provide useful information on dynamical systems, and may be useful in modeling transition between low-to high level processes.

The division between low and high-level cognitive functions is only a rough approximation to the processes in the brain. Embodied cognition has been intensively studied in the last decade, and developmental ideas showing how higher processes emerge from the lower ones have been embraced by robotics people. Even in linguistics it is now commonly acknowledged that real meaning comes from body-based metaphors [16], and the same is true for such abstract areas as mathematics [17]. New CI methods that go beyond pattern recognition and help to solve AI problems may eventually be developed, starting from distributed knowledge representation, graphical methods and spreading activations networks. The dynamics of such models will probably allow for reasonable symbolic approximations, although this is still an open problem.

It is instructive to think about the spectrum of CI problems and various approximations needed to solve them. Neural network models are inspired by brain processes and structures at quite low single-neuron level, while symbolic AI models are inspired by processes at the highest level. The brain has a very specific modular and hierarchical structure, it is not one huge neural network. Perceptron model of a neuron has only one internal parameter, the firing threshold, and a few synaptic weights that determine neuron-neuron interactions. Single neurons probably influence brain information processing in quite insignificant way. Larger neural structures, such as microcircuits or neural cell assemblies, could also be used as basic processors for neural modeling. They have more complex internal states and more complex interactions between elements, but connectionist systems are not trying to approximate these process in any systematic way [18]. A network of networks, hiding the complexity of its processors in a hierarchical way, with different emergent properties at each level, will have progressively more internal knowledge and more complex interactions with other such systems. At the highest level models of whole brains with an infinite number of potential internal states and very complex interactions may be obtained. Discussion of such transition from neurons to brains and to societies is presented in [19].

Computational intelligence is certainly more than just the study of the design of intelligent agents, it includes also study of all non-algoritmizable processes that humans (and sometimes animals) can solve with various degree of competence, and the engineeering approaches to solve such problems using hardware and software systems. CI should not be treated as a bag of tricks without deeper foundations. Competition to solve CI problems using approaches developed in other fields should be invited. Knowledge and search-based methods should complement the core CI techniques in problems requiring reasoning. Goldberg and Harik [20] see computational intelligence more as a way of thinking about problems, calling for a "broader view of the scope of the discipline". They have analyzed limitations to progress in computational manufacturing design, finding the models of human behaviors to be most useful. Although this is certainly worthwhile, defining clearly the problems that CI wants to solve and welcoming all methods that can be used in such solutions, independent of their inspirations, is even more important.

5 Grand Challenges to Computational Intelligence

AI has focused on many specific approaches to problem solving, useful for development of expert systems, neglecting its initial ambitious goals. Although a number of grand challenges for AI has been formulated, starting with the famous Turing Test for machine intelligence, these goals were perhaps reaching too far and thus were not realistic. Meaningful dialog in natural language requires a very-large knowledge base and efficient retrieval of knowledge structures. While the CyC project [21] has created huge knowledge base manually coding it over a period of more than 30 years the retrieval mechanisms that it offers are too inefficient to use it in large-scale dialog systems. A grand challenge for CI community is to propose more efficient knowledge representation and retrieval structures, perhaps modeled on the associative memory of the brain, perhaps using different knowledge representations for different purposes [22]. Vector and similarity-based models cannot yet replace complex frames in reasoning processes. Semantic networks have never been used in a large scale dialog systems, although in principle they could provide efficient association and inference mechanisms.

Feigenbaum [23] proposed a reasoning test which should be simpler for computers than the Turing Test. Instead of a general dialog that has to be based on extensive knowledge of the world, this test is based on the expert knowledge in a narrow domain. Reasoning in some field of mathematics or science by human expert and by an artificial system should be evaluated by another expert in the same field who will pose problems, questions, and ask for explanations. This could be achieved with super-expert systems in various domains, giving some measures of progress towards intelligent reasoning systems. The World Championship for 1st Order Automated Theorem Proving organized at the Conference on Automated Deduction (CADE) could be organized not only between computers, but could also involve humans, although much longer time to complete the proofs may be required. Other grand AI challenges [23] are concerned with large-scale knowledge bases, bootstraping on the knowledge resources from the Internet and creating semantic Internet. The 20-questions game could also be a good challenge for AI, much easier than the Turing test, requiring extensive knowledge about objects and their properties, but not about complex relations between objects. In fact some simple vector-space techniques may be used to play it [22], making it a good challenge not only for AI, but also for the broader CI community.

What would be a good grand challenge for non-AI part of computational intelligence? This has been the subject of a discussion panel on the challenges to the CI in the XXI century, organized at the World Congress on Computational Intelligence in Anchorage, Alaska, in 1998. The conclusion was that a grand challenge for CI is to build an artificial rat, an artificial animal that may survive in a hostile environment. The intermediate steps require solution to many problems in perception, such as object recognition, auditory and visual scene analysis, spatial orientation, memory, motor learning, behavioral control, but also some reasoning and planning. The ultimate challenge may be to build not only an animal, but a human-like system that in addition to survival will be able to pass the Turing test.

Imagine the future in which superintelligence based on some form of computations has been realized. In the long run everything seems to be possible, but what we would like it to do and like it to be? Computational intelligence should be human-centered, helping humans not only to solve their problems, but also to formulate meaningful goals, leading to a true personal fulfillment. It should protect us starting from birth, not only monitoring the health hazards, but also observing and guiding personal development, gently challenging children at every step to reach their full physical as well as mental potential. It should be a technology with access to extensive knowledge, but it also should help humans to make wise decisions presenting choices and their possible consequences. Although it may seem like a dangerous utopia perhaps deeper understanding of developmental processes, cognitive and emotional brain functions, real human needs, coupled with technology that can recognize behavioral patterns, make sense of observations, understand natural language, plan and reason with extensive background knowledge, will lead to a better world in which no human life is wasted. Intelligence with wisdom is perhaps an ultimate goal for human-oriented science. Such utopia is worth dreaming of, although we are still very far from this level (see some speculations on this topic in [24, 25, 26]).

A long-term goal for computational intelligence is to create cognitive systems that could compete with humans in large number of areas. So far this is possible only in restricted domains, such as recognition of specific patterns, processing of large amount of numerical information, memorization of numerous details, high precision control with small number of degrees of freedom, and reasoning in restricted domains, for example in board games. Brains are highly specialized in analysis of natural patterns, segmentation of auditory and visual scenes, and control of body movements, mapping perceptions to actions. Despite great progress in computational intelligence artificial systems designed to solve lower level cognitive functions are still far behind the natural ones. Situation is even worse when higher-level cognitive functions, involving complex knowledge structures necessary for understanding of language, reasoning, problem solving or planning, are considered. Human semantic and episodic memory is vastly superior to the most sophisticated artificial systems, storing complex memory patterns and rapidly accessing them in an associative way.

So far CI understood as a collection of different methods had no clear challenges of the AI magnitude. Improving clusterization, classification and approximation capabilities of CI systems is incremental and there are already so many methods that it is always possible to find alternative solutions. At the technical level fusion of different CI techniques is considered to be a challenge, but attempts to combine evolutionary and neural methods, to take just one example, have a long history and it is hard to find results that are significantly better than those achieved by competing techniques. The challenge is at the meta-level, to find all interesting solutions automatically, especially in difficult cases. Brains are flexible, and may solve the same problem in many different ways. Different applications – recognition of images, handwritten characters, faces, analysis of signals, mutimedia streams, texts, or various biomedical data – usually require highly specialized methods to achieve top performance. This is a powerful force that leads to compartmentalization of different CI branches. Creation of meta-learning systems competitive with the best methods in various applications is a great challenge for CI.

If we recognize that CI should be defined as the science of solving non-algorithmizable problems using computers or specialized hardware the whole field will be firmly anchored in computer and engineering sciences, many technical challenges may be formulated, and important biological sources of inspiration acknowledged. Focusing on broad, challenging problems instead of tools will enable competition with other methods for various applications, facilitating real progress towards even more difficult problems. It should also allow for greater integration of CI and AI communities working on real-world problems that require integration of perception with reasoning. Broad foundations for CI that go beyond pattern recognition need to be constructed, including solving problems related to the higher cognitive functions (see [27], this volume). Inspirations drawn from cognitive and brain sciences, or biology in general, will continue to be very important, but at the end of the road CI will become a solid branch of science on its own standing.

Acknowledgement. I am grateful for the support by the Polish Committee for Scientific Research, research grant 2005-2007.

References

[1] H.A. Simon, Artificial Intelligence: Where Has it Been, and Where is it Going? IEEE Transactions on Knowledge and Data Engineering 3(2): 128-136, 1991.
[2] A.K. Mackworth, The Coevolution of AI and AAAI, AI Magazine 26(4): 51-52, 2005.
[3] J.C. Bezdek, What is computational intelligence? In: Computational Intelligence Imitating Life, pp. 1–12, IEEE Press, New York, 1994.
[4] D. Poole, A. Mackworth and R. Goebel. Computational Intelligence – A Logical Approach. Oxford University Press, New York, 1998.
[5] Z. Chen, Computational Intelligence for Decision Support. CRC Press, Boca Raton, 2000.
[6] A.P. Engelbrecht, Computational Intelligence: An Introduction. Wiley, 2003.
[7] A. Konar, Computational Intelligence: Principles, Techniques and Applications. Springer 2005.

[8] A. Kusiak, Computational Intelligence in Design and Manufacturing. Wiley-Interscience, 2000.
[9] S. Dick and A. Kandel, Computational intelligence in software quality assurance. Series in Machine Perception and Artificial Intelligence, Vol. 63, World Scientific 2005.
[10] R.E. King, Computational intelligence in control engineering, Marcel Dekker Inc., NY, 1999.
[11] S.H. Chen, P. Wang, and P.P. Wang Computational Intelligence in Economics and Finance. Advanced Information Processing Series, Springer 2006.
[12] A. Newell and H.A. Simon, Computer science as empirical enquiry: Symbols and search. Communications of the ACM 19(3), 113–126, 1976.
[13] A. Newell, Unified Theories of Cognition. Cambridge, MA: Harvard University Press 1990.
[14] D. Lind, B. Marcus, Symbolic Dynamics and Coding, Cambridge University Press, 1995.
[15] H. Jacobsson, Rule extraction from recurrent neural networks: A taxonomy and review. Neural Computation, 17(6), 1223–1263, 2005.
[16] G. Lakoff, M. Johnson. Metaphors We Live By. University of Chicago Press, 2nd ed, 2003.
[17] G. Lakoff, R. Núnez, Where Mathematics Comes From: How the Embodied Mind Brings Mathematics into Being. Basic Books 2000.
[18] A. Clark, R. Lutz (eds), Connectionism in Context. Springer-Verlag, Berlin, 1992.
[19] W. Duch and J. Mandziuk, Quo Vadis Computational Intelligence? In: Machine Intelligence: Quo Vadis? Advances in Fuzzy Systems – Applications and Theory (eds. P. Sincak, J. Vascak, K. Hirota), World Scientific, pp. 3–28, 2004.
[20] D.E. Goldberg and G. Harik, A Case Study in Abnormal CI: The Design of Manufacturing and Other Anthropocentric Systems. International J. Computational Intelligence and Organizations, 1, 78-93, 1996.
[21] D. Lenat and R.V. Guha, Building Large Knowledge-Based Systems: Representation and Inference in the Cyc Project. Addison-Wesley 1990.
[22] J. Szymanski, T. Sarnatowicz and W. Duch, Towards Avatars with Artificial Minds: Role of Semantic Memory. Journal of Ubiquitous Computing and Intelligence (in print)
[23] E.A. Feigenbaum, Some Challenges and Grand Challenges for Computational Intelligence. Journal of the ACM 50(1), 32–40, 2003.
[24] R. Kurzweil. The age of spiritual machines: When computers exceed human intelligence. Penguin, New York, NY, 1999.
[25] J. McCarthy, The Future of AI–A Manifesto. AI Magazine 26, 39–40, 2005.
[26] L. Perlovsky. Knowledge Instinct. Basic Books, 2006.
[27] W. Duch, Towards comprehensive foundations of computational intelligence. In: Duch W, Mandziuk J, Eds, Challenges for Computational Intelligence. Springer, pp. 261–316, 2007.

New Millennium AI and the Convergence of History

Jürgen Schmidhuber

TU Munich, Boltzmannstr. 3, 85748 Garching bei, München, Germany &
IDSIA, Galleria 2, 6928 Manno (Lugano), Switzerland
juergen@idsia.ch - http://www.idsia.ch/~juergen

Summary. Artificial Intelligence (AI) has recently become a real formal science: the new millennium brought the first mathematically sound, asymptotically optimal, universal problem solvers, providing a new, rigorous foundation for the previously largely heuristic field of General AI and embedded agents. At the same time there has been rapid progress in practical methods for learning true sequence-processing programs, as opposed to traditional methods limited to stationary pattern association. Here we will briefly review some of the new results, and speculate about future developments, pointing out that the time intervals between the most notable events in over 40,000 years or 2^9 lifetimes of human history have sped up exponentially, apparently converging to zero within the next few decades. Or is this impression just a by-product of the way humans allocate memory space to past events?

1 Introduction

In 2003 we observed [84, 82] that each major breakthrough in computer science tends to come roughly twice as fast as the previous one, roughly matching a century-based scale: In 1623 the computing age started with the first mechanical calculator by Wilhelm Schickard (followed by machines of Pascal, 1640, and Leibniz, 1670). Roughly two centuries later Charles Babbage came up with the concept of a program-controlled computer (1834-1840). One century later, in 1931, Kurt Gödel layed the foundations of theoretical computer science with his work on universal formal languages and the limits of proof and computation. His results and Church's extensions thereof were reformulated by Turing in 1936, while Konrad Zuse built the first working program-controlled computers (1935-1941), using the binary system of Leibniz (1701) instead of the more cumbersome decimal system used by Babbage and many others. By then all the main ingredients of 'modern' computer science were in place. The next 50 years saw many less radical theoretical advances as well as faster and faster switches—relays were replaced by tubes by single transistors by numerous transistors etched on chips—but arguably this was rather predictable, incremental progress without earth-shaking events. Half a century

Jürgen Schmidhuber: *New Millennium AI and the Convergence of History*, Studies in Computational Intelligence (SCI) **63**, 15–35 (2007)
www.springerlink.com

later, however, Tim Berners-Lee triggered the most recent world-changing development by creating the World Wide Web (1990).

Extrapolating the trend, we should expect the next radical change to manifest itself one quarter of a century after the most recent one, that is, by 2015, when some computers will already match brains in terms of raw computing power, according to frequent estimates based on Moore's law, which suggests a speed-up factor of roughly 1000 per decade, give or take a few years. The remaining series of faster and faster additional revolutions should converge in an *Omega point* (term coined by Pierre Teilhard de Chardin, 1916) expected between 2030 and 2040, when individual machines will already approach the raw computing power of all human brains combined (provided Moore's law does not break down—compare Stanislaw Ulam's concept of an approaching *historic singularity* (cited in [40]) or Vinge's closely related technological singularity [113] as well as the subsequent speculations of Moravec [48] and Kurzweil [40]). Many of the present readers of this article should still be alive then.

Will the software and the theoretical advances keep up with the hardware development? We are convinced they will. In fact, the new millennium has already brought fundamental new insights into the problem of constructing theoretically optimal rational agents or universal Artificial Intelligences (AIs, more on this below). On the other hand, on a more practical level, there has been rapid progress in learning algorithms for agents interacting with a dynamic environment, autonomously discovering true sequence-processing, problem-solving programs, as opposed to the reactive mappings from stationary inputs to outputs studied in most traditional machine learning (ML) research. In what follows, we will briefly review some of the new results, then come back to the issue of whether or not history is about to "converge."

2 Overview

Since virtually all realistic sensory inputs of robots and other cognitive systems are sequential by nature, the future of machine learning and AI in general lies in sequence processing as opposed to processing of stationary input patterns. Most traditional methods for learning time series and mappings from sequences to sequences, however, are based on simple time windows: one of the numerous feedforward ML techniques such as feedforward neural nets (NN) [4] or support vector machines [112] is used to map a restricted, fixed time window of sequential input values to desired target values. Of course such approaches are bound to fail if there are temporal dependencies exceeding the time window size. Large time windows, on the other hand, yield unacceptable numbers of free parameters.

If we want to narrow the gap between learning abilities of humans and machines, then we will have to study how to learn general algorithms instead of such reactive mappings. In what follows we will first discuss very recent *universal* program learning methods that are optimal in various mathematical senses. For several reasons, however, these methods are not (yet) practically feasible. Therefore we will also discuss recent less universal but more feasible program learners based on recurrent neural networks.

Finally we will return to the introduction's topic of exponential speed-up, extending it to all of human history since the appearance of the Cro Magnon man roughly 40,000 years ago.

3 Notation

Consider a learning robotic agent with a single life which consists of discrete cycles or time steps $t = 1, 2, \ldots, T$. Its total lifetime T may or may not be known in advance. In what follows, the value of any time-varying variable Q at time t $(1 \leq t \leq T)$ will be denoted by $Q(t)$, the ordered sequence of values $Q(1), \ldots, Q(t)$ by $Q(\leq t)$, and the (possibly empty) sequence $Q(1), \ldots, Q(t-1)$ by $Q(< t)$.

At any given t the robot receives a real-valued input vector $x(t)$ from the environment and executes a real-valued action $y(t)$ which may affect future inputs; at times $t < T$ its goal is to maximize future success or *utility*

$$u(t) = E_\mu \left[\sum_{\tau=t+1}^{T} r(\tau) \;\middle|\; h(\leq t) \right], \tag{1}$$

where $r(t)$ is an additional real-valued reward input at time t, $h(t)$ the ordered triple $[x(t), y(t), r(t)]$ (hence $h(\leq t)$ is the known history up to t), and $E_\mu(\cdot \mid \cdot)$ denotes the conditional expectation operator with respect to some possibly unknown distribution μ from a set M of possible distributions. Here M reflects whatever is known about the possibly probabilistic reactions of the environment. For example, M may contain all computable distributions [104, 105, 42, 36]. Note that unlike in most previous work by others [38, 109], but like in much of the author's own previous work [97, 83], there is just one life, no need for predefined repeatable trials, no restriction to Markovian interfaces between sensors and environment [76], and the utility function implicitly takes into account the expected remaining lifespan $E_\mu(T \mid h(\leq t))$ and thus the possibility to extend it through appropriate actions [83, 86, 90, 87].

4 Universal But Incomputable AI

Solomonoff's theoretically optimal universal predictors and their Bayesian learning algorithms [104, 105, 42, 36] only assume that the reactions of the environment are sampled from an unknown probability distribution μ contained in a set M of all enumerable distributions—compare text after equation (1). That is, given an observation sequence $q(\leq t)$, we only assume there exists a computer program that can compute the probability of the next possible $q(t+1)$, given $q(\leq t)$. Since we typically do not know this program, we predict using a mixture distribution

$$\xi(q(t+1) \mid q(\leq t)) = \sum_i w_i \mu_i(q(t+1) \mid q(\leq t)), \tag{2}$$

a weighted sum of *all* distributions $\mu_i \in \mathcal{M}$, $i = 1, 2, \ldots$, where the sum of the positive weights satisfies $\sum_i w_i \leq 1$. It turns out that this is indeed the best one can possibly do, in a very general sense [105, 36]. The drawback is that the scheme is incomputable, since M contains infinitely many distributions.

One can increase the theoretical power of the scheme by augmenting M by certain non-enumerable but limit-computable distributions [80], or restrict it such that it becomes computable, e.g., by assuming the world is computed by some unknown but deterministic computer program sampled from the Speed Prior [81] which assigns low probability to environments that are hard to compute by any method. Under the Speed Prior the cumulative a priori probability of all data whose computation through an optimal algorithm requires more than $O(n)$ resources is $1/n$.

Can we use the optimal predictors to build an optimal AI? Indeed, in the new millennium it was shown we can. At any time t, the recent theoretically optimal yet uncomputable RL algorithm AIXI [36] uses Solomonoff's universal prediction scheme to select those action sequences that promise maximal future reward up to some horizon, typically $2t$, given the current data $h(\leq t)$. One may adapt this to the case of any finite horizon T. That is, in cycle $t+1$, AIXI selects as its next action the first action of an action sequence maximizing ξ-predicted reward up to the horizon, appropriately generalizing eq. (2). Recent work [36] demonstrated AIXI's optimal use of observations as follows. The Bayes-optimal policy p^ξ based on the mixture ξ is self-optimizing in the sense that its average utility value converges asymptotically for all $\mu \in \mathcal{M}$ to the optimal value achieved by the (infeasible) Bayes-optimal policy p^μ which knows μ in advance. The necessary condition that $\mathcal{M}$ admits self-optimizing policies is also sufficient. Furthermore, p^ξ is Pareto-optimal in the sense that there is no other policy yielding higher or equal value in *all* environments $\nu \in \mathcal{M}$ and a strictly higher value in at least one [36].

What are the implications? The first 50 years of attempts at "general AI" have been dominated by heuristic approaches [52, 69, 111, 47]. Traditionally many theoretical computer scientists have regarded the field with contempt for its lack of hard theoretical results. Things have changed, however. Although the universal approach above is practically infeasible due to the incomputability of Solomonoff's prior, it does provide, for the first time, a mathematically sound theory of AI and optimal decision making based on experience, identifying the limits of both human and artificial intelligence, and providing a yardstick for any future approach to general AI.

Using such results one can also come up with theoretically optimal ways of improving the predictive world model of a curious robotic agent [89]. The rewards of an optimal reinforcement learner are the predictor's improvements on the observation history so far. They encourage the reinforcement learner to produce action sequences that cause the creation and the learning of new, previously unknown regularities in the sensory input stream. It turns out that art and creativity can be explained as by-products of such intrinsic curiosity rewards: good observer-dependent art deepens the observer's insights about this world or possible worlds, connecting previously disconnected patterns in an initially surprising way that eventually becomes known and boring. While previous attempts at describing what is satisfactory art or music were informal, this work permits the first *technical, formal* approach to understanding the nature of art and creativity [89].

Using the Speed Prior mentioned above, one can scale the universal approach down such that it becomes computable [81]. In what follows we will mention ways of introducing additional optimality criteria that take into account the computational costs of prediction and decision making.

5 Asymptotically Optimal General Problem Solver

To take computation time into account in a general, optimal way [41] [42, p. 502-505], the recent asymptotically optimal search algorithm for *all* well-defined problems [35] allocates part of the total search time to searching the space of proofs for provably correct candidate programs with provable upper runtime bounds; at any given time it focuses resources on those programs with the currently best proven time bounds. The method is as fast as the initially unknown fastest problem solver for the given problem class, save for a constant slowdown factor of at most $1 + \epsilon$, $\epsilon > 0$, and an additive problem class-specific constant. Unfortunately, however, the latter may be huge.

Practical applications may not ignore the constants though. This motivates the next section which addresses all kinds of optimality (not just asymptotic optimality).

6 Optimal Self-Referential General Problem Solver

The recent Gödel machines [83, 86, 90, 87] represent the first class of mathematically rigorous, general, fully self-referential, self-improving, optimally efficient problem solvers. In particular, they are applicable to the problem embodied by objective (1), which obviously does not care for asymptotic optimality.

The initial software $\mathcal{S}$ of such a Gödel machine contains an initial problem solver, e.g., one of the approaches above [36] or some less general, typical sub-optimal method [38, 109]. Simultaneously, it contains an initial proof searcher (possibly based on an online variant of Levin's *Universal Search* [41]) which is used to run and test *proof techniques*. The latter are programs written in a universal programming language implemented on the Gödel machine within $\mathcal{S}$, able to compute proofs concerning the system's own future performance, based on an axiomatic system $\mathcal{A}$ encoded in $\mathcal{S}$. $\mathcal{A}$ describes the formal *utility* function, in our case eq. (1), the hardware properties, axioms of arithmetics and probability theory and string manipulation etc, and $\mathcal{S}$ itself, which is possible without introducing circularity [83].

Inspired by Kurt Gödel's celebrated self-referential formulas (1931), the Gödel machine rewrites any part of its own code in a computable way through a self-generated executable program as soon as its *Universal Search* variant has found a proof that the rewrite is *useful* according to objective (1). According to the Global Optimality Theorem [83, 86, 90, 87], such a self-rewrite is globally optimal—no local maxima!—since the self-referential code first had to prove that it is not useful to continue the proof search for alternative self-rewrites.

If there is no provably useful, globally optimal way of rewriting $\mathcal{S}$ at all, then humans will not find one either. But if there is one, then $\mathcal{S}$ itself can find and exploit it. Unlike previous *non*-self-referential methods based on hardwired proof searchers [36], Gödel machines not only boast an optimal *order* of complexity but can optimally reduce (through self-changes) any slowdowns hidden by the $O()$-notation, provided the utility of such speed-ups is provable at all.

Practical implementations of the Gödel machine do not yet exist though, and probably require a thoughtful choice of the initial axioms and the initial proof searcher. In the next sections we will deal with already quite practical, non-optimal and non-universal, but still rather general searchers in program space, as opposed to the space of reactive, feedforward input / output mappings, which still attracts the bulk of current machine learning research.

7 Supervised Recurrent Neural Networks

Recurrent neural networks (RNNs) are neural networks [4] with feedback connections that are, in principle, as powerful as any traditional computer. There is a very simple way to see this [74, 79]: a traditional microprocessor may be viewed as a very sparsely connected RNN consisting of very simple neurons implementing nonlinear AND and NAND gates, etc. Compare [100] for a more complex argument. Early RNNs [34, 1, 60] did not exploit this potential power because they were limited to static inputs. However, most interesting tasks involve processing sequences that consist of continually varying inputs. Examples include robot control, speech recognition, music composition, attentive vision, and numerous others.

The initial RNN enthusiasm of the 1980s and early 90s was fueled by the obvious theoretical advantages of RNNs: unlike feedforward neural networks (FNNs) [115, 70], Support Vector Machines (SVMs), and related approaches [112], RNNs have an internal state which is essential for many tasks involving program learning and sequence learning. And unlike in Hidden Markov Models

(HMMs) [125], internal states can take on continuous values, and the influence of these states can persist for many timesteps. This allows RNN to solve tasks that are impossible for HMMs, such as learning the rules of certain context-free languages [16].

Practitioners of the early years, however, have been sobered by unsuccessful attempts to apply RNNs to important real world problems that require sequential processing of information. The first RNNs simply did not work very well, and their operation was poorly understood since it is inherently more complex than that of FNNs. FNNs fit neatly into the framework of traditional statistics and information theory [99], while RNNs require additional insights, e.g., from theoretical computer science and *algorithmic* information theory [42, 80].

Fortunately, recent advances have overcome the major drawbacks of traditional RNNs. That's why RNNs are currently experiencing a second wave of attention. New architectures, learning algorithms, and a better understanding of RNN behavior have allowed RNNs to learn many previously unlearnable tasks. RNN optimists are claiming that we are at the beginning of a *"RN-Naissance"* [92, 85], and that soon we will see more and more applications of the new RNNs.

Sequence-processing, supervised, gradient-based RNNs were pioneered by Werbos [116], Williams [121], and Robinson & Fallside [65]; compare Pearlmutter's survey [56]. The basic idea is: a teacher-given training set contains example sequences of inputs and desired outputs or targets $d_k(t)$ for certain neurons y_k at certain times t. If the i-th neuron y_i is an input neuron, then its real-valued activation $y_i(t)$ at any given discrete time step $t = 1, 2, \ldots$ is set by the environment. Otherwise $y_i(t)$ is a typically nonlinear function $f(\cdot)$ of the $y_k(t-1)$ of all neurons y_k connected to y_i, where by default $y_i(0) = 0$ for all i. By choosing an appropriate $f(\cdot)$, we make the network dynamics differentiable, e.g., $y_i(t) = arctan(\sum_k (w_{ik} y_k(t-1))$, where w_{ik} is the real-valued weight on the connection from y_k to y_i. Then we use gradient descent to change the weights such that they minimize an objective function E reflecting the differences between actual and desired output sequences. Popular gradient descent algorithms for computing weight changes $\triangle w_{ik} \sim \frac{\partial E}{\partial w_{ik}}$ include *Back-Propagation Through Time* (BPTT) [116, 121] and *Real-Time Recurrent Learning* (RTRL) [65, 121] and mixtures thereof [121, 77].

The nice thing about gradient-based RNNs is that we can **differentiate our wishes with respect to programs,** e.g., [75, 74, 79]. The set of possible weight matrices represents a continuous space of programs, and the objective function E represents our desire to minimize the difference between what the network does and what it should do. The gradient $\frac{\partial E}{\partial w}$ (were w is the complete weight matrix) tells us how to change the current program such that it will be better at fulfilling our wish.

Typical RNNs trained by BPTT and RTRL and other previous approaches [50, 51, 107, 43, 61, 64, 10, 56, 55, 12, 13, 121, 77, 78, 122, 114, 45, 62], however, cannot learn to look far back into the past. Their problems were first rigorously analyzed in 1991 on the author's RNN long time lag project [29, 30]; also compare Werbos' concept of "sticky neurons" [117]. The error signals "flowing backwards in time" tend to either (1) blow up or (2) vanish: the temporal evolution of the back-propagated error exponentially depends on the weight magnitudes. Case (1) may lead to oscillating weights. In case (2), learning to bridge long time lags takes a prohibitive amount of time, or does not work at all. So then why bother with RNNs at all? For short time lags we could instead use short time-windows combined with non-recurrent approaches such as multi-layer perceptrons [4] or better *Support Vector Machines* SVMs [112].

An RNN called "Long Short-Term Memory" or LSTM (Figure S1) [32] overcomes the fundamental problems of traditional RNNs, and efficiently learns to solve many previously unlearnable tasks involving: Recognition of temporally extended patterns in noisy input sequences [32, 17]; Recognition of the temporal order of widely separated events in noisy input streams [32]; Extraction of information conveyed by the temporal distance between events [15]; Stable generation of precisely timed rhythms, smooth and non-smooth periodic trajectories; Robust storage of high-precision real numbers across extended time intervals; Arithmetic operations on continuous input streams [32, 14]. This made possible the numerous applications described further below.

We found [32, 17] that LSTM clearly outperforms previous RNNs on tasks that require learning the rules of regular languages (RLs) describable by deterministic finite state automata (DFA) [8, 101, 5, 39, 126], both in terms of reliability and speed. In particular, problems that are hardly ever solved by standard RNNs, are solved by LSTM in nearly 100% of all trials, LSTM's superiority also carries over to tasks involving context free languages (CFLs) such as those discussed in the RNN literature [108, 120, 106, 110, 67, 68]. Their recognition requires the functional equivalent of a stack. Some previous RNNs even failed to learn small CFL training sets [68]. Those that did not and those that even learned small CSL training sets [67, 6] failed to extract the general rules, and did not generalize well on substantially larger test sets. In contrast, LSTM generalizes extremely well. It requires only the 30 shortest exemplars ($n \leq 10$) of the context sensitive language $a^n b^n c^n$ to correctly predict the possible continuations of sequence prefixes for n up to 1000 and more.

Kalman filters can further improve LSTM's performance [58]. A combination of LSTM and the decoupled extended Kalman filter learned to deal correctly with values of n up to 10 million and more. That is, after training the network was able to read sequences of 30,000,000 symbols and more, one symbol at a time, and finally detect the subtle differences between *legal* strings such as $a^{10,000,000}b^{10,000,000}c^{10,000,000}$ and very similar but *illegal* strings such as $a^{10,000,000}b^{9,999,999}c^{10,000,000}$. This illustrates that LSTM networks can work in an extremely precise and robust fashion across very long time lags.

Speech recognition is still dominated by Hidden Markov Models (HMMs), e.g., [7]. HMMs and other graphical models such as Dynamic Bayesian Networks (DBN) *do* have internal states that can be used to model memories of previously seen inputs. Under certain circumstances they allow for learning the prediction of sequences of labels from unsegmented input streams. For example, an unsegmented acoustic signal can be transcribed into a sequence of words or phonemes. HMMs are well-suited for noisy inputs and are invariant to non-linear temporal stretching—they do not care for the difference between slow and fast versions of a given spoken word. At least in theory, however, RNNs could offer the following advantages: Due to their discrete nature, HMMs either have to discard real-valued information about timing, rhythm, prosody, etc., or use numerous hidden states to encode such potentially relevant information in discretized fashion. RNNs, however, can naturally use their real-valued activations to encode it compactly. Furthermore, in order to make HMM training tractable, it is necessary to assume that successive inputs are independent of one another. In most interesting sequence learning tasks, such as speech recognition, this is manifestly untrue. Because RNNs model the conditional probabilities of class occupancy directly, and do not model the input sequence, they do not require this assumption. RNN classifications are in principle conditioned on the entire input sequence. Finally, HMMs cannot even model the rules of context-free languages, while RNNs can [17, 16, 18, 94, 59].

LSTM recurrent networks were trained from scratch on utterances from the TIDIGITS speech database. It was found [3, 25, 26] that LSTM obtains results comparable to HMM based systems. A series of experiments on disjoint subsets of the database demonstrated that previous experience greatly reduces the network's training time, as well as increasing its accuracy. It was therefore argued that LSTM is a promising tool for applications requiring either rapid cross corpus adaptation or continually expanding datasets. Substantial promise also lies in LSTM-HMM hybrids that try to combine the best of both worlds, inspired by Robinson's hybrids based on traditional RNNs [66]. Recently we showed [28, 27] that LSTM learns framewise phoneme classification much faster than previous RNNs. Best results were obtained with a bi-directional variant of LSTM that classifies any part of a sequence by taking into account its entire past and future context.

Gradient-based LSTM also has been used to identify protein sequence motifs that contribute to classification [31]. Protein classification is important for extracting binding or active sites on a protein in order to develop new drugs, and in determining 3D protein folding features that can provide a better understanding of diseases resulting from protein misfolding.

Sometimes gradient information is of little use due to rough error surfaces with numerous local minima. For such cases, we have recently introduced a new, evolutionary/gradient-descent hybrid method for training LSTM and other RNNs called Evolino [96, 119, 93, 95]. Evolino evolves weights to the non-linear, hidden nodes of RNNs while computing optimal linear mappings from hidden state to output, using methods such as pseudo-inverse-based linear regression [57] or support vector machine-like quadratic programming [112], depending on the notion of optimality employed. Evolino-based LSTM can solve tasks that Echo State networks [37] cannot, and achieves higher accuracy in certain continuous function generation tasks than gradient-based LSTM, as well as other conventional gradient descent RNNs. However, for several problems requiring large networks with numerous learnable weights, gradient-based LSTM was superior to Evolino-based LSTM.

8 Reinforcement-Learning / Evolving Recurrent Neural Networks

In a certain sense, Reinforcement Learning (RL) is more challenging than supervised learning as above, since there is no teacher providing desired outputs at appropriate time steps. To solve a given problem, the learning agent itself must discover useful output sequences in response to the observations. The traditional approach to RL is best embodied by Sutton and Barto's book [109]. It makes strong assumptions about the environment, such as the Markov assumption: the current input of the agent tells it all it needs to know about the environment. Then all we need to learn is some sort of reactive mapping from stationary inputs to outputs. This is often unrealistic.

More general approaches search a space of truly sequence-processing programs with temporary variables for storing previous observations. For example, Olsson's ADATE [54] or related approaches such as *Genetic Programming (GP)* [9, 11, 73] can in principle be used to evolve such programs by maximizing an appropriate objective or fitness function. *Probabilistic Incremental Program Evolution (PIPE)* [71] is a related technique for automatic program synthesis, combining probability vector coding of program instructions [97] and Population-Based Incremental Learning [2] and tree-coded programs. PIPE was used for learning soccer team strategies in rather realistic simulations [72, 118].

A related, rather general approach for partially observable environments directly evolves programs for recurrent neural networks (RNN) with internal states, by applying evolutionary algorithms [63, 98, 33] to RNN weight matrices [46, 124, 123, 53, 44, 102, 49]. RNN can run general programs with memory / internal states (no need for the Markovian assumption), but for a long time it was unclear how to efficiently evolve their weights to solve complex RL tasks. Recent work, however, brought progress through a focus on reducing search spaces by co-evolving the comparatively small weight vectors of individual recurrent neurons [21, 22, 20, 23, 19, 24]. The powerful RNN learn to use their potential to create memories of important events, solving numerous RL / optimization tasks unsolvable by traditional RL methods [23, 19, 24]. As mentioned in the previous section, even supervised learning can greatly profit from this approach [96, 119, 93, 95].

9 Is History Converging? Again?

Many predict that within a few decades there will be computers whose raw computing power will surpass the one of a human brain by far (e.g., [48, 40]). We have argued that algorithmic advances are keeping up with the hardware development, pointing to new-millennium theoretical insights on universal problem solvers that are optimal in various mathematical senses (thus making *General AI* a real formal science), as well as practical progress in program learning through brain-inspired recurrent neural nets (as opposed to mere pattern association through traditional reactive devices).

Let us put the AI-oriented developments [88] discussed above in a broader context, and extend the analysis of past computer science breakthroughs in the introduction, which predicts that computer history will converge in an *Omega point* or historic *singularity* Ω around 2040.

Surprisingly, even if we go back all the way to the beginnings of modern man over 40,000 years ago, essential historic developments (that is, the subjects of the major chapters in history books) match a a binary scale marking exponentially declining temporal intervals, each half the size of the previous one, and even measurable in terms of powers of 2 multiplied by a human lifetime [91] (roughly 80 years—throughout recorded history many individuals have reached this age, although the average lifetime often was shorter, mostly due to high children mortality). Using the value $\Omega = 2040$, associate an error bar of not much more than 10 percent with each date below:

1. $\Omega - 2^9$ lifetimes: modern humans start colonizing the world from Africa
2. $\Omega - 2^8$ lifetimes: bow and arrow invented; hunting revolution
3. $\Omega - 2^7$ lifetimes: invention of agriculture; first permanent settlements; beginnings of civilization
4. $\Omega - 2^6$ lifetimes: first high civilizations (Sumeria, Egypt), and the most important invention of recorded history, namely, the one that made recorded history possible: writing
5. $\Omega - 2^5$ lifetimes: the ancient Greeks invent democracy and lay the foundations of Western science and art and philosophy, from algorithmic procedures and formal proofs to anatomically perfect sculptures, harmonic music, and organized sports. Old Testament written, major Asian religions founded. High civilizations in China, origin of the first calculation tools, and India, origin of the zero
6. $\Omega - 2^4$ lifetimes: bookprint (often called the most important invention of the past 2000 years) invented in China. Islamic science and culture start spreading across large parts of the known world (this has sometimes been called the most important event between Antiquity and the age of discoveries)
7. $\Omega - 2^3$ lifetimes: the largest and most dominant empire ever (perhaps including more than half of humanity and two thirds of the world economy) stretches across Asia from Korea all the way to Germany. Chinese

fleets and later also European vessels start exploring the world. Gun powder and guns invented in China. Rennaissance and Western bookprint (often called the most influential invention of the past 1000 years) and subsequent Reformation in Europe. Begin of the Scientific Revolution
8. $\Omega - 2^2$ lifetimes: Age of enlightenment and rational thought in Europe. Massive progress in the sciences; first flying machines; start of the industrial revolution based on the first steam engines
9. $\Omega - 2$ lifetimes: Second industrial revolution based on combustion engines, cheap electricity, and modern chemistry. Birth of modern medicine through the germ theory of disease; genetic and evolution theory. European colonialism at its short-lived peak
10. $\Omega - 1$ lifetime: modern post-World War II society and pop culture emerges. The world-wide super-exponential population explosion (mainly due to the Haber-Bosch process [103]) is at its peak. First commercial computers and first spacecraft; DNA structure unveiled
11. $\Omega - 1/2$ lifetime: 3rd industrial revolution based on personal computers and the World Wide Web. A mathematical theory of universal AI emerges (see sections above) - will this be considered a milestone in the future?
12. $\Omega - 1/4$ lifetime: This point will be reached in a few years. See introduction
13. ...

The following disclosure should help the reader to take this list with a grain of salt though. The author, who admits being very interested in witnessing the Omega point, was born in 1963, and therefore perhaps should not expect to live long past 2040. This may motivate him to uncover certain historic patterns that fit his desires, while ignoring other patterns that do not.

Others may feel attracted by the same trap. For example, Kurzweil [40] identifies exponential speedups in sequences of historic paradigm shifts identified by various historians, to back up the hypothesis that "the singularity is near." His historians are all contemporary though, presumably being subject to a similar bias. People of past ages might have held quite different views. For example, possibly some historians of the year 1525 felt inclined to predict a convergence of history around 1540, deriving this date from an exponential speedup of recent breakthroughs such as Western bookprint (around 1444), the re-discovery of America (48 years later), the Reformation (again 24 years later—see the pattern?), and other events they deemed important although today they are mostly forgotten.

In fact, could it be that such lists just reflect the human way of allocating memory space to past events? Maybe there is a general rule for both the individual memory of single humans and the collective memory of entire societies and their history books: constant amounts of memory space get allocated to exponentially larger, adjacent time intervals further and further into the past. For example, events that happened between 2 and 4 lifetimes ago get roughly as much memory space as events in the previous interval of twice the size. Presumably only a few "important" memories will survive the necessary compression. Maybe that's why there has never been a shortage of prophets predicting that the end is near - the important events according to one's own view of the past always seem to accelerate exponentially.

A similar plausible type of memory decay allocates $O(1/n)$ memory units to all events older than $O(n)$ unit time intervals. This is reminiscent of a bias governed by a time-reversed Speed Prior [81] (Section 4).

References

[1] L. B. Almeida. A learning rule for asynchronous perceptrons with feedback in a combinatorial environment. In *IEEE 1st International Conference on Neural Networks, San Diego*, volume 2, pages 609–618, 1987.

[2] S. Baluja and R. Caruana. Removing the genetics from the standard genetic algorithm. In A. Prieditis and S. Russell, editors, *Machine Learning: Proceedings of the Twelfth International Conference*, pages 38–46. Morgan Kaufmann Publishers, San Francisco, CA, 1995.

[3] N. Beringer, A. Graves, F. Schiel, and J. Schmidhuber. Classifying unprompted speech by retraining LSTM nets. In W. Duch, J. Kacprzyk, E. Oja, and S. Zadrozny, editors, *Artificial Neural Networks: Biological Inspirations - ICANN 2005, LNCS 3696*, pages 575–581. Springer-Verlag Berlin Heidelberg, 2005.

[4] C. M. Bishop. *Neural networks for pattern recognition.* Oxford University Press, 1995.

[5] Alan D. Blair and Jordan B. Pollack. Analysis of dynamical recognizers. *Neural Computation*, 9(5):1127–1142, 1997.

[6] M. Boden and J. Wiles. Context-free and context-sensitive dynamics in recurrent neural networks. *Connection Science*, 2000.

[7] H.A. Bourlard and N. Morgan. *Connnectionist Speech Recognition: A Hybrid Approach.* Kluwer Academic Publishers, 1994.

[8] M. P. Casey. The dynamics of discrete-time computation, with application to recurrent neural networks and finite state machine extraction. *Neural Computation*, 8(6):1135–1178, 1996.

[9] N. L. Cramer. A representation for the adaptive generation of simple sequential programs. In J.J. Grefenstette, editor, *Proceedings of an International Conference on Genetic Algorithms and Their Applications, Carnegie-Mellon University, July 24-26, 1985*, Hillsdale NJ, 1985. Lawrence Erlbaum Associates.

[10] B. de Vries and J. C. Principe. A theory for neural networks with time delays. In R. P. Lippmann, J. E. Moody, and D. S. Touretzky, editors, *Advances in Neural Information Processing Systems 3*, pages 162–168. Morgan Kaufmann, 1991.

[11] D. Dickmanns, J. Schmidhuber, and A. Winklhofer. Der genetische Algorithmus: Eine Implementierung in Prolog. Fortgeschrittenenpraktikum, Institut für Informatik, Lehrstuhl Prof. Radig, Technische Universität München, 1987.

[12] J. L. Elman. Finding structure in time. Technical Report CRL Technical Report 8801, Center for Research in Language, University of California, San Diego, 1988.

[13] S. E. Fahlman. The recurrent cascade-correlation learning algorithm. In R. P. Lippmann, J. E. Moody, and D. S. Touretzky, editors, *Advances in Neural Information Processing Systems 3*, pages 190–196. Morgan Kaufmann, 1991.

[14] F. A. Gers and J. Schmidhuber. Neural processing of complex continual input streams. In *Proc. IJCNN'2000, Int. Joint Conf. on Neural Networks*, Como, Italy, 2000.

[15] F. A. Gers and J. Schmidhuber. Recurrent nets that time and count. In *Proc. IJCNN'2000, Int. Joint Conf. on Neural Networks*, Como, Italy, 2000.

[16] F. A. Gers and J. Schmidhuber. LSTM recurrent networks learn simple context free and context sensitive languages. *IEEE Transactions on Neural Networks*, 12(6):1333–1340, 2001.

[17] F. A. Gers, J. Schmidhuber, and F. Cummins. Learning to forget: Continual prediction with LSTM. *Neural Computation*, 12(10):2451–2471, 2000.

[18] F. A. Gers, N. Schraudolph, and J. Schmidhuber. Learning precise timing with LSTM recurrent networks. *Journal of Machine Learning Research*, 3:115–143, 2002.

[19] F. Gomez and J. Schmidhuber. Evolving modular fast-weight networks for control. In W. Duch, J. Kacprzyk, E. Oja, and S. Zadrozny, editors, *Artificial Neural Networks: Biological Inspirations - ICANN 2005, LNCS 3697*, pages 383–389. Springer-Verlag Berlin Heidelberg, 2005.

[20] F. J. Gomez. *Robust Nonlinear Control through Neuroevolution.* PhD thesis, Department of Computer Sciences, University of Texas at Austin, 2003.

[21] F. J. Gomez and R. Miikkulainen. Incremental evolution of complex general behavior. *Adaptive Behavior*, 5:317–342, 1997.

[22] F. J. Gomez and R. Miikkulainen. Solving non-Markovian control tasks with neuroevolution. In *Proc. IJCAI 99*, Denver, CO, 1999. Morgan Kaufman.

[23] F. J. Gomez and R. Miikkulainen. Active guidance for a finless rocket using neuroevolution. In *Proc. GECCO 2003, Chicago*, 2003. *Winner of Best Paper Award in Real World Applications. Gomez is working at IDSIA on a CSEM grant to J. Schmidhuber.*

[24] F. J. Gomez and J. Schmidhuber. Co-evolving recurrent neurons learn deep memory POMDPs. In *Proc. of the 2005 conference on genetic and evolutionary computation (GECCO), Washington, D. C.* ACM Press, New York, NY, USA, 2005. *Nominated for a best paper award.*

[25] A. Graves, N. Beringer, and J. Schmidhuber. Rapid retraining on speech data with LSTM recurrent networks. Technical Report IDSIA-09-05, IDSIA, www.idsia.ch/techrep.html, 2005.

[26] A. Graves, S. Fernandez, F. Gomez, and J. Schmidhuber. Connectionist temporal classification: Labelling unsegmented sequence data with recurrent neural nets. In *ICML'06: Proceedings of the International Conference on Machine Learning*, 2006.

[27] A. Graves and J. Schmidhuber. Framewise phoneme classification with bidirectional LSTM and other neural network architectures. *Neural Networks*, 18:602–610, 2005.

[28] A. Graves and J. Schmidhuber. Framewise phoneme classification with bidirectional LSTM networks. In *Proc. Int. Joint Conf. on Neural Networks IJCNN 2005*, 2005.

[29] S. Hochreiter. Untersuchungen zu dynamischen neuronalen Netzen. Diploma thesis, Institut für Informatik, Lehrstuhl Prof. Brauer, Technische Universität München, 1991. See www7.informatik.tu-muenchen.de/~hochreit; advisor: J. Schmidhuber.

[30] S. Hochreiter, Y. Bengio, P. Frasconi, and J. Schmidhuber. Gradient flow in recurrent nets: the difficulty of learning long-term dependencies. In S. C. Kremer and J. F. Kolen, editors, *A Field Guide to Dynamical Recurrent Neural Networks.* IEEE Press, 2001.

[31] S. Hochreiter and K. Obermayer. Sequence classification for protein analysis. In *Snowbird Workshop*, Snowbird, Utah, April 5-8 2005. Computational and Biological Learning Society.

[32] S. Hochreiter and J. Schmidhuber. Long short-term memory. *Neural Computation*, 9(8):1735–1780, 1997.

[33] J. H. Holland. *Adaptation in Natural and Artificial Systems.* University of Michigan Press, Ann Arbor, 1975.

[34] J. J. Hopfield. Neural networks and physical systems with emergent collective computational abilities. *Proc. of the National Academy of Sciences*, 79:2554–2558, 1982.

[35] M. Hutter. The fastest and shortest algorithm for all well-defined problems. *International Journal of Foundations of Computer Science*, 13(3):431–443, 2002. (On J. Schmidhuber's SNF grant 20-61847).

[36] M. Hutter. *Universal Artificial Intelligence: Sequential Decisions based on Algorithmic Probability.* Springer, Berlin, 2004. (On J. Schmidhuber's SNF grant 20-61847).

[37] H. Jaeger. Harnessing nonlinearity: Predicting chaotic systems and saving energy in wireless communication. *Science*, 304:78–80, 2004.

[38] L. P. Kaelbling, M. L. Littman, and A. W. Moore. Reinforcement learning: a survey. *Journal of AI research*, 4:237–285, 1996.

[39] Y. Kalinke and H. Lehmann. Computation in recurrent neural networks: From counters to iterated function systems. In G. Antoniou and J. Slaney, editors, *Advanced Topics in Artificial Intelligence, Proceedings of the 11th Australian Joint Conference on Artificial Intelligence*, volume 1502 of *LNAI*, Berlin, Heidelberg, 1998. Springer.

[40] R. Kurzweil. *The Singularity is near.* Wiley Interscience, 2005.

[41] L. A. Levin. Universal sequential search problems. *Problems of Information Transmission*, 9(3):265–266, 1973.

[42] M. Li and P. M. B. Vitányi. *An Introduction to Kolmogorov Complexity and its Applications (2nd edition)*. Springer, 1997.

[43] T. Lin, B.G. Horne, P. Tino, and C.L. Giles. Learning long-term dependencies in NARX recurrent neural networks. *IEEE Transactions on Neural Networks*, 7(6):1329–1338, 1996.

[44] O. Miglino, H. Lund, and S. Nolfi. Evolving mobile robots in simulated and real environments. *Artificial Life*, 2(4):417–434, 1995.

[45] C. B. Miller and C. L. Giles. Experimental comparison of the effect of order in recurrent neural networks. *International Journal of Pattern Recognition and Artificial Intelligence*, 7(4):849–872, 1993.

[46] G. Miller, P. Todd, and S. Hedge. Designing neural networks using genetic algorithms. In *Proceedings of the 3rd International Conference on Genetic Algorithms*, pages 379–384. Morgan Kauffman, 1989.

[47] T. Mitchell. *Machine Learning*. McGraw Hill, 1997.

[48] H. Moravec. *Robot*. Wiley Interscience, 1999.

[49] D. E. Moriarty and R. Miikkulainen. Efficient reinforcement learning through symbiotic evolution. *Machine Learning*, 22:11–32, 1996.

[50] M. C. Mozer. A focused back-propagation algorithm for temporal sequence recognition. *Complex Systems*, 3:349–381, 1989.

[51] M. C. Mozer. Induction of multiscale temporal structure. In D. S. Lippman, J. E. Moody, and D. S. Touretzky, editors, *Advances in Neural Information Processing Systems 4*, pages 275–282. Morgan Kaufmann, 1992.

[52] A. Newell and H. Simon. GPS, a program that simulates human thought. In E. Feigenbaum and J. Feldman, editors, *Computers and Thought*, pages 279–293. McGraw-Hill, New York, 1963.

[53] S. Nolfi, D. Floreano, O. Miglino, and F. Mondada. How to evolve autonomous robots: Different approaches in evolutionary robotics. In R. A. Brooks and P. Maes, editors, *Fourth International Workshop on the Synthesis and Simulation of Living Systems (Artificial Life IV)*, pages 190–197. MIT, 1994.

[54] J. R. Olsson. Inductive functional programming using incremental program transformation. *Artificial Intelligence*, 74(1):55–83, 1995.

[55] B. A. Pearlmutter. Learning state space trajectories in recurrent neural networks. *Neural Computation*, 1(2):263–269, 1989.

[56] B. A. Pearlmutter. Gradient calculations for dynamic recurrent neural networks: A survey. *IEEE Transactions on Neural Networks*, 6(5):1212–1228, 1995.

[57] R. Penrose. A generalized inverse for matrices. In *Proceedings of the Cambridge Philosophy Society*, volume 51, pages 406–413, 1955.

[58] J. A. Pérez-Ortiz, F. A. Gers, D. Eck, and J. Schmidhuber. Kalman filters improve LSTM network performance in problems unsolvable by traditional recurrent nets. *Neural Networks*, (16):241–250, 2003.

[59] J. A. Pérez-Ortiz, F. A. Gers, D. Eck, and J. Schmidhuber. Kalman filters improve LSTM network performance in problems unsolvable by traditional recurrent nets. *Neural Networks*, 16(2):241–250, 2003.

[60] F. J. Pineda. Recurrent backpropagation and the dynamical approach to adaptive neural computation. *Neural Computation*, 1(2):161–172, 1989.

[61] T. A. Plate. Holographic recurrent networks. In J. D. Cowan S. J. Hanson and C. L. Giles, editors, *Advances in Neural Information Processing Systems 5*, pages 34–41. Morgan Kaufmann, 1993.

[62] G. V. Puskorius and L. A. Feldkamp. Neurocontrol of nonlinear dynamical systems with Kalman filter trained recurrent networks. *IEEE Transactions on Neural Networks*, 5(2):279–297, 1994.

[63] I. Rechenberg. Evolutionsstrategie - Optimierung technischer Systeme nach Prinzipien der biologischen Evolution. Dissertation, 1971. Published 1973 by Fromman-Holzboog.

[64] M. B. Ring. Learning sequential tasks by incrementally adding higher orders. In J. D. Cowan S. J. Hanson and C. L. Giles, editors, *Advances in Neural Information Processing Systems 5*, pages 115–122. Morgan Kaufmann, 1993.

[65] A. J. Robinson and F. Fallside. The utility driven dynamic error propagation network. Technical Report CUED/F-INFENG/TR.1, Cambridge University Engineering Department, 1987.

[66] Anthony J. Robinson. An application of recurrent nets to phone probability estimation. *IEEE Transactions on Neural Networks*, 5(2):298–305, March 1994.

[67] P. Rodriguez, J. Wiles, and J Elman. A recurrent neural network that learns to count. *Connection Science*, 11(1):5–40, 1999.

[68] Paul Rodriguez and Janet Wiles. Recurrent neural networks can learn to implement symbol-sensitive counting. In *Advances in Neural Information Processing Systems*, volume 10, pages 87–93. The MIT Press, 1998.

[69] P. S. Rosenbloom, J. E. Laird, and A. Newell. *The SOAR Papers*. MIT Press, 1993.

[70] D. E. Rumelhart and J. L. McClelland, editors. *Parallel Distributed Processing*, volume 1. MIT Press, 1986.

[71] R. P. Sałustowicz and J. Schmidhuber. Probabilistic incremental program evolution. *Evolutionary Computation*, 5(2):123–141, 1997.

[72] R. P. Sałustowicz, M. A. Wiering, and J. Schmidhuber. Learning team strategies: Soccer case studies. *Machine Learning*, 33(2/3):263–282, 1998.

[73] J. Schmidhuber. Evolutionary principles in self-referential learning. Diploma thesis, Institut für Informatik, Technische Universität München, 1987.

[74] J. Schmidhuber. Dynamische neuronale Netze und das fundamentale raumzeitliche Lernproblem. Dissertation, Institut für Informatik, Technische Universität München, 1990.

[75] J. Schmidhuber. An on-line algorithm for dynamic reinforcement learning and planning in reactive environments. In *Proc. IEEE/INNS International Joint Conference on Neural Networks, San Diego*, volume 2, pages 253–258, 1990.

[76] J. Schmidhuber. Reinforcement learning in Markovian and non-Markovian environments. In D. S. Lippman, J. E. Moody, and D. S. Touretzky, editors, *Advances in Neural Information Processing Systems 3 (NIPS 3)*, pages 500–506. Morgan Kaufmann, 1991.

[77] J. Schmidhuber. A fixed size storage $O(n^3)$ time complexity learning algorithm for fully recurrent continually running networks. *Neural Computation*, 4(2):243–248, 1992.

[78] J. Schmidhuber. Learning to control fast-weight memories: An alternative to recurrent nets. *Neural Computation*, 4(1):131–139, 1992.

[79] J. Schmidhuber. Netzwerkarchitekturen, Zielfunktionen und Kettenregel. Habilitationsschrift, Institut für Informatik, Technische Universität München, 1993.

[80] J. Schmidhuber. Hierarchies of generalized Kolmogorov complexities and nonenumerable universal measures computable in the limit. *International Journal of Foundations of Computer Science*, 13(4):587–612, 2002.

[81] J. Schmidhuber. The Speed Prior: a new simplicity measure yielding near-optimal computable predictions. In J. Kivinen and R. H. Sloan, editors, *Proceedings of the 15th Annual Conference on Computational Learning Theory (COLT 2002)*, Lecture Notes in Artificial Intelligence, pages 216–228. Springer, Sydney, Australia, 2002.

[82] J. Schmidhuber. Exponential speed-up of computer history's defining moments, 2003. http://www.idsia.ch/~juergen/computerhistory.html.

[83] J. Schmidhuber. Gödel machines: self-referential universal problem solvers making provably optimal self-improvements. Technical Report IDSIA-19-03, arXiv:cs.LO/0309048, IDSIA, Manno-Lugano, Switzerland, 2003.

[84] J. Schmidhuber. The new AI: General & sound & relevant for physics. Technical Report TR IDSIA-04-03, Version 1.0, cs.AI/0302012 v1, February 2003.

[85] J. Schmidhuber. Overview of work on robot learning, with publications, 2004. http://www.idsia.ch/~juergen/learningrobots.html.

[86] J. Schmidhuber. Completely self-referential optimal reinforcement learners. In W. Duch, J. Kacprzyk, E. Oja, and S. Zadrozny, editors, *Artificial Neural Networks: Biological Inspirations - ICANN 2005, LNCS 3697*, pages 223–233. Springer-Verlag Berlin Heidelberg, 2005. Plenary talk.

[87] J. Schmidhuber. Gödel machines: Towards a technical justification of consciousness. In D. Kudenko, D. Kazakov, and E. Alonso, editors, *Adaptive Agents and Multi-Agent Systems III (LNCS 3394)*, pages 1–23. Springer Verlag, 2005.

[88] J. Schmidhuber. Artificial Intelligence - history highlights and outlook: AI maturing and becoming a real formal science, 2006. http://www.idsia.ch/˜juergen/ai.html.

[89] J. Schmidhuber. Developmental robotics, optimal artificial curiosity, creativity, music, and the fine arts. *Connection Science*, 18(2):173–187, 2006.

[90] J. Schmidhuber. Gödel machines: fully self-referential optimal universal problem solvers. In B. Goertzel and C. Pennachin, editors, *Artificial General Intelligence*. Springer Verlag, in press, 2006.

[91] J. Schmidhuber. Is history converging? Again?, 2006. http://www.idsia.ch/˜juergen/history.html.

[92] J. Schmidhuber and B. Bakker. NIPS 2003 RNNaissance workshop on recurrent neural networks, Whistler, CA, 2003. http://www.idsia.ch/˜juergen/rnnaissance.html.

[93] J. Schmidhuber, M. Gagliolo, D. Wierstra, and F. Gomez. Evolino for recurrent support vector machines. In *ESANN'06*, 2006.

[94] J. Schmidhuber, F. Gers, and D. Eck. Learning nonregular languages: A comparison of simple recurrent networks and LSTM. *Neural Computation*, 14(9):2039–2041, 2002.

[95] J. Schmidhuber, D. Wierstra, M. Gagliolo, and F. Gomez. Training recurrent networks by EVOLINO. *Neural Computation*, 2006, in press.

[96] J. Schmidhuber, D. Wierstra, and F. J. Gomez. Evolino: Hybrid neuroevolution / optimal linear search for sequence prediction. In *Proceedings of the 19th International Joint Conference on Artificial Intelligence (IJCAI)*, pages 853–858, 2005.

[97] J. Schmidhuber, J. Zhao, and N. Schraudolph. Reinforcement learning with self-modifying policies. In S. Thrun and L. Pratt, editors, *Learning to learn*, pages 293–309. Kluwer, 1997.

[98] H. P. Schwefel. Numerische Optimierung von Computer-Modellen. Dissertation, 1974. Published 1977 by Birkhäuser, Basel.

[99] C. E. Shannon. A mathematical theory of communication (parts I and II). *Bell System Technical Journal*, XXVII:379–423, 1948.

[100] H. T. Siegelmann and E. D. Sontag. Turing computability with neural nets. *Applied Mathematics Letters*, 4(6):77–80, 1991.

[101] H.T. Siegelmann. *Theoretical Foundations of Recurrent Neural Networks*. PhD thesis, Rutgers, New Brunswick Rutgers, The State of New Jersey, 1992.

[102] K. Sims. Evolving virtual creatures. In Andrew Glassner, editor, *Proceedings of SIGGRAPH'94 (Orlando, Florida, July 1994)*, Computer Graphics Proceedings, Annual Conference, pages 15–22. ACM SIGGRAPH, ACM Press, jul 1994. ISBN 0-89791-667-0.

[103] V. Smil. Detonator of the population explosion. *Nature*, 400:415, 1999.

[104] R. J. Solomonoff. A formal theory of inductive inference. Part I. *Information and Control*, 7:1–22, 1964.

[105] R. J. Solomonoff. Complexity-based induction systems. *IEEE Transactions on Information Theory*, IT-24(5):422–432, 1978.
[106] M. Steijvers and P.D.G. Grunwald. A recurrent network that performs a contextsensitive prediction task. In *Proceedings of the 18th Annual Conference of the Cognitive Science Society*. Erlbaum, 1996.
[107] G. Sun, H. Chen, and Y. Lee. Time warping invariant neural networks. In J. D. Cowan S. J. Hanson and C. L. Giles, editors, *Advances in Neural Information Processing Systems 5*, pages 180–187. Morgan Kaufmann, 1993.
[108] G. Z. Sun, C. Lee Giles, H. H. Chen, and Y. C. Lee. The neural network pushdown automaton: Model, stack and learning simulations. Technical Report CS-TR-3118, University of Maryland, College Park, August 1993.
[109] R. Sutton and A. Barto. *Reinforcement learning: An introduction.* Cambridge, MA, MIT Press, 1998.
[110] B. Tonkes and J. Wiles. Learning a context-free task with a recurrent neural network: An analysis of stability. In *Proceedings of the Fourth Biennial Conference of the Australasian Cognitive Science Society*, 1997.
[111] P. Utgoff. Shift of bias for inductive concept learning. In R. Michalski, J. Carbonell, and T. Mitchell, editors, *Machine Learning*, volume 2, pages 163–190. Morgan Kaufmann, Los Altos, CA, 1986.
[112] V. Vapnik. *The Nature of Statistical Learning Theory.* Springer, New York, 1995.
[113] V. Vinge. The coming technological singularity, 1993. VISION-21 Symposium sponsored by NASA Lewis Research Center, and Whole Earth Review, Winter issue.
[114] R. L. Watrous and G. M. Kuhn. Induction of finite-state languages using second-order recurrent networks. *Neural Computation*, 4:406–414, 1992.
[115] P. J. Werbos. *Beyond Regression: New Tools for Prediction and Analysis in the Behavioral Sciences.* PhD thesis, Harvard University, 1974.
[116] P. J. Werbos. Generalization of backpropagation with application to a recurrent gas market model. *Neural Networks*, 1, 1988.
[117] P. J. Werbos. Neural networks, system identification, and control in the chemical industries. In D. A. Sofge D. A. White, editor, *Handbook of Intelligent Control: Neural, Fuzzy, and Adaptive Approaches*, pages 283–356. Thomson Learning, 1992.
[118] M. A. Wiering, R. P. Salustowicz, and J. Schmidhuber. Reinforcement learning soccer teams with incomplete world models. *Autonomous Robots*, 7(1):77–88, 1999.
[119] D. Wierstra, F. J. Gomez, and J. Schmidhuber. Modeling systems with internal state using Evolino. In *Proc. of the 2005 conference on genetic and evolutionary computation (GECCO), Washington, D. C.*, pages 1795–1802. ACM Press, New York, NY, USA, 2005. Got a GECCO best paper award.

[120] J. Wiles and J. Elman. Learning to count without a counter: A case study of dynamics and activation landscapes in recurrent networks. In *In Proceedings of the Seventeenth Annual Conference of the Cognitive Science Society*, pages pages 482 – 487, Cambridge, MA, 1995. MIT Press.

[121] R. J. Williams. Complexity of exact gradient computation algorithms for recurrent neural networks. Technical Report Technical Report NU-CCS-89-27, Boston: Northeastern University, College of Computer Science, 1989.

[122] R. J. Williams and J. Peng. An efficient gradient-based algorithm for on-line training of recurrent network trajectories. *Neural Computation*, 4:491–501, 1990.

[123] B. M. Yamauchi and R. D. Beer. Sequential behavior and learning in evolved dynamical neural networks. *Adaptive Behavior*, 2(3):219–246, 1994.

[124] Xin Yao. A review of evolutionary artificial neural networks. *International Journal of Intelligent Systems*, 4:203–222, 1993.

[125] S.J. Young and P.C Woodland. *HTK Version 3.2: User, Reference and Programmer Manual*, 2002.

[126] Z. Zeng, R. Goodman, and P. Smyth. Discrete recurrent neural networks for grammatical inference. *IEEE Transactions on Neural Networks*, 5(2), 1994.

The Challenges of Building Computational Cognitive Architectures

Ron Sun

Rensselaer Polytechnic Institute, Troy, New York, USA
rsun@rpi.edu
http://www.cogsci.rpi.edu/~rsun

Summary. The work in the area of computational cognitive modeling explores the essence of cognition through developing detailed understanding of cognition by specifying computational models. In this enterprise, a cognitive architecture is a domain-generic computational cognitive model that may be used for a broad, multiple-domain analysis of cognition. It embodies generic descriptions of cognition in computer algorithms and programs. Building cognitive architectures is a difficult task and a serious challenge to the fields of cognitive science, artificial intelligence, and computational intelligence. In this article, discussions of issues and challenges in developing cognitive architectures will be undertaken, examples of cognitive architectures will be given, and future directions will be outlined.

1 Introduction

While most work in computational intelligence takes an engineering approach, the field of cognitive modeling, with cognitive architectures in particular, takes a scientific approach — focusing on gathering empirical data and developing models that serve as scientific theories and scientific explanations of the data. It does so through an iterative hypothesis-test cycle. A cognitive architecture provides an essential framework to facilitate more detailed modeling and understanding of various components and processes of the mind.

Building cognitive architectures is a difficult challenge. In this article, discussions of issues and challenges in developing cognitive architectures will be undertaken, examples of cognitive architectures will be given, and future directions will be outlined. In the next section, the question of what a cognitive architecture is is answered. In section 3, the importance of cognitive architectures is addressed. In section 4, some background regarding the relation between artificial intelligence and cognitive science is provided. Then, in section 5, an example cognitive architecture is presented in detail and its applications to cognitive modeling and artificial intelligence described. In section 6, the significant challenges related to building cognitive architectures are articulated. Finally, section 7 concludes this article.

Ron Sun: *The Challenges of Building Computational Cognitive Architectures*, Studies in Computational Intelligence (SCI) **63**, 37–60 (2007)
www.springerlink.com

2 What is a Cognitive Architecture?

A cognitive architecture is a broadly-scoped, domain-generic computational cognitive model, capturing the essential structure and process of the mind, to be used for a broad, multiple-level, multiple-domain analysis of cognition [18, 35].

Let us explore this notion of architecture with an analogy. The architecture for a building consists of its overall framework and its overall design, as well as roofs, foundations, walls, windows, floors, and so on. Furniture and appliances can be easily rearranged and/or replaced and therefore they are not part of the architecture. By the same token, a cognitive architecture includes overall structures, essential divisions of modules, essential relations between modules, basic representations and algorithms within modules, and a variety of other aspects [37]. In general, an architecture includes those aspects of a system that are relatively invariant across time, domains, and individuals. It deals with processes of cognition in a structurally and mechanistically well defined way.

In relation to understanding the human mind (i.e., cognitive science), a cognitive architecture provides a concrete framework for more detailed modeling of cognitive phenomena, through specifying essential structures, divisions of modules, relations between modules, and so on. Its function is to provide an essential framework to facilitate more detailed modeling and exploration of various components and processes of the mind. Research in computational cognitive modeling explores the essence of cognition and various cognitive functionalities through developing detailed, process-based understanding by specifying computational models of mechanisms and processes. It embodies descriptions of cognition in computer algorithms and programs. That is, it produces runnable computational models. Detailed simulations are then conducted based on the computational models. In this enterprise, a cognitive architecture may be used for a broad, multiple-level, multiple-domain analysis of cognition.

In relation to building intelligent systems, a cognitive architecture specifies the underlying infrastructure for intelligent systems, which includes a variety of capabilities, modules, and subsystems. On that basis, application systems can be more easily developed. A cognitive architecture carries also with it theories of cognition and understanding of intelligence gained from studying the human mind. Therefore, the development of intelligent systems can be more cognitively grounded, which may be advantageous in many circumstances.

3 Why are Cognitive Architectures Important?

While there are all kinds of cognitive architectures in existence, in this article I am concerned specifically with psychologically oriented cognitive architectures (as opposed to software engineering oriented "cognitive" architectures): their importance and their applications. Psychologically oriented cognitive architectures are particularly important because (1) they are "intelligent" systems that are cognitively realistic (relatively speaking) and therefore they are more human-like in many ways, and (2) they shed new light on human cognition and therefore they are useful tools for advancing the understanding of cognition. Let us examine the importance of this type of cognitive architecture.

For cognitive science, the importance of such cognitive architectures lie in the fact that they are enormously useful in terms of understanding the human mind. In understanding cognitive phenomena, the use of computational simulation on the basis of cognitive architectures forces one to think in terms of process, and in terms of detail. Instead of using vague, purely conceptual theories, cognitive architectures force theoreticians to think clearly. They are therefore critical tools in the study of the mind. Researchers who use cognitive architectures must specify a cognitive mechanism in sufficient detail to allow the resulting models to be implemented on computers and run as simulations. This approach requires that important elements of the models be spelled out explicitly, thus aiding in developing better, conceptually clearer theories.

An architecture serves as an initial set of assumptions to be used for further modeling of cognition. These assumptions, in reality, may be based on either available scientific data (for example, psychological or biological observations), philosophical thoughts and arguments, or ad hoc working hypotheses (including computationally inspired such hypotheses). An architecture is useful and important precisely because it provides a comprehensive initial framework for further modeling in a variety of task domains.

Cognitive architectures also provide a deeper level of explanation. Instead of a model specifically designed for a specific task (often in an ad hoc way), using a cognitive architecture forces modelers to think in terms of the mechanisms and processes available within a generic cognitive architecture that are not specifically designed for a particular task, and thereby to generate explanations of the task that is not centered on superficial, high-level features of a task (as often happens with specialized, narrowly scoped models), that is, explanations of a deeper kind. To describe a task in terms of available mechanisms and processes of a cognitive architecture is to generate explanations centered on primitives of cognition as envisioned in the cognitive architecture, and therefore such explanations are deeper explanations. Because of the nature of such deeper explanations, this style of theorizing is also more likely to lead to unified explanations for a large variety of data and/or phenomena, because potentially a large variety of tasks, data, and phenomena can be explained on the basis of the same set of primitives provided by the same cognitive architecture. Therefore, using cognitive architectures leads to comprehensive theories of the mind [18, 2, 35].

In all, cognitive architectures are believed to be essential in advancing understanding of the mind [1, 18, 35]. Therefore, building cognitive architectures is an important enterprise for cognitive science.

On the other hand, for the fields of artificial intelligence and computational intelligence (AI/CI), the importance of cognitive architectures lies in the fact that they support the central goal of AI/CI — Building artificial systems that are as capable as human beings. Cognitive architectures help us to reverse engineer the only truly intelligent system around — the human mind. They constitute a solid basis for building truly intelligent systems, because they are well motivated by, and properly grounded in, existing cognitive research. The use of cognitive architectures in building intelligent systems may also facilitate the interaction between humans and artificially intelligent systems because of the similarity between humans and cognitively based intelligent systems.

It is also worth noting that cognitive architectures are the antithesis of expert systems: Instead of focusing on capturing performance in narrow domains, they are aimed to provide broad coverage of a wide variety of domains [11]. Business and industrial applications of intelligent systems increasingly require broadly scoped systems that are capable of a wide range of intelligent behaviors, not just isolated systems of narrow functionalities. For example, one application may require the inclusion of capabilities for raw image processing, pattern recognition, categorization, reasoning, decision making, and natural language communications. It may even require planning, control of robotic devices, and interactions with other systems and devices. Such requirements accentuate the importance of research on broadly scoped cognitive architectures that perform a wide range of cognitive functionalities across a variety of task domains.

4 AI and Cognitive Science

Artificial intelligence (or computational intelligence) and cognitive science have always been overlapping. Early in their history, that overlap was considerable. Herbert Simon once declared that “AI can have two purposes. One is to use the power of computers to augment human thinking. The other is to use a computer's artificial intelligence to understand how humans think.” Conversely, ever since the time when cognitive science was first created (that is, the time when cognitive science society was formed), artificial intelligence has been identified as one of the constituting disciplines. However, over time, the two disciplines have grown apart. The difficulty of many computational problems tackled by AI has led it to often adopt brute-force, optimality-oriented, or domain-specific solutions that are not cognitively realistic. Conversely, the need for experimental controllability, detail-orientedness, and precision has often led cognitive scientists to focus on some problems that AI would not consider as interesting problems.

One important question is whether cognitive science is relevant to solving AI/CI problems at all. Many cognitive scientists believe so. The human mind is one of the most flexible, general, and powerful intelligent systems in existence. Therefore, a good way to solving many AI/CI problems is a cognitive approach. Consequently, another important question is whether sub-optimal, "satisficing" methods or algorithms that are often revealed by cognitive science research are useful to AI/CI. Many cognitive scientists would say yes. It may be argued that in AI/CI, too much focus has been devoted to the search for domain-specific, brute-force, and/or optimal solutions [34]. Both robustness over a broad range of problems and computational efficiency (and tractability) are essential to long-term success of AI/CI as a field that generates general theories and general frameworks. Real-world problems are complex and they often include many different aspects, which require broad functionalities in order to be solved in a robust manner. All of these requirements above point to cognitively based approaches toward developing computational models, as human cognition is thus far the best example of intelligent systems that are robust and efficient.

In the reverse direction, can AI/CI contribute to our understanding of human cognition? The answer is clearly yes. AI/CI addresses many problems central to human cognition, usually in an mathematically/logically motivated or optimal way. AI/CI solutions thus often reflect fundamental mathematical/logical constraints and regularities underlying the problems, which should be relevant and applicable to all approaches to those problems, including those adopted in human cognition. Therefore, in that sense, they shed light on possible details of cognitive processes and mechanisms, and may lead to better understanding of these problems in general.

5 An Example of a Cognitive Architecture

5.1 An Overview

Below I will describe a cognitive architecture CLARION. It has been described extensively in a series of previous papers, including [43, 42], and [35, 36]. CLARION is an integrative architecture, consisting of a number of distinct subsystems, with a dual representational structure in each subsystem (implicit versus explicit representations). Its subsystems include the action-centered subsystem (the ACS), the non-action-centered subsystem (the NACS), the motivational subsystem (the MS), and the meta-cognitive subsystem (the MCS). The role of the action-centered subsystem is to control actions, regardless of whether the actions are for external physical movements or for internal mental operations. The role of the non-action-centered subsystem is to maintain general knowledge, either implicit or explicit. The role of the motivational subsystem is to provide underlying motivations for perception, action, and cognition, in terms of providing impetus and feedback (e.g., indicating whether outcomes are satisfactory or not). The role of the meta-cognitive subsystem is to monitor, direct, and modify the operations of the action-centered subsystem dynamically as well as the operations of all the other subsystems.

Each of these interacting subsystems consists of two levels of representation (i.e., a dual representational structure): Generally, in each subsystem, the top level encodes explicit knowledge and the bottom level encodes implicit knowledge. The distinction of implicit and explicit knowledge has been amply argued for before (see [23, 28, 5, 35]). The two levels interact, for example, by cooperating in actions, through a combination of the action recommendations from the two levels respectively, as well as by cooperating in learning through a bottom-up and a top-down process (to be discussed below). Essentially, it is a dual-process theory of mind [4]. See Figure 1.

This cognitive architecture is intended to satisfy some basic requirements as follows. It should be able to learn with or without a priori domain-specific knowledge to begin with [23, 42]. It also has to learn continuously from on-going experience in the world. As indicated by Medin et al. [13], Nosofsky et al [19], and others, human learning is often gradual and on-going. As suggested by Reber [23], Seger [28], Anderson [1], and others, there are clearly different types of knowledge involved in human learning (e.g., procedural vs. declarative, implicit vs. explicit, or sub-conceptual vs. conceptual). Moreover, different types of learning processes are involved in acquiring different types of knowledge [1, 10, 31, 42]. Finally, unlike ACT-R or SOAR, it should more fully incorporate motivational processes as well as meta-cognitive processes. Based on the above considerations, CLARION was developed.

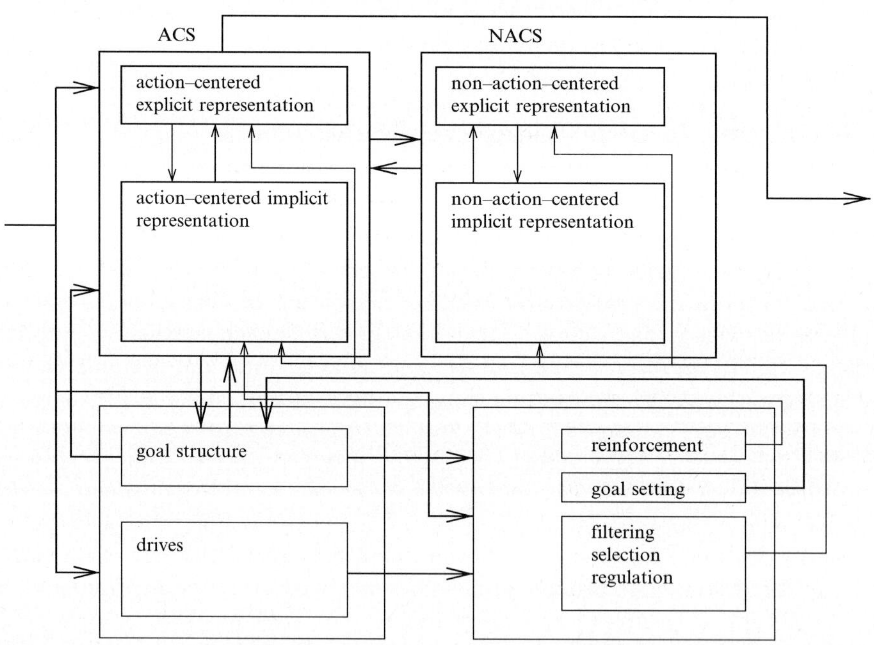

Fig. 1. The CLARION Architecture. ACS stands for the action-centered subsystem, NACS the non-action-centered subsystem, MS the motivational subsystem, and MCS the meta-cognitive subsystem

5.2 Some Details

The Action-Centered Subsystem

First, let us focus on the action-centered subsystem (the ACS) of CLARION (cf. [1, 31]). The overall operation of the action-centered subsystem may be described as follows:

1. Observe the current state x.
2. Compute in the bottom level the Q-values of x associated with each of all the possible actions a_i's: $Q(x, a_1)$, $Q(x, a_2)$, ..., $Q(x, a_n)$.
3. Find out all the possible actions $(b_1, b_2, ..., b_m)$ at the top level, based on the input x (sent up from the bottom level) and the rules in place.
4. Compare or combine the values of the selected a_i's with those of b_j's (sent down from the top level), and choose an appropriate action b.
5. Perform the action b, and observe the next state y and (possibly) the reinforcement r.
6. Update Q-values at the bottom level in accordance with the *Q-Learning-Backpropagation* algorithm.
7. Update the rule network at the top level using the *Rule-Extraction-Refinement algorithm.*
8. Go back to Step 1.

In the bottom level of the action-centered subsystem, implicit reactive routines are learned: A Q-value is an evaluation of the "quality" of an action in a given state: $Q(x, a)$ indicates how desirable action a is in state x (which consists of some sensory input). The agent may choose an action in any state based on Q-values. To acquire the Q-values, the Q-learning algorithm [48] may be used, which is a reinforcement learning algorithm. It basically compares the values of successive actions and adjusts an evaluation function on that basis. It thereby develops reactive sequential behaviors [36].

The bottom level of the action-centered subsystem is modular; that is, a number of small neural networks co-exist each of which is adapted to a specific modality, task, or group of input stimuli. This coincides with the modularity claim [6, 10, 8] that much processing is done by limited, encapsulated (to some extent), specialized processors that are highly efficient. These modules can be developed in interacting with the world (computationally, through various decomposition methods; e.g., [44]). Some of them, however, are formed evolutionarily, that is, given a priori to agents, reflecting their hardwired instincts and propensities [8].

In the top level of the action-centered subsystem, explicit conceptual knowledge is captured in the form of symbolic rules. See [36] for details. There are many ways in which explicit knowledge may be learned, including independent hypothesis-testing learning and "bottom-up learning" as discussed below.

Autonomous Generation of Explicit Conceptual Structures. Humans are generally able to learn implicit knowledge through trial and error, without necessarily utilizing a priori knowledge. On top of that, explicit knowledge can be acquired also from on-going experience in the world, through the mediation of implicit knowledge (i.e., bottom-up learning; see [35, 31, 10]). The basic process of bottom-up learning is as follows: if an action implicitly decided by the bottom level is successful, then the agent extracts an explicit rule that corresponds to the action selected by the bottom level and adds the rule to the top level. Then, in subsequent interaction with the world, the agent verifies the extracted rule by considering the outcome of applying the rule: if the outcome is not successful, then the rule should be made more specific and exclusive of the current case; if the outcome is successful, the agent may try to generalize the rule to make it more universal [14].[1]

After explicit rules have been learned, a variety of explicit reasoning methods may be used. Learning explicit conceptual representation at the top level can also be useful in enhancing learning of implicit reactive routines at the bottom level (e.g., [42]).

Assimilation of Externally Given Conceptual Structures. Although CLARION can learn even when no a priori or externally provided knowledge is available, it can make use of it when such knowledge is available [27, 1]. To deal with instructed learning, externally provided knowledge, in the forms of explicit conceptual structures such as rules, plans, categories, and so on, can (1) be combined with existent conceptual structures at the top level (i.e., internalization), and (2) be assimilated into implicit reactive routines at the bottom level (i.e., assimilation). This process is known as top-down learning. See [36] for more details.

The Non-Action-Centered Subsystem

The non-action-centered subsystem (NACS) may be used for representing general knowledge about the world (i.e., constituting the "semantic" memory as defined in, e.g., [21]), for performing various kinds of memory retrievals and inferences. Note that the non-action-centered subsystem is under the control of the action-centered subsystem (through its actions).

At the bottom level, associative memory networks encode non-action- centered implicit knowledge. Associations are formed by mapping an input to an output. The regular backpropagation learning algorithm may be used to establish such associations between pairs of inputs and outputs [26].

On the other hand, at the top level of the non-action-centered subsystem, a general knowledge store encodes explicit non-action-centered knowledge (cf. [32]). In this network, chunks are specified through dimensional values (features). A node is set up in the top level to represent a chunk. The chunk node connects to its corresponding features represented as individual

[1] The detail of the bottom-up learning algorithm can be found in [43].

nodes in the bottom level of the non-action-centered subsystem (see [32, 33] for details). Additionally, links between chunks encode explicit associations between pairs of chunks, known as associative rules. Explicit associative rules may be formed (i.e., learned) in a variety of ways [36].

During reasoning, in addition to applying associative rules, similarity-based reasoning may be employed in the non-action-centered subsystem. During reasoning, a known (given or inferred) chunk may be automatically compared with another chunk. If the similarity between them is sufficiently high, then the latter chunk is inferred (see [36] for details; see also [32, 33]).

As in the action-centered subsystem, top-down or bottom-up learning may take place in the non-action-centered subsystem, either to extract explicit knowledge in the top level from the implicit knowledge in the bottom level or to assimilate explicit knowledge of the top level into implicit knowledge in the bottom level.

The Motivational and the Meta-Cognitive Subsystem

The motivational subsystem (the MS) is concerned with drives and their interactions [47], which leads to actions. It is concerned with why an agent does what it does. Simply saying that an agent chooses actions to maximizes gains, rewards, reinforcements, or payoffs leaves open the question of what determines these things. The relevance of the motivational subsystem to the action-centered subsystem lies primarily in the fact that it provides the context in which the goal and the reinforcement of the action-centered subsystem are set. It thereby influences the working of the action-centered subsystem, and by extension, the working of the non-action-centered subsystem.

A bipartite system of motivational representation is in place in CLARION. The explicit goals (such as "finding food") of an agent (which is tied to the working of the action-centered subsystem) may be generated based on internal drive states (for example, "being hungry"). (See [36] for details.)

Beyond low-level drives (concerning physiological needs), there are also higher-level drives. Some of them are primary, in the sense of being "hard-wired". For example, Maslow [12] developed a set of these drives in the form of a "need hierarchy". While primary drives are built-in and relatively unalterable, there are also "derived" drives, which are secondary, changeable, and acquired mostly in the process of satisfying primary drives.

The meta-cognitive subsystem (the MCS) is closely tied to the motivational subsystem. The meta-cognitive subsystem monitors, controls, and regulates cognitive processes for the sake of improving cognitive performance [17, 30]. Control and regulation may be in the forms of setting goals for the action-centered subsystem, setting essential parameters of the action-centered subsystem and the non-action-centered subsystem, interrupting and changing on-going processes in the action-centered subsystem and the non-action-centered subsystem, and so on. Control and regulation can also be carried out through setting reinforcement functions for the action-centered subsystem. All of the above can be done on the basis of drive states in the motivational subsystem. The meta-cognitive subsystem is also made up of two levels: the top level (explicit) and the bottom level (implicit).

5.3 Accounting for Cognitive Data

CLARION has been successful in accounting for and explaining a variety of psychological data. For example, a number of well known skill learning tasks have been simulated using Clarion that span the spectrum ranging from simple reactive skills to complex cognitive skills. The simulated tasks include serial reaction time (SRT) tasks, artificial grammar learning (AGL) tasks, process control (PC) tasks, categorical inference (CI) tasks, alphabetical arithmetic (AA) tasks, and the Tower of Hanoi (TOH) task [35]. Among them, SRT and PC are typical implicit learning tasks (mainly involving implicit reactive routines), while TOH and AA are high-level cognitive skill acquisition tasks (with a significant presence of explicit processes). In addition, we have done extensive work on a complex minefield navigation (MN) task, which involves complex sequential decision making [42, 43]. We have also worked on an organizational decision task [38], and other social simulation tasks, as well as meta-cognitive tasks. While accounting for various psychological data, CLARION provides explanations that shed new light on cognitive phenomena. This point can be illustrated by the following example.

For instance, in [46], we simulated the alphabetic arithmetic task of [22]. In the task, subjects were asked to solve alphabetic arithmetic problems of the forms: $letter1 + number = letter2$ or $letter1 - number = letter2$, where $letter2$ is $number$ positions up or down from $letter1$ (depending on whether $+$ or $-$ was used; for example, $A + 2 = C$ or $C - 2 = A$). Subjects were given $letter1$ and $number$, and asked to produce $letter2$.

In experiment 1 of [22], during the training phase, one group of subjects (the consistent group) received 36 blocks of training, in which each block consisted of the same 12 addition problems. Another group (the varied group) received 6 blocks of training, in which each block consisted of the same 72 addition problems. While both groups received 432 trials, the consistent group practiced on each problem 36 times, but the varied group only 6 times. In the transfer phase, each group received 12 new addition problems (not practiced before), repeated 3 times. The findings were that, at the end of training, the consistent group performed far better than the varied group. However, during transfer, the consistent group performed worse than the varied group. The varied group showed perfect transfer, while the consistent group showed considerable slow-down. See Figure 2.

In experiment 2, the training phase was identical to that of experiment 1. However, during the transfer phase, both groups received 12 subtraction (not addition) problems, which were the reverse of the original addition problems, repeated 3 times. The findings were that, in contrast to experiment 1, during

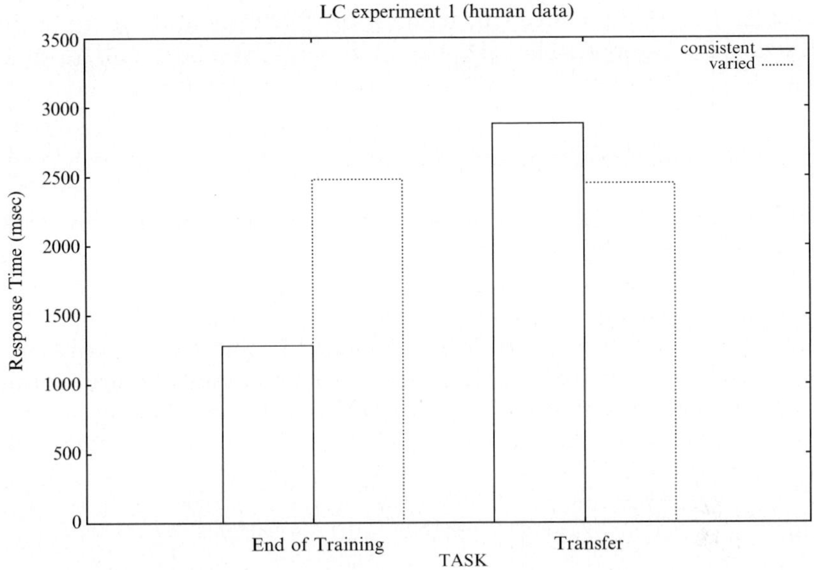

Fig. 2. Experiment 1 of the letter counting task (from [22])

transfer, the consistent group actually performed better than the varied group. Both groups performed worse than their corresponding performance at the end of training, but the varied group showed worse performance than the consistent group. See Figure 3.

How do we make sense of this complex pattern of data? Simulations were conducted based on CLARION. The CLARION simulation captured the data pattern, and provided detailed, process-based explanations for it. See the simulation data in Figure 4, which is to be compared with Figure 2. During the training phase of experiment 1, the simulated consistent group had a lower response time, because it had more practice on a smaller number of instances, which led to the better performing bottom level in the action-centered sub-system, as well as better performing instance retrieval from the top level of the non-action-centered subsystem.[2] Because they were better performing, the bottom level of the action-centered subsystem and the top level of the non-action-centered subsystem were more likely to be used in determining the overall outcome of the simulated consistent group, due to

[2] The bottom level in the action-centered subsystem of the simulated consistent group performed more accurately because of a more focused practice on a few instances by the consistent group (compared with the varied group). The top level of the non-action-centered subsystem of the simulated consistent group was more accurate for the same reason.

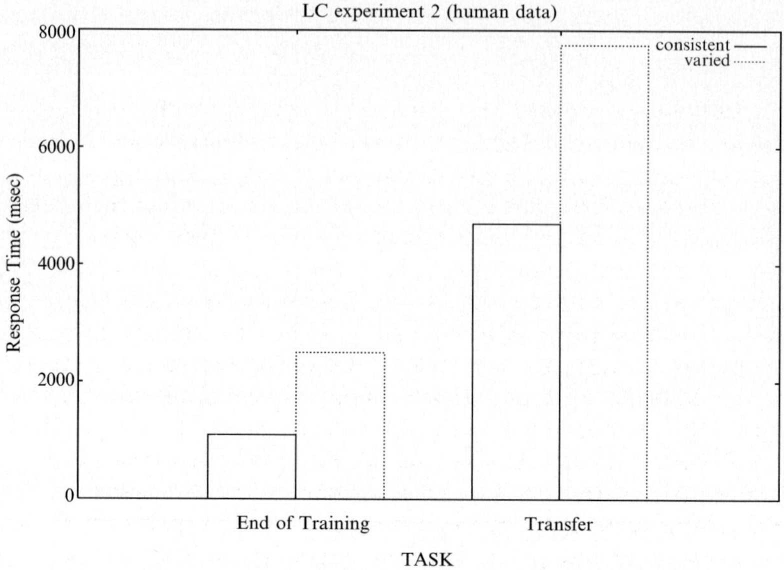

Fig. 3. Experiment 2 of the letter counting task (from [22])

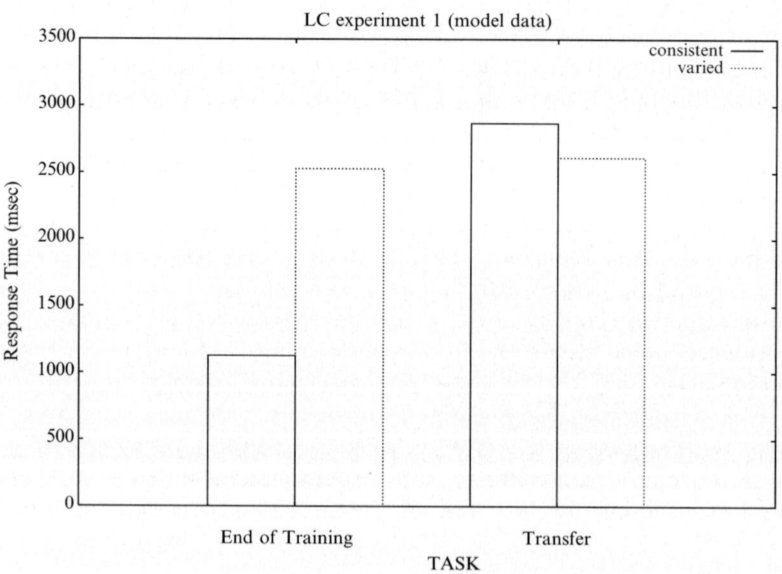

Fig. 4. Simulation of experiment 1 of the letter counting task

the competition among different components.[3] Because these two components had lower response times than other components,[4] a lower overall response time resulted for the simulated consistent group.

CLARION also matched the transfer performance difference between the two groups in experiment 1 (as shown in Figure 4). During the transfer phase of experiment 1, the performance of the simulated consistent group was worsened, compared with its performance at the end of training; the transfer performance of the simulated consistent group was in fact worse than that of the simulated varied group. This is because the simulated consistent group relied more on the bottom level of the action-centered subsystem and the non-action-centered subsystem during training and therefore, the activations of its counting rules (in the top level of the action-centered subsystem) were lower. As a result, it took more time to apply the counting rules during transfer, which it had to apply, due to the fact that it had to deal with a different set of instances during transfer. The performance of the simulated varied group hardly changed, compared with its performance at the end of training, because it relied mostly on the counting rules at the top level of the action-centered subsystem during training (which was equally applicable to both training and transfer). As a result, its counting rules had higher activations, and therefore it performed better than the simulated consistent group during transfer. See [46] for all the related statistical analysis.

As indicated by Figure 5, which is to be compared to Figure 3, the simulation also captured accurately the human data of experiment 2. During the transfer in experiment 2, due to the change in the task setting (counting down as opposed to counting up), the practiced rule for counting up was no longer useful. Therefore, both simulated groups had to use a new counting rule (for counting down), which had only the initial low activation for both groups. Similarly, both simulated groups had to use a new instance retrieval rule (for "reverse retrieval" of chunks such as "$A + 2 = C$"), which also had only the initial low activation in both cases.[5] Both simulated groups performed worse than at the end of training for the above reason.

Moreover, this simulation captured the fact that the varied group performed worse than the consistent group during transfer (Figure 5). This difference was explained by the fact that the simulated consistent group had

[3] We looked at the data, and indeed there were a lot more retrievals from the non-action-centered subsystem by the simulated consistent group than the simulated varied group [46]. The data also showed a higher selection probability of the bottom level of the action-centered subsystem in the simulated consistent group [46].

[4] It is either inherently so, as in the case of the bottom level of the action-centered subsystem, or due to more frequent use (and consequently higher activation), as in the case of the top level of the non-action-centered subsystem.

[5] Chunks were used in "reverse retrieval" during the transfer phase of experiment 2, because of the reverse relationship between the training and the transfer instances used in this experiment.

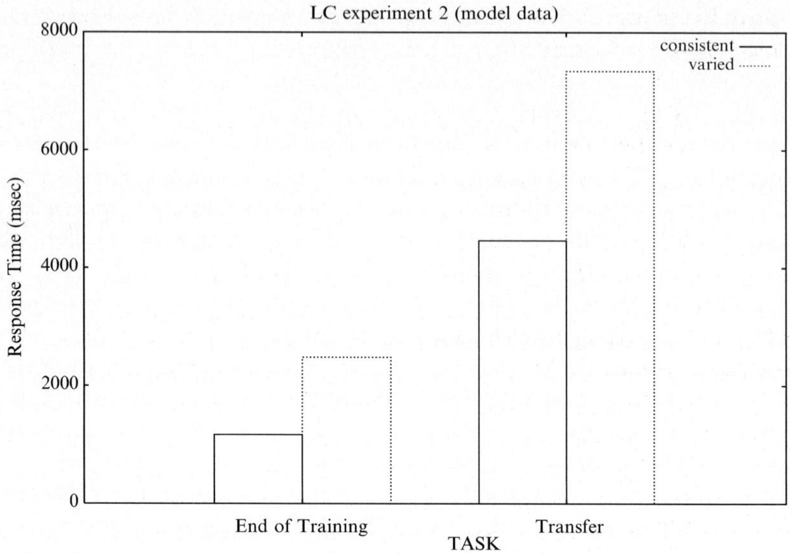

Fig. 5. Simulation of experiment 2 of the letter counting task

more activations associated with instance chunks (such as "$A + 2 = C$") than the simulated varied group, because the simulated consistent group had more practice with these chunks. Therefore, the simulated consistent group performed better than the simulated varied group in this phase.

CLARION provides some interesting interpretations of the human data. For example, it attributes (in part) the performance difference at the end of the training between the consistent and the varied group to the difference between relying on implicit knowledge and relying on explicit rules. Moreover, the CLARION simulation was far more accurate than the ACT-R simulation [9]. This fact suggests, to some extent, the advantage of CLARION.

In all of these simulations, the simulation with the CLARION cognitive architecture forced one to think in terms of process, and in terms of details. For instance, in the simulation described above, we investigated detailed computational processes involved in performing this task, in particular the roles of the two levels, and generated some conjectures in this respect.

CLARION also provides a deeper level of explanation. For example, in the case of simulating the alphabetic arithmetic task, explanations were provided in terms of action-centered knowledge or non-action-centered knowledge, in terms of explicit knowledge or implicit knowledge, or in terms of activations of representational units (in explaining response time differences), and so on. They were deeper because the explanations were centered on lower-level mechanisms and processes.

Due to the nature of such deeper explanations, this approach is also likely to lead to unified explanations, unifying a large variety of data and/or phenomena. For example, all the afore-mentioned tasks have been explained computationally in a unified way in CLARION.

5.4 A Model for Autonomous Intelligent Systems

CLARION also serves as a model for building autonomous intelligent systems (that is, intelligent agents). We tried to apply CLARION to a few reasonably interesting tasks in this regard, including learning minefield navigation.

As described in [43], CLARION was used to tackle a complex simulated minefield navigation task. The task setting was as shown in Figure 6. The CLARION based agent had to navigate an underwater vessel through a minefield to reach a target location. The agent received information only from a number of instruments, as shown in Figure 7. The sonar gauge showed how close the mines were in 7 equal areas that range from 45 degrees to the left of the agent to 45 degrees to the right. The fuel gauge showed the agent how much time was left before fuel ran out. The bearing gauge showed the direction of the target from the present direction of the agent. The range gauge showed how far the target was from the current location. Using only this sparse information, the agent decided (1) how to turn and (2) how fast to move. The agent, within an allotted time period, could either (a) reach the target (which is a success), (b) hit a mine (a failure), or (c) run out of fuel (a failure). The agent was under severe time pressure, so it had to be reactive in decision making, and it had no time for reasoning, episodic memory retrieval, and other slow cognitive processes. This task was thus quite

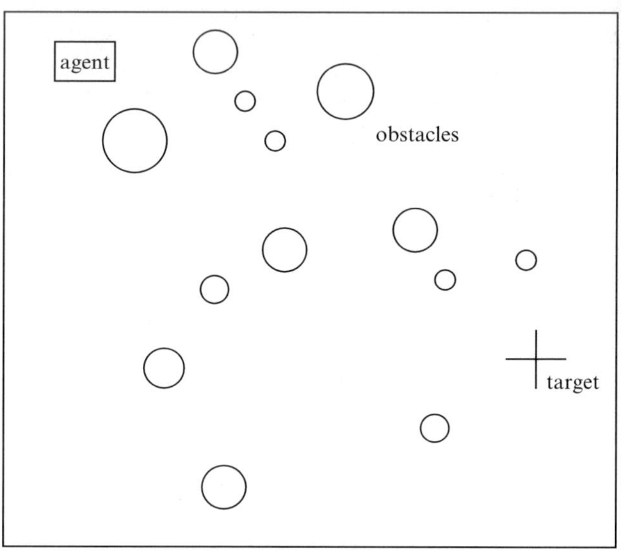

Fig. 6. Navigating through mines

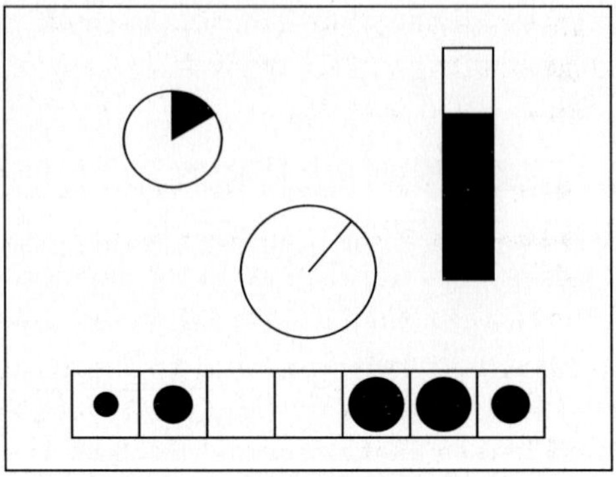

Fig. 7. The navigation input. The display at the upper left corner is the fuel gauge; the vertical one at the upper right corner is the range gauge; the round one in the middle is the bearing gauge; the 7 sonar gauges are at the bottom

difficult. CLARION based agents were applied to learning this task, starting from scratch, without any a priori knowledge. CLARION learned to reach a high level of performance through such autonomous learning. See Figure 8 for learning curves. As shown in the figure, CLARION based agents outperformed regular reinforcement learning (Q-learning) significantly, due to the synergy between the two levels of the ACS.

6 The Challenges Ahead

Let us look into some general and specific challenges in developing cognitive architectures in relation to cognitive science and AI/CI.

In general, building cognitive architectures is an extremely difficult task, because (1) a cognitive architecture needs to be compact but yet comprehensive in scope, (2) it needs to remain simple yet capture empirical data accurately, (3) it needs to be computationally feasible but also consistent with psychological theories, (4) it needs somehow to sort out and incorporate the myriad of incompatible psychological theories in existence, and so on.

6.1 The Challenges from Cognitive Science

Some have claimed that grand scientific theorizing have become a thing of the past. What remains to be done is filling in details and refining some minor points. Fortunately, many cognitive scientists believe otherwise. Many researchers in cognitive science are pursuing integrative approaches that attempt to explain data in multiple domains and functionalities [2, 35]. In cognitive science, as in many other scientific fields, significant advances may be made through discovering (hypothesizing and confirming) deep-level principles that unify superficial explanations across multiple domains, in a way

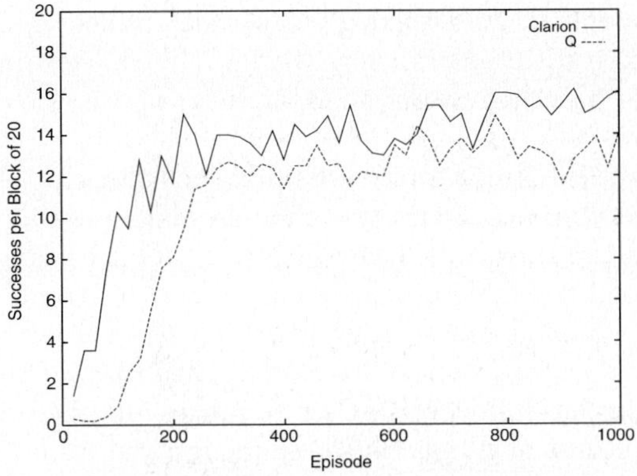

(a) The 30-mine learning curves.

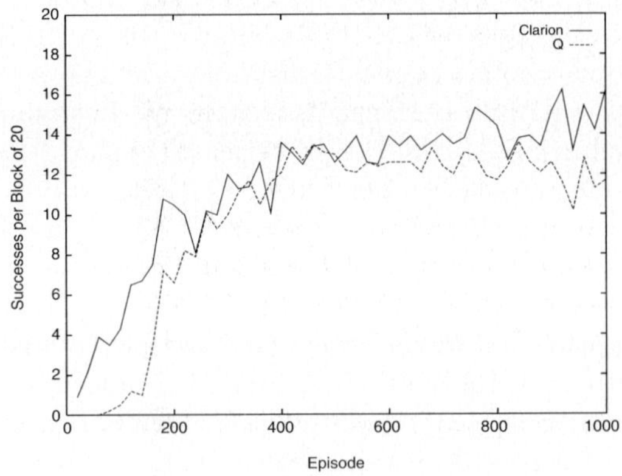

(b) The 60-mine learning curves.

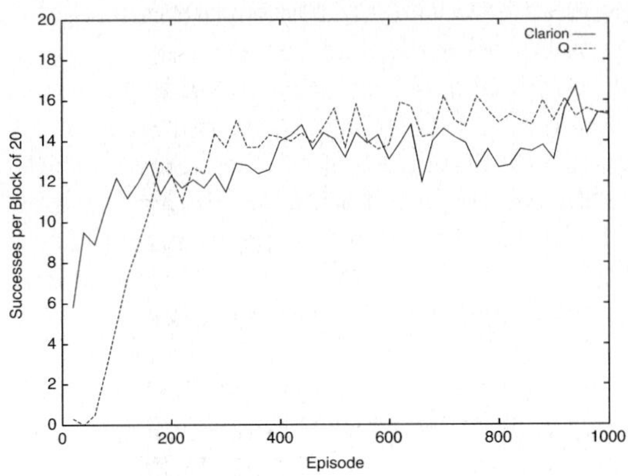

(c) The 10-mine learning curves.

Fig. 8. Learning curves for 10-mine, 30-mine, and 60-mine settings

somewhat analogous to Einstein's theory that unified electromagnetic and gravitational forces or String Theory that provides even further unifications [7]. Such theories, based on cognitive architectures, are exactly what cognitive science needs.

Cognitive scientists are actively pursuing such theories and, hopefully, will be increasingly doing so. Integrative models may serve as antidotes to the increasing specialization of research. Cognitive architectures that integrate a broad range of cognitive functionalities go against this trend of increasing specialization, and help to fit pieces together again. As voiced by many cognitive scientists, the trend of over-specialization is harmful and the reversal of this trend is a necessary step toward further advances [40]. Developing integrative cognitive architectures is thus a major challenge and a major opportunity.

In developing cognitive architectures, first of all, it is important to keep in mind a broad set of desiderata. For example, in [3] a set of desiderata proposed by Newell [18] was used to evaluate the ACT-R cognitive architecture versus conventional connectionist models. These desiderata include flexible behavior, real-time performance, adaptive behavior, vast knowledge base, dynamic behavior, knowledge integration, natural language, learning, development, evolution, and brain realization. These were considered functional constraints on cognitive architectures. In [37], another, broader set of desiderata was proposed and used to evaluate a wider set of cognitive architectures. These desiderata include ecological realism, bio-evolutionary realism, cognitive realism, and many others (see [37] for details). The advantages of coming up with and applying these sets of desiderata include (1) avoiding overly narrow models, (2) avoiding missing certain crucial functionalities, and (3) avoiding inappropriate approaches in implementing cognitive architectures. One can reasonably expect that this issue will be driving research in the field of cognitive architectures in the future. The challenge is to come up with better, more balanced sets of desiderata.

Related to that, some general architectural principles also need to be examined. For example, it was claimed that the strengths of ACT-R derive from its tight integration of the symbolic and the subsymbolic component [2]. On the other hand, it has been argued [45] that the strength of CLARION lies in its partitioning of cognitive processes into two broad categories: implicit and explicit processes (as well as other categories). It is important to explore such broad issues. It is a challenge to methodically explore such issues and reach some theoretically convincing and practically useful conclusions.

The validation of details of a cognitive architecture has been a difficult, but extremely important, issue. There have been too many instances in the past that research communities rushed into some particular model or some particular approach toward modeling cognition and human intelligence, without knowing exactly how much of the approach or the model was veridical. This often happens at the prompt of overly enthusiastic funding agencies or governmental panels. Often, without much effort at validation, claims are boldly made about the promise of a certain model or a certain approach.

Unfortunately, we have seen quite a few setbacks in the history of cognitive science as a result of this cavalier attitude. As in any other scientific fields, painstakingly detailed work needs to be carried out before sweeping claims can be made. This issue of validation of cognitive architectures poses a serious challenge, because a myriad of mechanisms are involved in cognitive architectures, and their complexity is high. Detailed validation has been extremely difficult. This challenge needs to be met by future research in this field.

Another challenge is how one can come up with a well constrained cognitive architecture with as few parameters as possible while accounting for a large variety of empirical data and observations [24]. Complex models have always invoked suspicion. An exaggerated argument against generic models was examined in [15]: "A good scientist can draw an elephant with three parameters, and with four he can tie a knot in its tail. There must be hundred of parameters floating around in this kind of theory and nobody will ever be able to untangle them". Counter-arguments to such objections can be advanced on the basis of the necessity of having complex models in understanding the mind, as argued in [15, 16, 18, 35], and so on. However, it should be clearly recognized that over-generality, beyond what is minimally necessary, is always a danger. Models may account well for a large set of data because of their extreme generality, rather than capturing any deep structures and regularities underlying cognitive processes in the human mind. This situation is to be avoided, by adopting a broad perspective, philosophical, psychological, biological, as well as computational, and by adopting a multi-level framework going from sociological, to psychological, to componential, and to biological levels, as argued in more detail in [39]. Techniques have been developed to accomplish this end and more work is needed (see, e.g., [24]).

While emphasizing the importance of being able to capture and explain the nuances of cognitive data, we also need to emphasize the importance of full functionality in cognitive architectures, which often does not get the attention it deserves in cognitive science. As mentioned before, cognitive architectures need to incorporate all of the following functionalities: perception, categorization and concepts, memory, reasoning, decision making, planning, problem solving, motor control, learning, language and communication, metacognition, and others. This issue has been raised very early on, but is still a major challenge for cognitive science [18, 3].

6.2 The Challenges from/to Artificial Intelligence

In the context of AI/CI, there have been various criteria proposed for evaluating cognitive architectures. Langley and Laird [11] proposed a set of useful criteria, including the following: (1) generality, versatility, and taskability, (2) both optimality and scalability (time/space complexity), (3) both reactivity and goal-directed behavior, (4) both autonomy and cooperation, (5) adaptation, learning, and behavioral improvements, and so on. To improve cognitive architectures in each of these aspects is a challenge by itself.

These are challenges from the perspective of AI/CI to the field of cognitive architectures.

To develop better cognitive architectures, we need better algorithms, which we may find in various subfields within AI/CI. It is extremely important that AI/CI researchers come up with better and better algorithms, for various functionalities such as information filtering, information retrieval, learning, reasoning, decision making, problem solving, communication, and so on. Only on the basis of such key algorithms that are continuously improving, better and better cognitive architectures may be built correspondingly.

In particular, we need better natural language processing capabilities, more powerful and more comprehensive learning algorithms, more efficient planning algorithms, and so on. Each of these types of algorithms could potentially significantly improve cognitive architectures, and cognitive architectures cannot be advanced without these algorithms. These are significant challenges from the field of cognitive architectures to AI/CI researchers.

AI/CI researchers also need to develop better computational methods (algorithms) for putting the pieces together to form better overall architectures. Various pieces have been, or are being, developed by various subfields of AI/CI (including neural networks, reinforcement learning, and so on). Now it is the time to put them together to form a more coherent, better integrated, and better functioning cognitive architecture. Better computational algorithms are needed for this purpose. That is another place where AI/CI researchers can come in. It will be a long and incremental process — the challenge is to continuously improving upon the state of the art and to come up with architectures that better and better mirror the human mind and serve a variety of application domains at the same time.

Cognitive architectures need to find both finer and broader applications, that is, both at lower levels and at higher levels. For example, some cognitive architectures found applications in large-scale simulation at a social, organizational level. For another example, some other cognitive architectures found applications in interpreting not only psychological data but also neuroimaging data (at a biological level). A review commissioned by the US National Research Council found that computational cognitive modeling had progressed to a degree that had made them useful in a number of application domains [20]. Another review [25] reached similar conclusions. Both reviews provided descriptions of some examples of applications of cognitive architectures. Inevitably, this issue will provide challenges for future research (applied, as well as theoretical) in cognitive architectures.

7 Concluding Remarks

Although progress has been made in recent decades in advancing the work on cognitive architectures, there is clearly a long way to go before we fully understand the architecture of the human mind and thereby develop computational cognitive architectures that successfully replicate the human mind.

An example cognitive architecture has been presented. However, it will be necessary to explore more fully the space of possible cognitive architectures [41], in order to further advance the state of the art in AI/CI and in cognitive science. It will also be necessary to enhance the functionalities of cognitive architectures so that they can be capable of the full range of intelligent behaviors. Tasks of gradually increasing complexity should be tackled. Many challenges and issues need to be addressed, including those stemming from cognitive science and from AI/CI, as discussed above.

We can expect that the field of cognitive architectures will have a profound impact, both in terms of better understanding cognition and in terms of developing better artificially intelligent systems. As such, it should be considered a grand challenge and correspondingly a significant amount of collective research effort should be put into it.

Acknowledgement

This work was carried out while the author was supported in part by ONR grant N00014-95-1-0440 and ARI grants DASW01-00-K-0012 and W74V8H-04-K-0002. Thanks are also due to Xi Zhang and Bob Mathews for their collaborations on related research.

References

[1] J. R. Anderson, (1983). The Architecture of Cognition. Harvard University Press, Cambridge, MA

[2] J. Anderson and C. Lebiere, (1998). The Atomic Components of Thought. Lawrence Erlbaum Associates, Mahwah, NJ.

[3] J. Anderson and C. Lebiere, (2003). The Newell Test for a theory of cognition. Behavioral and Brain Sciences, 26, 587-640

[4] S. Chaiken and Y. Trope (eds.), (1999). Dual Process Theories in Social Psychology. Guilford Press, New York.

[5] A. Cleeremans, A. Destrebecqz and M. Boyer, (1998). Implicit learning: News from the front. Trends in Cognitive Sciences, Volume 2, Issue 10, 406-416.

[6] J. Fodor, (1983). The Modularity of Mind. MIT Press, Cambridge, MA.

[7] G. Greene, (1999). The Elegant Universe. Norton, New York.

[8] L. Hirschfield and S. Gelman (eds.), (1994). Mapping the Mind: Domain Specificity in Cognition and Culture. Cambridge University Press, Cambridge, UK.

[9] T. Johnson, (1998). Acquisition and transfer of declarative and procedural knowledge. European Conference on Cognitive Modeling, pp. 15-22. Nottingham University Press, Nottingham, UK.

[10] A. Karmiloff-Smith, (1986). From meta-processes to conscious access: Evidence from children's metalinguistic and repair data. Cognition. 23. 95-147.
[11] P. Langley and J. Laird, (2003). Cognitive architectures: Research issues and challenges. Unpublished manuscript.
[12] A. Maslow, (1987). Motivation and Personality. 3rd Edition. Harper and Row, New York.
[13] D. Medin, W. Wattenmaker, and R. Michalski, (1987). Constraints and preferences in inductive learning: An experimental study of human and machine performance. Cognitive Science. 11, 299-339.
[14] R. Michalski, (1983). A theory and methodology of inductive learning. Artificial Intelligence. Vol.20, pp. 111-161.
[15] G. Miller, E. Galanter, and K. Pribram, (1960). Plans and the Structure of Behavior. Holt, Rinehart, and Winston, New York.
[16] M. Minsky, (1985). The Society of Mind. Simon and Schuster, New York.
[17] T. Nelson, (Ed.) (1993). Metacognition: Core Readings. Allyn and Bacon.
[18] A. Newell, (1990). Unified Theories of Cognition. Harvard University Press, Cambridge, MA.
[19] R. Nosofsky, T. Palmeri, and S. McKinley, (1994). Rule-plus-exception model of classification learning. Psychological Review. 101 (1), 53-79.
[20] R. W. Pew and A. S. Mavor (eds), (1998). Modeling Human and Organizational Behavior: Application to Military Simulations. National Academy Press, Washington, DC.
[21] M. R. Quillian, (1968). Semantic memory. In: M. Minsky (ed.), Semantic Information Processing. MIT Press, Cambridge, MA. pp. 227-270.
[22] M. Rabinowitz and N. Goldberg, (1995). Evaluating the structure-process hypothesis. In: F. Weinert and W. Schneider, (eds.) Memory Performance and Competencies. Lawrence Erlbaum, Hillsdale, NJ.
[23] A. Reber, (1989). Implicit learning and tacit knowledge. Journal of Experimental Psychology: General. 118 (3), 219-235.
[24] T. Regier, (2003). Constraining computational models of cognition. In: L. Nadel (ed.), Encyclopedia of Cognitive Science, pp. 611-615. MacMillan Reference Ltd. London.
[25] F. Ritter, Shadbolt, N., Elliman, D., Young, R., Gobet, F., and Baxter, G., (2003). Techniques for Modeling Human Performance in Synthetic Environments: A Supplementary Review. Human Systems Information Analysis Center, Wright-Patterson Air Force Base, Dayton, OH.
[26] D. Rumelhart, J. McClelland and the PDP Research Group, (1986). Parallel Distributed Processing: Explorations in the Microstructures of Cognition. MIT Press, Cambridge, MA.
[27] W. Schneider and W. Oliver (1991), An instructable connectionist/control architecture. In: K. VanLehn (ed.), Architectures for Intelligence, Erlbaum, Hillsdale, NJ.
[28] C. Seger, (1994). Implicit learning. Psychological Bulletin. 115 (2), 163-196.

[29] H. Simon, (1957), Models of Man, Social and Rational. Wiley, NY.
[30] J. D. Smith, W. E. Shields, and D. A. Washburn, (2003). The comparative psychology of uncertainty monitoring and metacognition. Behavioral and Brain Sciences, in press.
[31] W. Stanley, R. Mathews, R. Buss, and S. Kotler-Cope, (1989). Insight without awareness: On the interaction of verbalization, instruction and practice in a simulated process control task. Quarterly Journal of Experimental Psychology. 41A (3), 553-577.
[32] R. Sun, (1994). Integrating Rules and Connectionism for Robust Commonsense Reasoning. John Wiley and Sons, New York, NY.
[33] R. Sun, (1995). Robust reasoning: Integrating rule-based and similarity-based reasoning. Artificial Intelligence. 75, 2. 241-296.
[34] R. Sun, (2001). Does connectionism permit rule learning? INNS/ENNS/JNNS Newsletter, No.44, pp. 2-3. 2001.
[35] R. Sun, (2002). Duality of the Mind. Lawrence Erlbaum Associates, Mahwah, NJ.
[36] R. Sun, (2003). A Tutorial on CLARION. Technical report, Cognitive Science Department, Rensselaer Polytechnic Institute. http://www.cogsci.rpi.edu/rsun/sun.tutorial.pdf
[37] R. Sun, (2004). Desiderata for cognitive architectures. Philosophical Psychology, 17 (3), 341-373.
[38] R. Sun and I. Naveh, (2004). Simulating organizational decision making with a cognitive architecture CLARION. Journal of Artificial Society and Social Simulation, Vol.7, No.3, June, 2004. http://jasss.soc.surrey.ac.uk/7/3/5.html
[39] R. Sun, L. A. Coward, and M. J. Zenzen, (2005). On levels of cognitive modeling. Philosophical Psychology, 18 (5), 613-637.
[40] R. Sun, V. Honavar, and G. Oden, (1999). Integration of cognitive systems across disciplinary boundaries. Cognitive Systems Research, Vol.1, No.1, pp. 1-3.
[41] R. Sun and C. Ling, (1998). Computational cognitive modeling, the source of power and other related issues. AI Magazine. Vol.19, No.2, pp. 113-120.
[42] R. Sun, E. Merrill, and T. Peterson, (2001). From implicit skills to explicit knowledge: A bottom-up model of skill learning. Cognitive Science. Vol.25, No.2, 203-244.
[43] R. Sun and T. Peterson, (1998). Autonomous learning of sequential tasks: experiments and analyses. IEEE Transactions on Neural Networks, Vol.9, No.6, pp. 1217-1234.
[44] R. Sun and T. Peterson, (1999). Multi-agent reinforcement learning: Weighting and partitioning. Neural Networks, Vol.12, No.4-5. pp. 127-153.
[45] R. Sun, P. Slusarz, and C. Terry, (2005). The interaction of the explicit and the implicit in skill learning: A dual-process approach. Psychological Review, 112 (1), 159-192.

[46] R. Sun and X. Zhang, (2003). Accessibility versus action-centeredness in the representation of cognitive skills. Proceedings of the Fifth International Conference on Cognitive Modeling, pp. 195-200. Universitats-Verlag Bamberg, Bamberg, Germany.

[47] F. Toates, (1986). Motivational Systems. Cambridge University Press, Cambridge, UK.

[48] C. Watkins, (1989). Learning with Delayed Rewards. Ph.D Thesis, Cambridge University, Cambridge, UK.

Programming a Parallel Computer: The Ersatz Brain Project

James A. Anderson[1], Paul Allopenna[2], Gerald S. Guralnik[3], David Sheinberg[4], John A. Santini, Jr.[1], Socrates Dimitriadis[1], Benjamin B. Machta[3], and Brian T. Merritt[1]

[1] Department of Cognitive and Linguistic Sciences, Brown University, Providence, RI, 02912, USA
[2] Aptima, Inc., 12 Gill Road, Suite 1400, Woburn, MA, 01801, USA
[3] Department of Physics, Brown University, Providence, RI, 02912, USA.
[4] Department of Neuroscience, Brown University, Providence, RI, 02912, USA

Summary. There is a complex relationship between the architecture of a computer, the software it needs to run, and the tasks it performs. The most difficult aspect of building a brain-like computer may not be in its construction, but in its use: How can it be programmed? What can it do well? What does it do poorly? In the history of computers, software development has proved far more difficult and far slower than straightforward hardware development. There is no reason to expect a brain like computer to be any different. This chapter speculates about its basic design, provides examples of "programming" and suggests how intermediate level structures could arise in a sparsely connected massively parallel, brain like computer using sparse data representations.

1 Introduction

We want to design a suitable computer architecture – hardware and software – for the efficient execution of the applications now being developed that will display human-like cognitive abilities. We base our fundamental design on a few ideas taken from the neurobiology of the mammalian cerebral cortex, therefore we have named our project the Ersatz Brain Project. We gave it this name because, given our current state of knowledge, our goal can only be to build a shoddy second-rate brain. Even so, such a computer may be a starting point for projects that realize systems with the power, flexibility, and subtlety of the actual brain. We suggest that a "cortex-power" massively parallel computer is now technically feasible, requiring on the order of a million simple CPUs and a terabyte of memory for connections between CPUs.

We approach this problem from the point of view of cognitive computation. There is little attempt to make a detailed "biologically realistic" neurocomputer. We are proposing instead a "cortically inspired" computing system

James A. Anderson et al.: *Programming a Parallel Computer: The Ersatz Brain Project*, Studies in Computational Intelligence (SCI) **63**, 61–98 (2007)
www.springerlink.com

specifically for cognitive applications. Mammalian cerebral cortex is a remarkably uniform, large structure that seems to do much of the work of cognition, though other structures are involved. We know a good deal about the details of human cognition and the operations it performs. Our belief is that these operations will be performed most efficiently by a computer system based to at least some degree on the design of cerebral cortex.

However, the brain has severe intrinsic limitations on connectivity, speed, and accuracy and some computational strategies may simply be forced on the system by hardware limitations. In many respects we have a dual computational system, one system old, highly evolved, highly optimized, basically associative, perceptually oriented, memory driven, and alogical, and, a second system, recent, oddly contoured, unreliable, "symbolic" and "rule based". (See Sloman [1] for a cognitive science perspective.) We suggest a successful brain-like computer system should include aspects of both systems, since they are complementary and work effectively together, as the remarkable, very rapid success of our own species indicates.

2 Essentials of the Ersatz Approach

The human brain is composed of on the order of 10^{10} neurons, connected together with at least 10^{14} connections between neurons. These numbers are likely to be underestimates. Biological neurons and their connections are extremely complex electrochemical structures that require substantial computer power to model even in poor approximations. The more realistic the neuron approximation, the smaller is the network that can be modeled. Worse, there is very strong evidence that **a bigger brain is a better brain**, thereby increasing greatly computational demands if biology is followed closely. **We need good approximations to build a practical brain-like computer.**

2.1 The Ersatz Cortical Computing Module and the Network of Networks

Received wisdom has it that neurons are the basic computational units of the brain. However the Ersatz Brain Project is based on a different assumption. We will use the **Network of Networks [NofN]** approximation to structure the hardware and to reduce the number of connections required [2, 3, 4].

We assume that the basic neural computing units are not neurons, but small (perhaps $10^3 - 10^4$ neurons) attractor networks, that is, non-linear networks (modules) whose behavior is dominated by their attractor states that may be built in or acquired through learning [5, 6, 7]. Basing computation on module attractor states – that is, on intermediate level structure – and not directly on the activities of single neurons reduces the dimensionality of the system, allows a degree of intrinsic noise immunity, and allows interactions between networks to be approximated as interactions between attractor states. Interactions between modules are similar to the generic neural net unit except scalar connection strengths are replaced by state interaction matrices. (Fig. 1) The state interaction matrix gives the effect of an attractor state in one module upon attractor states in a module connected to it. Because attractors are derived from neuron responses, it is potentially possible to merge neuron-based preprocessing with attractor dynamics. The basic Network of Networks system is composed of very many of these basic modules arranged in a two-dimensional array. (Fig. 2).

2.2 Cortical Columns

The most likely physiological candidate for the basic component of a modular network is the cortical column. Cerebral cortex is a large two-dimensional layered sheet, with a repetitive structure. One of its most prominent anatomical features is the presence of what are called columns, local groups of cells oriented perpendicular to the cortical surface. There are several types of columns present at different spatial scales. A recent special issue of the journal *Cerebral Cortex* was devoted to cortical columns, their functions, and their connections. The introduction by Mountcastle [8] provides a useful summary of the two types of columns that will most concern us:

"The basic unit of cortical operation is the minicolumn ... It contains of the order of 80–100 neurons, except in the primate striate cortex, where

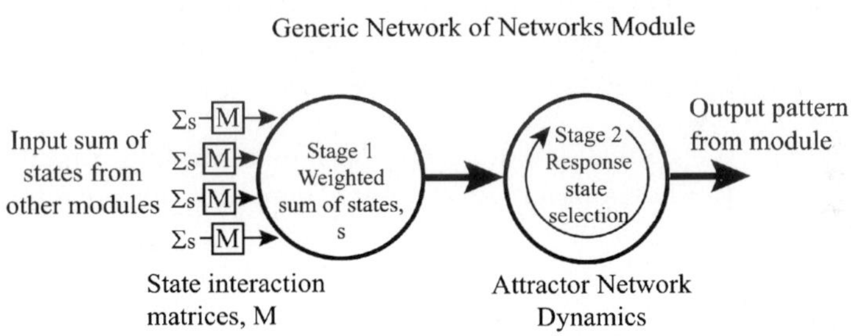

Fig. 1. Generic Network of Networks module structure

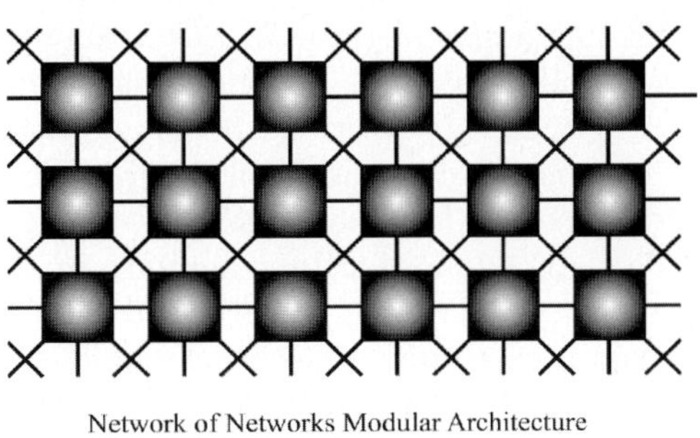

Fig. 2. Network of Networks 2-D modular architecture

the number is more than doubled. The minicolumn measures of the order of 40–50 μm in transverse diameter, separated from adjacent minicolumns by vertical cell-sparse zones which vary in size in different cortical areas. Each minicolumn has all cortical phenotypes, and each has several output channels. ... By the 26th gestational week the human neocortex is composed of a large number of minicolumns in parallel vertical arrays." ([8], p. 2)

Minicolumns form a biologically determined structure of stable size, form and universal occurrence. What are often called "columns" in the literature are collections of minicolumns that seem to form functional units. Probably the best-known examples of functional columns are the orientation columns in V1, primary visual cortex. Vertical electrode penetrations in V1, that is, parallel to the axis of the column, found numerous cells that respond to oriented visual stimuli with the same orientation [9]. The cells in a column are not identical in their properties and, outside of orientation, may vary widely in their responses to contrast, spatial frequency, etc. Clusters of minicolumns make up functional columns:

Mountcastle continues, "Cortical columns are formed by the binding together of many minicolumns by common input and short range horizontal connections. The number of minicolumns per column varies probably because of variation in size of the cell sparse inter-minicolumnar zones; the number varies between 50 and 80. Long-range, intracortical projections link columns with similar functional properties. Columns vary between 300 and 500 μm in transverse diameter, and do not differ significantly in size between brains that may vary in size over three orders of magnitude ... Cortical expansion in evolution is marked by increases in surface area with little change in thickness... ." ([8], p. 3).

If we assume there are 100 neurons per minicolumn, and roughly 80 minicolumns per functional column, this suggests there are roughly 8,000 neurons in a column.

2.3 Connectivity

Besides modular structure, an important observation about the brain in general that strongly influences how it works is its very sparse connectivity between neurons. Although a given neuron in cortex may have on the order of 100,000 synapses, there are more than 10^{10} neurons in the brain. Therefore, the fractional connectivity is very low; for the previous numbers it is 0.001 per cent, even if every synapse connects to a different cell. Connections are expensive biologically since they take up space, use energy, and are hard to wire up correctly. The connections that are there are precious and their pattern of connection must be under tight control. This observation puts severe constraints on the structure of large-scale brain models. One implication of expensive connections is that short local connections are relatively cheap compared to longer range ones. The cortical approximation we will discuss makes extensive use of local connections for computation in addition to the sparse, accurately targeted long-range connections.

2.4 Interactions between Modules

Let us discuss in a little more detail how to analyze interactions between small groups of modules. The attractor model we will use is the BSB network (Anderson, 1993) because it is simple to analyze using the eigenvectors and eigenvalues of its local connections.

The BSB model (Fig. 3.) was proposed several years ago as a simple feedback nonlinear neural network [6]. Its dynamics broke conveniently into a linear and a non-linear part. The analysis assumed it was a recurrent feedback network (See Fig. 3). An input pattern, **f**, appears on an interconnected group of neurons, say from a sensory input. There is vector feedback through a connection matrix, **A**, weighted by a constant α and an inhibitory decay constant, γ, with amount of inhibition a function of the amplitude of the activity. The state of the system is $\mathbf{x}(t)$. The system is linear up to the point where the LIMIT operation starts to operate. LIMIT($\mathbf{x}(t)$) is a hard limiter with an upper and lower threshold. Sometimes it is useful to maintain the outside input $\mathbf{f}(0)$ at some level; sometimes it is useful to remove the outside input. The constant δ performs this function.

The basic algorithm for BSB is:

$$x(t+1) = LIMIT(\alpha Ax(t) + \gamma x(t) + \delta f(0)) \tag{1}$$

In the nonlinear BSB network with growing activity, the state of the system will reach an attractor state based on the LIMIT function, usually the corner of a hypercube of limits. In practice, if **f** is an eigenvector the final BSB attractor state is close to the direction of **f**.

Activity can increase without bound or go to zero. The transition from increasing activity to decreasing activity is under control of α, γ and, the eigenvalues of **A**. These parameters provide a way of controlling network behavior.

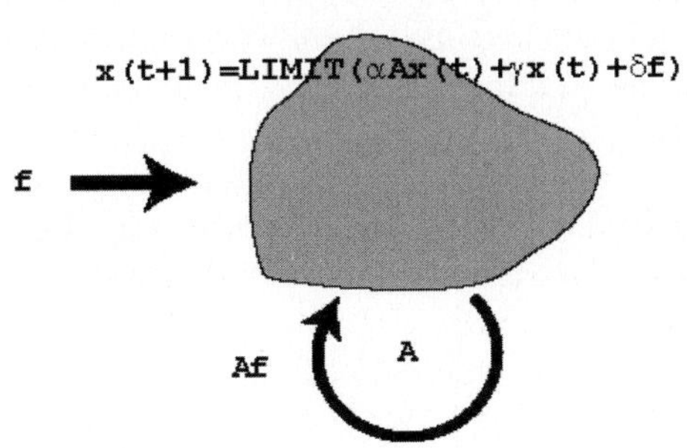

Fig. 3. Basic BSB module

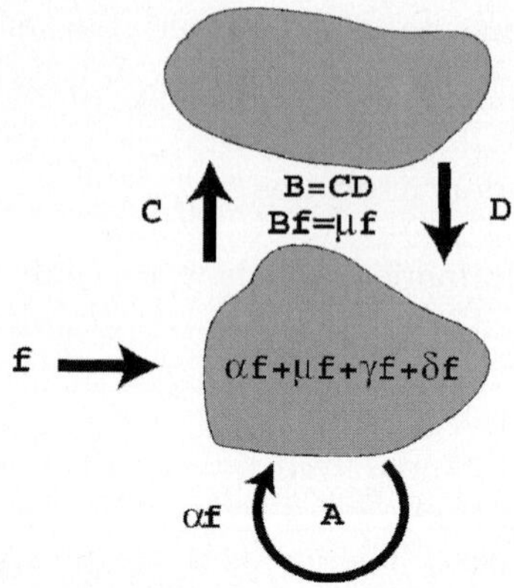

Fig. 4. Two modules linked by an associative pattern feedback loop. C and D are state interaction matrices between the two modules

Let us consider the implications of connections from other structures. In particular, we know two relevant facts about cortex: (1) one cortical region projects to others and, (2) there are back projections from the target regions to the first region that are comparable in size to the upward projections. Therefore let us consider the anatomy shown in Fig. 4. The upward association, **C**, can have almost any pattern as an output association. The downward association has an eigenvector of **A**, **f**, at its output, perhaps mixed with other eigenvectors. Hebbian learning operating at the start and finish of the loop will tend to learn **f** as an eigenvector of the entire loop because it is present at both input and output. These loops are reminiscent of the Bidirectional Associative Memory [BAM] of Kosko [10]. Analysis again is easy if **f** is an eigenvector. Let us assume δ is zero, as before. Analysis is the same as in the basic model, except that we now have a new term in the BSB equation corresponding to the contribution from the loop. (Fig. 4.)

We can also propose a computational mechanism for binding together a multi-module input pattern using local connections. If two modules are driven by two simultaneously presented patterns, **f** and **g**, associative links between **f** and **g** can be formed, increasing the gain of the module and therefore the likelihood that later simultaneous presentation of the patterns will lead to module activity reaching a limit. (Fig. 5) Local pattern co-occurrence will form local pattern associative bonds, letting larger groupings act as a unit, that is, a unit that increases and decreases in activity together. Large-scale patterns will tend to bind many module activities together since learning takes place embedded in a larger informational structure.

2.5 Interference Patterns and Traveling Waves

Because we have suggested many important connections are local, much information processing takes place by movement of information laterally from

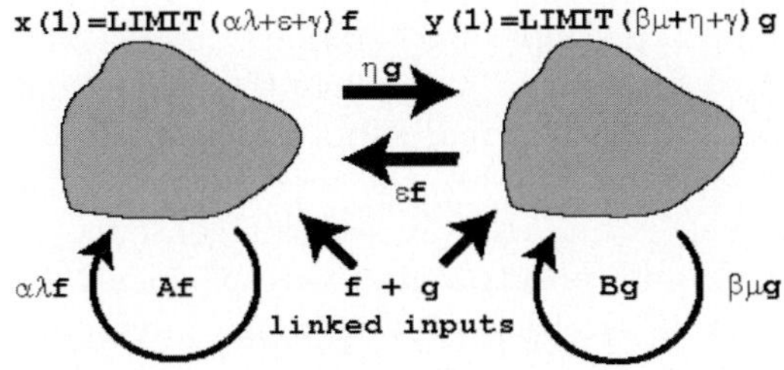

Fig. 5. Two coupled modules receiving a pair of linked input patterns, **f** and **g**

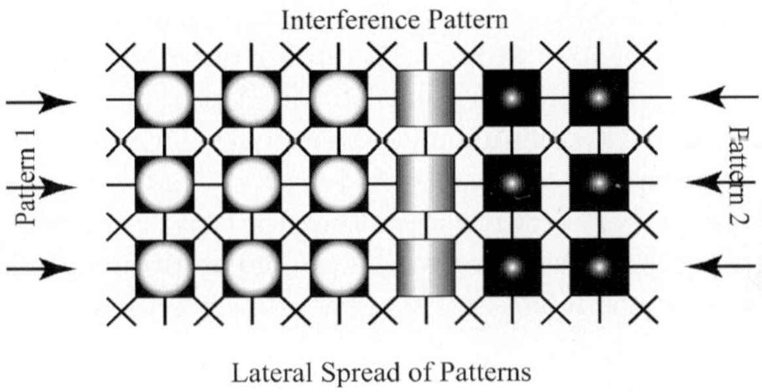

Fig. 6. Two patterns move laterally across an array and form an interference pattern, and can learn the resulting feature combination

module to module as shown in Fig. 6. This lateral information flow requires time and some important assumptions about the initial wiring of the modules. There is currently considerable experimental data supporting the idea of lateral information transfer in cerebral cortex over significant distances. The lateral information flow allows the potential for the formation of the feature combinations in the interference patterns, useful for pattern recognition. There is no particular reason to suggest that modules just passing information are in attractor states; for pattern transmission it is better if they re not.

Coincidences, where two patterns collide are of special interest. Since the individual modules are nonlinear learning networks, we have here the potential for forming new attractor states when an interference pattern forms, that is, when two patterns arrive simultaneously at a module over different pathways. (Fig. 6.)

There have been a number of related computational models specifically designed for vision that have assumed that image processing involves lateral spread of information. An early example is Pitts and McCulloch [11] who suggested, "A square in the visual field, as it moved in and out in successive constrictions and dilations in Area 17, would trace out four spokes radiating from a common center upon the recipient mosaic. This four-spoked form, not at all like a square, would the be the size-invariant figure of a square (p. 55)."

In the 1970's Blum [12] proposed the "grassfire" model where visual contours ignited metaphorical "grassfires" and where the flame fronts intersected produced a somewhat size invariant representation of an object. The propagating waves are computing something like the **medial axis representation**, that is, the point on the axis lying halfway between contours [13].

There are many examples of traveling waves in cortex. Bringuier, Chavane, Glaeser, and Fregnac [14] observed long range interactions in V1 with an inferred conduction velocity of approximately 0.1 m/sec. Lee, Mumford, Romero, and Lamme [15] discuss units in visual cortex that seem to respond to the medial axis. Particularly pertinent in this context is Lee [16] who discuses medial axis representations in the light of the organization of V1. In psychophysics, Kovacs and Julesz [17] and Kovacs, Feher, and Julesz [18] demonstrated threshold enhancement at the center of circle and at the foci of ellipses composed of oriented Gabor patches forming a closed contour. These models assume that an unspecified form of "activation" is being spread whereas the Network of Networks assumes that pattern information (a vector) related to module attractor states is being propagated. We feel that the traveling wave mechanism and its generalizations may have more general applications that vision.

2.6 Ersatz Hardware: A Brief Sketch

How hard would it be to implement such a cortex-like system in hardware? This section is a "back of the envelope" estimate of the numbers. We feel that there is a size, connectivity, and computational power sweet spot about the level of the parameters of the network of network model. If we equate an elementary attractor network with 10^4 actual neurons, that network might display perhaps 50 well-defined attractor states. Each elementary network might connect to 50 others through 50x50 state connection matrices. Therefore a cortex-sized artificial system might consist of 10^6 elementary units with about 10^{11} to 10^{12} (0.1 to 1 terabyte) total strengths involved to specify connections. Assume each elementary unit has the processing power of a simple CPU. If we assume 100 to 1000 CPU's can be placed on a chip there would be perhaps 1000 to 10,000 chips in a brain sized system. These numbers are within the capability of current technology.

Therefore, our basic architecture consists of a large number of simple CPUs connected to each other and arranged in a two dimensional array. (Fig. 7). A 2-D arrangement is simple, cheap to implement, and corresponds to the actual 2-D anatomy of cerebral cortex. An intrinsic 2-D topography can also make effective use of the spatial data representations used in cortex for vision, audition, skin senses and motor control.

2.7 Communications

The brain has extensive local and long-range communications. The brain is unlike a traditional computer in that its program, dynamics, and computations

Fig. 7. A 2-D array of modules with simple local connectivity. The basic Ersatz architecture

are determined primarily by strengths of its connections. Details of these relatively sparse interconnections are critical to every aspect of brain function.

Short-range connections

There is extensive local connectivity in cortex. An artificial system has many options. The simplest connections are purely to its neighbors. Expanding local connectivity to include modules two or three modules away, often seems to work better but is more complex to build.

Long-range connections

Many of the most important operations in brain computation involve pattern association where an input pattern is transformed to an associated output pattern that can be different from the input. One group of units connects to other groups with precisely targeted long-range connections.

CPU Functions

The CPUs must handle two quite different sets of operations. First, is communications with other CPUs. Much of the time and effort in brain-based computation is in getting the data to where it needs to be. Second, when the data arrives, it is can then be used for numerical computation.

3 Topographic Computation

The cortex, and, in imitation, our computational model is a 2-D sheet of modules. Connections between modules, slow and expensive in the brain, as we have noted, perform the computation. Therefore topographic relationships between modules and their timing relationships become of critical computational significance. We suggest that it is possible to use the topographic structure of the array to perform some interesting computations. Indeed, **topographic computation** may be a major mechanism used to direct a computation in practice, and, therefore, a tool that can be used to program the array.

The next three sections will describe some uses of topographic organization to (1) Arithmetic fact learning; (2) Information integration; (3) Filtering based on spatial organization; (4) Formation of spatial assemblies of modules within and across layers. This discussion is obviously highly speculative, but suggests the kinds of intermediate level structures that may be useful in a practical, programmable, and flexible brain-like computer.

4 Example: Arithmetic in Humans Using Topographic Control

A few years ago, we suggested a model for elementary arithmetic fact learning [19] that showed that it was possible to perform what appeared to be "symbolic" elementary arithmetic operations by using topographically organized spatial weighting functions. Because this kind of mechanism is suggestive of how an array of modules might be programmed topographically, we review it here even though the array of modules in the arithmetic model was one-dimensional. We will expand the array to two dimensions in the next sections. In addition the problem itself illustrates a few of the peculiarities of real-world cognitive computation.

When a computer multiplies two single digit integers, it performs a sequence of formal operations based on logic, the definitions of integers and the rules of binary arithmetic. Humans do multiplication very differently, using a combination of estimation and memory. The human algorithm might be formalized as something like "Choose the **product number** (that is, a number that is the answer to *some* multiplication problem) that is about the right size." For example, the most common error for 9x6 is 56. More careful inspection of the experimental data indicates a complex associative structure working along with estimation, Errors for 9x6 are frequently "off by one" multiples of 6 or of 9, for example, 48 or 63. These effects do not seem to arise because of errors in the application of formal logic to the computation.

The error rate for elementary multiplication facts in college-educated adults is as high as 8% in some studies. Such a high rate is remarkable considering that several years are devoted to practicing arithmetic facts in elementary school, at a time when learning of other complex material – language, for example – is fast and accurate. However, there are also some positive aspects to such an error prone algorithm. For example, errors in elementary arithmetic are usually "close" to the correct answer. Sometimes being "close" is good enough to do the job and more useful than the gross magnitude errors that malfunctioning computers can make.

Attempts by cognitive scientists to model human number performance have generally assumed the presence of an important "perceptual" part to the internal human number representation. In many experiments, numbers act more like "weights" or "light intensities" than abstract quantities. There is also evidence (Hadamard [20]) that higher mathematics as actually done by mathematicians or physicists is often more perceptual than abstract. The classic theorem-proof process is primarily used to check for errors and as a way of convincing others that results are correct. These ill-defined sensory and perceptual aspects of mathematics as it is practiced go by the name "mathematical intuition."

4.1 The Data Representation for Number

The most difficult problem in any application of neural networks is converting the input data into the state vectors that can be manipulated by network dynamics. Data representation often has more influence on system performance than the type of learning algorithm or the form of the network. There are few explicit rules that determine what the best data representations are for a particular problem. For example, there is constant intrinsic tension between accuracy and generalization since generalization to be useful requires an appropriate response to an unlearned pattern. Good generalization is critically dependent on the needs of the specific application, system history, and the data representation. In practice, generalization needs to be learnable and controllable so the same network can generalize one way at one time and in a different way with different task requirements.

A combination of computer simulations and inference from experimental data has suggested one useful data representation for number. The starting point for our formal system will therefore be a suggestion for the representation of number in a neural network or in a one-dimensional array of modules.

Topographic representations of parameters are very common in the nervous system. For example, in vertebrates, the early stages of the visual cortical representation are topographic in that the image on the retina is roughly mapped onto the two dimensional cortex. A topographic map of the visual field is used in vision in animals as diverse as humans and the horseshoe crab, *Limulus*. There are many other topographic maps in vertebrate cortex, for example, somatosensory maps of the body surface and frequency of sound in audition.

In a neural network, one useful form of such a topographic representation is called a **bar code**, as shown in Fig. 8. The value of the parameter represented depends on the location of a group of active units on a linear array of units. The price paid for this useful data representation is low precision and inefficiency in use of units. If a single parameter is represented only as a location then many units are required to represent a value that a single unit could

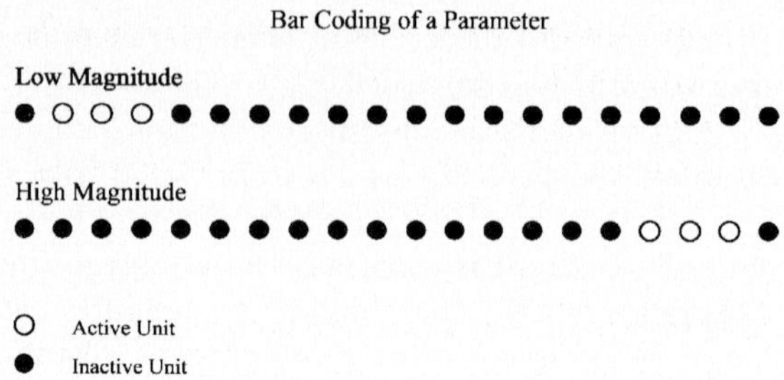

Fig. 8. Location of the active units on a linear array of units codes the magnitude of a parameter

represent by an activity level. Such a physical mapping is also inherently low precision, with precision determined by the number of units devoted to representing the parameter. Such a representation technique, and its variants are most naturally implemented in a system that is composed of many relatively inexpensive units and that is performing a function that is only secondarily concerned with high accuracy. Some nanocomponent based computing devices will fall into this category, as does, of course, the nervous system.

Such a representation can convey an essentially continuous range of values. However, in cognition we often use discrete entities – words, concepts, or numbers – to represent many different but somehow related examples. Much of the computational power of human cognition arises from its ability to keep only enough information about the world to be useful, and no more. For example, every physical example of a table is different in detail from every other one, but the word "table" often captures the essence of the commonalities of the group and the individual details can be discarded. With our network model, we can couple a continuous underlying data representation with a nonlinear attractor neural network with discrete attractors.

4.2 Simple Arithmetic Operations

Learning numbers is only the beginning of arithmetic. If we want to build a useful computational system, we have to be able to direct the computation to give answers to specific problems without further learning. Therefore, our first job is to specify the operations that we would reasonably expect an "arithmetic" network to perform. Let us suggest several primitive candidate operations for our computing system.

Counting seems to be present in all human societies of which we have knowledge and evidence for it goes far back into prehistory. There is good evidence that nonhuman primates and many higher mammals can count up to small integers. Formally, we can represent the counting function as two related operations: starting with a digit, add one to it, that is, **increment**, and, the symmetrical operation, subtract one, that is, **decrement**. Another valuable arithmetic related function is comparison of two numbers, that is, the equivalent of the formal operations **greater-than** or **lesser-than**. Another useful operation is **round-off**, that is, two and a bit more can be reported as "about two."

Therefore we will suggest that five operations form a useful starting point:

- **increment** (add 1)
- **decrement** (subtract 1)
- **greater-than** (given two numbers, choose the larger)
- **lesser-than** (given two numbers, choose the smaller)
- **round-off** to the nearest integer

4.3 Programming Patterns for the Simple Operations

The digits in order are represented as adjacent locations in the array, that is, the spatial locations follow the order of magnitudes: 1 next to 2, 2 next to 3, 3 next to 4, and so forth. If we use the bar coded data representation we have discussed (Fig. 9), it is surprisingly easy to find programming patterns that work. This robustness arises because we are dealing with qualitative properties of the geometry of representation, that is, **representational topology**.

In a previous paper [3] a way was suggested to control a network using a **vector programming pattern** that multiplied term by term the state vector derived from the input data. The number data representation we are using – the overlapping bar codes suggested earlier – contains enough information about the relations between digits to perform these operations.

The overall architecture of the system we have been describing is presented in Fig. 10. The system has two branches. One is connected to the physical world thorough the sensory systems. The other forms the "abstract" branch and chooses the desired arithmetic manipulation. An operation is chosen. This operation is associated with a programming pattern. In the other branch of the computation, information from the world is represented as a bar code. The two vectors are multiplied term by term. BSB attractor dynamics are

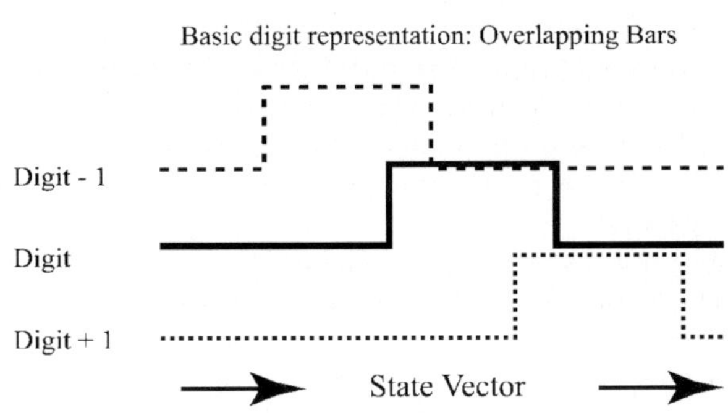

Fig. 9. Bar coded data representation for number magnitude. Bars are arranged spatially in order of magnitude as in the number line

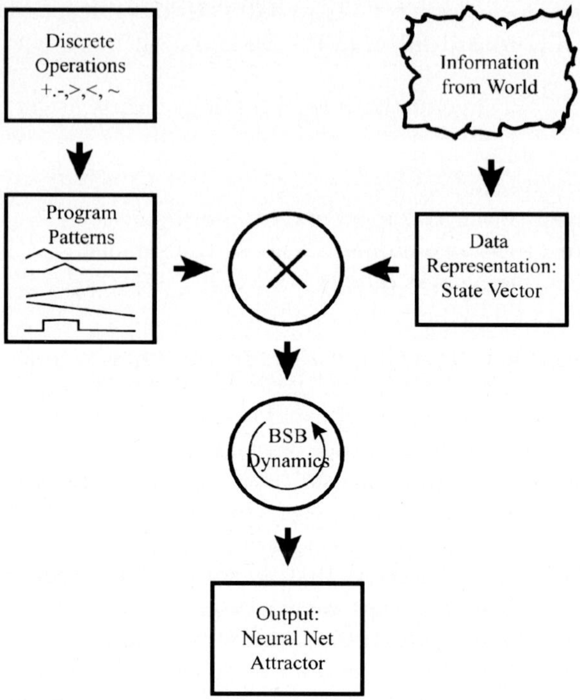

Fig. 10. Numerical computation involves merging two components. The left branch contains the program patterns and the right branch contains input quantities. The attractor network generates the output

then applied. The state vector then evolves to an attractor that contains the answer to the problem.

We can present intuitive arguments to support the particular programming patterns used. Consider counting, that is, the "Increment" operation. If we start from a particular location on the topographic map, we know that one direction on the map corresponds to larger numbers, the other to smaller. Therefore, if we differentially weight the map so that nearby larger numbers are weighted more strongly, the system state can be encouraged to move toward the attractor corresponding to the next largest digit. (The heavy dashed line in Fig. 11.)

The "greater-than" operation can also be done with differential weighting of the input data, so large digits reach attractors before small digits as shown in Fig. 12.

Round-off is performed slightly differently. A bar for an intermediate value is generated at a location between the two digits. The network then moves to the attractor with the largest overlap with the input data bar. The two other

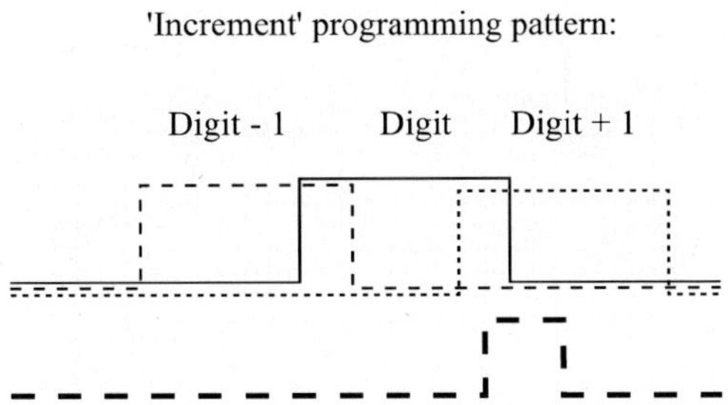

Fig. 11. The programming pattern for "Increment" weights the input digit and shifts it to the next attractor to the right

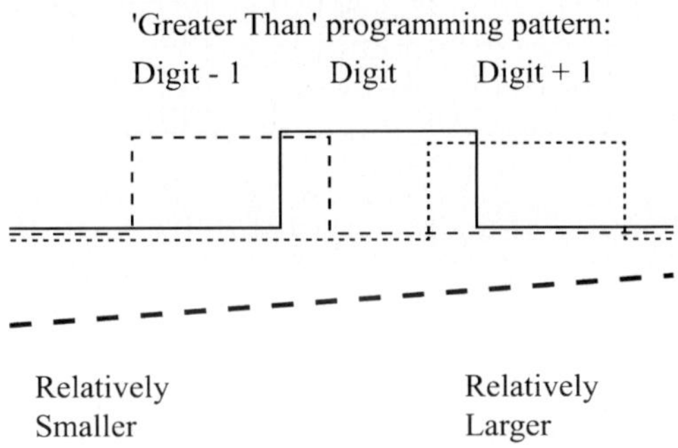

Fig. 12. The programming pattern for "Greater-Than" weights the larger digit more heavily and drives the state vector to that attractor

operations – less-than and decrement – are simple reflections of the patterns for greater than and increment.

There are some important properties of the operation of the program that are worth mentioning. For example, when the 'greater-than' program runs, it displays what is called a **symbolic distance** effect in cognitive psychology. The time to an attractor is a function of the difference in magnitude between the two numbers being compared, that is, the larger of the pair $(9, 1)$ is reaches its termination state faster than the larger of $(5, 4)$. In our model, this effect is driven by the magnitude representation in the data representation. The intrusion of statistical and perceptual components into what appears at first to be a purely abstract computation is to be expected from our approach, though perhaps a result not welcomed by purists since it would seem to have no place in logic. Notice also that the topographic map is effectively a realization of the number line analogy for integers.

Although these programming techniques work reliably for the digits one through ten, they are appallingly slow, limited in precision and dynamic range, and are clearly of no value for constructing a practical computing system that works with precise numbers. But, then, humans are not very good at arithmetic either.

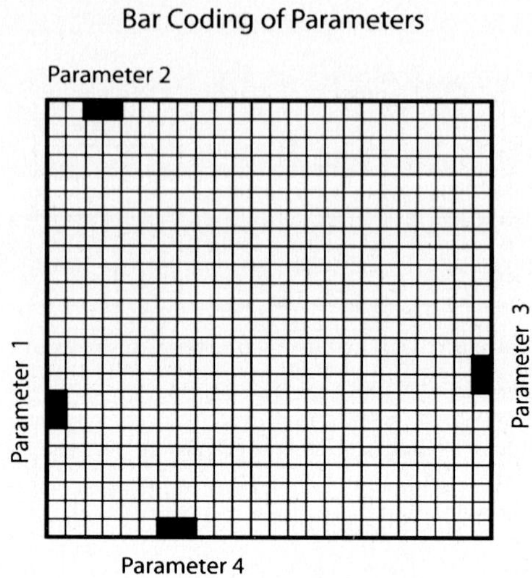

Fig. 13. Four bar codes

5 Example: Sensor Fusion Using Topographic Arrays

One potential application of our Ersatz Brain architecture is to **sensor fusion.** Sensor fusion involves integration of information from different types of sensor into a unified interpretation.

Suppose we have a set of four numerical sensor readings. We can do quite a bit with these values alone, for example, using them as a set of features for object recognition. However, this is not really sensor fusion since the sensor data is not integrated but used as information to determine a previously learned classification. This may not be possible if there is no previous classification, that is, in unsupervised learning.

Spatializing the data, that is letting it find a natural topographic organization that reflects the relationships between multiple data values, is a technique of great potential power, though an unfamiliar one. It is, however, a natural and effective way to compute in a two dimensional cortex and our two dimensional Ersatz architecture.

Assume we have four parameters that we want to represent in the activity pattern that describes a single entity (Fig. 13).

Our architecture assumes local transmission of patterns from module to module according to the Network of Networks model assumptions. Modules have multiple stable attractor states. Patterns are transmitted laterally from module to module. When two different patterns arrive at a module simultaneously, there is the possibility for a new pattern to be generated in the module, now representing the feature coincidence as a new part of the module's repertoire of attractor states. Standard learning rules plus the non-linear network can perform this operation. There may be many two parameter interference

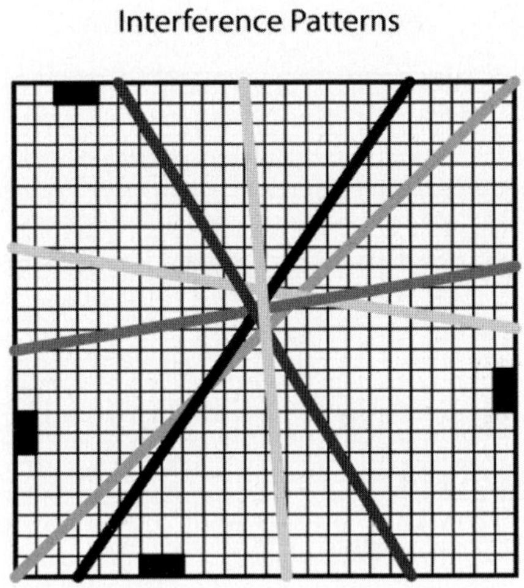

Fig. 14. Formation of simple interference patterns

patterns, the straight lines in Fig. 14. Each pair of input patterns gives rise to an interference pattern, a line perpendicular to the midpoint of the line between the pair of input locations.

5.1 Higher Level Features

In addition to the interference patterns representing coincidences between pairs of patterns there are often places where three or even four features coincide at a module. (Fig. 15) The higher-level combinations represent partial or whole aspects of the entire input pattern, that is, they respond to the Gestalt of the pattern. In this sense they have fused a number of aspects of the input pattern and represented it as a new activity pattern at a specific location.

5.2 Formation of Hierarchical "Concepts"

One intriguing aspect of this coding technique is the way it allows for the formation of what look a little like hierarchical concept representations. Suppose we have a set of "objects". For a simple demonstration, assume we have three parameter values that are fixed for each object and one value that varies widely from example to example. After a number of examples are seen, the system develops two different spatial codes. (Fig. 16).

In the first, a number of high order feature combinations are fixed since their three input "core" patterns never change. In the second, based on the additional spatial relations generated by the widely different examples, there is a varying set of feature combinations corresponding to the details of each specific example of the object. If the resulting spatial coding is looked at by an associative system, then two kinds of pattern can be learned.

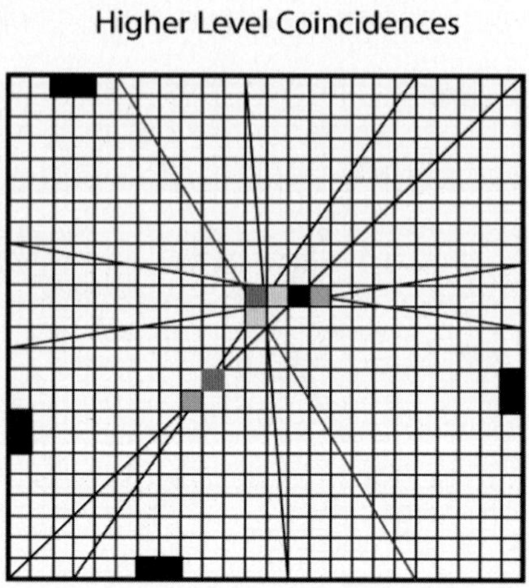

Fig. 15. Formation of higher-level feature combinations

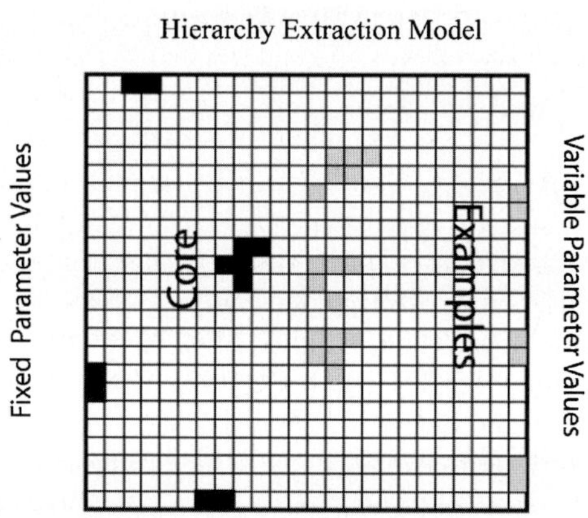

Fig. 16. Mechanism for formation of hierarchical structure

The first learned pattern corresponds to an unchanging **core** and might correspond to an abstraction of the commonalities of many examples. The second learned set of patterns corresponds to the core, plus the **examples** – patterns associated with each specific learned parameter set. All the specific examples are related by their common core pattern. This dichotomy has considerable similarity to the subordinate-superordinate relationships characterizing information stored in a hierarchical semantic network (Murphy [21]).

We have spatialized the logical structure of the hierarchy. Because the coincidences due to the "core" (three values) and to the "examples" (all four values) are spatially separated, we have the possibility of using the "core" as an abstraction of the examples and using it by itself as the descriptor of the entire set of examples Many of the control mechanisms suggested for cortex, most notably attention, are held to work by exciting or inhibiting regions of the cortical array.

6 Example: Selective Filtering of Vowels

Let us consider a test problem for some of our topographic based ideas on a real bit of biological signal processing. Let us consider how to build a geometric data representation inspired by one of the perceptual invariances seen in human speech.

6.1 Formant Frequencies in Speech Perception

The acoustic signals due to a vowel (and other phonemes) are dominated by the resonances of the vocal tract, called formants.

The resonant peaks are sometimes not very sharp and not very large. Vocal tracts come in different sizes, from men, to women, to children. Resonant peaks change their frequency as a function of vocal tract length.

Remarkably, this frequency shift – which can be substantial between a bass male voice and a small child – causes little problem for human speech perception. The important perceptual feature for phoneme recognition seems to be the based more on the ratios between the formant frequencies than on their absolute values. This relationship is what we would expect if a vocal tract simply increased in size without changing its shape. Then differences in vocal tract length approximately multiply the resonances by a constant.

We can see these properties using a classic set of data [22, 23] for the formant frequencies of vowels. Table 1 gives the average frequencies (in hz) of the formants F1, F2, and F3 for three vowels as spoken by men, women and children. The value of the formant for women is given the value 1.00. It can be seen the absolute values of the frequency of the formants for the same vowel vary about 30% between men women and children. If we look instead at the ratios of the formants the picture changes. Table 2 presents the values for the ratios of f1 and f2 and f2 and f3 for the same three vowels. Note the ratios vary only by a few percent over different vocal tract sizes. The ratio of formants is more constant than the absolute values of frequency.

6.2 Data Representation

We know from many sources there is a roughly logarithmic spatial mapping of frequency onto the surface of auditory cortex, what is sometimes called a tonotopic representation. A logarithmic spatial coding has the effect of translating the all the parameters multiplied by the constant by the same distance.

It might be easy to simply compute ratios of frequencies and use those values as inputs to a speech processor. However, there might be advantages to

Table 1. Average Formant Frequencies (Hz) and Ratios of Formant Frequencies for Men, Women and Children for Three Vowels. Data taken from Watrous [22] derived originally from Peterson and Barney [23]

		[i]		[æ]		[u]	
Men	f1	267	(0.86)	664	(0.77)	307	(0.81)
Women	f1	310	(1.00)	863	(1.00)	378	(1.00)
Children	f1	360	(1.16)	1017	(1.18)	432	(1.14)
Men	f2	2294	(0.82)	1727	(0.84)	876	(0.91)
Women	f2	2783	(1.00)	2049	(1.00)	961	(1.00)
Children	f2	3178	(1.14)	2334	(1.14)	1193	(1.24)
Men	f3	2937	(0.89)	2420	(0.85)	2239	(0.84)
Women	f3	3312	(1.00)	2832	(1.00)	2666	(1.00)
Children	f3	3763	(1.14)	3336	(1.18)	3250	(1.21)

Table 2. Ratios Between Formant Frequencies (Hz) for Men, Women and Children. Data taken from Watrous [22] derived originally from Peterson and Barney [23]

		[i]	[æ]	[u]
Men	f1/f2	0.12	0.38	0.35
Women	f1/f2	0.11	0.42	0.39
Children	f1/f2	0.11	0.43	0.36
Men	f2/f3	0.78	0.71	0.39
Women	f2/f3	0.84	0.72	0.36
Children	f2/f3	0.84	0.70	0.37

using a more brain-like approach. The transformations are not quite as simple as simple ratio invariance, the system is intrinsically noisy, the stimuli show substantial variability both within and between speakers (for example, there are many different accents), and the system is adaptive.

6.3 Representational Filtering

Our specific goal is to **enhance** the representation of ratios between "formant" frequencies, and to **de-emphasize** the exact values of those frequencies. That is, we wish to make a kind of filter using the data representation that responds to one aspect of the input data.

Let us start by assuming our usual information integration square array of modules with parameters fed in from the edges.

We start by duplicating the frequency representation three times along the edges of a square. Multiple frequency maps are common in the auditory system (See Talavage et al. [24] for an fMRI study of tonotopic maps in humans.).

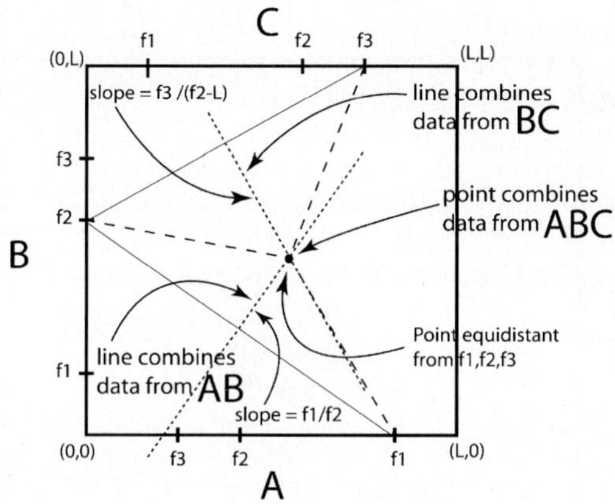

Fig. 17. The location of the point equidistant from f1, f2 and f3 contains information about the ratios between them

We will start by assuming that we are interested in locations where three frequency components come together at a single module at the same time. Figure 17 shows the point equidistant from f1 on the bottom edge, f2 on the side, and f3 on the top along with a few geometrical relations between the frequencies. We conjecture that this means the module may form a new internal representation corresponding to this triple coincidence. If we assume initially pattern transmission between modules is isotropic we then look for module locations equidistant from the locations of initial triple sets of frequencies. These points will be the modules that lie at the center of a circle whose circumferences contains the three points containing the three different frequency components, one from each side.

There are multiple combinations of the three formant frequencies. With three frequencies, the three possibilities are (f1,f2,f3), (f1,f3,f2), (f2,f1,f3), (f2,f3,f1), (f3,f1,f2), and (f3,f2,f1). Each triple combination produces a slightly different geometry.

Depending on what triple (location of the frequency combination) is used, different points are activated. Therefore a three "formant" system has six locations corresponding to possible triples. (We discuss in Section 8 the possibility of forming linked assemblies of active modules to function as a higher-level data representation.)

Figure 18 shows the points (marked with dots) that are equidistant from various combinations of frequencies. There are actually a lot more possible combinations possible involving both lower and higher order conjunctions, but we will start by only looking at triples.

The positions of the coincidences shift slightly if the frequency marks are translated by a uniform amount, corresponding in our model system to a change in vocal tract length giving rise to a shift in frequencies.

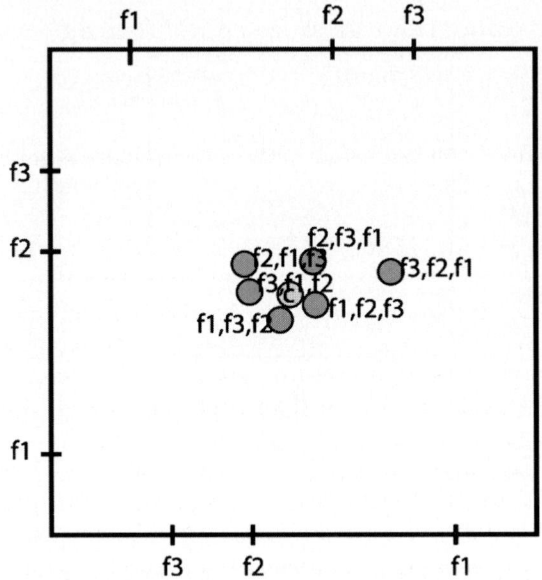

Fig. 18. Coincidence locations for various combinations of f1. f2. and f3

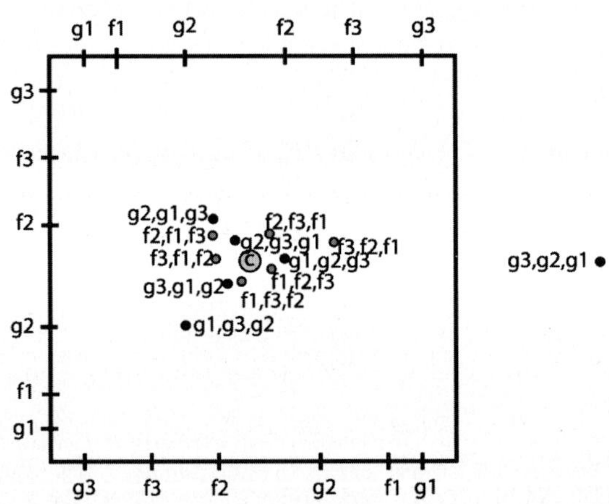

Fig. 19. (Top) Modules with triple coincidences for two sets of frequency patterns, f1, f2, and f3 and g1, g2, and g3

Because the geometry of the triple coincidence points varies with the location of the inputs along the edges, a different set of frequencies will give rise to a different set of coincidences. Figure 19 shows the set of triple coincidences for another set of frequencies. Note that one triple coincidence for the second set of frequencies lies outside the array.

7 Sparse Connectivity and Sparse Coding

7.1 Full Connectivity

Let us make some general, and highly speculative comments about the development of possible useful intermediate level structures for brain like computation.

Most neural network learning models assume full or nearly full connectivity between layers of units. Units are most often arranged in layers because it is an arrangement that pays homage to the neuroanatomy of mammalian cerebral cortex where cortical regions project to other cortical regions over long-range projection pathways through the white matter. Full connectivity means that a unit in a layer projects to, and it projected to, by all the units in layers above and below it.

Fully connected systems can be analyzed with standard mathematical techniques. They can perform a number of powerful information processing operations, and, combined with simple local learning algorithms such as the Hebb rule, they can be used to build adaptive systems with a number of useful applications in both engineering and science. The number of connections in fully connected systems grows very rapidly, order n^2, where n is the number of units.

7.2 Sparse Connectivity

Although many neural net models assume full or high connectivity, the actual cerebral cortex is **sparsely connected**, that is, each neuron projects to relatively few others given the potential number they could connect to, even in projections from one cortical region to another. (See Section 2.3).

7.3 Sparse Coding for Data Representation

Besides sparse connectivity, there is reasonably strong experimental evidence that **sparse coding** is used for data representation in the cortex, that is, information is represented by the activities of relatively few units. A review by Olshausen and Field [25] comments, "In recent years a combination of experimental, computational, and theoretical studies have pointed to the existence of a common underlying principle involved in sensory information processing, namely that information is represented by a relatively small number of simultaneously active neurons out of a large population, commonly referred to as 'sparse coding."' ([25], p. 481). Many of these ideas first emerged in a classic paper by Barlow [26].

There are numerous advantages to sparse coding. Olshausen and Field mention that sparse coding provides increased storage capacity in associative memories and is easy to work with computationally. Among other virtues, sparse coding also "makes structure in natural signals explicit" ([25], p. 481) and is energy efficient (see [27])

"Higher" regions (for example, inferotemporal cortex) seem to show a greater degree of sparse coding than lower ones (for example, V1). Cells in the higher levels of the visual system also have less spontaneous activity than lower regions, for example, cells in inferotemporal cortex are silent much of the time until they find a specific stimulus that piques their interest.

7.4 Sparse Coding Combined with Sparse Representation

Fig. 20 shows a cartoon version of a system that shows both **sparse coding** (three active units in input layer 1, four in layer 2, and two in output layer 3) and **sparse connectivity**.

Instead of trying to derive very general pattern association systems like back propagation, using high connectivity, let us see if we can make a learning system that starts from the assumption of both **sparse connectivity** and **sparse coding**.

Such extreme restrictions on connectivity and representation do not seem at first to form a very promising information processing system. We suspect this is why it has not been looked at seriously as a nervous system model, even though it seems to fit the qualitative structure of the cortex better than the high connectivity assumptions that underlie common neural network models.

7.5 Paths

In sparse systems, selectivity can come from other sources than a precise pattern of connection strengths. A useful notion in sparse systems is the idea

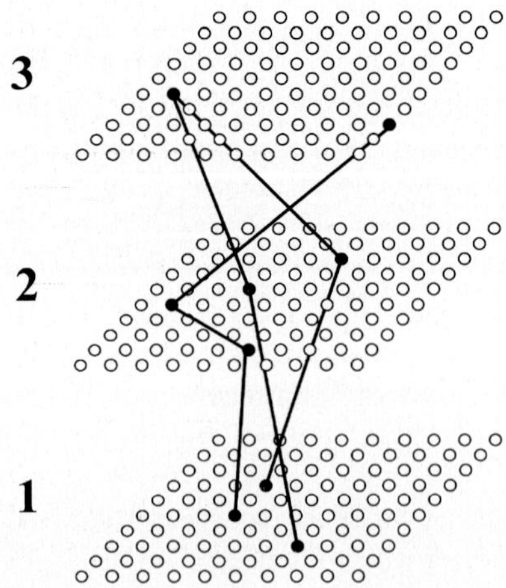

Fig. 20. A sparsely connected, sparsely coded three layer system

of a **path**. A **path** connects a sparsely coded input unit with a sparsely coded output unit. Paths have **strengths** just as individual connections do, but the strengths are based on the entire path, from beginning to end, which may involve several intermediate connections. A path may have initial strength zero or start with a nonzero strength due to initial connectivity. The concept of a path is related to useful ways to structure initial wiring of the nervous system. Connections that are likely to be valuable are sketched in initially. Learning can then tune and sharpen the pre-existing connections. This strategy seems to be present in sensory systems. For an example, see a recent discussion of the development of different aspects of the visual system [28].

Consider a path with several intermediate units. If there is no initial transmission of activity through the path, there will be no learning. If, however, when the input unit becomes active and gives rise to even low activity at the output unit, learning can take place and the path can become stronger. Continued learning will further strengthen that particular path. If the path is feedforward, connections earlier in the path cannot be changed properly without additional assumptions, for example, some form of backward information flow. Cortical regions generally project both forward and backwards.

7.6 Initial Biases in Connectivity

For a simple system with scalar weights by far the best, though unrealistic, strategy would be to somehow construct independent paths for each single unit association.

If many independent paths were desirable, then a useful initial construction bias for such a sparsely connected system would be to make available as **many** potential paths as possible. This bias differs significantly from back propagation, where there are almost always fewer units in the hidden layers than in the input and output layers, therefore **fewer** potential paths. (Many of the most important statistical properties of back propagation arise from the fact that there are a relatively small number of hidden layer units [7].) In a fully connected system, adding more units than contained in the input and output layers would be redundant. This is not so in sparser systems. For example, there is a huge expansion in number of units moving from retina to thalamus to cortex. A million input fibers drive 200 million V1 neurons.

7.7 Sparseness in Systems Using the Network of Networks

Suppose we assume that modules, not single units, are connected. Simple scalar activity is not, by nature, selective. Activity arising from different sources cannot be told apart; there is no "tag" giving its source. But if we use the Network of Networks approximations, activity along a path is not a scalar. Activity is now not simple unit activity but **module activity.** Patterned activity in a single module connected to another single module, can give rise to a selective pattern associator.

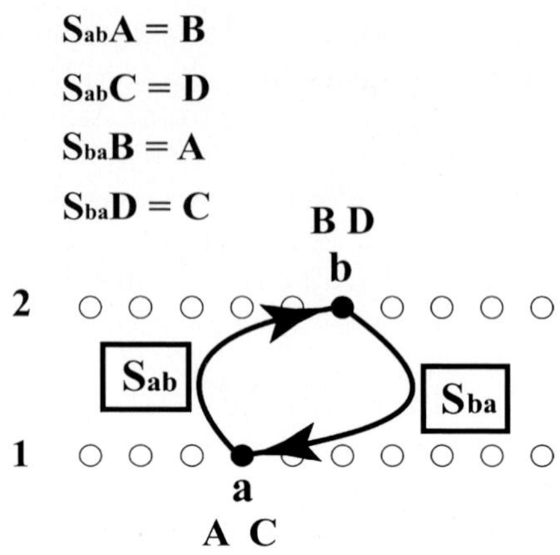

Fig. 21. About the simplest arrangement of interconnected modules. The **S**'s are matrices and **A**, **B**, etc. are patterns

For ease in analysis, let us assume the pattern associator is the simplest kind, the linear associator. Because we aiming for the development of correct associations over entire paths in a system using non-linear modules, we suspect the exact details of the network associators are not important.

If a pattern **A** on module **a** is input to Layer 1, (see Fig. 21) the activity, **B**, on **b** is given by the matrix-vector product, $\mathbf{S_{ab}A}$. We assume as part of the Network of Networks architecture that there are reciprocal connections between modules. If pattern **B** is present on module **b** of layer 2, the activity on **A** is given by $\mathbf{S_{ba}B}$. This loop between **a** and **b** is similar to a Bidirectional Associative Memory [10].

If pattern **A** on **a** and pattern **B** on **b** are simultaneously present, the increment in strengths are given by standard Hebb learning, that is,

$$\Delta S_{ab} \sim \mathrm{BA}^{\mathrm{T}} \; and \; \Delta S_{ba} \sim \mathrm{AB}^{\mathrm{T}} \qquad (2)$$

If no other pattern associations are stored in the connection matrices, after learning, if pattern **A** is present at module **a**, something like pattern **B** will appear at module **b**. The sparsely represented association (the only activity is on **a** and **b**) has been learned.

7.8 Multiple Patterns

Unlike scalar connections between units, a single connection between modules can learn multiple associations.

In the simplest situation, suppose we look at modules **a** and **b** as before, but assume that **a** and **b** can display two different patterns, **A** and **C** on **a** and **B** and **D** on **b**. Subject to the usual limitations of simple pattern associators, (that is, it works best if **A** and **B**, and **C** and **D**, are orthogonal)

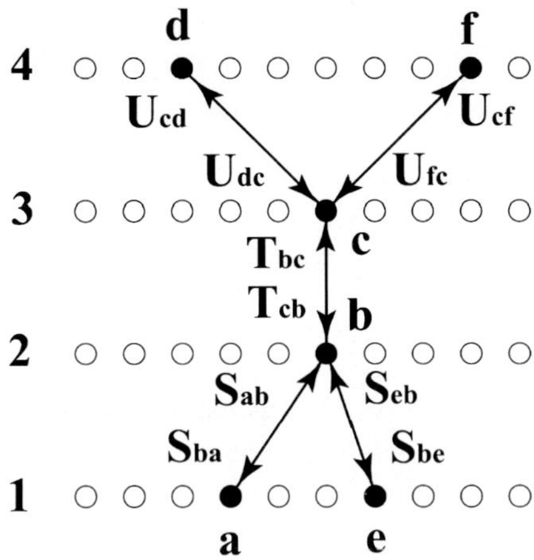

Fig. 22. Two network of networks paths with a common segment

such a network can easily learn to associate **A** with **B** and **C** with **D** using the same connections. We can use pattern selectivity to overcome some of the limitations of scalar systems.

We can also work with more complex systems. Consider a multilayer path with a common segment as in Fig. 22. We want to associate patterns on two paths, **a-b-c-d** and **e-b-c-f** with link **b-c** in common.

First, even though parts of the path are physically common they can be functionally separated if the paths use different module patterns.

Second, because the paths can be functionally separated by their pattern selectivity, pattern information propagating backwards can now be used to sharpen and strengthen one path without interfering with the strengths of another path. The next section gives a very simple example.

7.9 Backwards Projections

There is no reason to assume that the backward projections and the forward projections must have the same strength. All that may be required initially for sparse systems is that modules produce activity along all points in the path, backwards and forwards. We can analyze a simple arrangement of active, sparsely connected modules, part of a sparse representation, using elementary outer-product Hebb learning.

First consider the two associative chains, up and down, $\mathbf{a} > \mathbf{b} > \mathbf{c} > \mathbf{d}$ and $\mathbf{d} > \mathbf{c} > \mathbf{b} > \mathbf{a}$.

We assume supervised learning, that is, the desired patterns on **a** and **d** are present at input and output.

We will first consider outer-product learning in the weak signal linear range of operation of the system. As the path gets strong, we conjecture that modules in the path will develop internal states corresponding to the association activation patterns, that is, attractors.

Consider the two intermediate modules **b** and **c** on the path. Patterns on **b** and **c** are the result of information moving upwards from the input layer and downwards from the output layer, that is, initially with clamped patterns on **a** and **d**, with patterns **e** and **f** zero.

$$Activity\ on\ c = T_{bc}b + U_{dc}d \tag{3}$$
$$Activity\ on\ b = S_{ab}a + T_{cb} \tag{4}$$

If the activities on **b**, **c** and **d** are non-zero patterns, we have the ability to use simple Hebb learning to change the strengths of coupling between modules using their connection matrices so as to change the associative path strength connecting the pattern on **a** with the pattern on **d**.

There are two active modules connected to **b**, up from **a** and down from **c**. The change in upward coupling between **a** and **b**, through $\mathbf{S_{ab}}$ is given by

$$\Delta(coupling\ S_{ab}) \sim ba^{\mathrm{T}} \tag{5}$$

The change in upward coupling between **b** and **c** is

$$\Delta(coupling\ T_{bc}) \sim \mathrm{cb}^{\mathrm{T}} \tag{6}$$

and between **c** and **d** is

$$\Delta(coupling\ U_{cd}) \sim \mathrm{dc}^{\mathrm{T}} \tag{7}$$

We assume initial coupling is small, but sufficient to allow some patterns to emerge at **b** and **c**, that is, get through the path, weakly. After multiple learning events, the weak initial connections are small relative to later learning, and we can compute overall coupling between modules **a** and **d**.

If pattern **a** is presented at layer 1.and the pattern on **d** is zero, then the generated pattern on module

$$d \sim (\mathrm{U_{cd}})(T_{bc})(S_{ab})a \tag{8}$$
$$\sim (\mathrm{dc}^{\mathrm{T}})(cb^{T})(ba^{T})a \tag{9}$$
$$\sim \mathrm{d} \tag{10}$$

The downward associations learn in the same way so that if the supervised pattern **d** occurs it will produce a constant times pattern **a**. Note the potential for a BAM-like associative loop. These results are well known from the properties of simple associators. The properties of the system are subject to the capacity limitations of simple associators and the size of the vectors and matrices being associated.

The other leg of the association (Fig. 22), the pattern on **e** linked with the one on **f**, potentially can learn independently. It uses a different set of intermodule connections. One final state after extensive path learning might consist of a different set of activities on **b** and **c**, and, ultimately, different attractor states developing on **b** and **c** from learning on the other path.

Note that the somewhat arbitrary activities developing on **b** and **c** – the pattern sum of initial upward and downward connections – serves as a random 'hidden layer' link between input and output patterns. The argument holds for more intermediate layers and more associations, as long as sparse initial paths both up and down exist.

The system gets more complex as more patterns are learned. We know there are many regular topographic mappings in the cortex, particularly in the sensory systems. Our hunch is that some of the statistical abstraction and "concept forming" properties of associators may develop in the intermediate layers based on representational topography combined with learning in sparse systems. For example, it is possible, even likely, that sharing paths would be expected and computationally valuable in associations having a degree of sensory, perceptual or conceptual similarity. This issue remains to be explored.

7.10 Other Mechanisms

There are numerous other mechanisms that will be required to make this speculative, sparse system work in practice. For example, we have not considered here the important gain control mechanisms common in cortex that could limit the total number of active columns in a region. Besides generalized regional gain control, there is some evidence that active columns corresponding to a complex object in monkey inferotemporal cortex can inhibit other columns [29].

With sparse data representations and sparse connectivities, Hebb learning has the potential to form selective paths that can be used for general associative learning in Network of Networks systems. Both upward and downward paths can be used together. Because the modules are non-linear, with limits on activities and inhibition, stability of learning and dynamics is not likely to be a problem.

In both multilayer and single layer systems the potential exists for active feedback loops. Self-exciting loops may be useful intermediate level computational building blocks, as we suggest next.

8 Module Assemblies

The brain shows large differences in scale. Understanding how neurons work together in groupings of larger and larger size is perhaps the key to understanding brain function. We have already suggested one intermediate level of structure in the modules comprising the Network of Networks. We can take this idea a bit further where we speculate about the possibility of the formation of stable **module assemblies** as the next higher step in intermediate level representation.

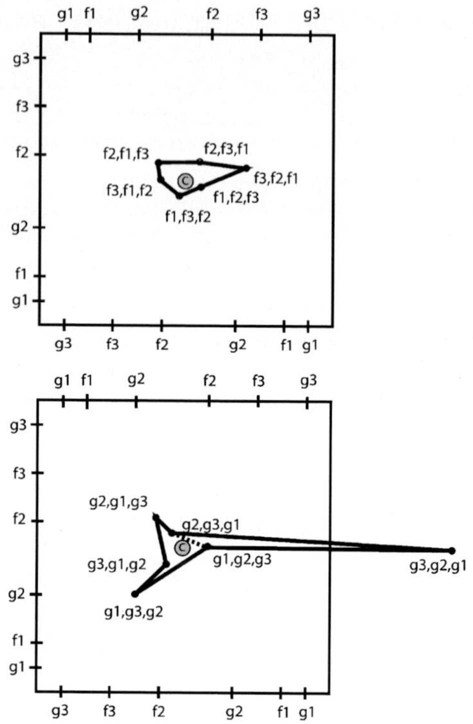

Fig. 23. Active modules for two different frequency patterns. (See Section 6.) These distinct frequency patterns have the potential to form distinct module assemblies

In Section 6, we showed that the information integration model as discussed in the context of vowel filtering, produced a stimulus dependent series of nearby excited modules. (See Fig. 23).

Suppose we bind together those nearby excited modules through associative linkages so they can learn to become mutually self excited in a group. An intermediate-level representation can arise we call a **module assembly**, which we discuss next.

8.1 Physiological Data from Cortex

We reviewed some of the properties of cortical columns in Section 2. It is hard to study detailed functional properties of columns. However, optical imaging of intrinsic cortical signals has now allowed visualization of structures of the size of cortical columns. The spatial resolution of this difficult technique can be a factor of 50 or more better than fMRI and gets into the size region where perhaps we can see some of the kinds of specialization for cortical computation that we need to see. A small body of intrinsic imaging work has been done on inferotemporal cortex [IT] in primates. As the highest visual area, IT contains the results of previous processing presumably in a form that is congenial for further use. Several groups have studied the organizational properties of inferotemporal cortex (area TE) from a computational point of view (See Tsunoda et al., [29]; Tanaka, [30, 31]). They proposed a model for inferotemporal cortex function that is in rough harmony with the basic architecture we assume for the Ersatz Brain.

Tanaka's early work on inferotemporal cortex used (1) computer simplified stimuli and (2) microelectrode recordings. Cells in area TE respond to few stimuli and have relatively little spontaneous activity. Once Tanaka found an image that drove a cell, the next step was to perform a series of abstractions of it until an optimal simplified image – a "critical feature" – was found that adequately drove the cell but further simplifications of it did not. Cells in a small region of TE tended to have similar critical features. For example, a number of cells in a small region might respond to various "T-junctions." The T-junctions would be different from each other in detail, but seemed to be examples of a basic 'T' structure. Regions of similar responses seemed to have roughly the size (300 μm) and character of the functional columns found in cortex. Nearby "columns" had quite different critical features. There was no sign of the continuous gradation of angle of best response found, for example, in orientation columns in V1.

From four to about a dozen columns were excited by presentation of complex objects and seemed to represent a complex percept. Such sets of observations led both Tanaka and Tsunoda et al. to propose a sparsely distributed, column based model of image recognition. From [29]:

"... objects are represented by combination of multiple columns in a sparsely distributed manner. The region activated by a single object image is only $3.3 \pm 2.5\%$ of the entire recording area (number of examined object images, 37). ... These results are consistent with feature-based representation, in which an object is represented by the combined activation of columns each responding to a specific visual feature of the object." ([29], p. 835).

8.2 "Cell Assemblies" and the Formation of Module Assemblies

If two modules are reciprocally associatively linked, we have a situation similar to the Bidirectional Associative Memory. If there are multiple interacting modules, we have the potential to form other interesting and complex associatively linked structures through Hebb learning, what we will call **module assemblies**, in harmony with what is seen in IT.

The source for this idea is the **cell assembly**, first proposed by Donald Hebb in his 1949 book *Organization of Behavior* [32]. What has become known as the **Hebb Rule** for synaptic coupling modification was originally proposed specifically to allow for the formation of cell assemblies.

A cell assembly is a closed chain of mutually self-exciting neurons. Hebb viewed the assembly as the link between cognition and neuroscience. When an assembly was active, it corresponded to a cognitive entity, for example, a concept or a word. Although the idea is an appealing one, it is hard to make it work in practice because it is difficult to form stable cell assemblies. Two common pathological situations are (a) no activity in the network after a period due to inadequate gain around the loop and, (b) spread of activity over the entire network since neurons in a realistic model system will participate in multiple assemblies and activity spread will widely. It is possible to control this behavior to some degree by making strong assumptions about inhibition, but the resulting systems are not robust.

As we mentioned, a key assumption of the Network of Networks model is that the basic computing elements are interacting groups of neurons. Module activity is not a scalar but a pattern of activity, that is, a high dimensional vector. Connections between modules are in the form of interactions between patterns. There is an intrinsic degree of selectivity. Patterns are less likely to spread promiscuously.

Because of this increased selectivity it might be possible that several nearby modules can become linked together to form loops through Hebb learning and can remain stable structures. We showed in Section 2 that associatively connecting modules together can increase the feedback coefficients in both modules (Figs. 4, 5.).

We can make speculations about the local density of connections in neocortex. Data from Lund et al. [33] suggests substantial connectivity over a region of one or two millimeters. Recurrent collaterals of cortical pyramidal cells form relatively dense projections around a pyramidal cell. The extent of lateral spread of recurrent collaterals in cortex seems to be over a circle of roughly 3 mm diameter [34]. If we assume that a column is roughly a third of a mm, there are perhaps 10 columns per mm^2. A 3 mm diameter circle has an area of roughly 10 mm^2, suggesting that a column could project locally to perhaps 100 nearby columns.

Let us assume that the intralayer connections are sufficiently dense so that active modules a little distance apart can become associatively linked.

Consider the set of four active modules, $\mathbf{a}, \mathbf{b}, \mathbf{c}, \mathbf{d}$, in Fig. 24. Assume they are densely connected locally. That will imply that $\mathbf{a}$ is connected to $\mathbf{b}$, $\mathbf{c}$ to $\mathbf{d}$, etc. Suppose that $\mathbf{a}, \mathbf{b}, \mathbf{c}, \mathbf{d}$ are forced to hold a particular pattern by some outside mechanism, say another brain region, an outside supervising input, or perhaps frequent co-occurance. Then the analysis we just performed on sparse paths with simple associators suggests that associative links between modules will develop.

However, the path closes on itself. If the modules $\mathbf{a}, \mathbf{b}, \mathbf{c}, \mathbf{d}$ are simultaneously active and are associatively linked, then if $\mathbf{a}$ is present, after traversing the linked path $\mathbf{a} > \mathbf{b} > \mathbf{c} > \mathbf{d} > \mathbf{a}$, the pattern arriving at $\mathbf{a}$ will be a constant times the pattern on $\mathbf{a}$. If the constant is large enough and the starting pattern is still present, there is the potential for feedback loop gain greater than one. The same analysis holds true for traversing the loop in the opposite direction, that is, $\mathbf{a} > \mathbf{d} > \mathbf{c} > \mathbf{b} > \mathbf{a}$. Limitations on maximum activity in each module will stabilize activity as it becomes large so activity in the loop

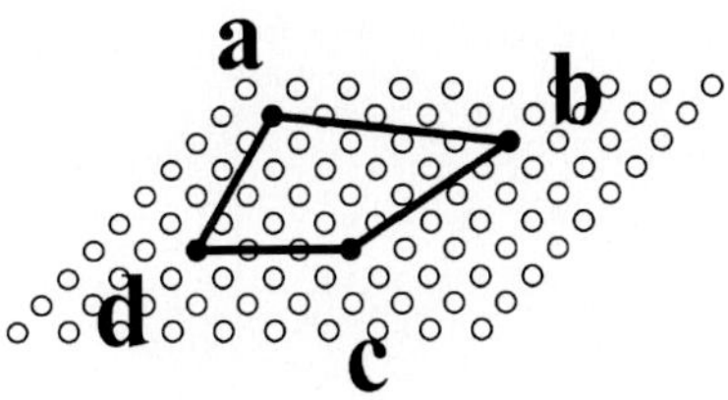

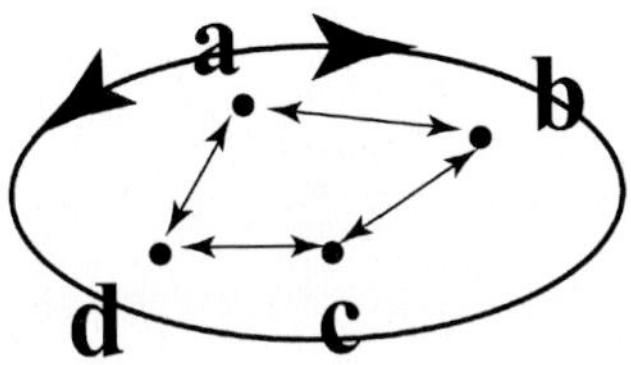

Fig. 24. Scheme for formation of a **module assembly**. Patterns flow around the loop based on local pairwise associations

will not increase without bounds. All the modules in the loop are likely to reach attractors in synchrony after extensive learning.

Timing considerations become of critical importance. Adjusting travel time through the strength of associative connections is a potentially promising way to control loop dynamics. Note that critical timing relationships also appeared earlier when we discussed for formation of high order feature coincidences to form selective filters. The parameters that control module dynamics are also a potential source of control.

We should note that the capacity of these seemingly impoverished sparse systems can be very large. For example, suppose columnar sized regions are about one third mm in diameter. Then one mm^2 contains roughly 10 columns and a cm^2 contains roughly 1,000 columns. If something like 10 columns become active to specify an object in a one cm^2 piece of cortex, there are roughly $(1000)^{10}$ or 10^{30} possible different combinations of 10 active columns. If we assume in addition that each column has 10 attractors, we increase the number of potential combinations of states shown by those 10 active columns by 10^{10} for a total number of states of a one cm^2 chunk of cortex to as much as 10^{40} distinct states, a large enough number to satisfy almost any conceivable need, especially when we consider that there may be on the order of a million columns in a human cortex.

8.3 Multi-layer Assemblies

We can speculate on the extension of idea of module assemblies to multiple layers. Now we can perhaps see the outlines of a dynamic, adaptive computational architecture that becomes both feasible and interesting. The basic notion is that local module assemblies form, perhaps at each layer. Sparse connections between each layer learn to link the assemblies through associative learning.

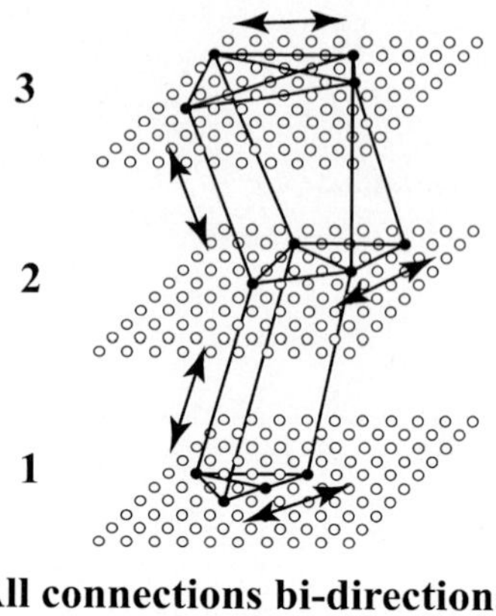

Fig. 25. Linkages between modules in a layer and from layer to layer may become bound into a more complex structure

The individual sparse paths now can work together to form complex multi-level entities. We might end up with something like that shown in Fig. 25, tightly connected module assemblies within a layer, sparsely linked together. This rings (loops in layers) and strings (sparse connections between layers) sparse architecture becomes a way to get multi-area activation patterns bound together through common dynamics.

Computation and learning from this point of view is based on the formation of sparsely interconnected between layers, less sparsely interconnected at a layer, module assemblies working with sparsely represented data.

8.4 Combinations

A natural way for learning to progress with this approach is to build combinations of module assemblies. Early learning, we conjecture, might form small module assemblies. Fig. 26 shows two such module assemblies that occur together. They can become bound together by learning through co-occurrence (Fig. 27).

As learning progresses, groups of module assemblies will bind together through the linkage mechanisms we have discussed. The small initial assemblies perhaps can act as the "sub-symbolic" substrate of cognition and the larger assemblies, symbols and concepts. Because of sparseness smaller components might be largely independent of each other and not interfere with each other. Cognitive learning would then come to have something of the aspect of an erector set or Legos, where the parts start by being independent and then get bound together with associative epoxy to form a sturdy higher-level spatially extended assembly. Note that there are more associative connections possible in the bound system than in the parts because there are many more possible paths. An interesting conjecture is that larger assemblies (words?

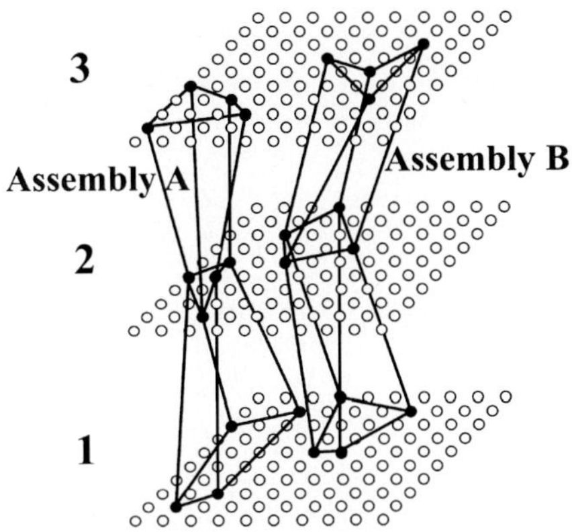

Fig. 26. Suppose we have two previously formed assemblies

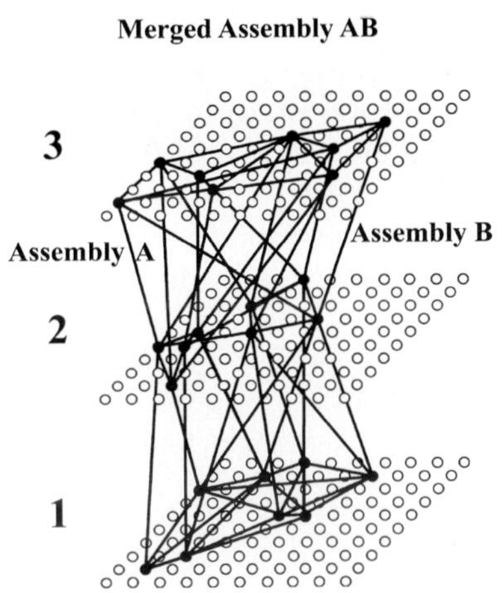

Fig. 27. Co-occurrence between assemblies can cause the structures to become associatively bound, forming a new more complex linked structure

concepts?) can be stable or more so than their component parts because of extensive cross-linkages.

This process looks somewhat like what is called compositionality. (Geman [35]). The virtues of compositionality are well known. It is a powerful and flexible way to build information processing systems because complex mental and cognitive objects can be built from previously constructed, statistically well-designed pieces.

What we are suggesting in this discussion is a highly speculative possible model for the dynamics, intermediate level structure, and learning in a potentially compositional system. Note however, that this system is built based on constraints derived from connectivity, learning, and dynamics and not as a way to do optimal information processing.

Perhaps compositionality as we see it manifested in cognitive systems is more like a splendid bug fix than a carefully chosen computational strategy.

References

[1] Sloman, S.: Causal Models: How People Think About the World and Its Alternatives. Oxford University Press, New York (2005)
[2] Anderson, J.A., Sutton, J.P.: If We Compute Faster, Do We Understand Better? Behavior Research Methods, Instruments, and Computers. **29** (1997) 67–77
[3] Anderson, J.A. Arithmetic on a Parallel Computer: Perception Versus Logic. Brain and Mind. **4** (2003) 169–188
[4] Strangman, G., Sutton, J.P.: The Behaving Human Neocortex as a Network of Networks. In: Hecht-Nielsen, R., McKenna, T. (eds.): Theories of the Cerebral Cortex. Springer, New York (2003) 205–219
[5] Hopfield, J.: Neural Networks and Physical Systems with Emergent Collective Computational Abilities. Proceedings of the National Academy of Sciences. **79** (1982) 2554–2558
[6] Anderson, J.A.: The BSB Network. In: Hassoun, M.H. (ed.): Associative Neural Networks. Oxford University Press, New York (1993) 77–103
[7] Anderson, J.A.: An Introduction to Neural Networks. MIT Press, Cambridge, MA (1995)
[8] Mountcastle, V.B. Introduction. Cerebral Cortex. **13** (2003) 2–4
[9] Hubel, D.H.,Wiesel, T.N.: Receptive Fields, Binocular Interactions, and Functional Architecture in the Cat's Visual Cortex. Journal of Physiology. **148** (1962) 106–154
[10] Kosko, B.: Bidirectional Associative Memory. IEEE Transactions on Systems, Man, and Cybernetics. **18** (1988) 49–60
[11] Pitts, W., McCulloch, W.S.: How We Know Universals: The Perception of Auditory and Visual Forms. In: McCulloch, W.S. (ed., 1965): Embodiments of Mind. MIT Press, Cambridge, MA (1947/1965) 46–66
[12] Blum, H.J.: Biological Shape and Visual Science (Part I). Journal of Theoretical Biology. **38** (1973) 205–287
[13] Kimia, B., Tannenbaum A., Zucker, S.W.: Shapes, Shocks and Deformations {I}: The Components of Shape and the Reaction-Diffusion Space. International Journal of Computer Vision. **15** (1995) 189–224
[14] Bringuier, V., Chavane, F., Glaeser, L., Fregnac, Y.: Horizontal Propagation of Visual Activity in the Synaptic Integration Field of Area 17 Neurons. Science **283** (1999) 695–699
[15] Lee, T.-S., Mumford, D., Romero, R., Lamme, V.A.F.: The Role of Primary Visual Cortex in Higher Level Vision. Vision Research. **38** (1998) 2429–2454

[16] Lee, T.-S.: Analysis and Synthesis of Visual Images in the Brain. In: Olver, P., Tannenbaum, A. (eds): Image Analysis and the Brain. Springer, Berlin (2002) 87–106
[17] Kovacs, I., Julesz, B.: Perceptual Sensitivity Maps Within Globally Defined Shapes. Nature. **370** (1994) 644–6
[18] Kovacs, I., Feher, A., Julesz, B.: Medial Point Description of Shape: A Representation for Action Coding and its Psychophysical Correlates. Vision Research. **38** (1998) 2323–2333
[19] Anderson, J.A.: Seven times seven is About Fifty. In: Scarborough, D., Sternberg, S. (eds.): Invitation to Cognitive Science, Volume 4. MIT Press, Cambridge, MA (1998) 255–299
[20] Hadamard, J.: The Psychology of Invention in the Mathematical Field. Dover, New York (1949)
[21] Murphy, G.L.: The Big Book of Concepts. MIT Press, Cambridge, MA (2002)
[22] Watrous, R.L.: Current Status of Peterson-Barney Vowel Formant Data. Journal of the Acoustical Society of America. **89** (1991) 2459–2460
[23] Peterson, G.E., Barney, H.L.: Control Methods Used in a Study of the Vowels. Journal of the Acoustical Society of America. **24** (1952) 175–184
[24] Talavage, T.M., Sereno, M.I., Melcher, J.R., Ledden, P.J., Rosen, B.R., Dale, A.M.: Tonotopic Organization in Human Auditory Cortex Revealed by Progressions of Frequency Sensitivity. Journal of Neurophysiology. **91** (2004) 1292–1296
[25] Olshausen, B.A., Field, D.J.: Sparse Coding of Sensor Inputs. Current Opinions in Neurobiology. **14** (2004) 481–487
[26] Barlow, H.B.: Single Units and Sensation: A Neuron Doctrine for Perceptual Psychology? Perception **1** (1972) 371–394
[27] Vincent, B.T., Baddeley, R.J., Troscianko, T., Gilchrist, I.D.: Is the Early Visual System Optimized to be Energy Efficient? Network: Computational neural systems. (In press)
[28] Cooper, L.N., Intrator, N., Blais, B.S., Shoval, H.Z.: Theory of Cortical Plasticity. World Scientific Publishing, Singapore (2004)
[29] Tsunoda, K., Yamane, Y., Nishizaki, M., Tanifuji, M.: Complex Objects are Represented in Macaque Inferotemporal Cortex by the Combination of Feature Columns. Nature Neuroscience. **4** (2003) 832–838
[30] Tanaka, K.: Inferotemporal Cortex and Object Vision. In: Cowan, W.M., Shooter, E.M., Stevens, C.F., Thompson, R.F. (eds.): Annual Review of Neuroscience. **19** (1996) 109–139
[31] Tanaka, K.: Columns for Complex Visual Object Features in Inferotemporal Cortex: Clustering of cells with similar but slightly different stimulus selectivities. Cerebral Cortex. **13** (2003) 90–99
[32] Hebb, D.O.: The Organization of Behavior. Wiley, New York (1949)
[33] Lund, J.S., Angelucci, A., Bressloff, P.C.: Anatomical Substrates for Functional Columns in Macaque Monkey Primary Visual Cortex. Cerebral Cortex. **13** (2003) 15–24

[34] Szentagothai, J.: Specificity Versus (Quasi-) Randomness in Cortical Connectivity. In: Brazier, M.A.B., Petsche, H. (eds,) Architectonics of the Cerebral Cortex. Raven, New York (1978) 77–97
[35] Geman, S.: Invariance and Selectivity in the Ventral Visual Pathway. Division of Applied Mathematics, Brown University, Providence, RI. (in preparation)

The Human Brain as a Hierarchical Intelligent Control System

JG Taylor

Department of Mathematics, King's College, Strand London WC2R2LS, UK
john.g.taylor@kcl.ac.uk

Summary. An approach to intelligence and reasoning is developed for the brain. The need for such an approach to Computational Intelligence is argued for on ethical grounds. The paper then considers various components of information processing in the brain, choosing attention, memory and reward as key (language cannot be handled in the space available). How these could then be used to achieve cognitive faculties and ultimately reasoning are then discussed, and the paper concludes with a brief analysis of reasoning tasks, including the amazing powers of Betty the Crow.

Contents

1 Human Intelligence

It is self-evident that the human brain is an intelligent system. But of what our faculty of human intelligence consists, and how it could be modelled in detail is still somewhat unclear. This lack of progress is in spite of the great strides being made by brain science to probe and define the processes the

JG Taylor: *The Human Brain as a Hierarchical Intelligent Control System*, Studies in Computational Intelligence (SCI) **63**, 99–122 (2007)
www.springerlink.com

brain supports, and how such support is achieved in neural terms. The neural networks being used to perform certain cognitive functions are becoming ever more delineated. But the main principles of information processing in the human brain are still uncertain. In this paper we present fragments of a general theory of human intelligence based on recent brain science results. The approach is to suggest how the functional neural architecture (modules, their functions and the functional networks they combine to produce) now being exposed in the brain could be joined together to lead to intelligence. The approach is based on detailed neural modelling of specific regions and their functions, but not yet on any model of the global brain. That would require much more extensive computation than able to be achieved presently, only then allowing a real proof of the resulting theory. However general principles are beginning to emerge from this approach which we explore.

There is great relevance of a brain-guided approach for the development of a more general theory of computational intelligence. The main thrust of the argument is that, given the long-term aim of computational intelligence being to construct a super-intelligent system, the ethical dangers that result from this aim necessitates use of brain guidance to help such a development proceed without danger to the future of the human race. I will come back to that in more detail shortly.

Human intelligence is hard to define. In action it is clear, as the following story indicates. Two work-men were sweating away digging a hole in the ground with the sun beating down on them, whilst their boss relaxed more comfortably in the nearby shade. One of the workmen turned to his colleague and asked 'Why are we out in the hot sun while he's so much better off in the shade'. 'Ask him' responded his colleague. So he did. The boss replied: 'Intelligence'. Here we see emphasis on the ability to use simple reasoning power to arrive at task solutions (the boss keeps cool, for example, by sheltering under the trees) given a set of constraints (but the workmen have to stay out in the sun where the work is sited).

Standard dictionary definitions of intelligence do not take us much farther. Thus Chambers dictionary defines intelligence most relevantly as intellectual skill or knowledge; mental brightness. It also defines intellect as the mind with reference to its rational power; the thinking principle. Similarly Collins dictionary defines intelligence most closely to our considerations as: the capacity for understanding; the ability to perceive and comprehend meaning. Both of these definitions involve the employment of some level of awareness, together with the possession of a good data-base of knowledge to help guide any intelligent reasoning process. More generally we may consider intelligence as the ability to use relevant past information to solve a given task by reasoning. So a more general definition of intelligence is as possessing the power to manipulate suitable information by other information so as to solve a specific task to obtain a reward; the more tasks being solvable in this way the more intelligent will be the system.

Computational intelligence need not be defined as in any way related to human intelligence. Such a position, of trying to develop an intelligent system with no relation to human powers is completely defensible. Yet it misses out on guidance from human intelligence as an existing solution, even if it is presently difficult to decipher. Even at a lower level there are animals with reasoning powers, such as crows or chimpanzees, which do not possess the linguistic powers of humans but are still able to perform what are intelligent actions, such as tool using or making.

Instead of going off into the unknown, computational intelligence can be suggested as the computer equivalent of human intelligence. Taking guidance from the definition of human intelligence we can define computational intelligence as the ability to solve tasks of increasing complexity using as much as is needed of the machine's data-base of knowledge (being close to the definition of computational intelligence in [15]).

There may be a variety of algorithms involved in the 'thinking' of an intelligent machine, but they can all be validly considered as its intelligence repertoire when they are used in attaining its goals. Such computational intelligence achieved in a machine can be seen as very similar to that of the human sort.

There is presently a clear gap between human and computational intelligence as arising from the presence (in humans) or not (in machines) of awareness. However that is difficult to understand due to the controversy presently surrounding the nature of any form of consciousness. That is expected to change in the future, but there is a more immediate gap between humans and machines based on the presence or absence of reasoning in solving a task. Humans have considerable reasoning powers, which have been the subject of much speculation. Machines are to be regarded as reasoning machines, but only through the symbolic calculus of language that may be encoded into suitable high-level codes. Humans have a relatively high level of linguistic reasoning powers, but also can reason non-linguistically. Other animals also possess non-linguistic reasoning powers, as I already mentioned, although it is controversial that they possess any linguistic reasoning capabilities. How animals (including humans) might reason non-linguistically is one of the problems to be dwelt on shortly in this paper. A better understanding of such non-linguistic reasoning is expected to lead also to a better understanding of creativity (which occurs at an unconscious level as well as expected to occur both at non-linguistic and linguistic levels). Creativity is a human power of great importance to understand and apply to machine intelligence (the epithet 'creative machine' is a contradiction in terms). This supports the strategy to be pursued here of considering reasoning, for initial modelling purposes, only at non-linguistic level.

The paper starts by discussing some problems facing computational intelligence, and in particular why the human brain is going to be important in the long run. Section 3 outlines the basic functionalities of the modules of networks of the human brain involved in solving tasks intelligently, and associated global information processing principles to which these components are related. The most important of these are extracted as: attention, memory and value. The following three sections cover each of those component functionalities respectively. A possible brain architecture to support reasoning is discussed in Section 7. The following three chapters are concerned with detailed applications to cognitive and most specifically some simple reasoning tasks. The final section contains concluding remarks.

2 Why the Brain is Important to Computational Intelligence (CI)

A set of important questions facing CI have been suggested recently [6] as:

a) What is the goal of CI?
b) What are the short term/long term technological challenges?
c) Is the field developing?
d) How can progress be measured?

The nature of answers to the above questions on CI need not be the same as those facing human intelligence. But the answers should be much closer if we consider human intelligence as what CI should at least in part try to emulate.

The answer to question a) (the goal of CI): is to develop a super-intelligent computational system. The success of this over-riding long-term goal of CI will depend on the definition of intelligence used by the assessor. From what has been said already, there is no doubt that the task facing CI is great. Even the reasoning powers of Betty the Crow are still beyond our understanding in terms of modelling. To neglect trying to model Betty's or chimpanzees' reasoning powers, and assume instead that it would be a small step in directly modelling human non-linguistic reasoning powers would be a misleading attitude to take. If we watch the development of an infant through the ages of the first few years of its life we see remarkable changes in its reasoning powers. These indicate that we are faced with a system – the developing human brain - of considerable complexity and one we can usefully study in what we expect to be a simpler 'frozen' form in lower animals such as crows or chimpanzees.

However before we leave this important first question we need to turn to an ethical question that stems from it and mentioned earlier: is a super-intelligent CI system meant to be autonomous? If so, then it does not need any further contact with its initial human creators. But such an autonomous super-intelligent system could as well be inimical as be supportive of humanity. The creation of such a super-intelligent system is thus suspect when looked at from a human-based morality. We should therefore ask how we can develop a super-intelligent CI system which is consistent with our own morality.

If we look at human society, we recognize that each time an infant is born this same question – how to create a self-sufficient super-intelligent CI (now regarded as the infant) – has to be faced. This has to be done initially by the infant's parents and then later by its teachers and ultimately by society at large. Various institutions have been created in human society to achieve a good solution to this basic moral problem: schools, Universities, religions, prisons, rules of law, and so on. There is therefore a solution, or even a number of solutions, each being appropriate in its own human culture.

We can arguably see this developmental/educative approach – presently employed by the majority of human cultures - as the optimal way to achieve a morally equable society for its participants. From this we can conclude that in order to be able to build a morally-defensible super-intelligent CI system we must take guidance not from a single human brain but from the methods of child-rearing that are basic in our societies. These have considerable variations across cultures, but in essence we can see that to produce members of human society that do not cause too much disruption we need a disciplined but loving environment. Elsewhere I have called this 'Guided Creative Fun' [22]. This consists of giving reward for good responses, but indicating by some form of penalty that a bad response has been made where relevant (the word NO is operative here). It is guided creative fun that needs to be incorporated into the development cycle of a super-intelligent CI. Thereby hopefully it would enable the super-intelligent CI to be become a valid and successful member of human society as well as being super-intelligent.

We are thereby drawn in trying to develop the super-intelligent CI by re-creating many of the important features of the human brain, especially those associated with the emotions. It will be through learning the rewards and penalties of the environments in which it could move and interact with the objects and humans, and learn to control its desires so as not to cause hurt and pain to others, that the requisite system would develop a basis for crucial emotional responses. These in humans help guide us to be able to be part of the complex web of human society. The same should occur for the super-intelligent CI. At a higher level, as ever more complex representations of its surroundings are created in the 'brain' of the super-intelligent CI system, it would begin to show emotion responses common in young children, as in the various social emotions of guilt, embarrassment, and so on. If the super intelligent system could be granted conscious experience then these would become feelings, helping guide it especially through empathy, to be able to take part effectively in human society.

It is clear that the incorporation of emotions into the super intelligent CI system is essential if it is to function in a morally defensible manner in human society. The example of psychopaths, with little empathy for the feelings of others, show how dangerous such super intelligent CI systems would be if unleashed on society. In any case CI systems must be trained to be able to perform tasks, and an effective approach is to train by mean of reward/penalty learning. There is already a broad body of knowledge about such reinforcement learning, under the heading of the Adaptive Critic Element/Adaptive Search Element or Temporal Difference learning, as well as the related machine-learning area of dynamic programming to learn an appropriate utility function to enable the CI system to perform a limited range of tasks. It may be that only through the use of ACE/ASE learning would a suitably flexible system be able to be developed as part of a super-intelligent CI system to enable it to function in the manner described above – as a society-compatible machine system.

The problem of society-compatibility is not a trivial one. It has taken us hundreds of millions of years of evolutionary pressure to arrive, as human beings, at a solution of intelligence which has enough of societal compatibility to enable societies to develop so as to live relatively equably together. Thus we, as humans, present a solution to the problem of creating super-intelligent systems able to live relatively equable together. There may be other solutions, but even in small experiments where human societies have tried to bring up portions of their members without guided creative fun, there has been failure of cohesion in the end. Thus we can conclude that it may be otherwise very difficult to succeed in creating a society of super-intelligent beings without an important emotional content in each. This would be used so that guided creative fun would enable these emotional powers to be harnessed to allow for societal compatibility between different elements of society brought up in that empathic manner.

This brings us back more strongly to the need for brain guidance in the creation of a super-intelligent CI system. Without that brain guidance we may not get some form of guided creative fun factor to be able to work, and hence may create a race of rampaging psychopathic monster CI systems, with no thought for humanity. Such a step would be racial suicide for us. The stamping of each of these machines by a set of 'rules of behaviour', such as Asimov's Three Laws of Robotics, would be difficult, since there are various situations where these rules are difficult to apply consistently. Only through use of an emotional preference for responses to be made, with the emotional basis being carefully trained along guided creative fun lines, can we ensure that human-like decision making would occur in moments of crisis. In particular we would expect that humanity should be no more at danger from the vagaries of such super-intelligent CU systems than they are from their own human neighbours. That is the best we can hope for from such intelligent systems.

The implications of this are that, in the long-term, as super-intelligence emerges, we must develop the super-intelligent CI systems as guided by the internal workings of the human brain. Not only will this be good in the long run for the super-intelligent CI systems, it will also be good for the human race.

This leads also to the realization that in any case much of what passes for artificial intelligence today is mainly algorithmics. These are designed by computer scientists and applied mathematicians so as to solve non-trivial problems such as difficult problems in algebra, differential equations, optimisation, or similar areas. But they are not intelligent in the way that we consider human intelligence. The data-bases of knowledge possessed by these algorithm solvers may be large, and the speeds they work at fast, But there are no such systems that have anything like human intelligence, This leads to further support to developing a super-intelligent CI system by looking at the structures used by the brain in its processing.

To conclude, I would strongly argue that the long-term goal of CI is to create a brain-guided super-intelligent CI system, trained initially in a developmental fashion so as to be able to function in an effective autonomous manner in human society.

At this point it could be argued against the brain-guided approach to CI I am advocating: the direction being followed so far would lead us to create a silicon-based version of humans. Why do that since there are already too many humans on the planet, polluting it in an increasingly dangerous manner. That is not the purpose of CI, nor my purpose in this article, although such an approach is an important but different process of understanding ourselves by the creation of a relatively good replica of ourselves. Here I am arguing that the reason to bring in the human brain is to allow development of a super-intelligent autonomous CI system along human lines so as to guide it to have empathy for its human creators. We do not have to go all the way towards making a human-replica brain. Instead we are probing the human brain so as to extract its secrets of overall design. It may be that alternate but better designs for some of the components can arise from different engineering approaches. But firstly we have to understand the basic principles underlying brain function and the functional architecture of global brain networks before we can begin to consider how to make improvements on the design of the human brain.

If we look at some areas of machine intelligence, such as machine vision or predicting the future value of stocks or shares, we have to accept that the human is still in front. That is why guidance from the brain is of importance even in these areas of heavy algorithmic processing; in both machine vision and prediction of the future values of financial assets, no algorithms have yet been created able to surpass the powers of the brain.

The short term challenges to CI (question b) are numerous: in machine vision, in robot control, in software agents, in creating systems able to handle complex domains, in affective computational systems (with emotions being included), in language understanding software systems, in the development of various tools enabling ever more complex computer programs to be validated, in developing ever more complex programs for running intelligent applications on cluster or grid computers, and so on.

Is the field developing (question c)? Undoubtedly it is, especially with ever increasing computer power. I will describe shortly some of the developments, from a brain guided point of view, that are thereby being made possible, although these are only still under development and have not yet arrived at specific CI applications.

How can progress be measured (question d)? One way, following brain guidance, is as to how much understanding is thereby generated about the brain itself. This requires relating software models built on the basis of the chosen CI design architecture to the experimental results of specific brain science experiments, say at a single cell or at a more global brain level.

However there are more applications-oriented criteria, such as the solutions that may be provided by brain-guided architectures for some of the present standard problems already mentioned under the answer to question c) above. Such criteria are thereby highly relevant to showing how the brain-guidance approach may be helpful in solving industrial–style problems.

I will now turn to consider a tentative set of processing principles that can now dimly be glimpsed as basic to the information processing powers of the human brain. It is these which will be explored more fully in the following.

3 Global Principles for the Human Brain

It is currently easy to get lost in the welter of brain science data now deluging us from all levels in the brain and thereby become unable to see the wood for the trees. This makes it hard to extract basic principles of information processing in the human brain. This has been almost made a tenet of brain science: difficulties in understanding the brain globally are constantly being claimed by distinguished neuroscientists, along the lines that 'It is not possible to attempt to model the whole brain since we do not know its basic underlying principles of operation' [3]. However that does not prevent us from trying to extract the principles upon which the brain operates, and hence embark on such an ambitious computational exercise. It will undoubtedly be a difficult exercise, but if we do not begin then we will never finish.

Other neuroscientists charge that unless every microscopic detail is included in any brain model it will fail. Such an approach indicates that no principles could be extracted at a global scale: the principles of the brain would involve ongoing activities at all levels and scales. That makes any attempt at the extraction of principles of information processing a very hard task. However when one asks 'And should superstrings be also included?' then the absurdity of such a position becomes apparent. The human brain works at a much higher level than superstrings or even at single neuron level, with much higher-level linguistic and concept activations and their manipulations as being basic to the information flow in the brain. Feature level descriptions are also tapped into by the brain, so we cannot neglect a variety of scales of coding there. But we will not need to go down to lower than synaptic level to begin to build an information-effective description of brain processing.

Understanding at the single neuron and at the lower synaptic level is of course proceeding apace. One important example of this is of spike-timing dependent potentiation (STDP for short [2]). This is now regarded as the most likely learning law at synaptic level, although there are still other forms of learning that may be present in the brain. But STDP is clearly an important learning component: it brings Hebbian learning into the temporal domain, so that only spikes which arrive at a cell causing a spike in the post-synaptic cell up to some tens of milliseconds later will have the relevant synaptic strength increased. If the pre-synaptic spike arrives tens of milliseconds later than the appearance of any post-synaptic spike, then there will be reduction of the relevant synaptic strength, since the incoming spike was not likely to be the cause of the outgoing one.

The STDP learning law allows the brain to capture causality in events involving external stimuli. Such causal sensitivity seems to be needed in a model of the brain so that it can begin to mirror the sensitivity present in the living brain. This would be best achieved by taking a model of spiking neurons, with their associated high sensitivity to temporal features of external stimuli.

This example shows that there are expected to be several constraints that enter in any approach to modelling the brain when such aspects as suitable sensitivity to the temporality of external events are to be included. However this does not necessarily imply that higher-level principles cannot be extracted: how these are implemented in detail (and most efficiently) in our brains need not impinge on less efficient implementations by models with graded neurons and other unrealistic features. Thus we will proceed to look for high level principles of brain processing, ignoring what may go on at synaptic and neural level.

At the highest level we have awareness, or consciousness. Awareness presents an enormous difficulty to model, since there is no universally accepted model of its nature, nor we do properly understand it. A model of awareness, developed from a control model of attention using engineering control ideas, however, has been developed elsewhere [18, 21, 22]; the claim that awareness is properly included in such a model has been argued in detail elsewhere [20, 21]. The model itself is based on attention, so let us consider that next.

Attention is a control system employed by the brain to handle the problem of the complexity of the world in which we live. Attention in the brain acts as a filter to reduce the processing load for the hard problems which invariably arise in the complex environments which we inhabit. This filtering process is achieved by a method used in engineering control, as also observed in motor control by the brain: the use of state estimator internal models. Attention acts to create an 'attended state estimate' of the external world, as represented in the brain. The external world exists in a lower level representation in the brain at a pre-attentive level. However that amount of information is too much for further processing, with many stimuli present, all coded at a variety of levels. The total amount of information is therefore cut down by attention being directed to the lower level brain activations for a single stimulus. It is such attended stimulus activity which is allowed to go forward to higher executive sites in the brain for further processing.

Attention thereby cuts down the computational complexity of the environment. To be able to use the results of the filter of attention at later times, the brain stores attended state estimates for later long-term memory (LTM), especially in the hippocampus acting as a recurrent neural network. This information is later transferred to nearby cortical sites to constitute either episodic memory (with the added tag of ownership) or to more distant sites to act as semantic memory (with no ownership tag). These memory structures are constantly developing through life, and form an important component of the knowledge base of the relevant brain.

Long-term memory can also exist in a non-declarative form, as procedural memory, able to be used in making responses and manipulating objects. Such memory is thought to be encoded in the frontal cortices, basal ganglia and related sites in cerebellum and parietal cortex. However to enter into effective procedural memory it is necessary that some form of attention control be exercised over the motor apparatus. This has been termed 'motor attention', and allows the selection of only certain most task-relevant responses as compared to many others. Thus attention can exist to control, select or filter out sensory input and also to control and select motor response. It is the motor attention control system, coupled with the sensory attention control system, which produces effective motor response in sensory-motor tasks, and whose continued learning (as over-learnt responses) enables the responses to become automatic.

Reward and drives are also important components of processing in the brain. The value of a stimulus is the level of reward received by the person when they interact with it (such as eat or drink it). Such reward as is thereby gained is used to build a so-called value map (in the orbito-frontal cortex and the related sub-cortical amygdala nucleus). Such values are used to guide responses. At the same time the level of motivation for the rewarding stimulus is used to modulate the level of drive to achieve that specific reward. Thus the reward value of a particular food is reduced if a period of eating much of the food has just occurred. In this way the motivation (hunger) is reduced and thereby the reward value of the relevant food is devalued.

Language is a crucial human ability. It involves assigning utterances (words) to external stimuli or to actions so as to build up a vocabulary enabling concept creation to be efficient, as well as concepts be universally recognised. Besides the one-stimulus-to-one-word component (semantics) there is also syntax in which rules for sentences are developed so that statements about sequences of events can be understood by others of the same culture.

Finally there is the process of reasoning, in which sequences of actions on stimuli are reasoned about by allowing imagination to take place. This involves the ability to replay previous action/ stimulus sequences, or make up imaginary new ones from previously remembered ones. The reasoning process is used to determine how a specific goal could be reached, which, if predicted to possess a high enough reward would then be achieved by means of taking the relevant actions in reality.

There are five functional components of the brain that have been singled out above: the attention control system, the memory system, the value map system, the language system and the reasoning system. These five aspects are fundamental in brain processing, but may not be completely independent.

The present evidence from the brain is that the executive processes involved in reasoning are those based heavily on the attention control system. It was argued earlier that language was not essential for some forms of reasoning. I will not consider it further here, in spite of its importance in human activity, so as to keep this contribution in length. In all, then, we have the following three global or high-level processing principles:

P1. Attention Control

Attention functions as a filter, to create an attended state of the world, which thereby reduces complexity, and allows for efficient analysis of the external world.

P2. Long-term Memory Databases

There are a variety of memory databases constructed during the activity of the brain: declarative (created under attention) and automatic (over-learned) involving especially motor responses.

P3. Value Maps and Their Guidance

Value maps are created in the brain of expected reward values of different stimuli in the outside world, and as coded in lower level cortical brain sites; these expected rewards are modulated by the level of any associated primary drive (such as hunger).

We claim that these three principles, properly implemented, provide the core functionalities to achieve the complexities achieved by the human brain (barring language). Non-linguistic reasoning is expected to arise from the first three principles, as will be discussed later.

More specific detail needs to be put in the structure of the attention control system, as well as bringing in memory and reward to make the overall system both autonomous and more powerful. Thus we must turn to how the brain fleshes out the functionality of attention, memory and reward.

4 Attention as a Filter

Attention is now known to arise from a control system in higher order cortex (parietal and prefrontal) generating a signal to amplify a specific target representation in posterior cortex. This is true for the both the endogenous (goal directed) and exogenous (externally directed) forms of attention. The higher cortical sites generating the control signal (inverse model for attention movement) use a bias from prefrontal goals (held in working memory) to amplify (by contrast gain) posterior activity in semantic memory sites (early occipital, temporal and parietal cortices). This leads to the following ballistic model of attention control:

$$\begin{gathered}\text{Goal bias (in Pre Frontal Cortex)} \rightarrow \\ \text{Inverse model controller (in Parietal lobe)} \rightarrow \\ \text{Lower Posterior Cortex} \\ \text{(in various modalities)}\end{gathered} \tag{1}$$

The amplified target activity is then able to access a buffer working memory site in posterior cortices (temporal and parietal) which acts as an attended state estimator for the modality and features that are being attended to. The access to the buffer has been modelled in the more extended CODAM model [18, 8] as a threshold process arising from two reciprocally-coupled neurons almost in bifurcation in models in [8]. The access to the sensory buffer is aided by an efference copy of the attention movement control signal generated by the inverse attention model sited in the brain in the Superior Parietal Lobe. The existence of an efference copy of attention was predicted as being observable by its effect on the sensory buffer signal (as represented by its P300 event-related potential) [19, 8]; this has just been observed in an experiment on the Attentional Blink. In this paradigm a rapid stream of visual stimuli are presented to a subject, who has to detect a specific first target T1 (such as a white X) and then a second target T2 (such as a black O). When the second target is about 250 milliseconds after the first there is least awareness of the second target. In that case the N200 (an event related brain potential arising about 200–250 milliseconds after stimulus onset) of the second target is observed to inhibit the P300 (a further event related brain potential related to the awareness of the stimulus) of the first when T2 is detected [17].

A possible architecture for the overall attention control circuit is shown in Figure 1, being based on the COrollary Discharge of Attention Model (or CODAM for short) model [18, 19, 21] of attention control. This makes important use of the efference copy of the attention movement control signal

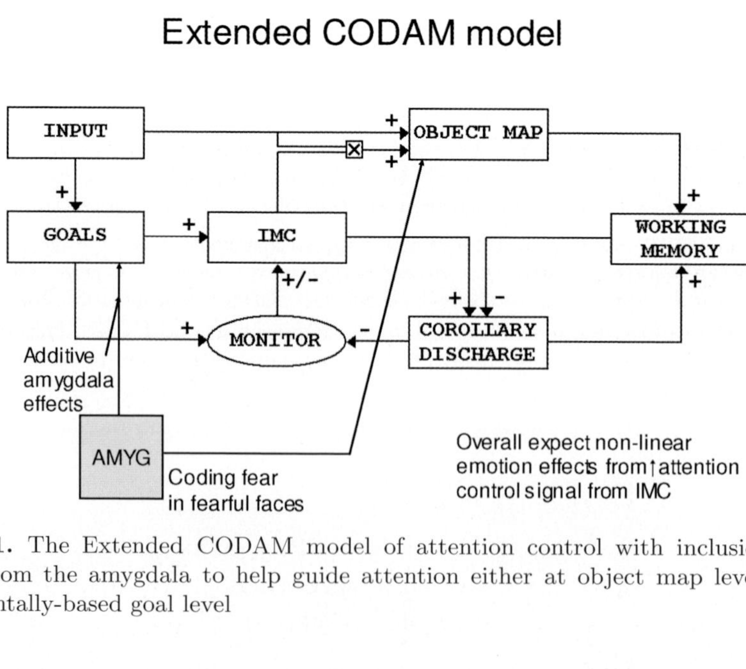

Fig. 1. The Extended CODAM model of attention control with inclusion of a bias from the amygdala to help guide attention either at object map level or at prefrontally-based goal level

to act as a predictor of the attended state, and so enhance the efficiency of attention movement. The model of Figure 1 contains an extension of the original CODAM model by the presence of an emotional component, the amygdala, which influences the activation of object stimuli on the object map as well as by further modulation of the activity of emotionally valued goals.

There is presently considerable controversy over the micro-structure of attention feedback. It could be one or more of three possible forms: 1) additive feedback, altering the overall threshold to an attended neuron; 2) a multiplicative effect on the output neuron to which attention is being paid; 3) contrast gain enhancing the inputs from a stimulus being attended to but not those from other stimuli that act as distracters. Present evidence supports the third of the above for the action of endogenous or top-down attention [21, 22], and only a small additional component from 2) above in addition in the case of exogenous or bottom-up attention [12].

5 Memory as a Data-Base

There are a variety of forms of memory: short-term, long-term, procedural, declarative, and so on, as already mentioned. We have already considered short-term (working) memory as part of CODAM (in both parietal lobes as the buffer system and in prefrontal cortex as part of the executive system). Declarative long-term memory is based on the hippocampus (HC). Numerous models of the HC cell fields exist, as well as models with more extensive connections to prefrontal cortex. HC stores memories in at least two ways: in changes of synaptic strengths of afferents to the module CA1, an output component of the hippocampus (and here acting as a feed-forward net) and in changes of the strengths of recurrent connections in the module CA3, which is an intermediate component of the hippocampus (thought to act like a Hopfield net in the rapid storage of permanent memories). These are both stored during theta wave activity in HC, and then observed to be played back in the HC in slow wave sleep. Finally playback to cortex is thought to occur during Rapid Eye Movement sleep, leading to a more permanent code both of episodic and semantic memories. These two types of memory form the basis of knowledge codes for sensory stimuli. There is also procedural memory based on pre-motor cortical sites and basal ganglia and thalamic sub-cortical sites, as well as involving parietal lobes and the cerebellum.

6 Rewards/Values as Biases

An organism tries to optimise its experience in the world by taking actions on stimuli which maximise its future rewards. The signal for reward, and in particular reward prediction error, has been observed as carried in the limbic brain by dopamine, with predicted future reward values of stimuli encoded in the orbito-frontal cortex (and very likely also in amygdala). A machine learning model of this has been developed under the guise of Temporal Difference learning and the Adaptive Critic Element/Adaptive Search Element system [1]. It allows a further bias to be exerted on the attention control system and resulting responses; it also provides a method of enhanced learning of stimulus representations as valued goals in prefrontal cortex.

Such a bias as brought about by the amygdala is shown in Figure 1 of the extended CODAM model.

7 An Architecture for Reasoning

The above components of attention, memory and reward are now to be put to work to attain reasoning. The memory component gives a data-base on which reasoning can be based, while reward biases the responses to be made. But whilst both of these components make for greater efficiency, they cannot function to provide intelligence and more specifically reasoning without attention as the work-horse driving the movement of information around, and attaining agreed goals. The engineering control approach has been developed for attention, as mentioned earlier [18, 19, 21, 22]. This will now be considered in more detail as to how it can provide a basis for the executive processes necessary in intelligent processing of information, ultimately leading to an approach to reasoning. Before we turn to the important component of attention let us consider how coupled forward/inverse models could allow some form of reasoning to be achieved.

The necessary component of engineering control to help develop reasoning is a forward model involving the process of prediction:

$$x(t+1) = F[x(t), u(t)] \tag{2}$$

where x(t) is the estimated or predicted state of the controlled system at time t, and u(t) is a control signal at that time, which through the forward model function F[,] in (2) leads to an estimate of the state at the next time step t + 1. Internal purely sensory 'thinking' can thus be envisaged as arising from the recurrent running of the above forward model, using the control signal u(t) from, say, the attention control system. The sequence would be concluded when a suitable goal state had been reached (as set up initially, for example, as part of a reasoning problem).

The ability to predict future states by the forward model of (2) still requires in detail a sequence of actions {u(t)} to be generated. This can be achieved by use of an inverse model controller (IMC), which is a map G from the actual state of the system x(t), and its desired state x(des, t) to the action to achieve that transformation form the one state to the other:

$$u(t+1) = G[x(t), x(des, t)] \tag{3}$$

We may use the coupled set of F and G in reasoning as follows. We suppose that an inverse model IMC of the form (3) has been trained so that given a sequence of visual state {x(n)} there is a sequence of actions {u(n)} generated by the IMC as

$$u(n+1) = G[x(n), x(n+1)] \tag{4}$$

In order to make the sequence of virtual actions take the internal states of the system through the sequence {x(n)} we need to use the forward model to make the virtual changes

$$x(n+2) = F[x(n+1), u(n+1)] \tag{5}$$

Thus at each step of the reasoning process (4) & (5), the result of the IMC is fed, along with the expected state, into the forward model (FM) to produce the next state. This is then fed, with the previous state into the IMC to produce the next action, and so on. That the desired sequence of states is produced by this process is expected only to be possible after a suitable learning process; as noted in Figure 2, where the error at each stage is used to retrain the FM (and to a lesser extent the IMC) so as to attain the correct sequence of states at each point. The error is thus, for example in terms of a squared error:

$$\sum [x'(n) - x(n)]^2 \tag{6}$$

where x'(n) is the predicted next state at the n'th step of the iteration, as compared to the actual value of the state x(n).

More general thinking and reasoning can be achieved using a sensory-motor coupled system, with state estimate x being the attended sensory and

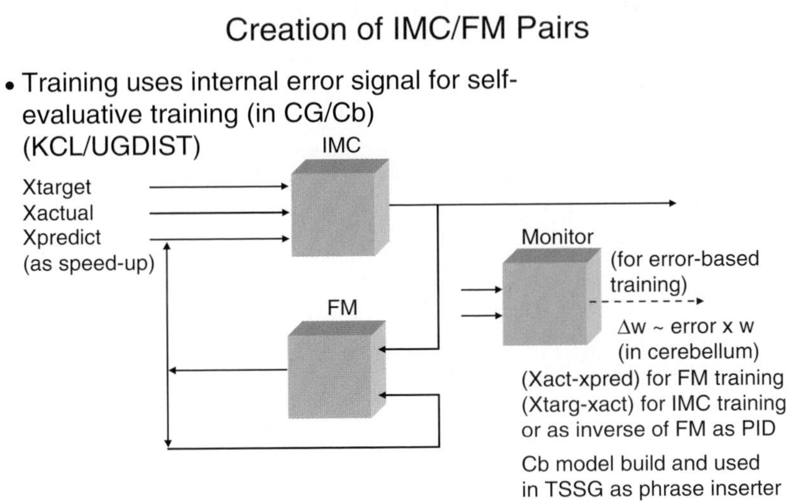

Fig. 2. The reasoning model and its learning component

motor state and the forward model run by joint sensory-motor attention control signals (which arise from separate parts of the brain, sensory attention mainly being in the right hemisphere, motor attention mainly in the left). The forward and related models will involve memory codes, so as to use previous information about the world more efficiently.

A possible architecture is shown in Figure 2 for reasoning, including a module to achieve learning through error-based feedback, and using the coupled IMC/Forward Model pair introduced above.

Without attention there is no filtering process on either the sensory inputs or the motor responses. This means that there will be considerable interference, especially in complex domains. To include attention the 'reasoning' model of Figure 2 is extended to include sensory and motor attention in Figure 3.

What has been added to the modules of Figure 2 are the sensory attention control components: goals module, inverse model controller (generating the attention movement control signal) and the 'plant' consisting of lower level cortical sites with activity to be attended to. There is also a working memory buffer, whose output has a threshold applied to it (so mimicking report); in the CODAM model of Figure 1 there is a further predictor of the attended state, achieved by using an efference copy of the IMC signal; this has been left out explicitly from Figure 3, although it can be regarded as a component of the WM buffer. There is also an IMC for motor attention, whose output is sent to a motor response system not shown explicitly in Figure 3. Finally there is a monitor or error module, as in Figure 2, allowing for error-based learning of the forward model as suggested there. Finally we note that the forward model FM is fed by output from the motor IMC (denoted IMCam) and not

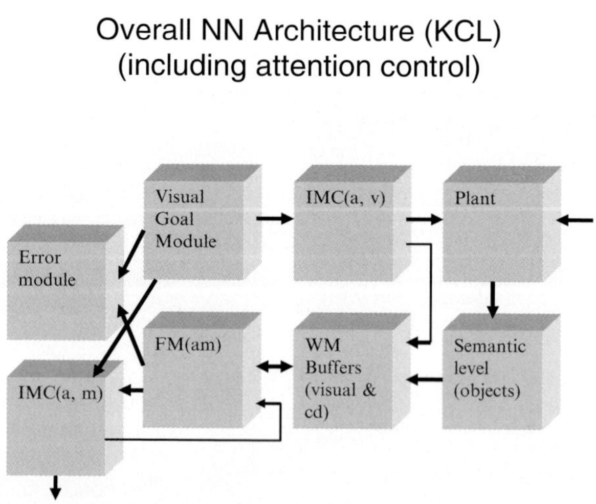

Fig. 3. The attention control components added to the model of figure 2, so as to include both sensory and motor attention control

from the lower-level motor output module not shown in Figure 3. The reason for this is related to the need for motor attention in the initial stages of motor learning, so that the IMCam must play a crucial role in the initial learning process, and only later can automatic processing be achieved in which the IMCam is expected to drop out.

We wish to apply the extended model of Figure 3 to various important reasoning tasks below, but we need to tease out various components of reasoning as part of the overall task.

8 Components of Reasoning

Cognitive Sub-Tasks: In reasoning, thinking and planning a variety of cognitive subtasks are carried out, including: 1) Storage and retrieval of memories in hippocampus (HC) and related areas; 2) Rehearsal of desired inputs in working memory; 3) Comparison of goals with new posterior activity; 4) Transformation of buffered material into a new, goal-directed form (such as spatial rotation of an image held in the mind); 5) Inhibition of pre-potent responses [9]; 6) The development of forward maps of attention in both sensory and motor modalities, so that possibly consequences of attended actions on the world can be imagined (along the lines of the architectures in Figs 2 and 3, and as discussed above); 7) Determination of the value of elements of sequences of sensory-motor states as they are being activated in recurrence through a forward model; 8) Learning of automatic sequences (chunks) so as to speed up the cognitive process.

The rehearsal, transformation, inhibition and retrieval processes are those that can be carried out already by a CODAM model [18, 19, 8] (with additional hippocampus activity for encoding & retrieval). CODAM can be used to set up a goal, such as the transformed state of the buffered image, or its preserved level of activity on the buffer, and transform what is presently on the buffer by the inverse attention controller into the desired goal state. Such transformations arise by use of the monitor in CODAM to enable the original image to be transformed or preserved under an attention feedback signal, generated by an error signal from the monitor and returning to the inverse model generating the attention movement control signal so as to modify (or preserve) attention and hence what is changed (or held) in the buffer, for later report. Longer term storage of material for much later use would proceed in the HC, under attention control. The comparison process involves yet again the monitor of CODAM. The use of forward models like (2) allows for careful planning of actions and the realisation and possible valuation of the consequences. Multiple recurrence through the forward models of the form (2) and the IMC model (3), as shown in Figure 2, allow further look-ahead, and prediction of consequences of several further action steps. Automatic process are created by sequence learning in the frontal cortex, using Frontal Cortex-> Basal Ganglia $\rightarrow$ Thalamus $\rightarrow$ Frontal Cortex, as well as with Cerebellum involvement, to obtain the recurrent architecture needed for learning chunks (although shorter chunks are also learnt in hippocampus). Attention agents based on these attention control principles have been constructed in software [10], and most recently combined with reward learning [11].

9 Extraction of Cognitive Principles

Cognitive Principles: We can deduce from these component functions some principles of cognitive processing:

CP1: There is overall control by attention of the cognitive process, using attention-based control signals to achieve suitable transformations to solve cognitive tasks;

CP2: Fusion of attention control (in parietal lobe and Prefrontal Cortex) and long-term learning in HC occurs to achieve an expanding state space of stimuli and actions, and of corresponding attention control;

CP3: The creation of a valuation of goals occurs in Prefrontal Cortex to handle reward prediction biasing of the processes 1) to 6) above;

CP4: Transformations on buffered stimuli is achieved by creation of suitable Prefrontal Cortex goals associated with the required transformations being carried out on existing buffered activities, under the control of attention;

CP5: Forward models (along the lines of equation (1) in section 2)) are created under attention control so that, by their recurrence, sequential passage through sequences of attended sensory-motor states is achieved (as in thinking), possibly to reach a desired goal (as in reasoning) or valued states that may even correspond to new ways of looking at a problem (as in creativity);

CP6: There is creation of, and ability to access, automated 'chunks' of knowledge, so they can be inserted into forward model sequences under CP4. These automated chunks are initially created by effortful attention at an earlier time (using error-based learning in cerebellum [16]) but are then gradually transferred to automatic mode by suitably long rehearsal (through reinforcement training by Dopamine from the brain stem);

CP7: Attention is employed as a gateway to consciousness, with consciousness providing an overarching control function of speed-up of attention (through the prediction model of the attended state in the WM buffer in the CODAM architecture), thereby giving consciousness overall guidance over ongoing processes (plus other features still to be defined).

Cognitive Architecture: A possible architecture is a) CODAM as an attention controller (with both sensory and motor forms and containing forward models) b) Extension of CODAM by inclusion of value maps and the reward error prediction delta, as begun in [23], and shown in Figure 1; c) Extension of CODAM to include a HC able to be attended to and to learn short sequences

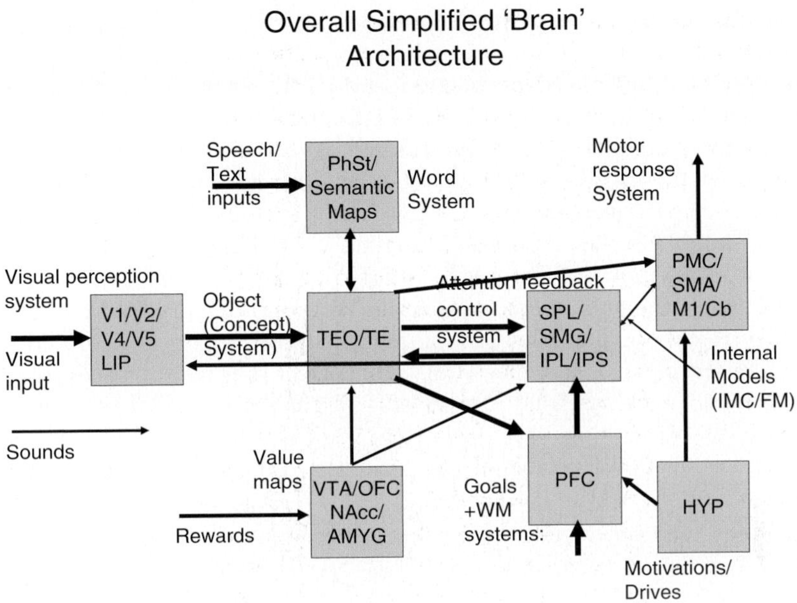

Fig. 4. Proposed Overall Cognitive Architecture for a Model Global Brain

d) Further extension of CODAM by addition of cerebellum to act as an error learner for 'glueing' chunked sequences together, with further extension to addition of basal ganglia so as to have the requisite automated chunks embedded in attended control of sequential progression. The goal systems in Prefrontal Cortex are composed of basal ganglia/thalamus architecture, in addition to prefrontal cortex, as in [24].

The overall cognitive architecture (not including the hippocampus) is shown in Figure 4.

We will now apply the above architecture and cognitive principles to some specific reasoning tasks.

10 Analysis of Specific Reasoning Tasks

The tasks I will consider first are: Wisconsin Card Sorting (WCST) and the Tower of Hanoi. These are valuable paradigms used to investigate prefrontal deficits in cognitive processing. I will then briefly describe logical reasoning in this brain-based framework, and finally turn to reasoning in animals, most specifically in Betty the Crow.

WCST:

The WCST task has a pack of cards with 1, 2, 3 or 4 shapes of 4 kinds, with each card being in one of 4 colours. Four test cards lie face up on the table. The subject's task is to take a card from the pack and place it on one of the four test cards according to a rule (following the matching of colour, shape or number). If the choice fits a rule chosen by the experimenter, unbeknownst to the subject, then the experimenter says 'yes', otherwise 'no'. Either way the subject has to try again. After (usually) 10 correct choices by the subject, the experimenter changes the rule, although without telling the subject except inthe 'yes' or 'no' response they give to the subject's choices.

Numerous neural models of this task exist, such as [13]. This uses a set of recurrent loops (Cortex → Basal Ganglia → Thalamus → Cortex) so as to store in working memory the rule presently in use by the subject. Receipt of a 'yes' continues the rule; a 'no' causes the Working Memory (WM) activation to be destroyed by the amygdala, acting as a punishment device, and a new rule is then formulated and tried until the correct new rule is discovered. This model captures the essential components of the process, and fits reasonably well with brain imaging data [13]. However the results in [14] indicate the need for further development of the recurrent prefrontal lobe loop structure to be more completely implemented.

Tower of Hanoi:

This task involves a set of vertical rods arranged, for example, in a line, with rings of increasing radius that can fit over each rod. It is not allowed to have a larger ring above a smaller one on any rod, and only one ring can be moved at any time. The purpose is for the subject to work out the smallest number of moves to take the rings from one allowed configurations of rings to another only making allowed moves. A standard initial goal state is all the rings on one rod (in a legal configuration), with the final goal state they being all on another rod (again in a legal configuration); the task usually has only three rods and 3–5 rings.

There are several executive processes involved in solving the problem: Working Memory encoding of the present position and of the ultimate goal state; transformation by a legal move to a further arrangement of rings from the initial one; further transformation from one configuration to the next by a legal move: Config(n) → Config(n + 1); achievement of the goal state by the final configuration reached, Config(N); if the desired final state is not achieved then start a new set of legal moves from the initial state; repeat until successful. There are more abstract approaches to this problem but we are considering here how the task would be solved by a person on first acquaintance with it. Thus the most crucial step is the manipulation of Config(n)→Config(n + 1) in Working Memory, using a suitable legal move as represented in Prefrontal Cortex. This corresponds to taking a mental image (in Working Memory) of Config(n), and making a mental legal move of one ring to the top of the set of rings on another rod, in which the final state Config(n+1) is also legal. Thus a check has to be made that such a move is legal by performing the move and then checking that there are no smaller ring below the moved ring on the new rod. If not, the move is legal and is made (and also stored in memory); if there is a smaller ring then the move is not taken and a new move is tried for its legality.

We note that in the above processes, attention and memory play a crucial role. Reward is involved in setting up the goal, but is not necessarily needed unless complete autonomy is needed for the agent to be suitable motivated.

Logical/Mathematical Reasoning:

There have been interesting new results on how this is achieved in the brain [9]. In particular the debate between use of either a linguistic or a spatial mechanism in reasoning seems to be settled in favour of either approach when it is most appropriate (as detected by the brain sites being activated). Thus in spatial reasoning sites in the brain are active known to support spatial processing; in linguistic or logical reasoning brain areas involved in language are observed active. Also an interesting result has arisen from the importance of inhibition in logical reasoning process, where perceptual reasoning methods give incorrect results and have to be inhibited by developing prefrontal cortical connections to prevent perceptual reasoning; it was placed as item 7) in the list of subtasks, and comes under principle CP3 in the previous section. Finally it is clear that logical reasoning (based on 'if-then' manipulations) is based on semantic processes; to understand the brain basis of this better, as pointed out in [9], it is needed to clarify the various functions being performed in left prefrontal regions centred on Broca's area (BA44-45) since that appears a crucial area involved in executive control in logical reasoning.

We can begin to model these processes most easily in the spatial case, based on initial models of numerosity. There are two number systems involved, integers and real numbers, the former involving language (so by some form of supervision by a teacher) and the latter by a Self Organising Map, so in an unsupervised manner; various models of these processes have been suggested recently.

Modelling Animal Reasoning:

Recent research has uncovered unexpected hidden cognitive powers in a group of animals, the corvids, of which the Caledonian crow is an important example, as further discussed in [4, 5, 7, 25]. We will also consider here, besides her powers of reasoning a low-level creative process present in Betty the Crow.

Betty was set the task by her handlers to remove a basket from the base of a transparent fixed vertical tube, open at the upper end; she was provided with a bent piece of wire which lay by the side of the vertical tube. Betty would view the situation for a few minutes, and then proceed to pick up the bent piece of wire in her beak and stick the bent end into the tube. She would then manoeuvre the bent end under the basket of food at the bottom of the tube, and thereby lift out the basket and its food. In a number of cases, however, Abel, Betty's partner, stole the food from the basket before Betty could eat it.

In one trial it was observed that Abel stole the bent wire before Betty was able to use it to extract the basket and food. After a few stationary minutes, Betty picked the straight wire up in her beak, pushed it under a piece of sticky tape on the bench, and bent the wire so that it was similar to her previous bent wire stolen by Abel. She had already had several successful retrievals of the food bucket using bent wires provided at each test. Thus these memories would still be relatively active, so the Forward Model/ Inverse Modal Controller pair would have some Working Memory activations for the 'bent-wire food-retrieval' schemata. She also had past experience (reported by the experimenters as occurring about a year previously) of playing with pipe-cleaners; it is to be assumed she developed a 'bent cleaner' schemata involving a further Forward Model/Inverse Model Controller pair (plus associated connections as needed in Figure 3). Thus we can conjecture that the processes going on in Betty's mind to enable her to take the creative 'tool making' step were:

1) The Forward Model/Inverse Model Controller pair for 'bent wire food retrieval' was still partly active in her brain;
2) The Forward Model/Inverse Model Controller pair for 'bent cleaner' was present in her memory, and especially was available as a reasoning system;
3) She had in her Working Memory (visual) images of both bent wires (from her recent successful trials) and a straight wire (this being straight in front of her);
4) Through some form of generalisation (unconscious, so outside any Working Memory modules) in the Forward Model or related modules, she must have been able to generalise from the straight wire to a straight pipe cleaner, and so to reason that she could use the Forward Model/Inverse Model Controller for bending pipe-cleaners to bend the wire;
5) She could check, through a further stage of reasoning, that this bent wire would then be usable to extract the food bucket;
6) She then proceeds to make the sequence of actions in reality: bend the wire, then lift the food bucket, then obtain the food.

The 'Aha' stage appears to be at stage 4), where some form of generalisation occurs. Such generalisation is common in neural modules with distributed representations of stimuli (so having overlap between relatively similar stimuli). In a module with dedicated nodes for inputs, it is also possible to have a similar spreading of activity for one input to activate sites normally activated by another if there is appropriately-designed lateral connectivity between similar nodes. It is not necessarily simple to achieve such 'semantic spreading' without carefully defined inputs (feature-based, for example), so that semantically similar nodes are both activated to a certain degree and corresponding learning can occur of appropriate lateral connection strengths between the nodes. It seems much easier to work directly with distributed representations, when such generalisation is well known to occur.

There is also a step left out of the above: the creation of a goal of a 'bent wire', in order to activate steps 5) and 6).

11 Conclusions

We started by comparing and contrasting the nature of computational and human intelligence. After a discussion of the nature of cognition in Section 2, we considered in the following sections some details of the brain basis of the three basic functions involved in cognition: attention, memory and reward. This led to elaboration of a set or more detailed component functions of cognition, and thence to a set of principles for cognitive processing in the brain and an associated cognitive architecture. In Section 10 we analyzed several basic cognitive tasks, the WCST and the Tower of Hanoi, leading to a brief discussion of the difference between linguistic and spatial reasoning. We finally considered the important creative steps in the reasoning of Betty the Crow, in which she created for herself a bent piece of wire from a straight one. This amazing process was used to extract a food reward; it involved several features of reasoning as well as the 'aha' process, very likely arising from some form of generalisation process in the neural networks of Betty's brain.

Our discussion is very incomplete. Further careful work, as well trying to bring in language, should allow the area of brain-based reasoning to be increasingly better understood and modelled. In such a manner it may be possible to create a new generation of computer programmes, able to generate reasoning results and even results based on creative processes. However the advances depend on the presence of increasingly sophisticated databases, as is clear for the foundations of the reasoning/creative components.

Acknowledgement

The author would like to acknowledge the support of the European Union through the FP6 IST GNOSYS project (FP6-003835) of the Cognitive Systems Initiative, and to his colleagues in this project for constant support. He would also like to thank Professor Kacelnik for a useful discussion, and his colleagues Dr M Hartley, Dr N Taylor, Dr N Fragopanagos and N Korsten for stimulating discussions on much of the area covered in the paper.

References

[1] Barto A (1995) Adaptive Critics and the Basal Ganglia. In: Models of Information Processing in the Basal Ganglia, JC Houk, J Davis & DC Beiser (Eds). Cambridge MA: MIT Press. pp 215–232

[2] Bi G & Poo B (1998) Synaptic Modifications in Cultured Hippocampal Neurons: Dependence on Spike Timing, Synaptic Strength and Postsynaptic Cell Type. Journal of Neuroscince 18:10464–72

[3] Blakemore C (2005) Private communication

[4] Boysen ST & Himes GT (1999) Current Issues and Emerging Theories in Animal Cognition. Annual Reviews of Psychology 50:683–705

[5] Chappell J & Kacelnik J (2002) Selection of tool diameter by New Caledonian crows Corvus moneduloides Animal Cognition 7:121–127.
[6] Duch W (2005) Private communication
[7] Emery NJ & Clayton NS (2004) The Mentality of Crows: Convergent Evolution of Intelligence in Corvids and Apes. Science 306: 1903–1907
[8] Fragopanagos N, Kockelkoren S & Taylor JG (2005) Modelling the Attentional Blink Brain Research Cogn Brain Research 24:568–586
[9] Houde O & Tzourio-Mazayer N (2003) Neural foundations of logical and mathematical cognition. Nat Rev Neuroscience 4:507–514
[10] Kasderidis S & Taylor JG (2004) Attentional Agents and Robot Control. International Journal of Knowledge-based & Intelligent Systems 8:69–89
[11] Kasderidis S & Taylor JG (2005) Rewarded Attentional Agents. ICANN2005 (Warsaw).
[12] Ling S & Carrasco M (2006) Sustained and Transient Covert Attention Enhance the Signal via Different Contrast Response Functions. Vision Research 46:210–220
[13] Monchi O, Taylor JG & Dagher A (2000) A neural model of working memory processes in normal subjects, Parkinson's disease and schizophrenia for fMRI design and predictions. Neural Networks 13(8–9):963–973
[14] Monchi O, Petrides M, Doyon J, Postuma RB, Worsley K & Dagher A (2004) Neural Bases of Set Shifting Deficits in Parkinson's Disease. Journal of Neuroscience 24:702–710
[15] Newell A (1990) Unified Theories of Cognition. Cambridge Mass: Harvard University Press
[16] Ohyama T, Nores WL, Murphy M & Mauk MD (2003) What the cerebellum computes. Trends in Neuroscience 26(4):222–6
[17] Sergent C, Baillet S & Dehaene S (2005) Timing of the brain events underlying access to consciousness during the attentional blink. Nature Neuroscience 8:1391–1400
[18] Taylor JG (2000) Attentional Movement: the control basis for Consciousness. Soc Neurosci Abstr 26:2231 #839.3
[19] Taylor JG (2003) Paying Attention to Consciousness. Progress in Neurobiology 71:305–335
[20] Taylor JG (2004) A Review of brain-based neuro-cognitive models. Cognitive processing 5(4):19–217
[21] Taylor JG (2005) From Matter to Consciousness: Towards a Final Solution? Physics of Life Reviews 2:1–44
[22] Taylor JG (2006) The Mind: A User's Manual. Wiley & Son
[23] Taylor JG & Fragopanagos N (2005) The interaction of attention and emotion. Neural Networks 18(4) 353–369
[24] Taylor N & Taylor JG (2000) Analysis of Recurrent Cortico-Basal-Ganglia-Thalamic Loops for Working Memory. Biological Cybernetics 82:415–432
[25] Weir AAS, Chappell J & Kacelnik A (2002) Shaping of Hooks in New Caledonian Crows. Science 297:981–3

Artificial Brain and *OfficeMate*^{TR} based on Brain Information Processing Mechanism

Soo-Young Lee

Korea Advanced Institute of Science and Technology, Korea

Summary. The Korean Brain Neuroinformatics Research Program has dual goals, i.e., to understand the information processing mechanism in the brain and to develop intelligent machine based on the mechanism. The basic form of the intelligent machine is called Artificial Brain, which is capable of conducting essential human functions such as vision, auditory, inference, and emergent behavior. By the proactive learning from human and environments the Artificial Brain may develop oneself to become more sophisticated entity. The OfficeMate will be the first demonstration of these intelligent entities, and will help human workers at offices for scheduling, telephone reception, document preparation, etc. The research scopes for the Artificial Brain and OfficeMate are presented with some recent results.

1 Introduction

Although people had tried to understand the mechanism of brain for a long time, still only a few are understood. Even with the limited knowledge on biological brain artificial neural network researchers have come up with powerful information processing models and developed useful applications in the real world. Here we report an ambitious research program to develop human-like intelligent systems, called *'Artificial Brain'*, based on the brain information processing mechanism.

The Korean Brain Neuroinformatics Research Program got into the third phase in July 2004 for 4 years, which is regarded as the final phase of Korean brain national research program started in November 1998 for 10 years [1]. The program was initially sponsored by Ministry of Science and Technology, and is now sponsored by Ministry of Commerce, Industry, and Energy. It is a joint effort of researchers from many different disciplines including neuroscience, cognitive science, electrical engineering, and computer science, and currently about 35 PhDs and about 70 graduate students are involved in the program.

The Korean Brain Neuroinformatics Research Program has two goals, i.e., to understand information processing mechanisms in biological brains and to

Soo-Young Lee: *Artificial Brain and OfficeMate*^{TR} *based on Brain Information Processing Mechanism*, Studies in Computational Intelligence (SCI) **63**, 123–143 (2007)
www.springerlink.com

develop intelligent machines with human-like functions based on these mechanisms. In the third phase we are developing an integrated hardware and software platform for the brain-like intelligent systems, which combine all the technologies developed for the brain functions in the second phase. With two microphones, two cameras (or retina chips), and one speaker, the *Artificial Brain* looks like a human head, and has the functions of vision, auditory, cognition, and behavior. Also, with this platform, we plan to develop a testbed application, i.e., "artificial secretary" alias *OfficeMate*, which will reduce the working time of human secretary by a half.

In this chapter the goals and current status of the Korean Brain Neuroinformatics Research Program will be presented with some recent developments. The research goals and scopes are first described, and recent developments are presented latter.

2 Research Goals

To set up the research goals we incorporated two approaches, i.e., the bottom-up and top-down approaches, and set common goals for them. The bottom-up approach we incorporated is to extrapolate technology development trends and foresee future technology. The prediction of technology demands in future society has always been the main force of technology developments, and we regard it as the top-down approach.

Scientific progress has a tendency to be greatly influenced by unforeseeable breakthroughs, and the reliability of long-term prediction in scientific discovery is always questionable. However, it is still safe to say that recent developments in high performance brain imaging and signal processing equipment will greatly speed up understanding of brain architecture and information processing mechanisms. By reducing resolution both in time and space, it may be possible to record neuronal signals with sufficient accuracy for precise mathematical modeling. Although there still exists a big gap between the molecular neuroscience and system neuroscience, the existing gap between microscopic cellular models and macroscopic behavior models may eventually be bridged resulting in a unified brain information processing model.

Prediction in technology developments is regarded as more reliable. Especially, there exists a well-accepted theory, called "Moore's Law", in semiconductor industry. In 1965 Gordon Moore realized that each new chip contained roughly twice as many structures as its predecessor and each chip was released within 18–24 months of the previous chip. This trend is still remarkably accurate, and an increase rate of about 50 times is expected for the next 10 years. With more than 10 billion transistors in a single chip one may be able to implement a small part of human brain functions. More powerful systems may be built with multiple chips. Even in conventional computing architectures, the communication bottleneck between processors and memories will become more serious, and distributed computing and storage architectures will be pursued. Therefore, neuro-chip technology will fit into the main stream of computer and semiconductor industries.

Another interesting law, called "Machrone's Law", says that the machine you want always costs US$5,000. Actually it seems the cost is going down to US$1,000. People always wanted more powerful systems, and engineers had always come up with powerful systems with the same or lower price range. Therefore, the bottom-up prediction says that the enormous computing power will be available with affordable price in the future.

In human history the Industrial Revolution is regarded as the first big step to utilize machines for human welfare. With the help of powerful energy-conversion machines such as steam engines, the Industrial Revolution paved a road to overcome the physical limitation of humans and result in mass-production of consumer products. The second big step may be the Computer Revolution, which is based on electronic technology with accurate number crunching and mass data storage. Nowadays we can not imagine the world without the mass-production machines and computers.

What the future human society will be? People always wanted to resolve present difficulties and found ways to overcome the difficulties for the better life. The Machrone's law may be a simple example. Although the Computer Revolution has provided better human life, it also creates problems, too. Computers are not yet sufficiently intelligent to work in a friendly way with humans. To make use of computers people must learn how to use it. In many cases it means learning tedious programming languages or memorizing the meaning of graphic user interface icons. Also, current computers do whatever they are programmed for, and do not have generalization and self-learning capabilities. Therefore, the programmers should take into account all possible cases for the specific application, and provide a solution for each case. Only a few people have these programming and logical-thinking abilities. To make computers useful to everybody, it is strongly recommended to make computers as human-friendly as possible. People shall be able to use computers as their friends or colleagues. Computers shall have human-like interfaces, self-learning capabilities, and self-esteem. The best way to accomplish this goal is to learn from the mother nature.

In Figure 1 information processing functions in brains are divided into 4 modules. A human has 5 sensors to receive information from environment, does some information processing based on these sensor signals, and provides motor controls. Among 5 sensors the vision and the auditory sensors provide the richest information, and complex information processing is performed. All the sensory information is integrated in the inference module, which provides learning, memory, and decision-making functions. The last module, Action Module, generates signals for required sensory motor controls. Although there may be many feedback pathways in biological brains, feed-forward signal pathways are mainly depicted here for simplicity.

Although the role of early vision systems is relatively well understood, we believe what we know about the brain is much less than what we do not

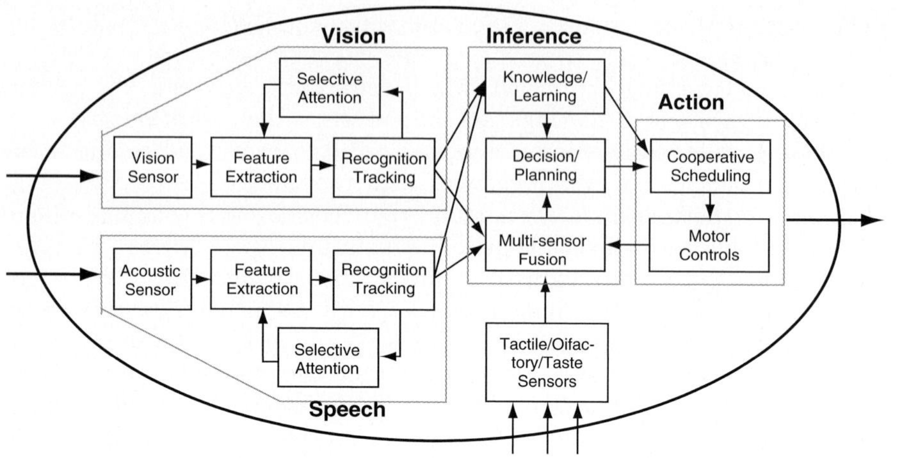

Fig. 1. Four functional modules in brain information processing systems. The *Artificial Brain* should also have 4 functional modules for vision, auditory, inference, and action systems

know. Compared to the vision and auditory modules the knowledge on the inference module is much more limited. However, even with a small hint from biological brains, we believe much more intelligent systems can be built. If neuroscientists concentrate on functions required to fill in the gaps of engineering functions, much faster progress may be achieved. Issues on invariant feature extraction, selective attention, adaptive dynamic ranges, sensory fusion, knowledge representation, generalization, self-learning, emotion, and cooperative behavior are only a few examples. Specific hardware implementations are also essential for the success. Therefore, a "system approach" is required to integrate efforts of researchers from many different disciplines for each module. Finally, the four modules need to be integrated as a single system, i.e., *Artificial Brain.*

The *Artificial Brain* may be trained to work for specific applications, and the *OfficeMate* is our choice of the application test-bed. Similar to office secretaries the *OfficeMate* will help users for office jobs such as scheduling, telephone calls, data search, and document preparation. The *OfficeMate* should be able to localize sound in normal office environment, rotate the head and cameras for visual attention and speech enhancement. Then it will segment and recognize the face. The lip reading will provide additional information for robust speech recognition in noisy environment, and both visual and audio features will be used for the recognition and representation of "machine emotion." The *OfficeMate* will use natural speech for communications with the human users, while electronic data communication may be used between *OfficeMates.* A demonstration version of the *Artificial Brain* hardware is shown in Figure 2.

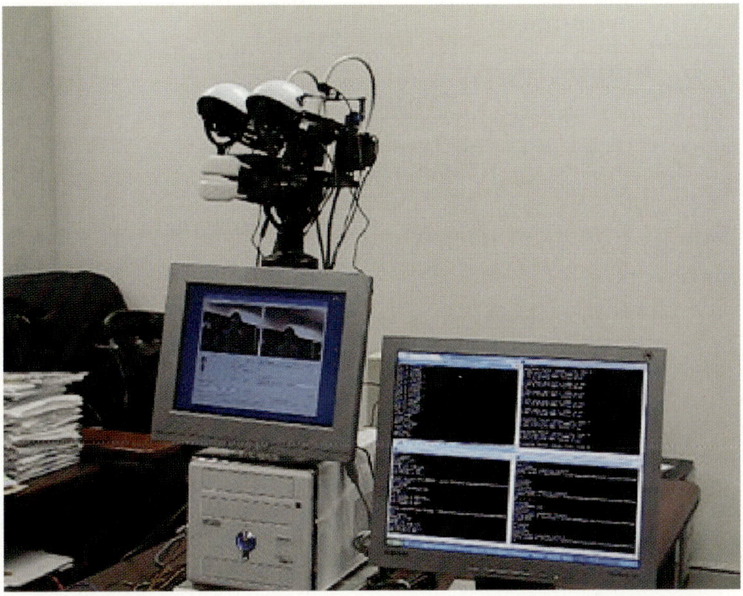

Fig. 2. *Artificial Brain* with two eyes, two ears, and one microphone. The lips are used for lip-sink, and 2 LCD displays are used for camera inputs and internal processor status

3 Research Scope

As shown in Figure 3 the *Artificial Brain* should have sensory modules for human like speech and visual capabilities, internal state module for the inference, and the output module for human-like behavioral control.

The sensory modules receive audio and video signals from the environment, and conduct feature extraction and recognition in the forward path. The backward path conducts top-down attention, which greatly improves the recognition performance of the real-world noisy speech and occluded patterns. The fusion of video and audio signals is also greatly influenced by this backward path.

The internal state module is largely responsible for intelligent functions and has a recurrent architecture. The recurrent architecture is required to model human-like emotion and self-esteem. Also, the user adaptation and proactive learning are performed at this internal state module.

The output module generates human-like behavior with speech synthesizer and facial representation controller. Also, it provides computer-based services for *OfficeMate* applications.

In the Korean Brain Research Program we are trying to develop detail mathematical models for the Artificial Brain. In the mathematical model the internal state value $\mathbf{H}[n+1]$ is defined as a function of sensory inputs, previous internal states, and system outputs, i.e.,

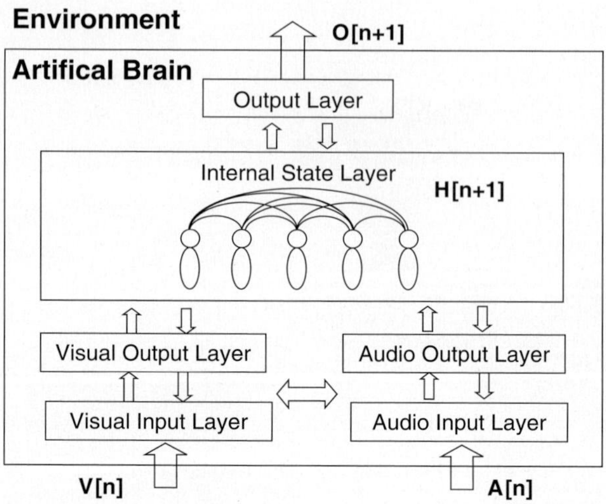

Fig. 3. Block diagram of *Artificial Brain* and its interaction with the environment

$$\boldsymbol{H}[n+1] = \boldsymbol{f}(\boldsymbol{V}[n], \boldsymbol{A}[n], \boldsymbol{H}[n], \boldsymbol{O}[n]), \tag{1}$$

where $\mathbf{V}[n], \mathbf{A}[n], \mathbf{H}[n]$, and $\mathbf{O}[n]$ denote video inputs, audio inputs, internal state values, and outputs at time n, respectively, and $f(\cdot)$ is a nonlinear function. The output is defined as a function of the internal states and sensory inputs, i.e.,

$$\boldsymbol{O}[n+1] = \boldsymbol{g}(\boldsymbol{H}[n+1],\ \boldsymbol{V}[n], \boldsymbol{A}[n]), \tag{2}$$

where $\boldsymbol{g}(\cdot)$ is a nonlinear function. It is also worth noting that the sensory inputs are functions of both the system outputs and environment states as

$$\boldsymbol{V}[n] = \boldsymbol{p}(\boldsymbol{O}[n],\ \boldsymbol{E}[n]),\ \boldsymbol{A}[n] = \boldsymbol{q}(\boldsymbol{O}[n], \boldsymbol{E}[n]), \tag{3}$$

where $\mathbf{E}[n]$ is the environment state value and $\boldsymbol{p}(\cdot)$ and $\boldsymbol{q}(\cdot)$ are nonlinear functions.

Although the basic technologies had been developed for the visual and audio perception during the last 8 years, the most challenging part is the development of the "Machine Ego" with human-like flexibility, self-learning performance, and emotional complexity. It will also have user-modeling capabilities for practical user interfaces. We believe the *Artificial Brain* should have active learning capability, i.e., the ability to ask "right" questions interacting with people. To ask right questions the *Artificial Brain* should be able to monitor itself and pinpoint what it may need to improve. Based on this observation we would like to develop a mathematical model of the Machine Ego, which is the most important component of the *Artificial Brain*. Research scopes for the four modules are summarized as follows.

3.1 Auditory Module

The research activities on the auditory module are based on the simplified diagram of the human auditory central nervous system. Detail functions currently under modeling are summarized in Figure 4. The object path, or "what" path, includes nonlinear feature extraction, time-frequency masking, and complex feature formation from cochlea to auditory cortex. These are the basic components of speech feature extraction for speech recognition. The spatial path, or "where" path, consists of sound localization and noise reduction with binaural processing. The attention path includes both bottom-up (BU) and top-down (TD) attention. However, all of these components are coupled together. Especially, the combined efforts of both BU and TD attention control the object and spatial signal paths.

The nonlinear feature extraction model is based on cochlear filter bank and logarithmic nonlinearity. The cochlear filter bank consists of many bandpass filters, of which center frequencies are distributed linearly in logarithmic scale. The quality factor Q, i.e., ratio of center frequency to bandwidth, of bandpass filters is quite low, and there are overlaps in frequency characteristics. The logarithmic nonlinearity provides wide dynamic range and robustness to additive noise. Time-frequency masking is a psychoacoustic phenomenon,

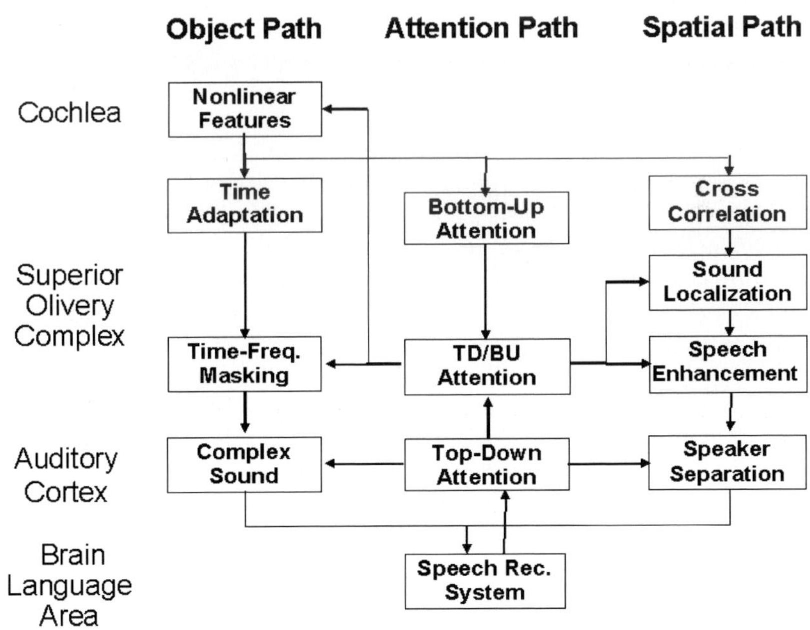

Fig. 4. Block diagram of auditory pathway model. The object path and spatial path deal with speech feature extraction and sound localization, respectively, and the attention path controls the other two paths for robust speech recognition

where a stronger signal suppresses weaker signals in nearby time and frequency domains.

For the binaural processing at the spatial path conventional models estimate interaural time delay, i.e., time-delay between signals from left and right ears, based on cross-correlation, and utilize the time-delay for sound localization and noise reduction. Interaural intensity difference is also utilized for advanced models. However, these models assume only direct sound paths from a sound source to two ears, which is not valid for many real-world environments with multipath reverberation and multiple sound sources (e.g., speech inside an automobile with external road and wind noise, and reverberation of speech mixed with music from the audio system). Therefore, it is required to incorporate deconvolution and separation algorithms in the binaural processing.

For the attention path, a model is being developed to combine both the bottom-up (BU) and top-down (TD) attention mechanisms. The BU attention usually results from strong sound intensity and/or rapid intensity changes in time, and is closely related to the time-frequency masking. However, TD attention comes from familiarity and importance of the sound, and relies on existing knowledge of each person. For example, a specific word or a person's voice may trigger TD attention for relevant people only. Therefore, TD attention originates from the higher-level brain areas that may be modeled in a speech recognition system.

3.2 Vision Module

The vision module also consists of submodules for feature extraction in the object path, stereo vision in the spatial path, and image recognition in the attention path. Also, it is closely coupled to the action module for the active vision and facial representation of emotion.

The object path starts from the bottom-up saliency map [2] to identify the area of interests, and pattern recognition with top-down attention is performed only at those areas. The saliency map consists of colors and orientation edges with several different scales. The recognition submodule will visit each area with high saliency one by one, and classify the images. In the first version of *Artificial Brain* the vision module mainly identifies the facial areas from background images, and recognizes the name and emotional status of the person. Similar to the auditory module the top-down attention greatly improves the recognition performance of occluded or confusing patterns in complex backgrounds.

An important research topic for this module is the color constancy with different illumination conditions.

In future lip-reading will be added for robust recognition in very noisy environment. Since the human perception of motion goes through two different pathways, i.e., the lateral fusiform gyrus for the invariant aspects and the superior temporal sulcus for the changeable aspects of faces [3], the dynamic video features may be different from static image features, and efficient unsupervised learning algorithm should be developed to extract the dynamic features.

3.3 Inference Module

The inference module performs knowledge acquisition, emotional transition, and user adaptation. Applications of inference functions for *OfficeMates* are also integrated in this module.

The knowledge acquisition should be autonomous and proactive. For the autonomous learning it should be able to learn without intervention of users. For example, if a textbook on medicine is provided, the *Artificial Brain* should be able to learn the knowledge of medical doctors. To accomplish this goal we develop unsupervised learning algorithms to extract the basic components of knowledge from the text. A hierarchical architecture may be adopted to build complex knowledge systems from these basic components. The proactive learning then improves the knowledge by asking proper questions. The module estimates what need to be learnt more, phrases proper questions, figures out appropriate person to ask, and incorporates the answers into its knowledge system.

Even with the proactive learning the inference module may experience new circumstances that it has never been exposed to before in the real world applications. Therefore, another important characteristic of the learning system is the generalization capability, which may be obtained by additional constraints on the cost function during learning [4].

The emotional transition is one important characteristic of human-like behavior. To incorporate the emotional transitions we use recurrent neural networks in the inference module, and one hidden neuron is assigned to each independent emotional axis. The transition may be triggered by sensory perception and its own actions to the environment. However, in the future the emotion assignment and transition will be learnt autonomously, and the efficient learning algorithm of this emotional network still need be investigated. If successful, it may lead us to the more complex topics of consciousness and self esteem.

The user adaptation has many different levels, and the simplest level may be implemented by adjusting some parameters of the inference system. However, we plan to implement the user adaptation as the training of another inference system for the user. In this framework both the *Artificial Brain* and users share the same inference architecture, and the two inference modules are learnt simultaneously.

The applications of the inference module include language understanding, meeting scheduling, and document preparation. Actually the language understanding is the fundamental function for efficient man-machine interface. Also, the extraction of emotional components from speech and texts is conducted during this process.

The *Artificial Brain* need to be trained for specific applications of the *OfficeMates*. We focus on two jobs of office secretaries, i.e., meeting scheduling and document preparation. Of course we do not expect perfect performance at the early stage, but hope to save time of human secretaries by a half.

3.4 Action Module

The action module consists of speech synthesizer and facial representation controller. Both are expected to have capabilities of emotional representation, which is very important for the natural interaction between the *Artificial Brain* and its users. The speech synthesizer is based on commercial TTS (Text-To-Speech) software, and we are just adding capability of emotional speech expressions. The emotional facial representation has been analyzed, and the robot head of the *Artificial Brain* is capable of representing simple emotions.

Another important function of the action module is the communication with other office equipments such as telephone, computer, fax machine, copier, etc. Although it does not require intelligence, it is needed to work as *OfficeMates*.

4 Research Results

In this section some research results are reported mainly for the auditory and vision modules. The inference and action modules are still at the early stage of research.

4.1 Self-Organized Speech Feature

The nonlinear feature extraction in auditory pathway is based on cochlear filter bank and logarithmic nonlinearity. The cochlear filter bank consists of many bandpass filters, of which center frequencies are distributed linearly in logarithmic scale. Based on the information-theoretic sparse coding principle we present the frequency-selective responses at the cochlea and complex time-frequency responses at the auditory cortex.

At cochlea we assume that speech signal is a linear combination of the independent basis features, and find these basis features by unsupervised learning from the observed speech. The Independent Component Analysis (ICA) minimizes the mutual information and extracts the statistically independent features [5]. For speech signals we assume the Laplacian distribution, of which sparsity was supported by an experiment on the dynamic functional connectivity in auditory cortex [6].

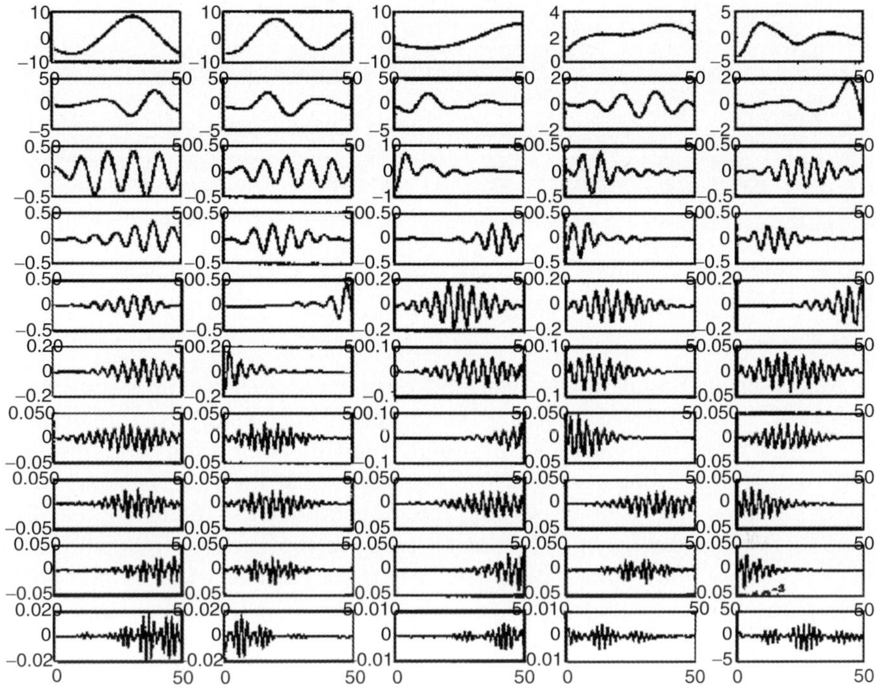

Fig. 5. Fifty simple speech features extracted by independent component analysis from raw speech signals

The training data consist of 60 sentences from six speakers in the TIMIT continuous speech corpus (http://www.ldc.upenn.edu/Catalog/docs/LDC93S2/timit.html), and speech segments composed of 50 samples, i.e., 10 ms time interval at 16 kHz sampling rates, are randomly generated.

As shown in Figure 5, the obtained 50 basis feature vectors are localized in both time and frequency [7]. Average normalized kurtosis of the extracted features is 60.3, which shows very high sparseness. By applying the topology-preserving ICA [8], the basis features are extracted in the order of the center frequency [9].

After the frequency-selective filtering at the cochlea, more complex auditory features are extracted at the latter part of the human auditory pathway, i.e., inferior colliculus and auditory cortex. This complex features may also be understood as the information-theoretic sparse coding. Here we model the earlier part of the human auditory pathway as a simple mel-scaled cochlear filterbank and the logarithmic compression. The time-frequency representation of the speech signal is estimated at each time frame with 10 msec intervals, and the ICA algorithm is applied to this two-dimensional data. The 23 mel-scaled filters and 5 time frames are selected with the local feature dimension of 115, which is reduced to 81 using the principal component analysis (PCA). Topology-preserving self-organization is also incorporated [8].

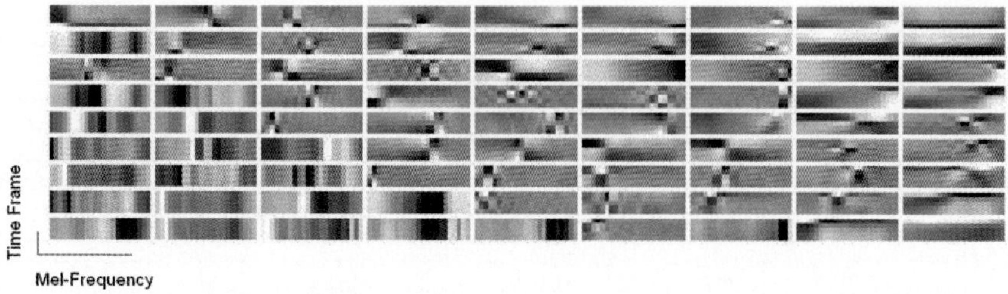

Fig. 6. Eighty one complex speech features extracted by independent component analysis from time-frequency spectrum

As shown in Figure 6, the resulting complex speech features show many aspects of the features extracted at the auditory cortex [10]. At the lower left side of the map, vertical lines represent frequency-maintaining components with complex harmonics structures. The horizontal lines at the upper right side of the map represent on-set and off-set components. In the center of the map, there exist frequency-modulation components such as frequency-rising and frequency-falling components. In fact, there exist neurons responding to these three basic sound components in the human auditory pathways, i.e., the steady complex harmonics, on/off-sets, and frequency modulation. Many auditory cortical areas are tonotopically organized, and are specialized to specific sound features [11].

4.2 Time-Frequency Masking

Another important characteristic of the signal processing in the human auditory pathway is the time-frequency masking, which had been successfully modeled and applied to the noise-robust speech recognition [12]. Time-frequency masking is a psychoacoustic phenomenon, where the stronger signal suppresses the weaker signals in nearby time and frequency domains [13]. It also helps to increase frequency selectivity with overlapping filters.

As shown in Figure 7, the frequency masking is modeled by the lateral inhibition in frequency domain, and incorporated at the output of the Mel-scale filterbank. The time masking is also implemented as lateral inhibition, but only the forward (progressive) time masking is incorporated.

The developed time-frequency masking model is applied to the isolated word recognition task. Frequency masking reduces the misclassification rates greatly, and the temporal masking reduces the error rate even further [12].

4.3 Binaural Speech Separation and Enhancement

For the binaural processing the usual model estimates interaural time delay based on cross-correlation, and utilizes the time-delay for sound localization and noise reduction. Interaural intensity difference is also utilized for advanced

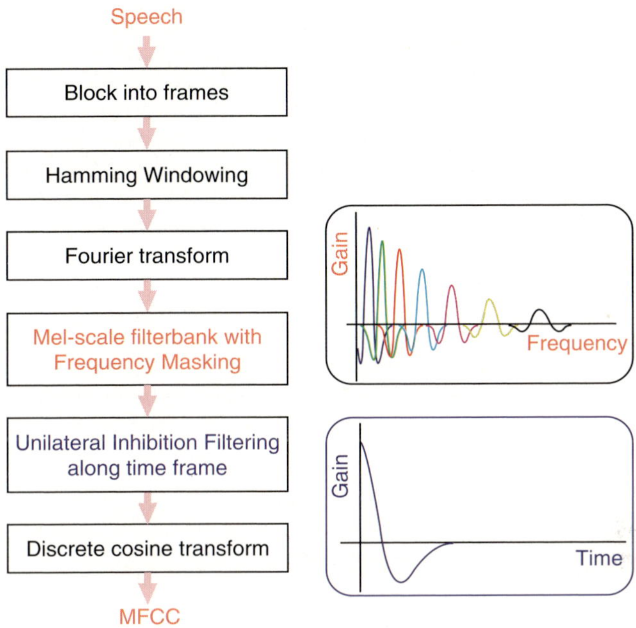

Fig. 7. Block diagram of time-frequency masking and their lateral interconnections

models [13]. However, these models assume only direct sound paths from one sound source to two ears, which is not true for many real-world environments with multipath reverberation and multiple sound sources. Therefore, it is required to incorporate deconvolution and separation algorithms, and an extended binaural processing model has been developed based on information theory. We have extended the convolutive ICA algorithm [14] to multiple filterbanks [15] and further extended the cochlea filterbank.

As shown in Figure 8, the signals from the left ear and the right ear first go through the same filterbank, and the outputs of each filter are de-mixed by separate ICA networks. Then, the clean signals are recovered through inverse filterbanks. If two signal sources exist, each signal can be recovered. If only one signal source exists, the signal and a noise will be recovered.

In ref. [15] the ICA-based binaural signal separation with uniform filterbank results in much higher final SIR than the fullband time-domain approach and the frequency-domain approach. The poor performance of the frequency-domain approach comes from the boundary effects of the frame-based short-time Fourier transform as well as the permutation problem of the ICA algorithm. Although the permutation problems still needs to be solved, compared to the standard time-domain approach without the filterbank, the filterbank approach converges much faster giving better SNR. Basically the filterbank approach divides the complex problem into many easier problems. Due to the decimation at each filter the computational complexity is also

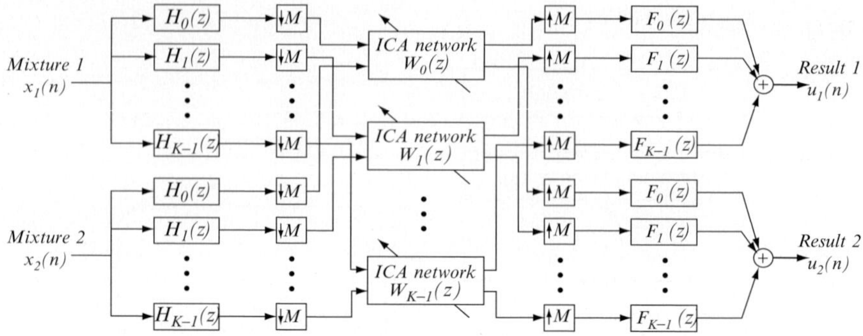

Fig. 8. Binaural processing model with filterbank and independent component analysis. Two microphone signals first go through bandpass filterbank, and separate ICA network is applied to each filtered signals. The filtered signals may be combined by inverse filters

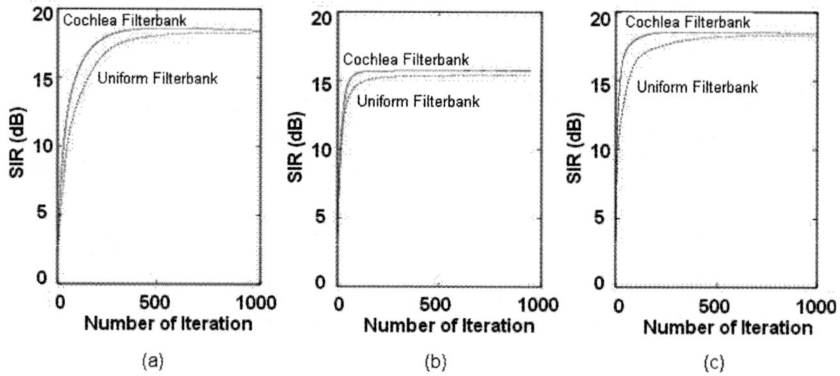

Fig. 9. Performance of ICA-based binaural signal separation methods from convolutive mixtures. The ICA with cochlea filterbank converges much faster than uniform filterbank approach. (a) Speech and music; (b) two speeches; (c) speech and car noise

reduced. Also, it is more biologically plausible. As shown in Figure 9, the utilization of cochlea filterbank greatly improves the convergence.

The de-mixing matrices include information on the relative positions of the sources from the microphones, and the sound localization is also achievable from the de-mixing coefficients. The filterbank approach is quite advantageous for the accurate estimation of the sound direction, especially for noisy multi-source cases. Also, the estimated sound direction may also be utilized to solve the permutation problem [16].

4.4 Top-Down Selective Attention

In the cognitive science literature two different processes are presented with the word "selective attention", i.e., the bottom-up (BU) and top-down (TD) attention mechanisms. The BU attention usually incurs from strong sound intensity and/or fast intensity changes in time. However, the TD attention comes from familiarity and perceptual importance of the sound, and relies on existing knowledge of each person. For example, a specific word or a person's voice may trigger TD attention for relevant people only.

The TD attention originates from the higher brain areas, which may be modeled as a speech recognition system. A simple but efficient TD attention model has been developed with a multilayer perceptron classifier for the pattern and speech recognition systems [17][18]. As shown in Figure 10, the sensory input pattern is fed to a multi-layer Perceptron (MLP), which generates a classified output. Then, an attention cue may be generated either from the classified output or from an external source. The attended output class estimates an attended input pattern based on the top-down attention. It may be done by adjusting the attention gain coefficients for each input neuron by error backpropagation. For unattended input features the attention gain may become very small, while those of attended features remains close to 1. Once a pattern is classified, the attention shifting may occurs to find the remaining patterns. In this case the attention gain coefficients of highly-attended features may be set to 0, while the other may be adapted.

The main difficulty of this top-down expectation comes from the basic nature of the pattern classification. For pattern classification problems many input patterns may belong to the same output class, and the reverse is not unique. However, for many practical applications, one only needs to find the closest input pattern to the attended class, and the gradient-descent algorithm does just that.

Figure 11 shows examples of selective attention and attention switching algorithm in action for confusing patterns [19] and overlapped numerals [17].

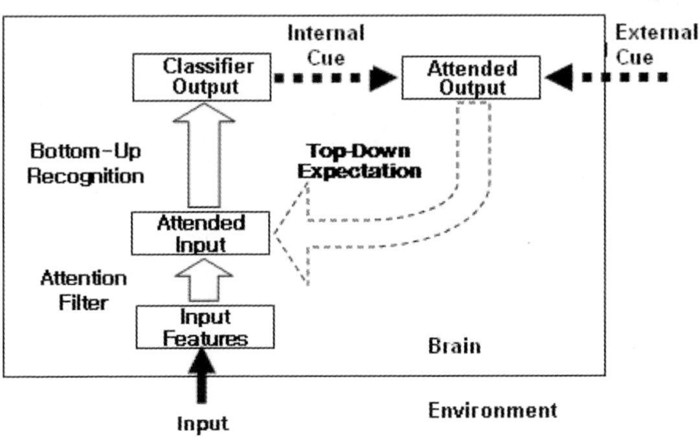

Fig. 10. Block diagram of top-down attention mechanism. The top-down expectation is estimated from the attended output class by the multi-layer perceptron classifier, which mimics the previous knowledge on the words and sounds

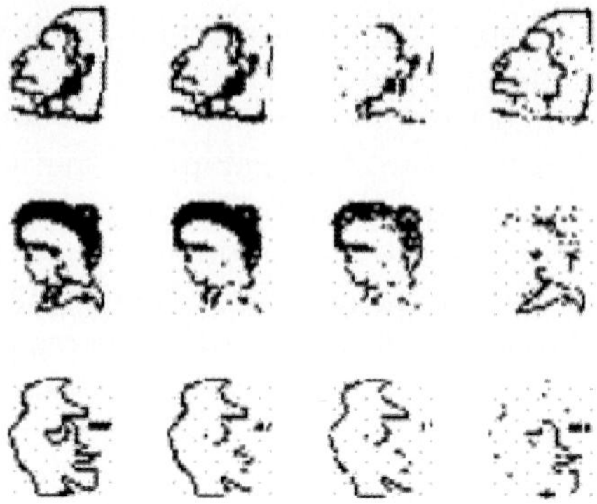

(a)

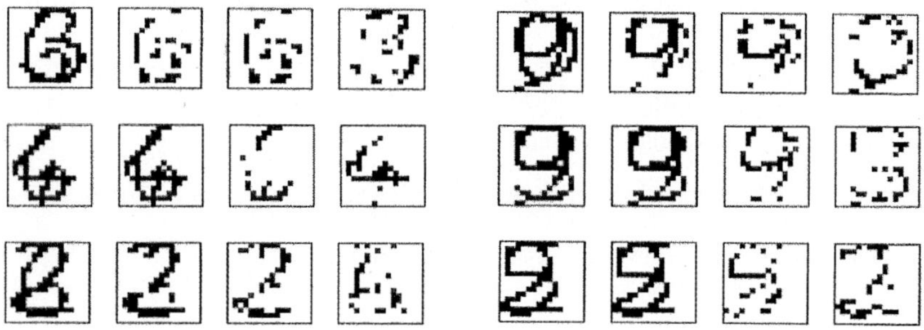

(b)

Fig. 11. Examples of selective attention and attention switching. The four images in each row show from the left the test input, attended input, attention-switched input, and the second-round input, respectively. (a) Results for 3 confusing images, i.e. Eskimo and the facial side view, lady face and old man face, and trumpet player and facial front view; (b) results from overlapped 10 numeric characters

The four images on the horizontal sequences show results on one test. The first image shows the confusing or overlapped test pattern. The second image shows the attended input for the first round classification. The third image shows the masking pattern for attention switching. The fourth image shows the residual input pattern for the second round classification. Figure 11 clearly shows that selective attention and attention switching are performed effectively, and the remaining input patterns for the second round classification are quite visible. The top-down attention algorithm recognized much better than standard MLP classifier, and the attention shifting successfully recognized two superimposed patterns in sequence. It also achieved much better recognition rates for speech recognition applications in real-world noisy environment [18].

We also combined the ICA-based blind signal separation and top-down attention algorithms [20]. The ICA algorithm assumes that the source signals are statistically independent, which is not true for many real-world speech signals. Therefore, the ICA-based binaural signal separation algorithm results in non-exact source signals. By incorporating attention layer at the output of

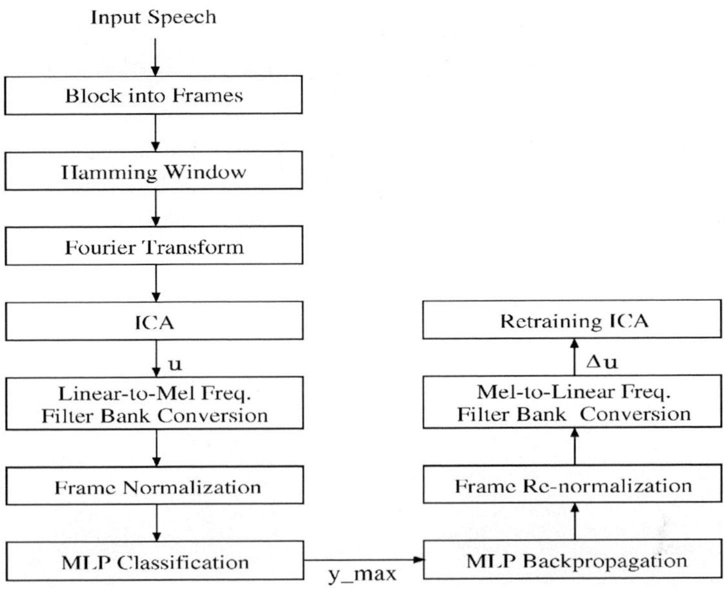

Fig. 12. Block diagram of ICA-based signal separation with deviation correction from top-down attention. It may be understood as a BU-TD combined approach, in which the ICA network serves for the bottom-up attention

the ICA network, this deviation may be compensated for the reference signal provided by the top-down attention. For speech recognition tasks the Mel-Frequency Cepstral Coefficient (MFCC) feature is the popular choice, and the backward evaluation becomes complicated. However, as shown in Figure 12, it is still applicable. Basically one may consider the calculation steps of the MFCC as another layer of a nonlinear neural network, and apply the error backpropagation with the specific network architecture.

4.5 Dynamic Features for Lip-reading

In previous studies the lip-motion features are extracted from single-frame images and the sequential nature of the motion video is not utilized. However, it is commonly understood that the human perception of static images and motion go through different pathways. The features of motion video may be different from the features for the face recognition, and requires more representation from consecutive multiple frames.

Figure 13 shows the dynamic features extracted by 3 decomposition techniques, i.e., Principal Component Analysis (PCA), Non-negative Matrix Factorization (NMF), and Independent Component Analysis (ICA), from multi-frame lip videos [21]. While the PCA results in global features, the ICA results in local features with high sparsity. The sparsity of the NMF-based features resides between those of the PCA and ICA-based features. The

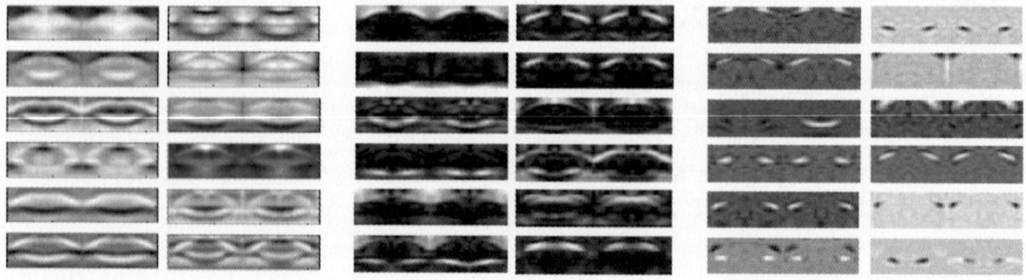

Fig. 13. Extracted lip motion features by PCA (left figures), NMF (center figures), and ICA (right figures) algorithms. Only features from 2-frames are shown for simplicity

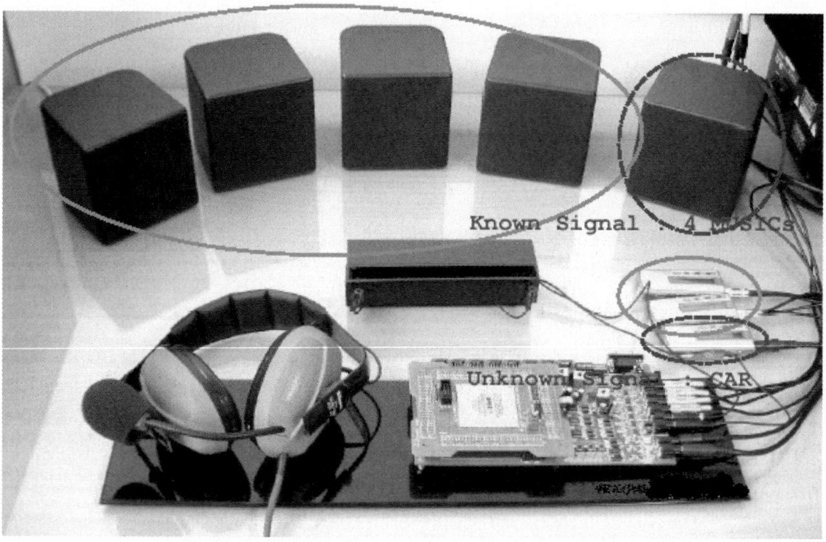

Fig. 14. Demonstration system for the blind signal processing and adaptive noise canceling. Two microphones received 6 signals, i.e., one human speech, one car noise from the right speaker, and 4 background music signals from the remaining 4 speakers

probability density functions and kurtosis of these features are almost independent on the number of the consecutive frames, and the multiple-frame features require less number of coefficients to represent video clips than the single-frame static features. It was also found that the ICA-based features result in the best recognition performance for the lip-reading.

4.6 Hardware Implementations

Many auditory models require intensive computing, and special hardware has been developed for real-time applications. A speech recognition chip had been developed as a System-On-Chip, which consists of circuit blocks for AD conversion, nonlinear speech feature extraction, programmable processor for recognition system, and DA conversion. Also, the extended binaural processing model has been implemented in FPGAs [22].

The developed FPGA-chip was tested with a board with two microphones and 5 speakers. Four of these speakers mimic car audio signals, of which original waveforms are available from electric line jacks. The other speaker generates car noise signal. Also, there is another human speaker. Therefore, the two microphones receive 6 audio signals as shown in the upper part of Figure 14. The developed chip and board demonstrated great signal enhancement, and result in about 19 dB final SNR or 18 dB enhancements. The performance of the FPGA-chip is tested for speech recognition tasks, and the achieved recognition rates are almost the same as those of a clean speech.

5 The Future

The intelligent machines will help human as friends and family members in the early 21st century, and provide services for the prosperity of human beings. In 2020 each family will have at-least one intelligent machine to help their household jobs. At offices intelligent machines, such as the *OfficeMates*, will help human to work efficiently for the organizations. We expect the number of working people may be reduced by a half with the help of *OfficeMates*, and the other half may work on more intelligent jobs. Or, they may just relax and enjoy their freedom.

Intelligence to machines, and freedom to mankind!

Acknowledgment

This research has been supported by the Brain Neuroinformatics Research Program by Korean Ministry of Commerce, Industry, and Commerce. The author also would like to represent his sincere thanks to his former and current students who had contributed to the researches.

References

[1] Lee, S.Y.: Korean Brain Neuroinformatics Research Program: The 3rd Phase. International Joint Conference on Neural Networks, Budapest, Hungary (2004).

[2] Itti L., Koch, C.: Computational model of visual attention. Nature Reviews Neuroscience 2 (2001) 194–203.

[3] Haxby, J.V., Hoffman, E.A., Gobbini, M.I.: The distributed human neural system for face perception. Trends in Cognitive Sciences 4 (2000) 223–233.
[4] Jeong, S.Y., Lee, S.Y.: Adaptive learning algorithm to incorporate additional functional constraints into neural networks. Neurocomputing 35 (2000) 73–90.
[5] Olshausen, B., Field, D.: Emergence of simple-cell receptive field properties by learning a sparse code for natural images. Nature 381 (1996) 607–609.
[6] Clement, R.S., Witte, R.S., Rousche, P.J., Kipke, D.R.: Functional connectivity in auditory cortex using chronic, multichannel unit recordings. Neurocomputing 26 (1999) 347–354.
[7] Lee, J.H., Lee, T.W., Jung, H.Y., Lee, S.Y.: On the Efficient Speech Feature Extraction Based on Independent Component Analysis. Neural Processing Letters 15 (2002) 235–245.
[8] Hyvarinen, A., Hoyer, P.O., Inki, M.: Topographic independent component analysis. Neural Computation 13 (2001) 1527–1558.
[9] Jeon, H.B., Lee, J.H., Lee, S.Y.: On the center-frequency ordered speech feature extraction based on independent component analysis. International Conference on Neural Information Processing, Shanghai, China (2001) 1199–1203.
[10] Kim, T., Lee, S.Y.: Learning self-organized topology-preserving complex speech features at primary auditory cortex. Neurocomputing 65-66 (2005) 793–800.
[11] Eggermont, J.J.: Between sound and perception: reviewing the search for a neural code. Hearing Research 157 (2001) 1–42.
[12] Park, K.Y., Lee, S.Y.: An engineering model of the masking for the noise-robust speech recognition. Neurocomputing 52-54 (2003) 615–620.
[13] Yost, W.A.: Fundamentals of hearing – An introduction. Academic Press (2000).
[14] Torkkola, T.: Blind separation of convolved sources based on information maximization. In Proc. IEEE Workshop on Neural Networks for Signal Processing, Kyoto (1996) 423–432.
[15] Park, H.M., Jeong, H.Y., Lee, T.W., Lee, S.Y.: Subband-based blind signal separation for noisy speech recognition. Electronics Letters 35 (1999) 2011–2012.
[16] Dhir, C.S., Park, H.M., Lee, S.Y.: Permutation Correction of Filter Bank ICA Using Static Channel Characteristics. Proc. International Conf. Neural Information Processing, Calcutta, India (2004) 1076–1081.
[17] Lee, S.Y., Mozer, M.C.: Robust Recognition of Noisy and Superimposed Patterns via Selective Attention. Neural Information Processing Systems 12 (1999) MIT Press 31–37.
[18] Park, K.Y., and Lee, S.Y.: Out-of-Vocabulary Rejection based on Selective Attention Model. Neural Processing Letters 12 (2000) 41–48.

[19] Kim, B.T., and Lee, S.Y.: Sequential Recognition of Superimposed Patterns with Top-Down Selective Attention. Neurocomputing 58-60 (2004) 633–640.

[20] Bae, U.M., Park, H.M., Lee, S.Y.: Top-Down Attention to Complement Independent Component Analysis for Blind Signal Separation. Neurocomputing 49 (2002) 315–327.

[21] Lee, M., and Lee, S.Y.: Unsupervised Extraction of Multi-Frame Features for Lip-Reading. Neural Information Processing – Letters and Reviews 10 (2006) 97–104.

[22] Kim, C.M., Park, H.M., Kim, T., Lee, S.Y., Choi, Y.K.: FPGA Implementation of ICA Algorithm for Blind Signal Separation and Active Noise Canceling. IEEE Transactions on Neural Networks 14 (2003) 1038–1046.

Natural Intelligence and Artificial Intelligence: Bridging the Gap between Neurons and Neuro-Imaging to Understand Intelligent Behaviour

Stan Gielen

Dept. of Biophysics Radboud University Nijmegen

Summary. The brain has been a source of inspiration for artificial intelligence since long. With the advance of modern neuro-imaging techniques we have the opportunity to peek into the active brain in normal human subjects and to measure its activity. At the present, there is a large gap in knowledge linking results about neuronal architecture, activity of single neurons, neuro-imaging studies and human cognitive performance. Bridging this gap is necessary before we can understand the neuronal encoding of human cognition and consciousness and opens the possibility for Brain-Computer Interfaces (BCI). BCI applications aim to interpret neuronal activity in terms of action or intention for action and to use these signals to control external devices, for example to restore motor function after paralysis in stroke patients. Before we will be able to use neuronal activity for BCI-applications in an efficient and reliable way, advanced pattern recognition algorithms have to be developed to classify the noisy signals from the brain. The main challenge for the future will be to understand neuronal information processing to such an extent that we can interpret neuronal activity reliably in terms of cognitive activity of human subjects. This will provide insight in the cognitive abilities of humans and will help to bridge the gap between natural and artificial intelligence.

1 Introduction

In July 2005 the journal Science celebrated its 125 years of existence by publishing a series of ten "hard questions". These questions were posed to set new goals for science: "The pressures of the great, hard questions bend and even break well-established principles, which is what makes science forever self-renewing—and which is what demolishes the nonsensical notion that science's job will ever be done". Most of these hard questions were related to major problems in astronomy, physics, neurobiology, and only one problem ("What Are the Limits of Conventional Computing?") was directly related to Computational Science. Yet, several of the questions, that were posed from the perspective of Neurobiology, are directly related to computer science and

Stan Gielen: *Natural Intelligence and Artificial Intelligence: Bridging the Gap between Neurons and Neuro-Imaging to Understand Intelligent Behaviour*, Studies in Computational Intelligence (SCI) **63**, 145–161 (2007)
www.springerlink.com

artificial/computational intelligence. Questions like "What Is the Biological Basis of Consciousness?", "How Are Memories Stored and Retrieved?", and "How Did Cooperative Behavior Evolve?" are equally crucial to computer science, where problems related to autonomous unsupervised decision making and information retrieval in large, complex data bases, and emergent intelligence are at the heart of Computer Science and Artificial/Computational Intelligence. In this chapter, I will shortly address some aspects of the "hard" problems in Neuroscience regarding consciousness, storage and retrieval of information and the evolution of cooperative behaviour. Then I will explain how these questions relate to major problems in the field of artificial/computational intelligence.

Since long, the basic principles of neuronal information processing have served as a source of inspiration for advanced applications in computer science. In this context it is remarkable that after the booming of the neural network hype in the nineteen eighties, the neural network community has become separated in two streams : one community (frequently called artificial neural network, AI or machine learning community) focussed on algorithms for advanced applications in real-world problems. The other community, called the computational neuroscience community, focussed on more-or-less realistic models to describe the behaviour and function of biological neurons and networks of neurons. The development of these two research lines along separated tracks is somewhat surprising and undesirable since these research lines have many problems in common. Examples of fields of common interest are decision making in complex situations with incomplete information, the design of intelligent autonomous agents and advanced data analysis/retrieval. Presumably the most interesting problems of common interest are related to the phenomena of emergent intelligence and consciousness and their implementation in the neural/computer hardware, which belong to the major long-term goals in both computational intelligence and in AI/machine learning. To some extent, the separate evolution of both fields reflects the different implementation (neural wetware versus computer hardware) and the different background of researchers in both communities. The AI/machine learning community has many researchers with a background in computer science or statistical physics, whereas the computational neuroscience has mainly researchers with a background in neurobiology and theoretical physics.

A particular example of a complex problem that is of relevance to both research communities is the topic of Brain-Computer Interfaces (BCI). BCI tries to extract meaningful signals from neuronal signals, among others for the diagnosis and rehabilitation of patients with neurological disorders. This topic deals with many problems that belong to the core of the AI/machine learning community and of the computational neuroscience community. It deals with pattern recognition and classification of highly complex data with a very poor signal to noise ratio. Moreover, the neural code, i.e. the meaning and function of the neuronal signals is hardly known, since we do not know the detailed functional role of various brain structures, nor do we know how information is encoded in the temporal properties of the parallel distributed pathways and how attention or dysfunction modifies the neuronal signals. A reliable classification of neuronal signals not only requires advanced data-analysis techniques; knowledge about the neural code and about the involvement of neuronal structures in neuronal information processing is equally important for successful BCI applications.

2 Brain Computer Interfaces

Any type of goal-directed behaviour is reflected in a characteristic sequence of neuronal activities in various parts of the brain. In the past decade it has become clear that not only the decision to start a task and the performance of the task (which should be taken very general as it can imply both perceptual tasks as well as motor performance), but even the intention to take action, is reflected in the neuronal activity. In large parts of the human brain the neuronal activity is very similar to that for subjects who intend or plan actions and for subjects, who really perform these actions [25]. This feature is used for BCI applications. A Brain-Computer Interface (BCI) generally aims to provide a communication channel from a human to a computer that directly translates brain activity into sequences of control commands. Such a device may give disabled people direct control over a neuro-prosthesis or over computer applications as tools for communicating solely by their intentions that are reflected in their brain signals (e.g. [27, 26, 53, 3, 38]). The hope is that BCI can possibly reach this goal in the near future by recording brain activity and by using these signals to control a device. For example, this device could be the limb of a subject when the subject is paralysed (for example after a stroke or in case of Amyotrophic Lateral Sclerosis (ALS), which leads to complete paralysis of all muscles for an otherwise intact central nervous system). In that case the recorded brain signals could be used for artificial electrical stimulation of muscles, which leads to muscle contraction and limb movements. BCI tools are thought to become an important tool when normal functioning of the brain (such as after a stroke) is limited.

Not only does BCI address the issue of translating brain activity into control commands, it also deals with interfering with brain activity by electrical stimulation of the brain. At present deep-brain stimulation is used in severe Parkinson patients, when the traditional treatment of patients with Parkinson's disease with levodopa (a drug which replaces the neurotransmitter dopamine which is no longer produced in the basal ganglia in Parkinson Patients) fails. Deep brain stimulation does not provide a cure for Parkinson's Disease, but greatly alleviates the symptoms [18]. Other applications can be found in the sensory domain. The artificial cochlea, which provides hearing to deaf people with a disorder in the peripheral auditory system [5, 6, 21, 54], has become a standard medical treatment. Another application, that is expected to become equally successful as the "artificial cochlea", is the "artificial retina" [13, 14, 39].

In most BCI applications (both realised as well as planned applications), brain activity is measured by means of a multi-electrode (typically 128) electroencephalogram (EEG), which is a measure for the weighed activity of many cells in the brain. The temporal resolution of EEG is excellent (typically in the range of a millisecond). The main problem of this technique is the poor signal-to-noise ratio (SNR), which makes it hard to distinguish the location and contribution of multiple sources of activity in a normal functioning brain. The accuracy of source localisation is typically in the order of a centimetre if the number of sources is limited to three, but becomes rather hopeless if more than 5 sources of neuronal activity are involved. High-resolution EEG is non-invasive as opposed to invasive work by e.g. Nicolelis [30, 31] who used implanted electrodes in the brain to record the activity in various brain structures. Sub-dural electrodes have a much better signal-to-noise ratio, but have the disadvantage of being invasive. Non-invasive data acquisition is a requirement for most applications, but has the disadvantage that the signals of interest are 'hidden' in a highly 'noisy' environment as EEG signals consist of a superposition of a large number of simultaneously active brain sources that are typically distorted by artefacts and even subject to non-stationarity. The non-stationarity is the consequence of modulation of neuronal signals by attention or by competition between multiple sensory stimuli. Moreover, the brain is highly adaptive and can even involve new pathways to compensate for lesions in the brain. An ideal BCI application should be adaptive to the task and the subject and should adapt rapidly. Actually, it should be the algorithm, which adapts itself to the subject, rather than the subject who adapts to the BCI device. Moreover, BCI should have short yield high information transfer rates. Therefore, advanced data-analysis techniques are absolutely necessary.

3 The Neural Code

Traditionally the easiest and most accurate method to measure neuronal activity is to record the action potentials of single neurons. The long tradition of single-unit recording has revealed that information in the brain is coded in firing rate of a neuron (i.e. the number of action potentials per unit of time) and in recruitment: the orderly recruitment of neurons as a function of stimulus intensity or motor output. Single-unit recordings have been very successful and have revealed many secrets about neuronal information processing. However, single-unit recordings are not suitable to measure correlations in neuronal activity of various neurons within a neuronal ensemble. Such correlations might be due to common input or to lateral neuronal interactions. Knowledge about correlations in neuronal activity is important to understand the nature and amount of information that is encoded by an ensemble of neurons. For an ensemble of N neurons the firing rate of this ensemble to a stimulus s can be represented by the vector $\vec{r}$ with r_j representing the firing rate of neuron j. If the probability for firing rate r_j of neuron j given the stimulus s is $p(r_j|s)$, the information encoded by this ensemble is $I(s) = -\int p(\vec{r}|s) \ln p(\vec{r}|s) d\vec{r}$. If neurons do not have neuronal interactions, such that firing rates are independent, we have $p(\vec{r}|s) = \prod_j p(r_j|s)$, such that the information I(s) can be written as $I(s) = -\sum_j \int p(r_j|s) \ln p(r_j|s) dr_j$. This implies that the information in the activity of the ensemble of neurons is simply the sum of information encoded by each of the neurons. If neuronal information encoded by different neurons is correlated, it is not longer true that $p(\vec{r}|s) = \prod_j p(r_j|s)$. From a theoretical point of view, this would imply that the amount of information encoded by the ensemble of neurons is in general less than the sum of information encoded by the individual neurons (see e.g. [33]). However, since the generation of an action potential is a stochastic process, correlated firing allows elimination of noise by averaging. Therefore, correlated firing may reflect a compromise to obtain optimal information transfer by reducing noise. More detailed information about the information in neuronal ensembles can be found in [42].

In order to explore the temporal correlation between firing of neurons, the next logical step was the development of multi-unit recording techniques by arrays of electrodes. These multi-unit recordings have revealed a third coding mechanism for neuronal information: coding by temporal correlation of action potential firing [9, 47]. At any moment in time, many neurons are active in the brain. Multi-unit recordings have shown that active neurons can be subdivided in subgroups of neurons, where neurons in the same subgroup reveal a high temporal correlation of firing. Microelectrode recordings in monkeys, as well as neuroimaging studies in man have revealed that these ensembles of neurons can (re)organize rapidly and in a flexible way into subgroups, where activity of neurons in the same subgroup reveals a high temporal correlation of firing without changes in mean firing rate (see e.g. [46, 47]. The functional significance of the temporal locking is not known, but there are various hypotheses about its functional significance (see e.g. [44]).

4 Recording Neuronal Signals from the Brain

Each neuron receives spike-input from many other neurons. The input of action potentials to a neuron arrives at the dendrites of the neuron where each action potential induces the release of a specific neurotransmitter. This neurotransmitter opens ion-channels, which allows ions to move through the cell membrane into the neuron. These ion currents cause a local change in the membrane potential (the so-called post-synaptic potential). Changes of the membrane potential of a neuron are the result of the many post-synaptic potentials due to input by action potentials from other neurons. The flow of these currents from the dendrites to the cell body explains why a neuron can be modelled as a dipole.

Typically a neuron receives input from 10^3 to 10^4 neurons. The amount of synaptic input modulates the strength of the dipole. The EEG activity, recorded on the skin above the skull, reflects the contribution of the many dipoles. If all neurons would receive uncorrelated input, the EEG on the skull would be nothing more than noise. However, the input to ensembles of neighbouring neurons is not uncorrelated. This is particularly true for so called Evoked-Brain potentials, where simultaneous onset of neuronal activity is triggered by the sudden onset of a stimulus. Well-known examples are the EEG activity above visual cortex due to onset of neuronal activity at the presentation of a checkerboard pattern or the evoked potentials in the auditory pathway due to sudden onset of a sound. In addition to these transient components of EEG, the temporal correlation of synchronized neuronal activity is reflected in rapid oscillations in EEG activity. These oscillations have been reported at various frequencies, such as the alpha (8–12 Hz) or theta (5–10 Hz) rhythm and the frequently reported beta (12–28 Hz) and gamma oscillations (29–80 Hz). EEG activity reflects the activity of neurons with a dipole orientation orthogonal to the skull. However, since the cortex folds with various sulci, many neurons have an orientation parallel to the skull, rather than orthogonal. These neurons do not or hardly contribute to EEG activity on the skull. However, the ion currents of the neurons parallel to the skull give rise to tiny magnetic fields with an amplitude smaller than that of the earth magnetic field. These small magnetic field components can be measured using SQUIDS in the so-called magneto-encephalogram (MEG). Therefore, MEG is complementary to EEG. Both reflect neuronal activity, but of different groups of neurons.

Another important measure of neuronal activity is obtained by functional Magnetic Resonance Imaging (fMRI). The metabolism related to neuronal activity causes differences in oxygen consumption. Oxygen is transported through the blood vessels by means of hemoglobin molecules. Hemoglobin with oxygen (oxyhemoglobin) is diamagnetic, whereas deoxyhemoglobin (hemoglobin after release of oxygen) is paramagnetic, causing microscopic magnetic field inhomogeneities that affect the transverse relaxation time (called T_2) of the MRI. Since increased neuronal activity leads to an increased blood flow, actually overcompensating for the neuronal oxygen need, the oxygen concentration increases in the blood vessels. Hence the relaxation time T_2 of brain tissue to a radio pulse, which deflects the atom spins oriented along the major magnetic field, is larger for active neuronal tissue than for neuronal tissue at rest. fMRI measures the magnetic relaxation signal due to the perturbing radio pulse.

EEG and MEG both have a high temporal resolution. The disadvantage is that the inverse problem (the problem of finding the location of the electric or magnetic sources that gave rise to the measured EEG or MEG activity) is an ill-posed problem since many different sources of activity can provide the same

EEG or MEG activity on the skull. Therefore, source estimation (estimating the temporal properties and the location of the electric or magnetic sources) is possible only if prior knowledge is available about the number of sources (which should be limited) or if prior knowledge is available about the position and temporal modulation of the sources. fMRI typically has a high spatial resolution (typically a few tenths of a millimeter). However, the temporal resolution (tenth of a second) is way above a millisecond, which is the time constant to characterise neuronal activity. Therefore, a combination of both techniques is typically used in advanced neuroimaging research.

5 Basic Questions Regarding the Interpretation of Neuronal Oscillations

5.1 Functional Role of Neuronal Oscillations

Although no one will deny the existence of neuronal oscillations nowadays, their functional significance is yet a topic of debate and few hypotheses exist to explain why and how various ensembles of neurons develop in a flexible way, each with a high temporal correlation structure. These two issues are related and reflect two important problems in neuroscience. Understanding the functional role and the origin of synchronized neural activity is crucial for research on neuronal information processing with large implications for BCI. As pointed out before, correlated firing may be a way to obtain more accurate information coding by eliminating noise. However, other hypotheses have been put forward that attribute other functional roles to correlated firing. In order to explain this we will first discuss the various hypotheses about the functional significance (see also [44]) before we discuss the possible neuronal mechanisms that can explain the initiation and disappearance of neuronal oscillations.

The first hypothesis to provide a functional significance to synchronized neuronal activity is that synchronization plays a role in the representation of sensory information. The most well-known example is the hypothetical role to solve the binding problem. Visual information comes from the retina and passes along the LGN (Lateral Geniculate Nucleus) in the thalamus to the visual cortex (V1). After V1, different features of visual information are processed along different parallel channels. Each channel encodes a particular feature of visual objects, such as color, position of the object, nature of object, and object velocity. For a single object in the visual environment, each channel carries information about a single feature of the object. However, since the visual environment contains multiple objects, each channel carries information about features from multiple objects and the question is how the central nervous system knows which feature belongs to which object. For example, if we have a red pencil and a blue coffee cup, how does the brain know that the label "blue" belongs to the coffee cup and not to the pencil. The idea has been proposed (see [47]) that the temporal correlation might serve as a label for all features that belong to the same object (however, see [45]). Previous work hypothesized that neuronal coherence (or phase-locking or synchronization) could provide a tag that binds those neurons that represent the same perceptual object. This binding tag would be a flexible code for linking neurons into assemblies and thereby would greatly enlarge the representational capacity of a given pool of neurons. In line with this hypothesis, it has been suggested that object features represented by spatially distinct neural assemblies are dynamically linked to a coherent percept by synchronized activity in the gamma range [10]. This hypothesis can explain why information in the brain is processed, transferred, and stored by flexible cell assemblies, defined as distributed networks of neuronal groups that are transiently synchronized by dynamic connections [10, 52]. A particular example of temporal locking is the observation of phase-encoding in hippocampal place cells [35]. When a rodent moves around in a limited area, the phase of firing in the theta-rhythm carries more information about location of the rodent within this space than does firing rate [49].

Another hypothesis is that synchrony enhances the saliency of neural responses. This can be understood from the fact that two action potentials, arriving simultaneously at the dendrites of a neuron are much more effective in eliciting an action potential than two action potentials which arrive with a time interval. This is particularly clear if the neuronal time constant is small, such that the neuron operates as a coincidence detector [23]. Therefore, correlated discharges have a much stronger impact on neuronal populations than temporally disorganized inputs [11, 41]. The regulation of interaction with target neurons by coherent firing has been reported in corticospinal projections from motor cortex to the spinal cord [43]. Thus, the oscillatory activity might serve as a dynamic filter, which selects the salient and significant inputs to the network. Along these lines, similar coherent oscillations have also been reported for recordings in monkey motor cortex (see e.g. [2, 1, 16, 19, 20, 36, 37], who studied the cross-correlation and coherence between local field potentials and neural spike trains in monkey primary motor cortex, and [40]).

5.2 Neuronal Mechanisms for Neuronal Synchronization

The role of tight neuronal synchronization has raised the question how "noisy" neurons are able to fire in close synchrony with millisecond accuracy. The explanation is that the time constant of the neuron can be modified by balanced excitatory and inhibitory input [23]. Changing the amount of balanced excitation and inhibition changes the time constant of the neuron without changes in firing rate of the neuron. This can be understood using a popular, but simplified representation of neuronal dynamics: the leaky integrate-and-fire model. According to this model, the dynamics of the membrane potential of the neuron is given by

$$C\frac{dV(t)}{dt} = -\frac{1}{R}V(t) + \sum_i \{G_i(t)^* s_i(t)\}(V(t) - V_i) \qquad (1)$$

where C represents the capacitance of the cell membrane, R represents the resistance of the cell membrane, $s_i(t)$ represents the spike input from neuron i, $G_i(t)$ represents the conductance of the synaptic contact between the cell and the input from neuron i, and V_N represents the Nernst potential. The symbol * represents convolution. The neurotransmitter released by an incoming action potential opens ion channels and thereby modifies the local conductance G_i of the post-synaptic membrane. The last term in Eq. (1) represents the synaptic input current. Eq. (1) can also be written as

$$\tau \frac{dV(t)}{dt} = -V(t) + R \sum_i I_i(t) \tag{2}$$

where the resistance R is the resistance of the cell membrane which is modulated by the synaptic input and with $\tau = RC$. A large amount of input implies a large conductance and a small resistance R. Therefore, input affects the time constant τ. Obviously, Eq. (2) clearly explains the name of the leaky integrate-and-fire model. For large values of the time constant τ, the neuron integrates the input until it reaches a threshold (typically near $-40\,\mathrm{mV}$). Then, an action potential is generated and the membrane potential is reset to the membrane potential at rest, typically near $-70\,\mathrm{mV}$. For small values of τ, the membrane potential decays rapidly to its rest value, such that a small value of τ turns the neuron into a coincidence detector: the neuron only fires an action potential if the input from different neurons arrives within a small time interval. This explains why balanced excitation and inhibition changes the behaviour of the neuron from a (leaky) integrator into a coincidence detector, which fires only for tightly synchronized input. Although this can explain the propagation of synchronized neuronal activity from one brain structure to another [7], it does not explain the initiation of synchronized activity.

As mentioned above, many studies have reported synchronized oscillations between various neuronal ensembles. The amount of temporal synchronization between neuronal ensembles is generally expressed by the coherence function in the frequency domain. The coherence between two signals x(t) and y(t) is defined by

$$\gamma(\omega) = \frac{\langle R_{xy}(\omega) \rangle}{\left\langle \sqrt{R_{xx}(\omega)} \right\rangle \left\langle \sqrt{R_{yy}(\omega)} \right\rangle} \tag{3}$$

where $< \cdot >$ represents ensemble average over many corresponding time segments for x(t) and y(t). $R_{xy}(\omega)$ represents the cross-covariance function between x(t) and y(t) in the frequency domain. Usually, one will find the squared coherence function $|\gamma(\omega)|^2$ in the literature to explore the relation between two signals. This squared coherence is a real-valued function of frequency in the range between 0 and 1. If the signal y(t) can be obtained from the signal x(t) by convolution by a linear system in the absence of noise, the squared coherence has value 1. This value becomes smaller when noise or nonlinearities are involved. The more noise or the more complex (nonlinear) the relation between x(t) and y(t), the more the squared coherence approaches the lower limit value of zero. This explains why the squared coherence, in addition to the mutual-information, has often been used to explore the relation between input and output of an ensemble of neurons or to explore the similarity between signals in different parts of the brain (see e.g. [29, 17]).

The coherence function $\gamma(\omega)$ has a special property in that it captures the frequency-dependent phase relation between x(t) and y(t) by the complex-valued function $R_{xy}(\omega)$. The variability of the relative phase provides information about the coupling strength between two signals. If two signals are tightly coupled, the variability of relative phase will be small. This property is highly important in neuronal synchronization (see also [7]). Moreover, when information goes from x to y, any time delay Δt will cause a frequency dependent phase shift $\Delta\phi = \omega\Delta t$. One might expect that if one brain structure provides input to another brain structure, the phase difference between synchronized activities in these two brain structures will reflect at least the effect of finite conduction velocity of signals between the two brain structures. These differences can be quite large in the motor system, where signals from motor cortex project to neurons in the spinal cord, approximately one meter away. With a typical axonal conduction velocity of 60 m/s, this gives rise to as pure time delay of $\Delta t = 16\,\text{ms}$ and to a frequency-dependent phase shift of $\omega\Delta t$ (see [43]). Quite remarkably, oscillatory neuronal activity in different parts of the brain appears to be almost synchronous, without significant time delays. Significant time delays should be expected for serial processing in several brain structures due to the conduction velocity of neuronal signals in the brain. The absence of time delays is what one should expect for a highly connected network of neuronal ensembles with multiple feedback loops. Such highly connected networks operate as a functional unit and cannot be separated into a set of subsystems with clear unambiguous causal relationships between these subsystems. This finding argues against the simple view of neuronal information processing as a serial process from sensory cortices to motor cortex, for example in the case of sensory information about position of objects, which is translated into motor commands to grasp an object.

Comparison of the relative phase of synchronized neuronal oscillations in two functionally related parts of the nervous system has suggested that the excitability of neurons is modulated such that excitability is maximal at the time of arrival of periodic oscillatory activity [43]. This is highly remarkable: how can the receiving neurons adjust their excitability such that it is optimal at the time of arrival of the synchronized input? Based on the findings by Schoffelen et al. [43], Fries [12] hypothesized that neuronal communication is mechanistically subserved by neuronal coherence. The idea is that activated neuronal groups oscillate and thereby undergo rhythmic excitability fluctuations that produce temporal windows for communication. A recent modelling study showed that coherence is processed accurately between subsequent groups of neurons [57]. Coherence by coherently oscillating neuronal groups is a requirement for effective interaction, because they ensure that the communication windows for input and for output at the interacting neuronal groups are open at the same times. A recent study [8] suggested a mechanism for modulation of excitation such that the neuronal excitability is optimal at the arrival of a period synchronized input. Thus, a flexible pattern of coherence defines a flexible communication structure.

6 Interpreting EEG/MEG Data : Reading Out the Brain

Triggered by the increased knowledge about the neuronal information processing in the central nervous system, the past five years have shown an exponential increase in publications on BCI. These publications mainly referred to new algorithms for classification of EEG/MEG signals and for transforming these signals into mechanical or electronic output (for a recent overview see [32]). Although the aim is to use BCI for the human brain, most experimental data have been obtained in animal experiments using neuronal activity recorded invasively in multiple brain areas (see e.g. [30, 31]). Patterns of spike trains and local field potentials from multi-electrode recordings represent astonishingly well imagined or intended movements. As explained before, the spatial resolution of source localisation estimation based on EEG or MEG is rather poor. This causes a great problem in recording the activity in a particular brain structure with non-invasive EEG electrodes in humans. A recent study in epileptic patients using invasive presurgically implanted subdural electrodes over frontal regions [24] has shown a good performance in classification of neuronal activity, suggesting that it would be a good BCI tool. With these patients, it was possible in just one session to differentiate without any training imagination of hand-, tongue-, and mouth movement from the electrocorticogram (ECoG). However, invasive recordings cannot be used in standard clinical applications.

These results have created enormous public interest and hope for a rapid solution to critical clinical problems such as communication in locked-in patients and movement restoration in patients with spinal cord lesions and chronic stroke. Unfortunately, there are many problems that have to be solved and standard clinical use of BCI seems out of the question for the near future (see a recent review by Birbaumer [4]) illustrating the complexity of the problem with great technical and conceptual problems.

Further progress in this field depends on several developments. It will be of great help, if more detailed knowledge will become available on the precise role of various brain structures in normal human perception, action, and decision making. Knowledge about the role of various brain structures in sensori-motor tasks will provide insight in the spatial and temporal properties of activity in the brain. Prior information about the source location will enable the extension of the temporal filtering, which is currently used in BCI-applications, to spatio-temporal filters that act as templates for classifying EEG/MEG signals. This will improve the signal to noise ratio of EEG/MEG signals considerably.

Along these lines, the use of advanced data-analysis tools like multi-taper techniques (see [29, 57]) will be necessary to reduce the signal-to-noise ratio. Moreover, more information about the neuronal code will be necessary. What is the functional role of various rhythms of neuronal activity? How are these rhythms created and what modulates their amplitude? It is well known that synchronization (especially in the β and γ range) depends on attention and expectation [40, 43]. Knowledge of the task-dependent functional role of neuronal oscillations might be useful to extract particular frequency bands in EEG/MEG for BCI applications in particular tasks. A proper interpretation of EEG/MEG patterns will also require a better insight in the plasticity of the brain. Plasticity in the brain takes place on a large range of time scales. Some processes of plasticity develop on a time scale of seconds, whereas other processes, such as the activation of some brain region to compensate for damage or dysfunction in another part of the brain, become effective only after days or weeks. This is particularly important for elderly people, when brain function deteriorates, where good tools to diagnose symptoms of dementia and other neurological pathologies might help to alleviate symptoms and to save expenses by timely and effective treatment.

7 Implications for Artificial/Computational Science

A better understanding of neuronal information processing will have large implications for artificial/computational science and for BCI in particular. Although the emergence of intelligent behaviour will remain one of the mysteries of the human brain for quite a while, there are many other aspects that already have an impact.

One example concerns the design of an autonomous system. How can such a system distinguish irrelevant stimuli from relevant stimuli when operating in a complex environment. The problem is that the sensory input in a normal environment contains a huge amount of information. Detailed processing of all sensory information would require large amounts of time and would prohibit rapid responses to relevant stimuli. This is where attention starts to play a role. If prior knowledge is available about the possible relevance of stimuli, attention might help to focus and select the relevant stimuli to speed up sensory processing. Indeed, attention has been shown to reduce reaction times and a recent study [43] has shown that the attention-related probability for the stimulus is highly correlated to the amount of gamma-activity in the EEG, giving rise to shorter reaction times. Several other studies on neuronal information processing have shown that sensory processing is not just a bottom-up process, driven by peripheral stimuli. Rather, neuronal information processing of sensory stimuli is governed by Bayes' law, which says that the sensory interpretation of neuronal activity is determined both by the log-likelihood of the stimulus given the neuronal activity and by the prior probability for the stimulus [15, 22, 34, 50, 51, 55, 56].

Classical theories of sensory processing view the brain as a passive, stimulus-driven device. By contrast, more recent approaches emphasize the constructive nature of perception, viewing it as an active and highly selective process. Indeed, there is ample evidence that the processing of stimuli is controlled by top–down influences that strongly shape the intrinsic dynamics of thalamocortical networks and constantly create predictions about forthcoming sensory events. Coherence among subthreshold membrane potential fluctuations could be exploited to express selective functional relationships during states of expectancy or attention, and these dynamic patterns could allow the grouping and selection of distributed neuronal responses for further processing. Top-down driven selection and processing of sensory information has become one of the basic concepts in robotics and in multi-agent technology, although the implementation is very different from that in the brain. Without any doubt this is to large extent determined by the differences in hardware/wetware.

But how do groups of neurons communicate? And how do top-down influences modify the communication structure within a range of hundred milliseconds while anatomical connections stay unchanged on that time scale? In very general terms, the dominant model of neuronal communication is that a neuron sends its message (encoded in e.g. firing rate or in the degree of action potential synchronization) down its axons to all neurons to which it is anatomically connected. Those receiving neurons combine (e.g. sum and threshold) all the inputs and send their output to neurons to which they have connections. An important aspect of this model is that both the distribution and the reception of neuronal signals is governed solely by the structure of the anatomical connections, i.e. there is no further communication structure beyond the one imposed by anatomical connectedness. However, cognitive functions require flexibility in the routing of signals through the brain. They require a flexible effective communication structure on top of the anatomical communication structure that is fixed, at least on the time scale at which cognitive demands change.

Fries [12] hypothesized that this effective communication structure is mechanistically implemented by the pattern of coherence among neuronal groups, i.e. the pattern of phase-locking among oscillations in the communicating neuronal groups. As explained before, the key factor in this model is that neuronal communication between two neuronal groups mechanistically depends on coherence between them while the absence of neuronal coherence prevents communication. Although this idea has been proposed as a working hypothesis, which needs firm experimental testing, the idea may be a crucial step to understand the biological basis of consciousness [28]. If we understand the neurobiological basis of consciousness, this may serve as an example for the implementation of "consciousness" in artificial systems. However, the diversity of definitions for consciousness hamper progress on this topic both in neurobiology and in AI.

The main challenge for the near future will be to understand the neuronal code and to understand the role of various brain structures in memory, sensorimotor processing and decision making. It would be a tremendous achievement if this information could be used for successful BCI applications. On a longer time scale, we need to understand how self-organization in the brain results in emergent intelligent behaviour. What are the underlying principles for the autonomous development of intelligence and can we find where and how these processes take place in the brain? If so, could we measure this brain activity for advanced BCI applications? BCI applications so far allow only binary decisions with an information flow of just a few bits per second at best. Will we be able to implement models for emergent intelligence and will we be able to use these models to solve complex real-world problems? This information will be crucially important to develop advanced adaptive algorithms to interpret EEG/MEG activity, which can then be used for the diagnosis and therapy of patients with neurological disorders.

References

[1] Baker SN, Spinks R, Jackson A, and Lemon RN (2001) Synchronization in monkey motor cortex during a precision grip task. I. Task-dependent modulation in single-unit synchrony. J Neurophysiol 85: 869–885

[2] Baker SN, Pinches EM, and Lemon RN (2003) Synchronization in monkey motor cortex during a precision grip task. II. Effect of oscillatory activity on corticospinal output. J Neurophysiol 89: 1941–1953

[3] Birbaumer N, Ghanayim N, Hinterberger T, Iversen I, Kotchoubey B, Kübler A, Perelmouter J, Taub E, and Flor H (1999) A spelling device for the paralysed. Nature 398: 297–298

[4] Birbaumer N (2006) Brain–computer-interface research: Coming of age. Clin Neurophysiol 117: 479–483

[5] Carlyon RP, van Wieringen A, Long CJ, Deeks JM, and Wouters J (2002) Temporal pitch mechanisms in acoustic and electric hearing. J Acoust Soc Am 112: 621–633

[6] Chen HB, Ishihara YC, and Zeng FG (2005) Pitch discrimination of patterned electric stimulation. J Acoust Soc America 118: 338–345

[7] Diesmann M, Gewaltig MO, and Aertsen A (1999) Stable propagation of synchronous spiking in cortical neural networks. Nature 402: 529–533

[8] Dodla R, Svirskis, G and Rinzel J (2006) Well-timed, brief inhibition can promote spiking: postinhibitory facilitation. J Neurophysiol 95: 2664–2677

[9] Eckhorn R, Bauer R, Jordan W, Brosch M, Kruse W, Munk M, and Reitboeck HJ (1988) Coherent Oscillations – a Mechanism of Feature Linking in the Visual-Cortex – Multiple Electrode and Correlation Analyses in the Cat. Biol Cybern 60: 121–130

[10] Engel AK, Fries P, and Singer W (2001) Dynamic predictions: oscillations and synchrony in top-down processing. Nature Rev Neurosci 2: 704–711
[11] Fellous JM, Rudolph M, Destexhe A, and Sejnowski TJ (2003) Synaptic background noise controls the input/output characteristics of single cells in an in vitro model of in vivo activity. Neuroscience 122: 811–829
[12] Fries P (2005) A mechanism for cognitive dynamics: neuronal communication through neuronal coherence. Trends in Cognitive Science 9: 474–480
[13] Funatsu E, Kuramochi S, Nagafuchi Y, Kage H, Sakashita N, Murao F, and Kyuma K (2002) Artificial retina large scale integration with on-sensor projection function for high-speed motion detection. Optical Engineering 41: 2709–2718
[14] Gekeler F, and Zrenner E (2005) Status of the subretinal implant project. An overview. Ophthalmologe 102: 941–945
[15] Gielen CCAM, Hesselmans GHFM, and Johannesma PIM (1988) Sensory interpretation of neural activity patterns. Math Biosciences 88: 15–35
[16] Jackson A, Gee VJ, Baker SN, and Lemon RN (2003) Synchrony between neurons with similar muscle fields in monkey motor cortex. Neuron 38: 115–125
[17] Jarvis MR, and Mitra, PP (2001). Sampling properties of the spectrum and coherence of sequences of action potentials. Neural Comp. 13: 717–749
[18] Jenkinson N, Nandi D, Miall RC, Stein JF, and Aziz TZ (2004) Pedunculopontine nucleus stimulation improves akinesia in a Parkinsonian monkey. Neuroreport 15: 2621–2624
[19] Kilner JM, Baker SN, and Lemon RN (2002) A novel algorithm to remove electrical cross-talk between surface EMG recordings and its application to the measurement of short-term synchronisation in humans. J Physiol Lond 538: 919–930
[20] Kilner JM, Salenius S, Baker SN, Jackson A, Hari R, and Lemon RN (2003) Task-dependent modulations of cortical oscillatory activity in human subjects during a bimanual precision grip task. Neuroimage 18: 67–73
[21] Kong YY, Stickney GS, and Zeng FG (2005) Speech and melody recognition in binaurally combined acoustic and electric hearing. J Acoust Soc Am 117: 1351–1361
[22] Kording KP, and Wolpert DM (2004) Bayesian integration in sensorimotor learning. Nature 427: 244–247
[23] Kuhn A, Aertsen A, and Rotter S (2004) Neuronal integration of synaptic input in the fluctuation-driven regime. J Neurosci 24: 2345–2356
[24] Leuthardt EC, Schalk G, Wolpaw J, Ojeman JG, and Moran DW (2004) Invasive BCI in presurgically implanted patients. J Neural Eng 1:63–71.
[25] Lotze M, Montoya P, Erb M, Hülsmann E, Flor H, Klose U, Birbaumer N, and Grodd W (1999) Activation of cortical and cerebellar motor areas

during executed and imagined hand movements: an fMRI study. J Cogn Neurosci 11: 491–501
[26] McFarland DJ, Sarnacki WA, and Wolpaw JR (2003) Brain-computer interface (BCI) operation: optimizing information transfer rates. Biol Psychol 63: 237–251
[27] Millán JR, Renkens F, Mouriñ J, and Gerstner W (2004) Noninvasive brain-actuated control of a mobile robot by human EEG. IEEE Trans Biomed Eng 51: 1026–33.
[28] Miller G (2005) What is the biological basis of consciousness. Science 309:79
[29] Mitra PP, and Pesaran B (1999). Analysis of dynamic brain imaging data. Biophys J 76: 691–708
[30] Nicolelis MAL (2001) Actions from thoughts. Nature 409:403–407.
[31] Nicolelis MAL (2003) Brain-machine interfaces to restore motor function and probe neural circuits. Nature Rev Neurosci 4: 417–422.
[32] Nicolelis MAL, Birbaumer N, Mueller KL (Eds) (2004) Special issue on brain–machine-interfaces. IEEE Trans Biomed Eng 51:877–1087
[33] Nirenberg S, Carcieri SM, Jacobs AL and Latham PE (2001) Retinal ganglion cells act largely as independent encoders. Nature 411: 698–701
[34] Nundy S, and Purves D (2002) A probabilistic explanation of brightness scaling. Proc Natl Acad Sci USA 99: 14482–14487
[35] O'Keefe J. and Recce M.L. (1993) Phase relationship between hippocampal place units and the EEG theta rhythm. Hippocampus 3: 317–330
[36] Olivier E, Baker SN, and Lemon RN (2002) Comparison of direct and indirect measurements of the central motor conduction time in the monkey. Clin Neurophysiol 113: 469–477
[37] Olivier E, Baker SN, Nakajima K, Brochier T, and Lemon RN (2001) Investigation into non-mono synaptic corticospinal excitation of macaque upper limb single motor units. J Neurophysiol 86: 1573–1586
[38] Pfurtscheller G, Neuper C, Guger C, Harkam W, Ramoser R, Schlögl A, Obermaier KB, and Pregenzer M (2000) Current trends in Graz brain-computer interface (BCI), IEEE Trans Rehab Eng 8: 216–219
[39] Piedade M, Gerald J, Sousa LA, Tavares G, and Tomas P (2005) Visual neuroprosthesis: A non invasive system for stimulating the cortex. IEEE Trans Circuits and Systems I-Regular Papers 52: 2648–2662
[40] Riehle A, Grun S, Diesmann M, and Aertsen A (1997) Spike synchronization and rate modulation differentially involved in motor cortical function. Science 278: 1950–1953
[41] Rudolph M, and Destexhe A (2003) Tuning neocortical pyramidal neurons between integrators and coincidence detectors. J Comp Neurosci 14: 239–251
[42] Samengo I, and Treves A (2000) Representational capacity of a set of independent neurons. Phys Rev E 63: 011910
[43] Schoffelen JM, Oostenveld R, and Fries P (2005) Neuronal coherence as a mechanism of effective corticospinal interaction. Science 308: 111–113

[44] Sejnowski T.J., and Paulsen O. (2006) Network oscillations: emerging computational principles. J Neurosci 26: 1673–1676

[45] Shadlen MN, and Movshon JA (1999) Synchrony unbound: A critical evaluation of the temporal binding hypothesis. Neuron 24: 67–77

[46] Singer W (1999) Neuronal synchrony: A versatile code for the definition of relations? Neuron 24: 49–65

[47] Singer W, and Gray CM (1995) Visual feature integration and the temporal correlation hypothesis. Annu Rev Neurosci 18: 555–586

[48] Smith E.C. and Lewicki M.S. (2006) Efficient auditory coding. Nature, 439: 978–982

[49] Tsodyks MV, Skaggs WE, Sejnowski TJ, and MacNaughton BL (1996) Population dynamics and theta rhythm phase precsession of hippocampal place cell firing: a spiking neuron model. Hippocampus 6: 271–280

[50] van Beers RJ, Sittig AC, and van der Gon JJD (1999) Integration of proprioceptive and visual position-information: An experimentally supported model. J Neurophysiol 81: 1355–1364

[51] van Beers RJ, Wolpert DM, and Haggard P (2002) When feeling is more important than seeing in sensorimotor adaptation. Current Biology 12: 834–837

[52] Varela F, J.-P. Lachaux, E. Rodriguez, J. and Martinerie (2001) The brainweb: Phase synchronization and large-scale integration. Nature Rev Neurosci 2: 229–239.

[53] Wolpaw JR, Birbaumer N, McFarland DJ, Pfurtscheller G, and Vaughan TM (2002) Brain-computer interfaces for communication and control. Clin. Neurophysiol. 113: 767–791

[54] Weber DJ, Stein RB, Chan KM, Loeb GE, Richmond FJR, Rolf R, James K, Chong SL, Thompson AK, and Misiaszek J (2004) Functional electrical stimulation using microstimulators to correct foot drop: a case study. Canadian J of Physiol Pharmacol 82: 784–792

[55] Yang ZY, and Purves D (2003) A statistical explanation of visual space. Nature Neurosci 6: 632–640

[56] Yang Z and Purves D (2004) The statistical structure of natural light patterns determines perceived light intensity. Proc. Natl. Ac. Sci. USA 101: 8745–8750

[57] Zeitler M, Fries P, and Gielen S (2006) Assessing neuronal coherence with single-unit, multi-unit, and local field potentials. Neural Comp, in press.

Computational Scene Analysis

DeLiang Wang

Department of Computer Science & Engineering and Center for Cognitive Science
The Ohio State University
Columbus, OH 43210-1277, U.S.A.
dwang@cse.ohio-state.edu

Summary. A remarkable achievement of the perceptual system is its scene analysis capability, which involves two basic perceptual processes: the segmentation of a scene into a set of coherent patterns (objects) and the recognition of memorized ones. Although the perceptual system performs scene analysis with apparent ease, computational scene analysis remains a tremendous challenge as foreseen by Frank Rosenblatt. This chapter discusses scene analysis in the field of computational intelligence, particularly visual and auditory scene analysis. The chapter first addresses the question of the goal of computational scene analysis. A main reason why scene analysis is difficult in computational intelligence is the binding problem, which refers to how a collection of features comprising an object in a scene is represented in a neural network. In this context, temporal correlation theory is introduced as a biologically plausible representation for addressing the binding problem. The LEGION network lays a computational foundation for oscillatory correlation, which is a special form of temporal correlation. Recent results on visual and auditory scene analysis are described in the oscillatory correlation framework, with emphasis on real-world scenes. Also discussed are the issues of attention, feature-based versus model-based analysis, and representation versus learning. Finally, the chapter points out that the time dimension and David Marr's framework for understanding perception are essential for computational scene analysis.

1 Introduction

Human intelligence can be broadly divided into three aspects: Perception, reasoning, and action. The first is mainly concerned with analyzing the information in the environment gathered by the five senses, and the last is primarily concerned with acting on the environment. In other words, perception and action are about input and output, respectively, from the viewpoint of the intelligent agent (i.e. a human being). Reasoning involves higher cognitive functions such as memory, planning, language understanding, and decision making, and is at the core of traditional artificial intelligence [49]. Reasoning also serves to connect perception and action, and the three aspects interact with one another to form the whole of intelligence.

DeLiang Wang: *Computational Scene Analysis*, Studies in Computational Intelligence (SCI) **63**, 163–191 (2007)
www.springerlink.com

This chapter is about perception - we are concerned with how to analyze the perceptual input, particularly in the visual and auditory domains. Because perception seeks to describe the physical world, or scenes with objects located in physical space, perceptual analysis is also known as scene analysis. To differentiate scene analysis by humans and by machines, we term the latter *computational scene analysis*[1]. In this chapter I focus on the analysis of a scene into its constituent objects and their spatial positions, not the recognition of memorized objects. Pattern recognition has been much studied in computational intelligence, and is treated extensively elsewhere in this collection.

Although humans, and nonhuman animals, perform scene analysis with apparent ease, computational scene analysis remains an extremely challenging problem despite decades of research in fields such as computer vision and speech processing. The difficulty was recognized by Frank Rosenblatt in his 1962 classic book, "Principles of neurodynamics" [47]. In the last chapter, he summarized a list of challenges facing perceptrons at the time, and two problems in the list "represent the most baffling impediments to the advance of perceptron theory" (p. 580). The two problems are figure-ground separation and the recognition of topological relations. The field of neural networks has since made great strides, particularly in understanding supervised learning procedures for training multilayer and recurrent networks [48, 2]. However, progress has been slow in addressing Rosenblatt's two chief problems, largely validating his foresight.

Rosenblatt's first problem concerns how to separate a figure from its background in a scene, and is closely related to the problem of scene segregation: To decompose a scene into its comprising objects. The second problem concerns how to compute spatial relations between objects in a scene. Since the second problem presupposes a solution to the first, figure-ground separation is a more fundamental issue. Both are central problems of computational scene analysis.

In the next section I discuss the goal of computational scene analysis. Section 3 is devoted to a key problem in scene analysis - the binding problem, which concerns how sensory elements are organized into percepts in the brain. Section 4 describes oscillatory correlation theory as a biologically plausible representation to address the binding problem. The section also reviews the LEGION[2] network that achieves rapid synchronization and desynchronization, hence providing a computational foundation for the oscillatory correlation theory. The following two sections describe visual and auditory scene analysis separately. In Section 7, I discuss a number of challenging issues facing computational scene analysis. Finally, Section 8 concludes the chapter.

Note that this chapter does not attempt to survey the large body of literature on computational scene analysis. Rather, it highlights a few topics that I consider to be most relevant to this book.

1 This is consistent with the use of the term Computational Intelligence.

2 LEGION stands for Locally Excitatory Globally Inhibitory Oscillator Network [68].

2 What is the Goal of Computational Scene Analysis?

In his monumental book on computational vision, Marr makes a compelling case that understanding perceptual information processing requires three different levels of description. The first level of description, called computational theory, is mainly concerned with the goal of computation. The second level, called representation and algorithm, is concerned with the representation of the input and the output, and the algorithm that transforms from the input representation to the output representation. The third level, called hardware implementation, is concerned with how to physically realize the representation and the algorithm.

So, what is the goal of computational scene analysis? Before addressing this question, let us ask the question of what purpose perception serves. Answers to this question have been attempted by philosophers and psychologists for ages. From the information processing perspective, Gibson [21] considers perception as the way of seeking and gathering information about the environment from the sensory input. On visual perception, Marr [30] considers that its purpose is to produce a visual description of the environment for the viewer. On auditory scene analysis, Bregman states that its goal is to produce separate streams from the auditory input, where each stream represents a sound source in the acoustic environment [6]. It is worth emphasizing that the above views suggest that perception is a private process of the perceiver even though the physical environment may be common to different perceivers.

In this context, we may state that *the goal of computational scene analysis is to produce a computational description of the objects and their spatial locations in a physical scene from sensory input.* The term 'object' here is used in a modality-neutral way: An object may refer to an image, a sound, a smell, and so on. In the visual domain, sensory input comprises two retinal images, and in the auditory domain it comprises two eardrum vibrations. Thus, the goal of visual scene analysis is to extract visual objects and their locations from one or two images. Likewise, the goal of auditory scene analysis is to extract streams from one or two audio recordings.

The above goal of computational scene analysis is strongly related to the goal of human scene analysis. In particular, we assume the input format to be similar in both cases. This assumption makes the problem well defined and has an important consequence: It makes the research in computational scene analysis perceptually relevant. In other words, progress in computational scene analysis may shed light on perceptual and neural mechanisms. This restricted scope also differentiates computational scene analysis from engineering problem solving, where a variety and a number of sensors may be used.

With common sensory input, we further propose that computational scene analysis should aim to achieve human level performance. Moreover, we do not consider the problem solved until a machine system achieves human level performance in *all* perceptual environments. That is, computational scene analysis should aim for the versatile functions of human perception, rather than its utilities in restricted domains.

3 Binding Problem and Temporal Correlation Theory

The ability to group sensory elements of a scene into coherent objects, often known as perceptual organization or perceptual grouping [40], is a fundamental part of perception. Perceptual organization takes place so rapidly and effortlessly that it is often taken for granted by us the perceivers. The difficulty of this task was not fully appreciated until effort in computational scene analysis started in earnest. How perceptual organization is achieved in the brain remains a mystery.

Early processing in the perceptual system clearly involves detection of local features, such as color, orientation, and motion in the visual system, and frequency and onset in the auditory system. Hence, a closely related question to perceptual organization is how the responses of feature-detecting neurons are bound together in the brain to form a perceived scene? This is the well-known *binding problem*. At the core of the binding problem is that sensory input contains multiple objects simultaneously and, as a result, the issue of which features should bind with which others must be resolved in objection formation. I illustrate the situation with two objects - a triangle and a square - at two different locations: The triangle is at the top and the square is at the bottom. This layout, shown in Figure 1, was discussed by Rosenblatt [47] and used as an instance of the binding problem by von der Malsburg [60]. Given feature detectors that respond to triangle, square, top, and bottom, how can the nervous system bind the locations and the shapes so as to perceive that the triangle is at the top and the square is at the bottom (correctly), rather than the square is on top and the triangle is on bottom (incorrectly)? We should note that object-level attributes, such as shape and size, are undefined before the more fundamental problem of figure-ground separation is solved. Hence, I will refer to the binding of local features to form a perceived object, or a percept, when discussing the binding problem.

How does the brain solve the binding problem? Concerned with shape recognition in the context of multiple objects, Milner [32] suggested that different objects could be separated in time, leading to synchronization of firing activity within the neurons activated by the same object. Later von der Malsburg [59] proposed a correlation theory to address the binding problem. The correlation theory asserts that the temporal structure of a neuronal signal provides the neural basis for correlation, which in turn serves to bind neuronal responses. In a subsequent paper, von der Malsburg and Schneider [61] demonstrated the temporal correlation theory in a neural model for segregating two auditory stimuli based on their distinct onset times - an example of auditory scene analysis that I will come back to in Section 6. This paper

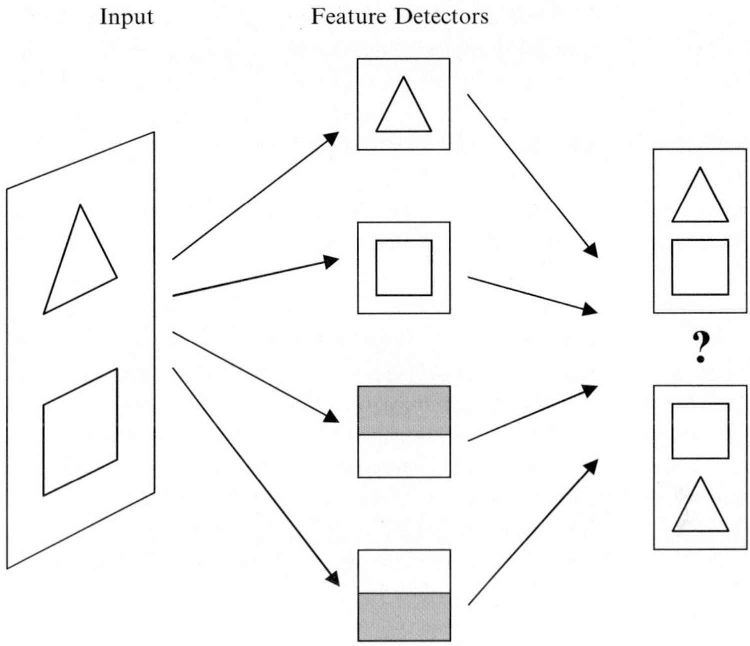

Fig. 1. Illustration of the binding problem. The input consists of a triangle and a square. There are four feature detectors for triangle, square, top, and bottom. The binding problem concerns whether the triangle is on top (and the square at bottom) or the square is on top (and the triangle at bottom)

proposed, for the first time, to use neural oscillators to solve a figure-ground separation task, whereby correlation is realized by synchrony and desynchrony among neural oscillations. Note that the temporal correlation theory is a theory of representation, concerned with how different objects are represented in a neural network, not a computational algorithm; that is, the theory does not address how multiple objects in the input scene are transformed into multiple cell assemblies with different time structures. This is a key computational issue I will address in the next section.

The main alternative to the temporal correlation theory is the hierarchical coding hypothesis, which asserts that binding occurs through individual neurons that are arranged in some cortical hierarchy so that neurons higher in the hierarchy respond to larger and more specialized parts of an object. Eventually, individual objects are coded by individual neurons, and for this reason hierarchical coding is also known as the cardinal cell (or grandmother cell) representation [3]. Gray [23] presented biological evidence for and against the hierarchical representation. From the computational standpoint, the hierarchical coding hypothesis has major drawbacks, including the need to encode a prohibitively large number of scenarios by cells [59, 65].

It should be clear from the above discussion that the figure-ground separation problem is essentially the same as the binding problem. A layered perceptron may be viewed as a computational implementation of the hierarchical coding hypothesis, and the problems challenging Rosenblatt's perceptrons underline the limitations of hierarchical coding.

4 Oscillatory Correlation Theory

A special form of temporal correlation - oscillatory correlation [52] - has been studied extensively. In oscillatory correlation, feature detectors are represented by oscillators and binding is represented by synchrony within an assembly of oscillators and desynchrony between different assemblies. The notion of oscillatory correlation is directly supported by the substantial evidence of coherent oscillations in the brain. In addition, the activity of a neuron or a local group of neurons can be accurately modeled by an oscillator. It is worth pointing out here that a mathematical oscillator need not always produce periodic behavior; indeed an oscillator in response to a time varying input often exhibits a variety of aperiodic responses.

Like the temporal correlation theory, the oscillatory correlation theory is a representation theory, not a computational mechanism. A computational mechanism for the oscillatory correlation theory needs to exhibit three key features [65]. First, the mechanism can synchronize a locally coupled assembly of oscillators. Second, it can desynchronize different assemblies of oscillators that are activated by multiple, simultaneously present objects. Third, both synchrony and desynchrony must occur rapidly in order to deal with the changing environment.

The first neural network that successfully met the above requirements is the LEGION mechanism proposed in 1995 by Terman and Wang [52, 62]. LEGION builds on relaxation oscillators characterized by two time scales [58]. Formally, a relaxation oscillator, i, is defined as a pair of an excitatory unit x_i and an inhibitory unit y_i [52]:

$$\dot{x}_i = 3x_i - x_i^3 + 2 - y_i + I_i + S_i + \rho \tag{1a}$$

$$\dot{y}_i = \varepsilon(\alpha(1 + \tanh(x_i/\beta)) - y_i) \tag{1b}$$

In the above equation, I_i denotes the external stimulation to the oscillator and S_i the input from the rest of the network, to be specified below. ρ denotes the amplitude of intrinsic noise (e.g. Gaussian noise) which assists the process of desynchronization, and α and β are parameters. ε is a small positive parameter, and it is this parameter that induces the two time scales with y on a slower one.

Figure 2 illustrates the oscillator defined in (1). As shown in Fig. 2A, the x-nullcline (i.e. $\dot{x} = 0$) is a cubic function and the y-nullcline is a sigmoid function. When $I > 0$, the two nullclines intersect at a single point on the middle

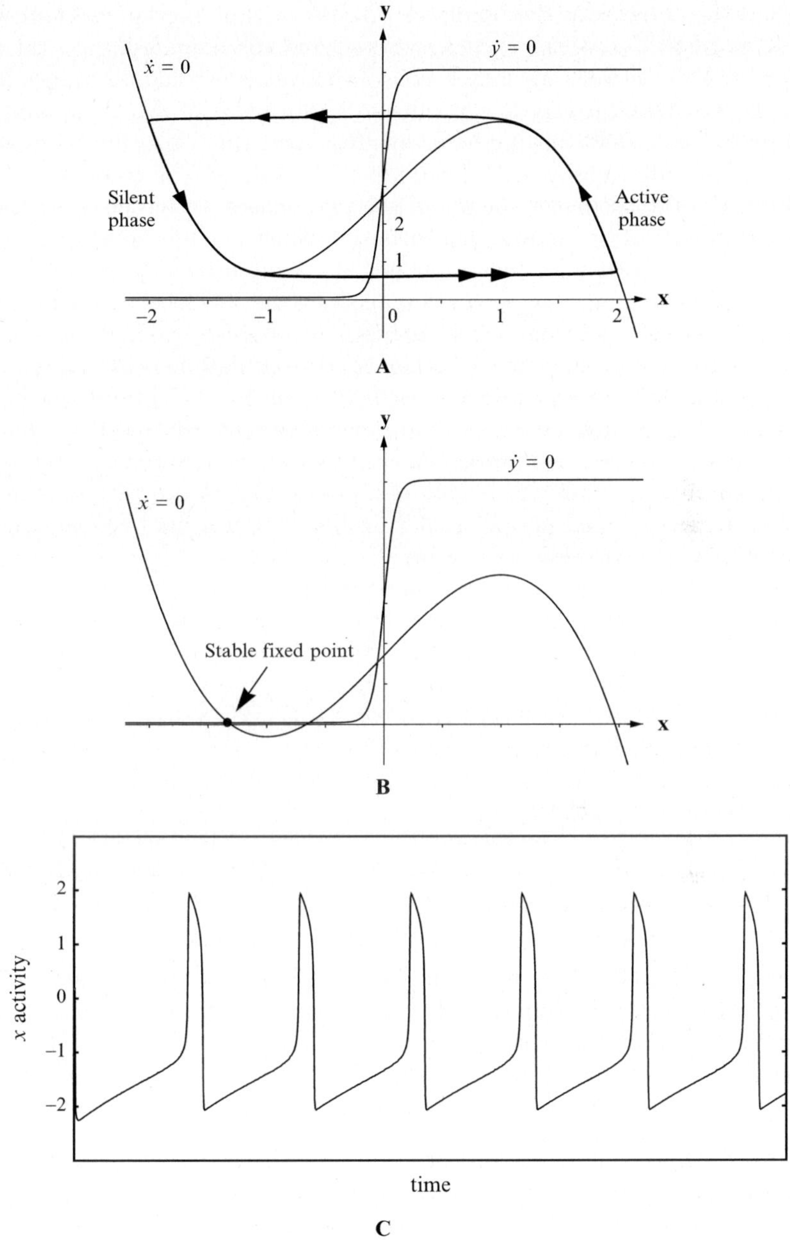

Fig. 2. Behavior of a relaxation oscillator. A. Enabled state of the oscillator. This state produces a limit cycle shown as the bold curve. The direction of the trajectory is indicated by the arrows, and jumps are indicated by double arrows. B. Excitable state of the oscillator. This state produces a stable fixed point. C. Temporal activity of the oscillator in the enabled state. The curve shows the x activity

branch of the cubic, and the oscillator produces a stable limit cycle shown in Figure 2A. In this case, the oscillator is referred to as *enabled*, and the limit cycle alternates between an *active phase* with relatively high x values and a *silent phase* with relatively low x values. Within each of the two phases the oscillator exhibits slow-varying behavior. However, the transition between the two phases occurs rapidly, called *jumping*. The role of α is to determine the relative times the oscillator spends in the two phases - a larger α produces a relatively shorter active phase. The situation when $I < 0$ is shown in Fig. 2B. In this case, the two nullclines intersect at a stable fixed point on the left branch of the cubic, and no oscillation occurs - the oscillator is referred to as *excitable*. Whether the state of an oscillator is enabled or excitable depends solely on external stimulation; in other words, oscillation is stimulus dependent. The x activity of an enabled state is given in Fig. 2C, and it resembles a spike train. Indeed, relaxation oscillators have been widely used as models of single neurons, where x is interpreted as the membrane potential of a neuron and y the activation state of ion channels [19, 36, 35]. A relaxation oscillation may also be interpreted as an oscillating burst of neuronal spikes, where x corresponds to the envelope of the burst.

In a LEGION network an oscillator is excitatorily coupled with other oscillators in its neighborhood, and excites a global inhibitor which then inhibits every oscillator in the network. Specifically, S_i in (1a) is defined as

$$S_i = \sum_{k \in N(i)} W_{ik} H(x_k - \theta_x) - W_z H(z - \theta_z) \tag{2}$$

where $N(i)$ denotes a set of neighbors of i, and W_{ik} the connection weight from oscillator k to i. H stands for the Heaviside step function, and θ_x and θ_z are thresholds. W_z is the weight of inhibition from the global inhibitor z, which is defined as

$$\dot{z} = \phi(\sigma_\infty - z) \tag{3}$$

where ϕ is a parameter. $\sigma_\infty = 1$ if at least one oscillator is in the active phase and $\sigma_\infty = 0$ otherwise. From (3) it is easy to see that $z \to 1$ when σ_∞ equals 1.

On the basis of the earlier analysis by Somers and Kopell [51] on two coupled relaxation oscillators, Terman and Wang [52] conducted an extensive analysis on LEGION networks. They showed that LEGION exhibits the mechanism of *selective gating* as follows. When an oscillator jumps to the active phase, its activity spreads to its neighboring oscillators, their neighbors, and so on. This leads to fast synchronization within the oscillator assembly that contains the oscillator. In addition, the oscillator activates the global inhibitor which prevents the oscillators of different assemblies from jumping up. This leads to desynchronization among different oscillator assemblies. They proved the following theorem: A LEGION network can reach both synchronization within each assembly and desynchronization between different assemblies, and does so in no greater than m cycles of oscillations, where m is the number of the oscillator assemblies. In other words, both synchronization and desynchronization are achieved rapidly.

The selective gating mechanism of LEGION successfully meets the three computational requirements stated at the beginning of this section. Subsequent research has shown that rapid synchronization and desynchronization can also be achieved using other types of oscillators, such as Wilson-Cowan and spike (integrate-and-fire) oscillators, although conclusions are typically drawn from numerical simulations. See [65] for a broad review on this topic.

As a concrete application of LEGION dynamics, I describe a solution to a classic problem in neural computation - the connectedness problem [64]. The connectedness problem, first described by Minsky and Papert in 1969, is the centerpiece of their consequential critique on perceptrons [33]. The connectedness predicate is innocuously simple: To classify whether an input pattern is connected or not. To appreciate the significance of this predicate, I need to give some context on perceptron theory. Rosenblatt's perceptrons [46, 47] are classification networks. A typical perceptron, illustrated in Figure 3, computes a predicate. It consists of a binary input layer R, which symbolizes retina, a layer of binary feature detectors, and a response unit that signals the result of a binary classification. A feature detector is activated if and only if all the pixels within the area of R sensed by the detector are black. The response unit outputs 1 if a weighted sum of all the feature detectors exceeds a threshold, and outputs 0 otherwise.

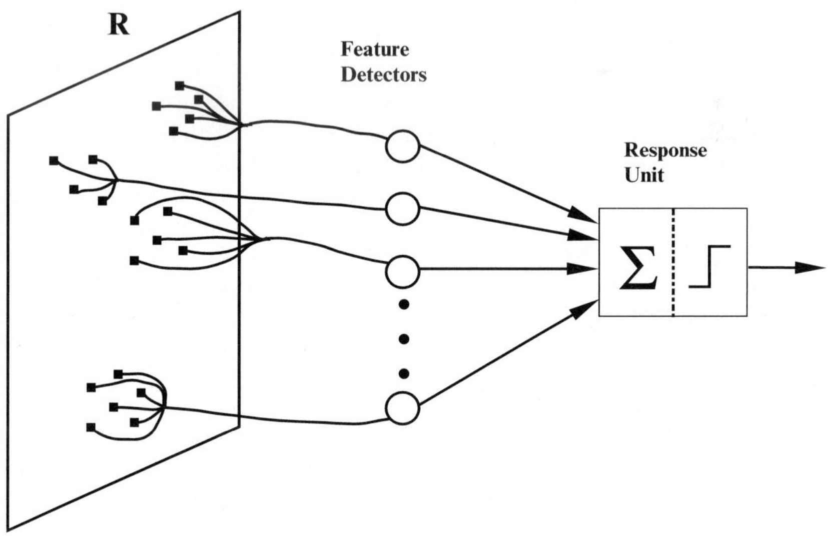

Fig. 3. Diagram of a perceptron. R denotes the input layer, which projects to a layer of feature detectors. The response unit takes a weighted sum of the responses of all the detectors, and outputs 1 if the sum passes a certain threshold and 0 otherwise

Minsky and Papert [33] define the *order* of a predicate as the smallest number of pixels in R that must be sensed by some feature detector in order to compute the predicate. With this notion, they prove that the order of the connectedness predicate increases at least as fast as $|R|^{1/2}$. That is, the order of this predicate is unbounded. What does this result mean? It means that, to compute a predicate of an unbounded order requires feature detectors with too large receptive fields (relative to R) and too many of detectors to be computationally feasible [33]. It is important to understand that the result is not about *computability*, or *whether* a perceptron exists to solve the problem. With a finite size of R, the number of connected patterns is finite, and we can simply find a perceptron to solve the problem, in which each connected pattern is sensed by a single feature detector. However, the number of connected patterns grows exponentially except for one-dimensional R [65], and this way of computing the connectedness predicate is computationally intractable. Hence, their result is about the *scalability* or *computational complexity*.

Thanks to recurrent connectivity and oscillatory dynamics, LEGION solves the connectedness problem in general form [64]. To explain the solution, Figure 4 shows the response of a two-dimensional (2-D) LEGION network to two binary images: The first one is a connected figure showing a cup (Fig. 4A) and the second one is a disconnected figure showing the word 'CUP' (Fig. 4D). To ensure that the network has no binding preference, we randomize oscillator phases at the beginning. The random initial conditions are illustrated in Fig. 4B, where the diameter of a circle represents the x activity of the corresponding oscillator. A snapshot of the network activity shortly after the beginning is shown in Fig. 4C where the oscillator assembly representing the cup is synchronized and other oscillators are in the excitable state. The response of the same network to 'CUP' is depicted in Figures 4E-G at different times. In this case, the network forms three assemblies corresponding to each of the three letters. Figure H shows the temporal activity of all the enabled oscillators for the connected cup image, where excitable oscillators are omitted. The upper panel of Fig. 4H shows the combined activity of the assembly representing the cup, and the middle panel shows the activity of the global inhibitor. Despite randomized phases to begin with, the assembly reaches synchrony in the first oscillation cycle. The temporal response to the disconnected 'CUP' is shown in Fig. 4I, where synchrony within each of the three assemblies and desynchrony between them are both achieved in the first two cycles. As illustrated in Figs. 4H and 4I, every time an assembly jumps to the active phase the global inhibitor is triggered. Thus, how many assemblies - or put differently how many connected patterns in the input image - is revealed by the ratio of the response frequency of the global inhibitor to the oscillation frequency of an enabled oscillator. A ratio of 1 indicates there is one pattern in the input figure, and thus the figure is connected. A ratio greater than 1 indicates there are more than one pattern in the input figure, and thus the figure is disconnected. Hence the solution to the predicate is given by a simple test of whether the ratio exceeds a threshold θ, chosen in

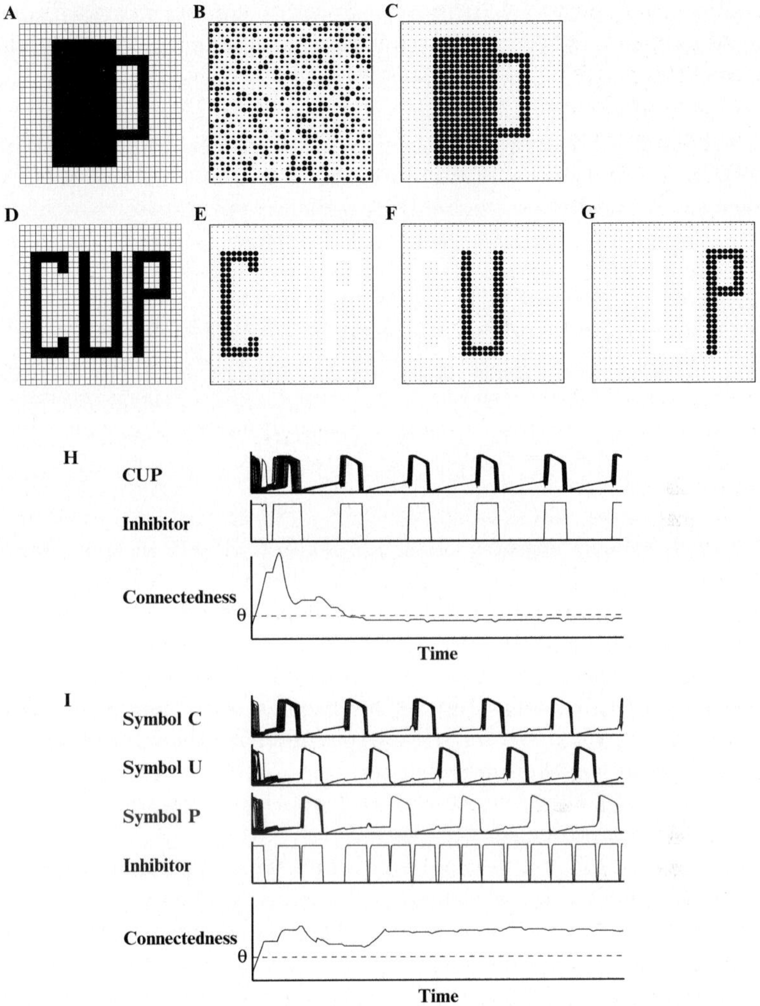

Fig. 4. Oscillatory correlation solution to the connectedness problem (from [64]). A. An input image with 30x30 binary pixels showing a connected cup figure. B. A snapshot from corresponding LEGION network showing the initial conditions of the network. C. A subsequent snapshot of the network activity. D. Another input image depicting three connected patterns forming the word 'CUP'. E.-G. Snapshots of the LEGION network at three different times. H. The upper trace shows the temporal activity of the oscillator assembly representing the connected cup image, the middle trace the activity of the global inhibitor, and the bottom trace the ratio of the global inhibitor's frequency to that of enabled oscillators. The threshold is indicated by the dash line. I. The upper three traces show the temporal activities for the three assemblies representing the three connected patterns in the disconnected 'CUP' image, the next-to-bottom trace the activity of the global inhibitor, and the bottom one the ratio of the global inhibitor's frequency to that of enabled oscillators along with

the range $2 > \theta > 1$. The bottom traces of Fig. 4H and Fig. 4I show the ratios, where θ is chosen to be 1.6. As shown in the figure, the connectedness predicate is correctly computed beyond a beginning period that corresponds to the process of assembly formation.

The oscillatory correlation theory provides a general framework to address the computational scene analysis problem. The following two sections deal with visual and auditory scene analysis respectively.

5 Visual Scene Analysis

For computational scene analysis, some measure of similarity between features is necessary. What determines if local sensory elements should be grouped into the same object or separated apart? This is the main subject of Gestalt psychology [71, 27]. Major Gestalt grouping principles are summarized below [40]:

- *Proximity.* Sensory elements that are located closely in space tend to be grouped.
- *Similarity.* Elements with similar attributes, such as color, depth, or texture, tend to group.
- *Common fate.* Elements that move together, or show common motion, likely belong to the same object. Common fate is an instance of similarity in a sense, and it is listed separately to emphasize the importance of visual dynamics in perceptual organization.
- *Good continuity.* Elements that form smooth continuations of each other tend to be bound together.
- *Connectedness* and *common region.* Connected elements or elements inside the same connected region have the tendency to group.
- *Familiarity.* Elements that belong to the same memorized pattern tend to group.

To apply the above grouping principles requires a process of feature extraction, which may be a complex operation for certain features such as motion and texture. With extracted features, oscillatory correlation represents a general approach to visual scene analysis. In this approach, an oscillator typically corresponds to a spatial location and connection weights between neighboring oscillators are determined by feature extraction. The oscillator network then evolves autonomously. After assembly formation takes place, different assemblies representing different objects will pop out from the network at different times. It is *segmentation in time* that is unique of this approach to scene analysis. As a result, such segmentation gives rise to the notion of a *segmentation capacity* [69] - at least for networks of relaxation oscillators with a non-instantaneous active phase - that refers to a limited number of oscillator assemblies that may be formed. The segmentation capacity corresponds to the integer ratio of the oscillation period to the duration of the active phase for relaxation oscillators.

The first application of the oscillatory correlation approach to real image segmentation was made by Wang and Terman [69]. Their segmentation system is based on LEGION dynamics. Unlike synthetic images, real images are often noisy. Image noise may result in many fragments and deteriorate the result of oscillatory correlation. To address the problem of fragmentation, a lateral potential is introduced for each oscillator in order to distinguish between major assemblies and noisy fragments. A major assembly should contain at least one oscillator that lies at the center of a sizable homogeneous region. Such an oscillator, called a leader, has a high lateral potential because it receives a large amount of lateral excitation from its neighborhood. On the other hand, a fragment does not have a leader. All fragments, forming a background, will cease oscillating after a few periods. Another issue that has to be addressed is computing time required for integrating a large oscillator network. To alleviate the computational burden, Wang and Terman abstracted an algorithm from oscillatory dynamics that retains key features of LEGION dynamics, such as jumping and rapid spread of excitation and inhibition. The abstracted algorithm has the option to eliminate the segmentation capacity in order to segment a large number of regions. In the Wang-Terman system, each oscillator is mutually connected with its 8-nearest neighbors, and the connection weight between oscillators i and j is set proportional to $1/(1+|I_i+I_j|)$, where I_i and I_j represent the corresponding pixel intensities. W_z in (2) is a key parameter that controls the granularity of segmentation, whereby smaller values of W_z produce fewer and larger segments.

In a subsequent study, Chen et al. [10] suggested the idea of weight adaptation to perform feature-preserving smoothing before segmentation. In addition, they proposed to add a logarithmic normalization factor in excitatory coupling (cf. (2)):

$$S_i = \frac{\sum_{k\in N(i)} W_{ik} H(x_k - \theta_x)}{log(\sum_{k\in N(i)} H(x_k - \theta_x) + 1)} - W_z H(z - \theta_z) \quad (4)$$

The resulting algorithm produces robust segmentation results. An example is given in Figure 5, where the task is to extract hydrographic objects from satellite images from the United States Geological Survey (USGS). Figure 5A gives the original image containing water bodies, and Figure 5B shows the corresponding extraction results, where the extracted waterbodies are displayed as white and overlaid on the original image. For reference, Figure 5C provides the corresponding map from the USGS. A careful comparison should reveal that the extracted waterbodies match the image better than the map, since the latter is often not up to date.

Cesmeli and Wang [8] applied LEGION to motion-based segmentation that considers motion as well as intensity for analyzing image sequences (see also [75]). In their system, two pathways perform an initial optic flow estimation and intensity-based segmentation in parallel. A subsequent network

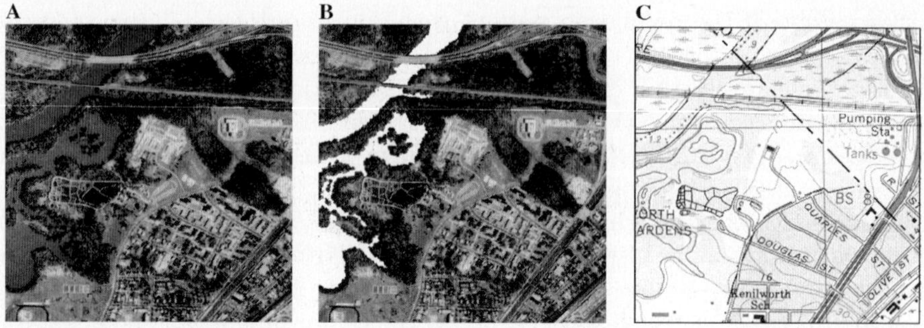

Fig. 5. Extraction of hydrographic regions (from [10]). A. Input satellite image consisting of 640x606 pixels. B. Extraction result, where segmented waterbodies are indicated by white. C. Corresponding 1:24,000 topographic map

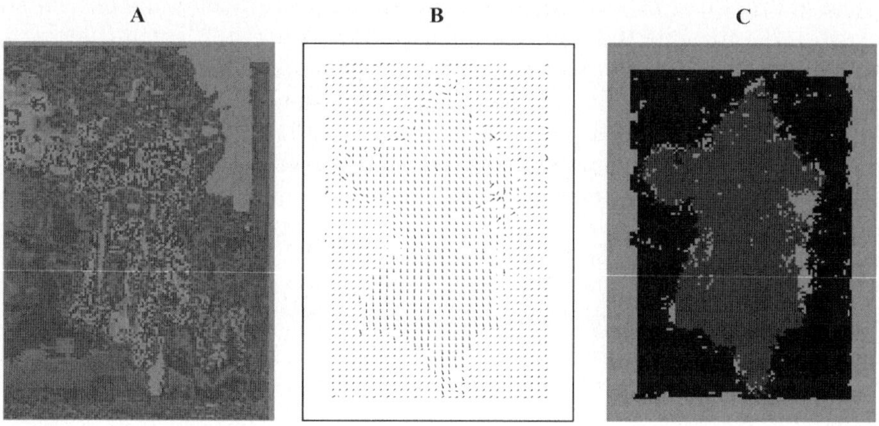

Fig. 6. Motion segmentation (from [8]). A. A frame of a motion sequence. B. Estimated optic flow. C. Result of segmentation

combines the two to refine local motion estimates. Motion analysis and intensity analysis complement each other since the former tends to be reliable for inhomogeneous, textured regions while the latter is most effective for homogeneous regions. The use of LEGION for segmentation allows for multiple motions at the same location, as in the case of motion transparency. The resulting system significantly reduces erroneous motion estimates and improves boundary localization. A typical example is given in Figure 6. A frame of a motion sequence is shown in Fig. 6A, where a motorcycle rider jumps to a dry canal with his motorcycle while the camera is tracking him. Due to the camera motion, the rider and his motorcycle have a downward motion with a small rightward component and the image background has an upright diagonal motion. Figure 6B shows the estimated optic flow after integrating motion and brightness analyses, and it is largely correct. The rider with his motorcycle is then segmented from the image background as depicted in Figure 6C. Their oscillator model has been favorably compared with a number of motion segregation algorithms including the one by Black and Anandan [5] based on robust statistics.

A large number of studies have applied the oscillatory correlation approach to visual scene analysis tasks, including segmentation of range and texture images, extraction of object contours, and selection of salient objects. See [65] for a recent review on this subject.

6 Auditory Scene Analysis

What grouping principles govern auditory scene analysis? Bregman systematically addresses this question in a comprehensive book [6]. According to Bregman, grouping principles analogous to Gestalt laws revealed in the visual domain are responsible for the segregation of auditory input into streams. Displaying the acoustic input in a 2-D time-frequency (T-F) representation such as a spectrogram, major grouping principles for auditory scene analysis (ASA) are given below [6, 13]:

- *Proximity in frequency and time.* Two tones that are close in frequency or time tend to be grouped into the same stream (an auditory object).
- *Periodicity.* Harmonically related frequency components tend to grouped.
- *Onset and offset.* Frequency components that onset or offset at the same time tend to be organized into the same stream.
- *Amplitude and frequency modulation.* Frequency components that have common temporal modulation tend to be grouped together. This principle applies to both amplitude modulation and frequency modulation.
- *Continuous/smooth transition.* Tones that form a continuous, or discontinuous but smooth, trajectory tend to be fused.
- *Familiarity.* Sound components belonging to the same learned pattern, such as a syllable, have the tendency to group.

Auditory scene analysis takes place in two stages in the brain according to Bregman [6]. The first stage, known as the segmentation stage [66], decomposes the acoustic mixture reaching the ears into a collection of time-frequency segments, each corresponding to a contiguous region in a T-F representation. The second stage groups the segments into streams.

The first study on auditory segregation using oscillatory correlation was made by von der Malsburg and Schneider [61]. As discussed in Section 3, their segregation is based on common onsets. However, their model relies on global connectivity to achieve synchronization among the oscillators that are stimulated at the same time. Desynchronization is obtained with a global inhibitor. Subsequently Wang [63] studied stream segregation by employing a

2-D LEGION network, where one dimension represents time and another one represents frequency. With appropriate connectivity, the LEGION network exhibits a set of psychophysical phenomena, such as dependency of stream segregation on spectrotemporal proximity and competition among different perceptual organizations (see [38, 9] for recent extensions).

Wang and Brown [66] studied a more complex problem: Segregation of voiced speech from its acoustic interference. After feature extraction using a model of auditory periphery that comprises cochlear filtering and mechanical to neural transduction, they compute a number of mid-level representations from peripheral responses, including a correlogram (autocorrelation) and a summary correlogram. The core of their model is an oscillator network with two layers performing auditory segmentation and grouping, respectively. The two-layer structure is designed to embody Bregman's two-stage notion. Auditory segmentation is based on cross-channel correlation in the frequency domain and temporal continuity in the time domain. Specifically, the first layer is a LEGION network where each oscillator is connected with its four nearest neighbors in time and frequency. The connection weight along the frequency axis is set to one if the corresponding cross-channel correlation exceeds a certain threshold and zero otherwise. The connection weight along the time axis is set to one uniformly. In response to an input mixture, the segmentation layer produces oscillator assemblies, representing regions of acoustic energy such as harmonics or formants. The second layer groups the segments that emerge from the first layer. Specifically, this layer contains lateral connections with both excitation and inhibition but no global inhibitor. Grouping in this layer is based on the dominant pitch extracted from the summary correlogram within each time frame. The extracted dominant pitch is used to divide the oscillators of the frame into two classes: One is consistent with the pitch frequency and the other is not. Then excitatory connections are formed between the oscillators of the same class and inhibitory connections are formed between the two classes. This pattern of connectivity within the grouping layer promotes synchronization among a group of segments that have common periodicity.

Figure 7 gives an example of segregating a mixture of a male utterance and telephone ringing. Figure 7A displays the peripheral response to the mixture. The 2-D response is generated by a filterbank with 128 channels over 150 time frames. Figure 7B shows a snapshot of the grouping layer, where active oscillators, indicated by white pixels, primarily correspond to the speech utterance. Figure 7C shows another snapshot of the grouping layer taken at a different time. At this time, active oscillators mostly correspond to the background, i.e. the telephone ringing.

Wrigley and Brown [72] recently proposed an oscillator network to model auditory selective attention. Their model first performs peripheral processing and then auditory segmentation. A unique part of model is an interactive loop between an oscillator layer that performs stream segregation and a leaky integrator that simulates the attentional process. The weights of the connections

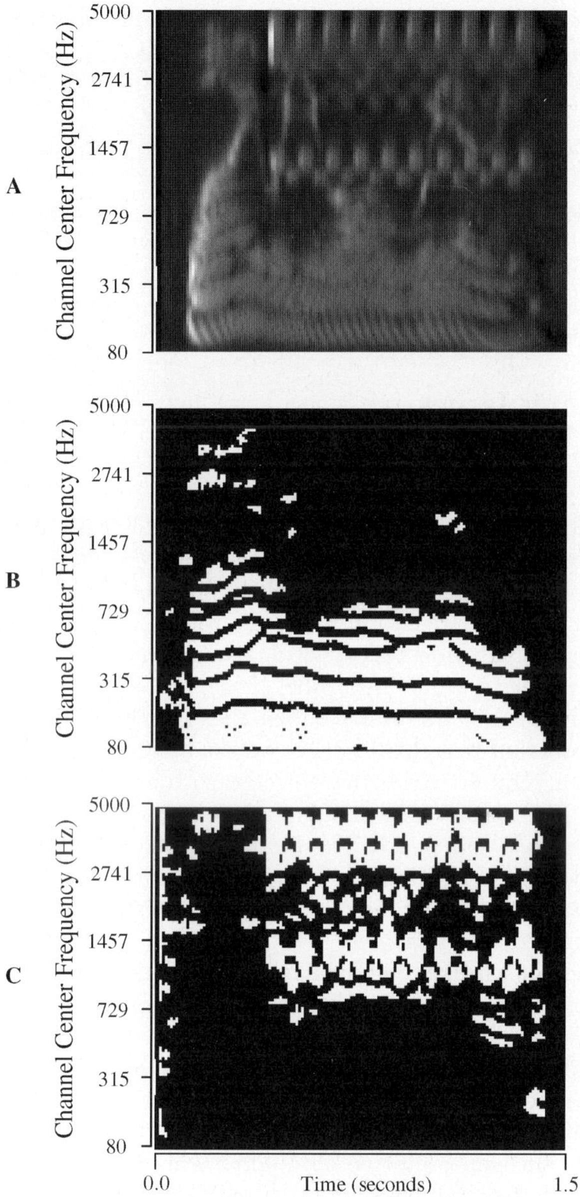

Fig. 7. Segregation of voiced speech from telephone ringing (from [66]). A. Peripheral response to an auditory stimulus consisting of a male utterance mixed with telephone ringing. A bank of 128 filters having center frequencies ranging from 80 Hz to 5 kHz is employed in peripheral processing. B. A snapshot of the grouping layer. Here, white pixels denote active oscillators that represent the segregated speech stream. C. Another snapshot showing the segregated background

between the oscillator layer and the leaky integrator are subject to modulation by the attentional interest of the model. Through this interaction, the attentional leaky integrator selects one dominant stream from the stream segregation layer. Their network successfully simulates a number of auditory grouping phenomena, including two-tone streaming with distracted attention and sequential capturing. At a conceptual level, a major difference between this model and Wang's model [63] concerns whether attention can be directed to more than one stream: In the Wrigley and Brown model only one stream may be attended to at a time whereas in Wang's model attention may be divided by more than one stream. This issue will be revisited in Sect. 7.1.

7 Challenging Issues

7.1 Attention

The importance of attention for scene analysis can hardly be overstated. In a way, to perceive is to attend.

The issues of binding and attention are related. It has been frequently suggested that selective attention plays the role of binding. According to the popular feature integration theory of Treisman and Gelade [57], the visual system first analyzes a scene in parallel by separate retinotopic feature maps and focal attention then integrates the analyses within different feature maps to produce a coherent percept. In other words, attention provides a 'spotlight' on the location map to bind and select an object [55]. Arguing from the neurobiological perspective, Reynolds and Desimone [43] also suggested that attention provides a solution to the binding problem. An alternative view - object-based theories of attention [42, 41] - claims that selective attention operates on the result of binding. So the key question is whether attention precedes or succeeds binding.

A visual object can have an arbitrary shape and size. This consideration creates the following dilemma for the feature integration theory. On the one hand, it is a location-based theory of attention that binds at the same location individual analyses from different feature maps. On the other hand, to select an object, attention spotlight must also have arbitrary shape and size, adapting to a specific object and thus object-based. Without a binding process, what produces such an adaptive spotlight? This is an intrinsic difficulty if focal attention, rather than perceptual organization, is to bind features across different locations. The difficulty is illustrated by the finding of Field et al. [18] that a path of curvilinearly aligned (snake-like) orientation elements embedded in a background of randomly oriented elements can be readily detected by observers, whereas other paths cannot. This is illustrated in Fig. 8, which shows a snake-like pattern (left panel) in a cluttered background (right panel). Note that there are virtually an infinite number of snake patterns that can

Fig. 8. Detection of snake-like patterns (from [18] with permission). Human observers can easily detect the occurrence of a snake-like pattern - shown on the left - that is embedded in a background of random orientation elements shown on the right. The snake pattern consists of 12 aligned orientation elements

be constructed from orientation elements. Grouping seems to be required to yield organized patterns for attentional spotlight.

This difficulty, however, does not occur in object-based theories, in which binding provides multiple segments for focal attention to perform sequential analysis. Though sometimes difficult to tear object-based attention apart from location-based attention, since the former implicitly provides the information for the latter, psychophysical and neurobiological studies show increasing support for the object-based view [37, 31, 15]. For example, a recent study demonstrates that the visual search for a target item in the presence of many distractors is very efficient if the distractors can be organized into a small number of groups on the basis of feature similarity, suggesting that visual attention examines organized groups rather than individual items [67]. Indeed, the Field et al. results have been successfully simulated by the oscillation model of Yen and Finkel [74].

The notion of a segmentation capacity (see Sect. 5) is a basic characteristic of the oscillatory correlation theory. A limited capacity naturally arises from relaxation oscillations because of their non-instantaneous active phase. On the other hand, networks of spiking neurons [7] or chaotic maps [76] do not exhibit a limited capacity. Although such a capacity is sometimes treated as a computational weakness [70, 76, 14], capacity limitation is a fundamental property of attention. Also it has been argued that a limited capacity is advantageous for information processing (e.g. [29, 25]).

Assuming a limited capacity, a related question is: Can we attend to more than one object at a time? A direct answer was offered by Cowan [12] after reviewing a large body of literature. His answer is that the attentional capacity is about four. Furthermore, the attentional capacity underlies the well-documented capacity in short-term memory. How to reconcile between a capacity limit of more than one and the phenomenological impression that we can focus on only one thing at a time? A capacity limit represents an upper bound on the number of items held by attention, and it does not necessarily mean that the attention span is constantly full. It may be possible for a subject to selectively attend to one thing in order to extract information from it. Even in the case of selective attention, however, unselected items still receive some analysis. In the classic experiment of Cherry [11], for example, listeners can detect the change of the speaker gender from the 'unattended' ear.

Another important question regarding attention is what to attend when faced with a myriad of stimuli? This decision can be a matter of survival for an animal. Attention can be either goal-driven or stimulus-driven [73]. When the perceiver seeks to find something, e.g. when it looks for a pen in an office, its attention is goal-driven (also called active attention). In contrast, when the perceiver's attention is captured by some salient stimulus in the input scene, such as a red pen on a gray desk, attention is said to be stimulus-driven (or passive attention). It is important to realize that these two modes of attention likely occur simultaneously in a given act of attention. Goal-driven attention is controlled by the perceiver's intentions at the moment. Stimulus-driven attention, on the other hand, can be studied by varying stimulus properties of the input scene. Perceptual studies [73, 42] suggest that stimuli that differ from the rest of the scene in one or more feature dimensions, e.g. color, depth, and motion for vision, tend to capture attention. In other words, salient objects draw attention. The saliency of a stimulus has two aspects. The first is the difference between the stimulus and its surround and the second is the homogeneity of the surround [16]; a stimulus is highly salient when it is different from its surrounding stimuli that are similar to each other. Visual feature dimensions include luminance, color, orientation, motion, and depth. Auditory feature dimensions include loudness, pitch, temporal modulation, and location. In addition to feature saliency, abrupt changes to the scene tend to capture attention [73], including the onset of a new stimulus in the scene and the abrupt change in a feature dimension of an existing stimulus. In other words, novel objects draw attention.

7.2 Feature-based Analysis versus Model-based Analysis

Scene analysis can be performed on the basis of the features of the objects in the input scene or the models of the objects in the memory. Feature-based versus model-based analysis is also metaphorically characterized as bottom-up versus top-down analysis. Familiarity has been acknowledged as an organizing principle in scene analysis so the issue is not whether memorized objects influence scene analysis. What's at issue is how much model-based analysis contributes to scene analysis, or whether binding should be part of a recognition process.

According to some, binding is a byproduct of recognition, which is typically coupled with some selection mechanism that brings the pattern of interest into focus, and there is really no binding problem so to speak [44]. For example, Fukushima and Imagawa [20] proposed a model that performs recognition and segmentation simultaneously by employing a search controller that selects a small area of the input image for processing. Their model is based on Fukushima's neocognitron model for pattern recognition, which is a hierarchical multilayer network. The neocognitron model is a prominent example of the hierarchical coding approach to the binding problem. The model contains a cascade of layers with both forward and backward connections. The forward path performs pattern recognition that is robust to a range of variations in position and size, and the last layer stores learned patterns. When a scene of multiple patterns is presented, a rough area selection is performed based on feature density of the input, and further competition in the last layer leads to a winner. The winning unit of the last layer, through backward connections, reinforces the pattern of the input image that is consistent with the stored template. This, in a way, segments that part of the input image from its background. After a while, the network switches to another area of high feature density and continues the analysis process. Their model has been evaluated on binary images of connected characters. Olshausen et al. [39] proposed a model that also combines pattern recognition and a model of selective attention. Their attention model is implemented by a shifting circuit that routes information in a hierarchical network while preserving spatial relations between visual features, and recognition is based on a Hopfield model of associative memory. The location and size of an attention blob are determined by competition in a feature saliency map, producing potential regions of interest on an image. This model is highlighted by Shadlen and Movshon [50] as an alternative to the temporal correlation theory. The model is evaluated on binary images with well-separated patterns. A later model along a similar line was proposed by Riesenhuber and Poggio [44], and it uses a hierarchical architecture similar to the neocognitron. Their model has been tested on two-object scenes: One is a stored pattern and another is a distractor.

In my opinion, model-based analysis has clear limits. Perceiving an object, e.g. a frog, with all its vivid details such as location, shape, color, orientation, and size, is more than simply recognizing that the object is a frog [24, 56]. Indeed, if feature analysis played no role in scene analysis, camouflage would not have emerged from animal evolution as a universal strategy of blending with the environment. This point is illustrated in Figure 9 which shows two frogs in a pond. It is effortful to spot a frog in its natural habitat even for an experienced observer. Also, perception operates on both familiar and unfamiliar objects, and model-based analysis is not applicable to the latter objects. Besides these and other conceptual difficulties with the hierarchical coding discussed in Section 3, it is unclear how the above model-based systems can be extended to analyze scenes where complex objects are arranged in arbitrary

Fig. 9. A natural image that contains two frogs in their natural habitat

ways. As mentioned in Sect. 7.1, the number of possible snake patterns (see Fig. 8) seems too large to search in a top-down manner.

7.3 Learning versus Representation

Learning - both supervised and unsupervised - is central to neural networks (and computational intelligence in general). The development of neural networks can be characterized, to a large extent, by the development of learning algorithms. Nowadays, much activity in neural networks is popularly called machine learning. There is also increasing interest in the research community to apply machine learning techniques to scene analysis. Some even argue that data-driven learning can do away with the need to search for appropriate representations for computational scene analysis.

While certain regularities of input data can be discovered by a learning system, the applicability of learning-based approaches to computational scene analysis is likely bounded. As pointed out by Konen and von der Malsburg [28], such approaches tend to be plagued by the problem of combinatorial explosion when dealing with realistically complex scenes. It is perhaps revealing to consider the connectedness problem in this context. The failure of perceptrons to solve this problem is rooted in the lack of a proper representation, not the lack of a powerful learning method. According to Minsky and Papert [34], "no machine can learn to recognize X unless it possesses, at least potentially, some scheme for representing X." (p. xiii). Indeed, modern multilayer perceptrons with the powerful backpropagation algorithm fare no better on the connectedness problem [65]. No learning machine, to my knowledge, has succeeded in solving this problem. The cause, as discussed in Sect. 4, is computational complexity - learning the connectedness predicate would require far too many training samples and too much learning time. The success of LEGION stems from the oscillatory correlation representation and the network structure.

The brain of a newborn possesses genetic knowledge resulting from millions of years of evolution. It is relevant to note that success is quite limited in teaching chimpanzees, the closest relatives of humans, basic language or arithmetic despite considerable effort by animal behaviorists. While in theory all is learnable, including genetic knowledge, evolution operates at much greater time scales. Even if evolutionary computing someday succeeded in uncovering the computational principles of evolution, the challenge would be insurmountable of simulating billions of years of environmental change that resulted in the flourishing of life on the earth. Furthermore, even if major evolutionary processes were totally reproducible on a computer there would still be no assurance that the result will a human rather than an amoeba.

Computational complexity should be of principal concern in computational intelligence. The essence of intelligence is the efficiency of information processing. Although stunning progress in computer speed and memory has enabled the execution of very complex algorithms, we should keep in mind that a slower algorithm will always be slower no matter how fast the computer is.

For those who are concerned with biological plausibility, the speed of human scene analysis has strong implications on the kind of processing employed. For visual analysis, it has been empirically shown that object identification in a visual scene takes less than 150 ms [4, 54]. Interestingly, the visual system categorizes novel scenes just as fast as highly familiar ones [17]. After estimation of processing time at various stages of the visual pathway, Thorpe and Fabre-Thorpe [53] conclude that such analysis must be based primarily on feedforward processing as there is little time left for iterative feedback. Coincidentally, a comprehensive analysis on noise robustness, time course, and language context led Allen [1] to essentially the same conclusion, that is, human speech analysis is primarily a bottom-up process. These results challenge the biological validity of the contemporary emphasis on statistical, model-based approaches [26] that typically boil down to a time-consuming search in a large probability space.

A major distinction between perception and reasoning is that the process of perception is rapid and automatic, whereas the process of reasoning is consciously deliberative and generally slow. Even when faced with certain ambiguous figures that permit multiple interpretations, such as the famous Necker cube shown in Fig. 10, perception seems to quickly dwell on one of the plausible interpretations and would require a slow, conscious effort to switch to a competing interpretation. This is not to say that perception never involves conscious deliberations. We do, on occasion, debate in our head what to make of an input scene. But such an experience is more of the exception rather

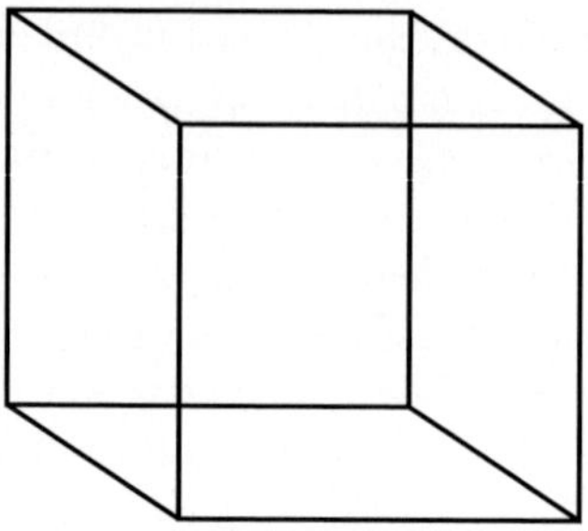

Fig. 10. Necker cube. This figure can be seen as a cube that is viewed either from above or from below

than the rule. From an ecological point of view, perception needs to figure out what is out there quickly as the scene changes constantly due to both environmental change and locomotion. The speed of perception is critical to the survival of an organism.

The emphasis on representations contrasts that on learning. A representation is a formal system that encodes certain types of information. Marr's pioneering study on computational vision exemplifies representational approaches to scene analysis [30]. In my view, the Marrian framework for computational perception provides the most promising roadmap for understanding scene analysis. Appropriate representations embody key insights and constraints in an information processing domain. How information is represented can profoundly affect how information is processed. For instance, the cepstral representation[3] separates voice excitation from vocal tract filtering [22], and the discovery of this representation pays huge dividends to speech processing tasks including automatic speech recognition where cepstral features are an indispensable part of any state-of-the-art system.

Since a good representation often captures the current state of scientific understanding on human perception, it does not seem to make sense to let a computer program 'discover' it through machine learning. For example, cochlear processing of the acoustic information is well understood and amounts to an elaborate filterbank. Why not codify such understanding, which is nontrivial to figure out from scratch, in a representation of auditory periphery?

The above discussion makes it plain that the investigation of computational scene analysis can be characterized in large part as the pursuit of appropriate representations. This vision implies that the research in computational scene analysis is an interdisciplinary enterprise, as psychophysics as well as cognitive neuroscience contributes to uncovering effective representations.

So what is the role of learning in computational scene analysis? A representation provides a framework, a skeleton, but it is by no means all that is needed to solve the computational scene analysis problem. Learning plays

[3] A cepstrum corresponds to the logarithm of the magnitude spectrum.

an important role within a representational framework to adjust parameter values and precisely model the distribution of input data in relation to the system. A recent study by Roman et al. [45] offers an example in this regard. Their binaural system for speech segregation builds on an auditory peripheral model and the notion of binary time-frequency masks for speech separation, and computes the binaural cues of interaural time difference and interaural intensity difference which are known to be employed by the auditory system. However, their use of supervised training is responsible for high-quality estimates of binary masks, which in turn lead to good segregation results. In other words, effective learning can substantially enhance system performance.

8 Concluding Remarks

In this chapter I have made an effort to define the goal of computational scene analysis explicitly. The challenges facing Rosenblatt's perceptrons are fundamentally related to the binding problem. Temporal correlation provides a biologically plausible framework to address the binding problem. Advances in understanding oscillatory dynamics lead to the development of the oscillatory correlation approach to computational scene analysis with promising results.

Perhaps the most important characteristic of natural intelligence compared to artificial intelligence is its versatility. Natural intelligence ranges from sensation, perceptual organization, language, motor control, to decision making and long-term planning. The substrate for all these functions is a brain - an immense network of neurons - whose structure is largely fixed after development. I have argued recently that time provides a necessary dimension to the versatility of brain function [65]. Temporal structure is shared by neuronal responses in all parts of the brain, and the time dimension is flexible and infinitely extensible.

Computational scene analysis is an extremely challenging problem. The bewildering complexity of perception makes it necessary to adopt a compass to guide the way forward and avoid many pitfalls along the way. I strongly recommend Marr's posthumous book to anyone who is attempted by the challenge.

Acknowledgments

I thank G. Hu, S. Srinivasan, and Y. Li for their assistance in formatting and figure preparation. This research was supported in part by an AFOSR grant (FA9550-04-1-0117) and an AFRL grant (FA8750-04-1-0093).

References

[1] Allen JB (2005) Articulation and intelligibility. Morgan & Claypool
[2] Arbib MA ed (2003) Handbook of brain theory and neural networks. 2nd ed, MIT Press, Cambridge MA
[3] Barlow HB (1972) Single units and cognition: A neurone doctrine for perceptual psychology. Percept 1:371-394
[4] Biederman I (1987) Recognition-by-component: A theory of human image understanding. Psychol Rev 94:115-147
[5] Black MJ, Anandan P (1996) The robust estimation of multiple motions: parametric and piecewise-smooth flow fields. CVGIP: Image Understanding 63:75-104
[6] Bregman AS (1990) Auditory scene analysis. MIT Press, Cambridge MA
[7] Campbell SR, Wang DL, Jayaprakash C (1999) Synchrony and desynchrony in integrate-and-fire oscillators. Neural Comp 11:1595-1619
[8] Cesmeli E, Wang DL (2000) Motion segmentation based on motion/brightness integration and oscillatory correlation. IEEE Trans Neural Net 11:935-947
[9] Chang P (2004) Exploration of behavioral, physiological, and computational approaches to auditory scene analysis. MS Thesis, The Ohio State University Department of Computer Science and Engineering (available at http://www.cse.ohio-state.edu/pnl/theses.html)
[10] Chen K, Wang DL, Liu X (2000) Weight adaptation and oscillatory correlation for image segmentation. IEEE Trans Neural Net 11:1106-1123
[11] Cherry EC (1953) Some experiments on the recognition of speech, with one and with two ears. J Acoust Soc Am 25:975-979
[12] Cowan N (2001) The magic number 4 in short-term memory: a reconsideration of mental storage capacity. Behav Brain Sci 24:87-185
[13] Darwin CJ (1997) Auditory grouping. Trends Cogn Sci 1:327-333
[14] Domijan D (2004) Recurrent network with large representational capacity. Neural Comp 16:1917-1942
[15] Driver J, Baylis GC (1998) Attention and visual object recognition. In: Parasuraman R (ed) The attentive brain. MIT Press Cambridge MA, pp. 299-326
[16] Duncan J, Humphreys GW (1989) Visual search and stimulus similarity. Psychol Rev, 96:433-458
[17] Fabre-Thorpe M, Delorme A, Marlot C, Thorpe S (2001) A limit to the speed of processing in ultra-rapid visual categorization of novel natural scenes. J Cog Neurosci 13:1-10
[18] Field DJ, Hayes A, Hess RF (1993) Contour integration by the human visual system: Evidence for a local "association field". Vis Res 33:173-193
[19] FitzHugh R (1961) Impulses and physiological states in models of nerve membrane. Biophys J 1:445-466
[20] Fukushima K, Imagawa T (1993) Recognition and segmentation of connected characters with selective attention. Neural Net 6:33-41

[21] Gibson JJ (1966) The senses considered as perceptual systems. Greenwood Press, Westport CT
[22] Gold B, Morgan N (2000) Speech and audio signal processing. Wiley & Sons, New York
[23] Gray CM (1999) The temporal correlation hypothesis of visual feature integration: still alive and well. Neuron 24:31-47
[24] Kahneman D, Treisman A, Gibbs B (1992) The reviewing of object files: object-specific integration of information. Cognit Psychol 24:175-219
[25] Kareev Y (1995) Through a narrow window: Working memory capacity and the detection of covariation. Cognition 56:263-269
[26] Knill DC, Richards W eds (1996) Perception as Bayesian inference. Cambridge University Press, New York
[27] Koffka K (1935) Principles of Gestalt psychology. Harcourt, New York
[28] Konen W, von der Malsburg C (1993) Learning to generalize from single examples in the dynamic link architecture. Neural Comp 5:719-735
[29] MacGregor JN (1987) Short-term memory capacity: Limitation or optimization? Psychol Rev 94:107-108
[30] Marr D (1982) Vision. Freeman, New York
[31] Mattingley JB, Davis G, Driver J (1997) Preattentive filling-in of visual surfaces in parietal extinction. Science 275:671-674
[32] Milner, PM (1974) A model for visual shape recognition. Psychol Rev 81(6):521-535
[33] Minsky ML, Papert SA (1969) Perceptrons. MIT Press, Cambridge MA
[34] Minsky ML, Papert SA (1988) Perceptrons (Expanded ed). MIT Press, Cambridge MA
[35] Morris C, Lecar H (1981) Voltage oscillations in the barnacle giant muscle fiber. Biophys J 35:193-213
[36] Nagumo J, Arimoto S, Yoshizawa S (1962) An active pulse transmission line simulating nerve axon. Proc IRE 50:2061-2070
[37] Nakayama K, He ZJ, Shimojo S (1995) Visual surface representation: A critical link between lower-level and higher-level vision. In: Kosslyn SM, Osherson DN (eds) An invitation to cognitive science. MIT Press, Cambridge MA, pp. 1-70
[38] Norris M (2003) Assessment and extension of Wang's oscillatory model of auditory stream segregation. PhD Dissertation, University of Queensland School of Information Technology and Electrical Engineering
[39] Olshausen BA, Anderson CH, Van Essen DC (1993) A neurobiological model of visual attention and invariant pattern recognition based on dynamic routing of information. J Neurosci 13:4700-4719
[40] Palmer SE (1999) Vision science. MIT Press, Cambridge MA
[41] Parasuraman R ed (1998) The attentive brain. MIT Press, Cambridge MA
[42] Pashler HE (1998) The psychology of attention. MIT Press, Cambridge MA

[43] Reynolds JH, Desimone R (1999) The role of neural mechanisms of attention in solving the binding problem. Neuron 24:19-29
[44] Riesenhuber M, Poggio T (1999) Are cortical models really bound by the "binding problem"? Neuron 24:87-93
[45] Roman N, Wang DL, Brown GJ (2003) Speech segregation based on sound localization. J Acoust Soc Am 114:2236-2252
[46] Rosenblatt F (1958) The perceptron: a probabilistic model for information storage and organization in the brain. Psychol Rev 65:386-408
[47] Rosenblatt F (1962) Principles of neural dynamics. Spartan, New York
[48] Rumelhart DE, McClelland JL eds (1986) Parallel distributed processing 1: Foundations. MIT Press, Cambridge MA
[49] Russell S, Norvig P (2003) Artificial intelligence: A modern approach. 2nd ed Prentice Hall, Upper Saddle River, NJ
[50] Shadlen MN, Movshon JA (1999) Synchrony unbound: a critical evaluation of the temporal binding hypothesis. Neuron 24:67-77.
[51] Somers D, Kopell N (1993) Rapid synchrony through fast threshold modulation. Biol Cybern, 68:393-407
[52] Terman D, Wang DL (1995) Global competition and local cooperation in a network of neural oscillators, Physica D 81:148-176
[53] Thorpe S, Fabre-Thorpe M (2003) Fast visual processing. In: Arbib MA (ed) Handbook of Brain Theory and Neural Networks. MIT Press, Cambridge MA, pp. 441-444
[54] Thorpe S, Fize D, Marlot C (1996) Speed of processing in the human visual system. Nature 381:520-522
[55] Treisman A (1986) Features and objects in visual processing. Sci Am, November, Reprinted in The perceptual world, Rock I (ed). Freeman and Company, New York, pp. 97-110
[56] Treisman A (1999) Solutions to the binding problem: progress through controversy and convergence. Neuron 24:105-110
[57] Treisman A, Gelade G (1980) A feature-integration theory of attention. Cognit Psychol 12:97-136
[58] van der Pol B (1926) On "relaxation oscillations". Phil Mag 2(11):978-992
[59] von der Malsburg C (1981) The correlation theory of brain function. Internal Report 81-2, Max-Planck-Institute for Biophysical Chemistry, Reprinted in Models of neural networks II, Domany E, van Hemmen JL, Schulten K, eds (1994) Springer, Berlin
[60] von der Malsburg C (1999) The what and why of binding: the modeler's perspective. Neuron 24:95-104
[61] von der Malsburg C, Schneider W (1986) A neural cocktail-party processor. Biol Cybern 54:29-40
[62] Wang DL (1995) Emergent synchrony in locally coupled neural oscillators. IEEE Trans Neural Net 6(4):941-948
[63] Wang DL (1996) Primitive auditory segregation based on oscillatory correlation. Cognit Sci 20:409-456

[64] Wang DL (2000) On connectedness: a solution based on oscillatory correlation. Neural Comp 12:131-139
[65] Wang DL (2005) The time dimension for scene analysis. IEEE Trans Neural Net 16:1401-1426
[66] Wang DL, Brown GJ (1999) Separation of speech from interfering sounds based on oscillatory correlation. IEEE Trans Neural Net 10:684-697
[67] Wang DL, Kristjansson A, Nakayama K (2005) Efficient visual search without top-down or bottom-up guidance. Percept Psychophys 67:239-253
[68] Wang DL, Terman D (1995) Locally excitatory globally inhibitory oscillator networks. IEEE Trans Neural Net 6(1):283-286
[69] Wang DL, Terman D (1997) Image segmentation based on oscillatory correlation. Neural Comp 9:805-836 (for errata see Neural Comp 9:1623-1626)
[70] Wersing H, Steil JJ, Ritter H (2001) A competitive-layer model for feature binding and sensory segmentation. Neural Comp 13:357-388
[71] Wertheimer M (1923) Untersuchungen zur Lehre von der Gestalt, II. Psychol Forsch 4:301-350
[72] Wrigley SN, Brown GJ (2004) A computational model of auditory selective attention. IEEE Trans Neural Net 15:1151-1163
[73] Yantis S (1998) Control of visual attention. In: Pashler H (ed) Attention. Psychology Press, London, pp. 223-256
[74] Yen SC, Finkel LH (1998) Extraction of perceptually salient contours by striate cortical networks. Vis Res 38:719-741
[75] Zhang X, Minai AA (2004) Temporally sequenced intelligent block-matching and motion-segmentation using locally coupled networks. IEEE Trans Neural Net 15:1202-1214
[76] Zhao L, Macau EEN (2001) A network of dynamically coupled chaotic maps for scene segmentation. IEEE Trans Neural Net 12:1375-1385

Brain-, Gene-, and Quantum Inspired Computational Intelligence: Challenges and Opportunities

Nikola Kasabov

Knowledge Engineering and Discovery Research Institute, KEDRI Auckland University of Technology, Auckland, New Zealand
nkasabov@aut.ac.nz
www.kedri.info

Summary. This chapter discusses opportunities and challenges for the creation of methods of computational intelligence (CI) and more specifically – artificial neural networks (ANN), inspired by principles at different levels of information processing in the brain: cognitive-, neuronal-, genetic-, and quantum, and mainly, the issues related to the integration of these principles into more powerful and accurate CI methods. It is demonstrated how some of these methods can be applied to model biological processes and to improve our understanding in the subject area, along with other – being generic CI methods applicable to challenging generic AI problems. The chapter first offers a brief presentation of some principles of information processing at different levels of the brain, and then presents brain-inspired, gene-inspired and quantum inspired CI. The main contribution of the chapter though is the introduction of methods inspired by the integration of principles from several levels of information processing, namely: (1) a computational neurogenetic model, that combines in one model gene information related to spiking neuronal activities; (2) a general framework of a quantum spiking neural network model; (3) a general framework of a quantum computational neuro-genetic model. Many open questions and challenges are discussed, along with directions for further research.

Key words: Artificial neural networks, Computational Intelligence, Neuro-informatics, Bionformatics, Evolving connectionist systems, Gene regulatory networks, Computational neurogenetic modeling, Quantum information processing.

1 Introduction

The brain is a dynamic information processing system that evolves its structure and functionality in time through information processing at different levels – Fig. 1: quantum-, molecular (genetic)-, single neuron-, ensemble of neurons-, cognitive-, evolutionary.

Nikola Kasabov: *Brain-, Gene-, and Quantum Inspired Computational Intelligence: Challenges and Opportunities*, Studies in Computational Intelligence (SCI) **63**, 193–219 (2007)
www.springerlink.com

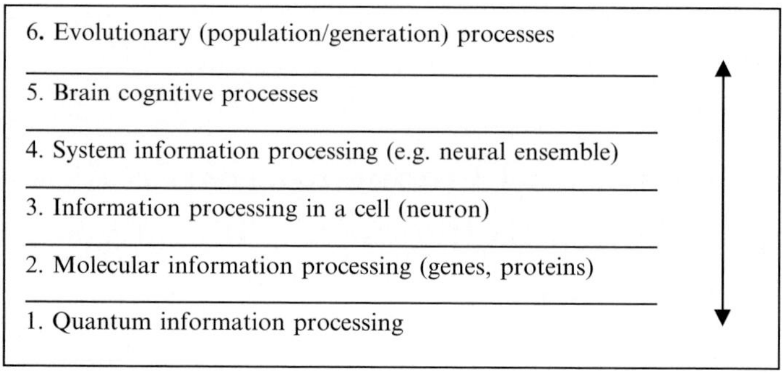

Fig. 1. Levels of information processing in the brain and the interaction between the levels

Principles from each of these levels have been already used as inspiration for computational intelligence (CI) methods, and more specifically – for methods of artificial neural networks (ANN). The chapter focuses on the interaction between these levels and mainly – on how this interaction can be modeled, and how it can be used in principle to improve existing CI methods and for a better understanding of brain-, gene-, and quantum processes.

At the quantum level, particles (atoms, ions, electrons, etc.), that make every molecule in the material world, are moving continuously, being in several states at the same time that are characterised by probability, phase, frequency, energy.

At a molecular level, RNA and protein molecules evolve in a cell and interact in a continuous way, based on the stored information in the DNA and on external factors, and affect the functioning of a cell (neuron) under certain conditions.

At the level of a neuron, the internal information processes and the external stimuli cause the neuron to produce a signal that carries information to be transferred to other neurons.

At the level of neuronal ensembles, all neurons operate in a "concert", defining the function of the ensemble, for instance perception of a spoken word.

At the level of the whole brain, cognitive processes take place, such as language and reasoning, and global information processes are manifested, such as consciousness.

At the level of a population of individuals, species evolve through evolution, changing the genetic DNA code for a better adaptation.

The information processes at each level from Fig. 1 are very complex and difficult to understand, but much more difficult to understand is the interaction between the different levels. It may be that understanding the interaction through its modeling would be a key to understanding each level of information processing in the brain and perhaps the brain as a whole. Using principles from different levels in one ANN CI model and modeling their relationship can lead to a next generation of ANN as more powerful tools to understand the brain and to solve complex problems.

Some examples of CI models that combine principles from different levels shown in fig. 1 are: computational neuro-genetic models [45, 4, 40]; quantum inspired CI and ANN [15, 47]; evolutionary models [17, 74]. Suggestions are made that modeling higher cognitive functions and consciousness in particular can be achieved if principles from quantum information processing are considered [48, 49]. There are many issues and open questions to be addressed when creating CI methods that integrate principles from different levels, some of them presented in this chapter.

The chapter discusses briefly in section 2 models inspired by information processes in the brain, that include local learning evolving connectionist systems (ECOS) and spiking neural networks (SNN). Section 3 presents CI methods inspired by genetic information processes, mainly models of gene regulatory networks (GRN). In section 4, the issue of combining neuronal with genetic information processing is discussed and the principles of computational neuro-genetic modeling (CNGM) are presented. Section 5 presents some ideas behind the quantum inspired CI. Section 6 presents a model of a quantum inspired SNN and offers a theoretical framework for the integration of principles from quantum-, -genetic- and neuronal information processing. Section 7 concludes the chapter with more open questions and challenges for the future.

2 CI and ANN Models Inspired by Neuronal and Cognitive Processes in the Brain

Many CI methods, in particular ANN, are brain-inspired (using some principles from the brain), or brain-like (more biologically plausible models, usually developed to model a brain function) [1, 2, 3, 4, 5, 8, 9]. Examples are: models of single neurons and neural network ensembles [41, 56, 57, 58, 59, 60, 34]; cognitive ANN models [23, 8, 9, 64] etc.

These models have been created with the goals of:

- Modeling and understanding brain functions;
- Creating powerful methods and systems of computational intelligence (CI) for solving complex problems in all areas of science and the humanity.

In this section we present only two groups of models, namely evolving connectionist systems (ECOS) and spiking neural networks (SNN) as they will be used in other sections to create models that incorporate principles from other levels of information processing.

2.1 Local, Knowledge–based Learning Evolving Connectionist Systems (ECOS) – Weakly Brain Inspired Models

ECOS are adaptive, incremental learning and knowledge representation systems, that evolve their structure and functionality, where in the core of a system is a connectionist architecture that consists of neurons (information processing units) and connections between them [35]. An ECOS is a CI system based on neural networks, but using other techniques of CI, that operates continuously in time and adapts its structure and functionality through a continuous interaction with the environment and with other systems. The adaptation is defined through:

1. A set of evolving rules.
2. A set of parameters ("genes") that are subject to change during the system operation.
3. An incoming continuous flow of information, possibly with unknown distribution.
4. Goal (rationale) criteria (also subject to modification) that are applied to optimize the performance of the system over time.

ECOS learning algorithms are inspired by brain-like information processing principles, e.g.:

1. They evolve in an open space, where the dimensions of the space can change.
2. They learn via incremental learning, possibly in an on-line mode.
3. They learn continuously in a lifelong learning mode.
4. They learn both as individual systems and as an evolutionary population of such systems.
5. They use constructive learning and have evolving structures.
6. They learn and partition the problem space locally, thus allowing for a fast adaptation and tracing the evolving processes over time.
7. They evolve different types of knowledge representation from data, mostly a combination of memory-based and symbolic knowledge.

Many ECOS have been suggested so far, where the structure and the functionality of the models evolve through incremental, continuous learning from incoming data, some times in an on-line mode, and through interaction with other models and the environment. Examples are: growing SOMs [41], growing gas [20], RAN [51], growing RBF networks[18, 52], FuzzyARTMAP [8], EFuNN [35, 36, 37], DENFIS [38] and many more.

A block diagram of EFuNN is given in fig. 2. It is used to model gene regulatory networks in section 5. At any time of the EFuNN continuous incremental learning, rules can be derived from the structure, which rules represent clusters of data and local functions associated with these clusters:

$$\begin{aligned}&\text{IF} < \text{data is in cluster Ncj, defined by a cluster center Nj, a cluster radius}\\&\text{Rj and a number of examples Njex in this cluster} > \\&\text{THEN} < \text{the output function is Fc} >\end{aligned} \tag{1}$$

In case of DENFIS, first order local fuzzy rule models are derived incrementally from data, for example:

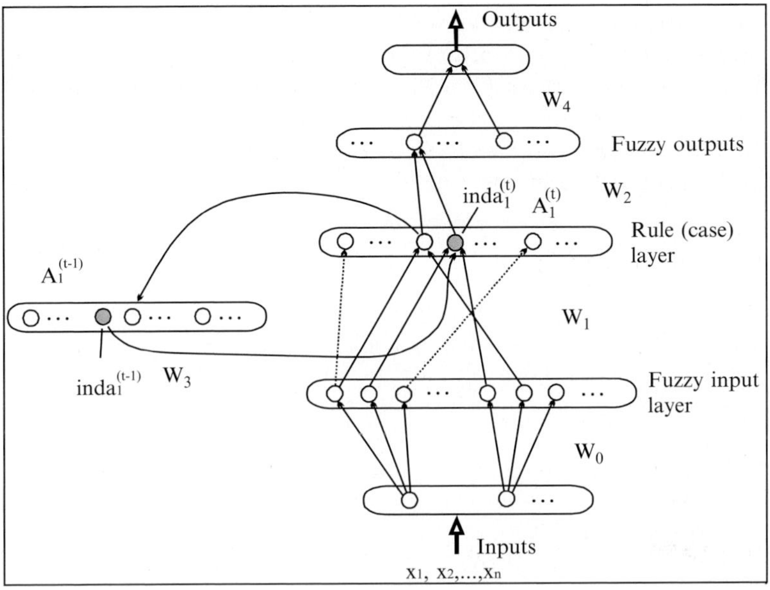

Fig. 2. An EFuNN architecture with a short term memory and feedback connections [36]. It is used in section 5 to model GRN with inputs being the expression of genes at a time (t) and the outputs being the expression of genes/proteins at time (t + dt)

IF $<$ the value of x1 is in the area defined by a Gaussian membership function with a center at 0.1 and a standard deviation of 0.05, AND $<$ the value of x2 is in the area defined by a Gaussian function with parameters $(0.25, 0.1)$ respectively $>$
THEN $<$ the output y is calculated by the formula:
$$y = 0.01 + 0.7x1 + 0.12x2 > \tag{2}$$

In case of EFuNN, local simple fuzzy rule models are derived, for example:

$$\begin{aligned}&\text{IF x1 is (Medium 0.8) and x2 is (Low 0.6)}\\&\text{THEN y is (High 0.7), radius R} = 0.24;\ \text{Nexamp} = 6,\end{aligned} \tag{3}$$

where: Low, Medium and High are fuzzy membership functions defined for the range of each of the variables x1, x2, and y; the number and the type of the membership functions can either be deduced from the data through learning algorithms, or it can be predefined based on human knowledge [75, 73]; R is the radius of the cluster and Nexamp is the number of examples in the cluster.

Further development of the EFuNN and the DENFIS local ECOS models is the Transductive Weighted Neuro-Fuzzy Inference Engine (TWNFI) [62, 37]. In this approach, for every new vector (sample/example S) a "personalized" model is developed from existing nearest samples, where each of the variables is normalised in a different sub-range of [0,1] so that they have a different influence on the Euclidean distance from (1), therefore they are ranked in terms of their importance to the output calculated for any new sample individually. Samples are also weighted in the model based on their distance to the new sample, where in the Euclidean distance formula variables are also weighted. Each personalized model can be represented as a rule (or a set of rules) that represents the personalized profile for the new input vector. The TWNFI model is evolving as new data samples, added to a data set, can be used in any further personalized model development. That includes using different sets of variables, features [62, 37].

ECOS have been applied to model brain functions and as general CI tools [37]. In one application, an ECOS was trained to classify EEG data, measured from a single person's brain, into four classes representing four perceptual states – hearing, seeing, both, and nothing [37]. In another application, ECOS were used to model emerging acoustic clusters, when multiple spoken languages are learned [37].

ECOS have been applied to a wide range of CI applications, such as adaptive classification of gene expression data, adaptive robot control, adaptive financial data modeling, adaptive environmental and social data modeling [37].

ECOS are used in section 3 for building GRN models.

2.2 Spiking Neural Networks – Strongly Brain-Inspired Models

Spiking models of a neuron and of neural networks – spiking neural networks (SNN), have been inspired and developed to mimic more biologically the spiking activity of neurons in the brain when processing information.

One model – the spike response model (SRM) of a neuron [44, 21] is described below and extended in section 4 to a computational neuro-genetic model (CNGM).

A neuron i receives input spikes from pre-synaptic neurons $j \in \Gamma i$, where Γi is a pool of all neurons pre-synaptic to neuron i. The state of the neuron i is described by the state variable $u_i(t)$ that can be interpreted as a total postsynaptic potential (PSP) at the membrane of soma – fig. 3. When $u_i(t)$

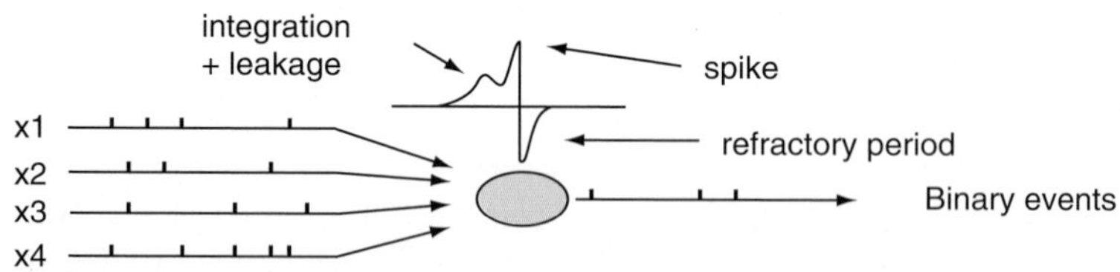

Fig. 3. A general representation of a spiking neuron model (from [4])

reaches a firing threshold $\vartheta i(t)$, neuron i fires, i.e. emits a spike. The value of the state variable $u_i(t)$ is the sum of all postsynaptic potentials, i.e.

$$u_i(t) = \sum_{j \in \Gamma_i} \sum_{t_j \in F_j} J_{ij} \varepsilon_{ij} \left(t - t_j - \Delta_{ij}^{ax}\right) \tag{4}$$

The weight of synaptic connection from neuron j to neuron i is denoted by J_{ij}. It takes positive (negative) values for excitatory (inhibitory) connections, respectively. Depending on the sign of J_{ij}, a presynaptic spike generated at time t_j increases (or decreases) $u_i(t)$ by an amount $\varepsilon_{ij}\left(t - t_j - \Delta_{ij}^{ax}\right)$. Δ_{ij}^{ax} is an axonal delay between neurons i and j which increases with Euclidean distance between neurons.

The positive kernel $\varepsilon_{ij}\left(t - t_j - \Delta_{ij}^{ax}\right) = \varepsilon_{ij}(s)$ expresses an individual postsynaptic potential (PSP) evoked by a pre-synaptic neuron j on neuron i. A double exponential formula can be used

$$\varepsilon_{ij}^{synapse}(s) = A^{synapse}\left(\exp\left(-\frac{s}{\tau_{decay}^{synapse}}\right) - \exp\left(-\frac{s}{\tau_{rise}^{synapse}}\right)\right) \tag{5}$$

The following notations are used above: $\tau_{decay/rise}^{synapse}$ are time constants of the rise and fall of an individual PSP; A is the PSP's amplitude; *synapse* represents the type of the activity of the synapse from the neuron j to neuron i, that can be measured and modeled separately for a *fast_excitation, fast_inhibition, slow_excitation, and slow_inhibition,* all integrated in formula [4]. These types of PSPs are based on neurobiology [13] and will be the basis for the development of the computational neuro-genetic model in section 4, where thye different synaptic activities are represented as functions of different proteins (neuro-transmitters and neuro-receptors).

External inputs from the input layer are added at each time step, thus incorporating the background noise and/or the background oscillations. Each external input has its own weight $J_{ik}^{ext\_input}$ and amount of signal $\varepsilon_k(t)$, such that:

$$u_i^{ext\_inpu}(t) = J_{ik}^{ext\_input} \varepsilon_{ik}(t) \tag{6}$$

It is optional to add some degree of Gaussian noise to the right hand side of the equation above to obtain a stochastic neuron model instead of a deterministic one.

SNN models can be built with the use of the above spiking neuron model. Spiking neurons within a SNN can be either excitatory or inhibitory. Lateral connections between neurons in a SNN may have weights that decrease in value with distance from neuron i for instance, according to a Gaussian formula while the connections between neurons themselves can be established at random.

SNN can be used to build biologically plausible models of brain functions. Examples are given in [4, 13, 21, 44]. Fig. 4 shows graphically an application

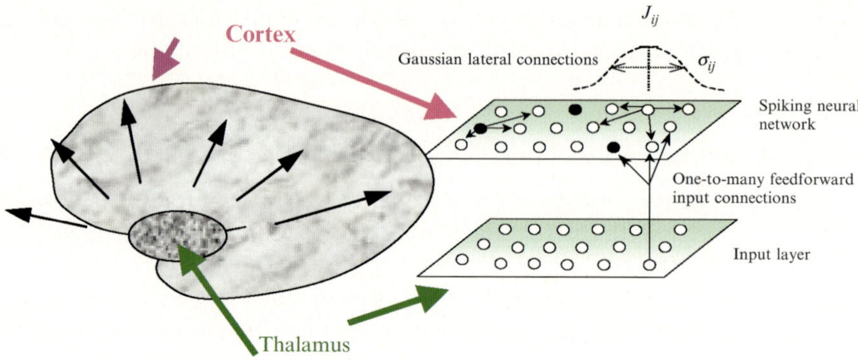

Fig. 4. An example of a SNN to model a function of the cortex with internal inputs from the thalamus and external input stimuli. About 20% of N = 120 neurons are inhibitory neurons that are randomly positioned on the grid (filled circles). External input is random with a defined average frequency (e.g. between 10–20 Hz) (from [4])

of a SNN to model brain functions that connect signals from the thalamus to the temporal cortex (from [4]).

Other applications of SNN include image recognition. In [71] adaptive *SNN* model is developed where new SNN sub-modules (maps) are created incrementally to accommodate new data samples over time. For example, a new sub-module of several spiking neurons and connections is evolved when a new class of objects (e.g. a new face in case of face recognition problem) is presented to the system for learning at any time of this process. When there are no active inputs presented to the system, the system merges close spiking neuronal maps depending on their similarity.

Developing new methods for learning in evolving SNN is a challenging direction for future research with a potential for applications in both computational neuroscience and pattern recognition, e.g. multimodal information processing – speech, image, odor, gestures, etc.

SNN are extended to computational neuro-genetic models (CNGM) in section 4.

2.3 Open Questions

Further development of brain-like or brain-inspired ANN requires some questions to be addressed:

- How much should an ANN mimic the brain in order to become an efficient CI model?
- How is a balance between structure definition and learning achieved in ANN?
- How can ANN evolve and optimize their parameters and input features over time in an efficient way?

- How incremental learning in ANN can be applied without a presentation of an input signal (e.g., "sleep" learning)?

3 Gene-Inspired Methods of Computational Intelligence

3.1 The Central Dogma in Molecular Biology and GRN

The central dogma of molecular biology states that DNA, which resides in the nucleus of a cell or a neuron, transcribes into RNA, and then translates into proteins, which process is continuous, evolving, so that proteins, called transcription factors, cause genes to transcribe, etc. [14, 7] (Fig. 5).

The DNA is a long, double stranded sequence (a double helix) of millions or billions of 4 base molecules (nucleotides) denoted as A, C, T and G, that are chemically and physically connected to each other through other molecules. In the double helix, they make pairs such that every A from one strand is connected to a corresponding T on the opposite strand, and every C is connected to a G. A gene is a sequence of hundreds and thousands of bases as part of the DNA that is translated into protein. Only less than 5% of the DNA of the human genome constitutes genes, the other part is a non-coding region that contains useful information as well.

The DNA of each organism is unique and resides in the nucleus of each of its cells. But what make a cell alive are the proteins that are expressed from the genes and define the function of the cell. The genes and proteins in each cell are connected in a dynamic *gene regulatory network* consisting of regulatory pathways.

Normally, only few hundreds of genes are expressed as proteins in a particular cell. At the transcription phase, one gene is transcribed in many RNA copies and their number defines the expression level of this gene [14, 7] Some

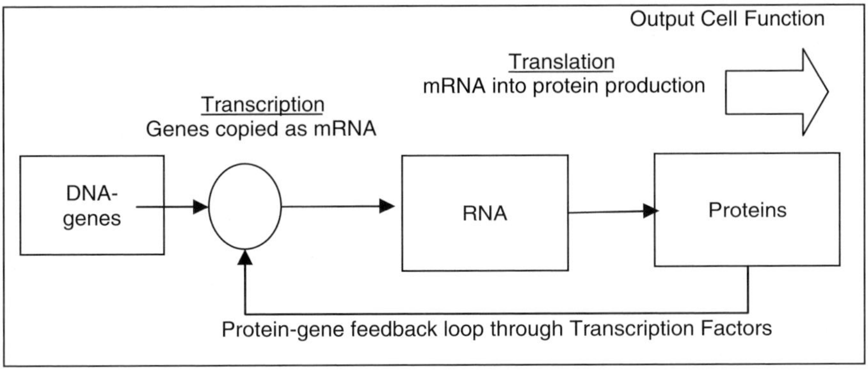

Fig. 5. The genes in the DNA transcribe into RNA and then translate into proteins that define the function of a cell (The central dogma of molecular biology)

genes may be over-expressed, resulting in too much protein in the cell, some genes may be under-expressed resulting in too little protein, in both cases the cell may be functioning in a wrong way that may be causing a disease. Abnormal expression of a gene can be caused by a gene mutation – a random change in the code of the gene, where a base molecule is either inserted, or – deleted, or – altered into another base molecule. Drugs can be used to stimulate or to suppress the expression of certain genes and proteins, but how that will affect indirectly the other genes related to the targeted one, has to be evaluated and that is where computational modeling of gene regulatory networks (GRN) can help.

It is always difficult to establish the interaction between genes and proteins. The question "What will happen with a cell or the whole organism if one gene is under-expressed or missing?" is now being attempted by using a technology called "Knock-out gene technology". This technology is based on a removal of a gene sequence from the DNA and letting the cell/organism to develop, where parameters are measured and compared with these parameters when the gene was not missing.

3.2 GRN-ANN Models

Modeling gene regulatory networks (GRN) is the task of creating a dynamic interaction network between genes that defines the next time expression of genes based on their previous time expression. A detailed discussion of the methods for GRN modeling can be found in [14, 11, 39]. Models of GRN, derived from gene expression RNA data, have been developed with the use of different mathematical and computational methods, such as: statistical correlation techniques; evolutionary computation; ANN; differential equations, both ordinary and partial; Boolean models; kinetic models; state-based models and others [14].

A model of GRN, trained on time-course data is presented in [10] where the human response to fibroblast serum data is used (Fig. 6) and a GRN is extracted from it (Fig. 7). The method uses a genetic algorithm to select the initial cluster centers of the time course clustered gene expression values, and then applies a Kalman filter to derive the gene connecting equations.

In [39] a GRN-ECOS is proposed and applied on a small scale cell line gene expression data. An ECOS is evolved with inputs being the expression level of a certain number of selected genes (e.g. 4) at a time moment (t) and the outputs being the expression level of the same or other genes/proteins at the next time moment (t + dt). After an ECOS is trained on time course gene expression data, rules are extracted from the ECOS and linked between each other in terms of time-arrows of their creation in the model, thus representing the GRN. The rule nodes in an ECOS capture clusters of input genes that are related to the output genes/proteins at next time moment. An example of EFuNN used for modeling GRN is shown in [39, 36].

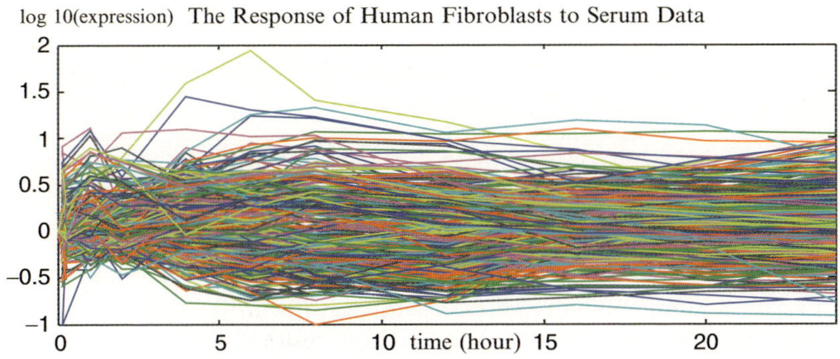

Fig. 6. Time-course gene expression data representing the response of thousands of genes of fibroblast to serum (from [10])

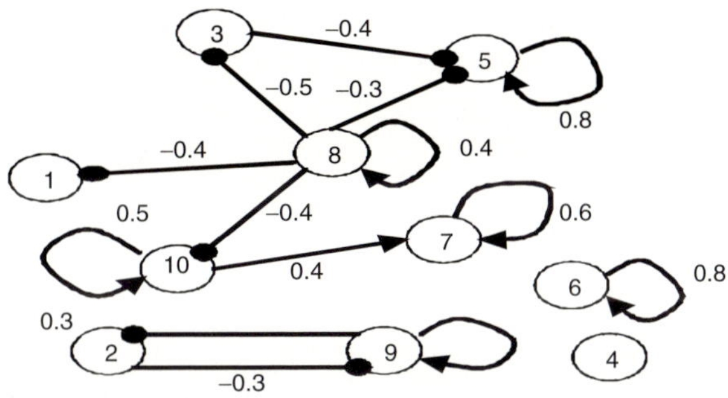

Fig. 7. A GRN obtained with the use of the method from [10] on the data from Fig. 6 after the time gene expression series are clustered into 10 clusters. The nodes represent gene clusters while the arcs represent the dynamic relation (interaction) between these gene groups over consecutive time moments

The extracted rules from an EFuNN model for example represent the relationship between the gene expression of a group of genes G(t) at a time moment t and the expression of the genes at the next time moment G(t + dt), e.g.:

$$\begin{aligned}&\text{IF } g13(t) \text{ is High } (0.87) \text{ and } g23(t) \text{ is Low } (0.9)\\&\text{THEN } g87\ (t+dt) \text{ is High } (0.6) \text{ and } g103(t+dt) \text{ is Low}\end{aligned} \tag{7}$$

Through modifying a threshold for rule extraction one can extract stronger or weaker patterns of dynamic relationship.

Adaptive training of an ECOS makes it possible for incremental learning of a GRN as well as adding new inputs/outputs (new genes) to the GRN.

A set of DENFIS models can be trained, one for each gene g_i, so that an input vector is the expression vector G(t) and the output is a single variable $g_i(t + dt)$. DENFIS allows for a dynamic partitioning of the input space. Takagi-Sugeno fuzzy rules, that represent the relationship between gene g_i

with the rest of the genes, are extracted from each DENFISi model, e.g.:

$$\begin{aligned}&\text{IF g1 is 0.63, 0.70, 0.76) and g2 is (0.71, 0.77, 0.84) and g3 is (0.71,}\\&\quad\text{0.77, 0.84) and g4 is (0.59, 0.66, 0.72)}\\&\text{THEN g5} = 1.84 - 1.26\text{g1} - 1.22\text{g2} + 0.58\text{g3} - 0.03\text{g4}\end{aligned} \tag{8}$$

The ECOS structure from fig. 2 can be used in a multilevel, hierarchical way, where transcription process is represented in one ECOS and translation – in another ECOS which inputs are connected to the outputs of the first one, using feedback connections to represent transcription factors.

Despite of the variety of different methods used so far for modeling GRN and for systems biology in general, there is not a single method that will suit all requirements to model a complex biological system, especially to meet the requirements for adaptation, robustness, information integration.

In the next section GRN modeling is integrated with SNN to model the interaction between genes/proteins in relation to activity of a spiking neuron and a SNN as a whole.

4 Computational Neuro-Genetic Models (CNGM)

4.1 General Notions

With the advancement of molecular and brain research technologies more and more data and information is being made available about the genetic basis of some neuronal functions (see for example: the brain-gene map of mouse http://alleninstitute.org; the brain-gene ontology BGO at http://www.kedri.info).

This information can be utilized to create biologically plausible ANN models of brain functions and diseases that include models of gene interaction. This area integrates knowledge from computer and information science, brain science, molecular genetics and it is called here computational neurogenetic modeling (CNGM) [40].

A CNGM integrates genetic, proteomic and brain activity data and performs data analysis, modeling, prognosis and knowledge extraction that reveals relationship between brain functions and genetic information. Let us look at this process as a process of building mathematical function or a computational algorithm as follows.

A future state of a molecule M′ or a group of molecules (e.g. genes, proteins) depends on its current state M, and on an external signal Em:

$$\mathrm{M}' = \mathrm{Fm}\ (\mathrm{M}, \mathrm{Em}) \tag{9}$$

A future state N′ of a neuron, or an ensemble of neurons, will depend on its current state N and on the state of the molecules M (e.g. genes) and on external signals En:

$$\mathrm{N}' = \mathrm{Fn}\ (\mathrm{N}, \mathrm{M}, \mathrm{En}) \tag{10}$$

And finally, a future neuronal state C′ of the brain will depend on its current state C and also on the neuronal-N, and the molecular-M state and on the external stimuli Ec:

$$\mathrm{C}' = \mathrm{Fc}\ (\mathrm{C}, \mathrm{N}, \mathrm{M}, \mathrm{Ec}) \tag{11}$$

The above set of equations (or algorithms) is a general one and in different cases it can be implemented differently, e.g.: one gene – one neuron/brain function; multiple genes – one neuron/brain function, no interaction between genes; multiple genes – multiple neuron/brain functions, where genes interact in a gene regulatory network (GRN) and neurons also interact in a neural network architecture; multiple genes – complex brain/cognitive function/s, where genes interact within GRN and neurons interact in several hierarchical neural networks.

Several CNGM models have been developed so far varying from modeling a single gene in a biologically realistic ANN model [45], to modeling a set of genes forming an interaction gene regulatory network (GRN) [11, 4]. In the next section we give an example of a CNGM that combines SNN and GRN into one model [4].

4.2 A Computational Neuro-Genetic Model (CNGM) that Integrates GRN within a SNN Model

The main idea behind the model, proposed in [40], is that interaction of genes in neurons affect the dynamics of the whole ANN through neuronal parameters, which are no longer constant, but change as a function of gene/protein expression. Through optimization of the GRN, the initial gene/protein expression values, and the ANN parameters, particular target states of the ANN can be achieved, so that the ANN can be tuned to model real brain data in particular.

This idea is illustrated in fig. 8. The behavior of the SNN is evaluated by means of the local field potential (LFP), thus making it possible to attempt modeling the role of genes in different brain states, where EEG data is available to test the model. A standard FFT signal processing technique is used to evaluate the SNN output and to compare it with real human EEG data. Broader theoretical and biological background of CNGM construction is given in [4].

In general, we consider two sets of genes – a set $\mathrm{G_{gen}}$ that relates to general cell functions, and a set $\mathrm{G_{spec}}$ that defines specific neuronal information-processing functions (receptors, ion channels, etc.). The two sets form together a set $\mathbf{G} = \{G_1, G_2, \ldots, G_n\}$. We assume that the expression level of each gene is a nonlinear function of expression levels of all the genes in $\mathbf{G}$:

$$g_j(t + \Delta t') = \sigma\left(\sum_{k=1}^{n} w_{jk} g_k(t)\right) \tag{12}$$

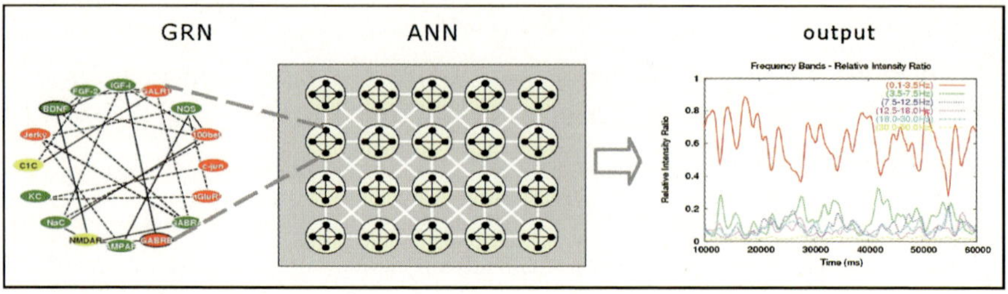

Fig. 8. A CNGM, where a GRN is used to represent the interaction of genes, and a SNN is employed to model a brain function. The model output is compared against real brain data for validation of the model and for verifying the derived gene interaction GRN after model optimization is applied [4]

In [4] it is assumed that: (1) one protein is coded by one gene; (2) relationship between the protein level and the gene expression level is nonlinear; (3) protein levels lie between the minimal and maximal values. Thus, the protein level is expressed by

$$p_j(t + \Delta t) = \left(p_j^{\max} - p_j^{\min}\right) \sigma \left(\sum_{k=1}^{n} w_{jk} g_k(t) \right) + p_j^{\min} \tag{13}$$

The delay constant introduced in the formula corresponds to the delay caused by the gene transcription, mRNA translation into proteins and post-translational protein modifications, and also the delay caused by gene transcription regulation by transcription factors.

Some proteins and genes are known to be affecting the spiking activity of a neuron represented in a SNN model by neuronal parameters, such as *fast_excitation, fast_inhibition, slow_excitation, and slow_*inhibition (see section 2). Some neuronal parameters and their correspondence to particular proteins are summarized in Table 1.

Besides the genes coding for the proteins mentioned above and directly affecting the spiking dynamics of a neuron, a GRN model can include other genes relevant to a problem in hand, e.g. modeling a brain function or a brain disease. In [4] these genes/proteins are: c-jun, mGLuR3, Jerky, BDNF, FGF-2, IGF-I, GALR1, NOS, S100beta [4].

The goal of the CNGM from fig. 8 is to achieve a desired SNN output through optimization of the model parameters. The LFP of the SNN, defined as LFP = $(1/N)\Sigma\ u_i(t)$, by means of FFT is evaluated in order to compare the SNN output with the EEG signal analyzed in the same way. It has been shown that brain LFPs in principle have the same spectral characteristics as EEG [19].

In order to find an optimal GRN within the SNN model, so that the frequency characteristics of the LFP of the SNN model are similar to the brain EEG characteristics, the following evolutionary computation procedure is used:

Table 1. Neuronal Parameters and Related Proteins
ABBREVIATIONS: PSP = POSTSYNAPTIC POTENTIAL, AMPAR = (AMINO-METHYLISOXAZOLE-PROPIONIC ACID) AMPA RECEPTOR, NMDAR = (N-METHYL-D-ASPARTATE ACID) NMDA RECEPTOR, GABRA = (GAMMA-AMINOBUTYRIC ACID) $GABA_A$ RECEPTOR, GABRB = $GABA_B$ RECEPTOR, SCN = SODIUM VOLTAGE-GATED CHANNEL, KCN = KALIUM (POTASSIUM) VOLTAGE-GATED CHANNEL, CLC = CHLORIDE CHANNEL, PV = PARVALBUMIN.

Neuronal parameter Amplitude and time constants of	Protein
Fast excitation PSP	AMPAR
Slow excitation PSP	NMDAR
Fast inhibition PSP	GABRA
Slow inhibition PSP	GABRB
Firing threshold	SCN, KCN, CLC
Late excitatory PSP through GABRA	PV

1. Generate a population of CNGMs, each with randomly, but constrained, generated values of coefficients for the GRN matrix W, initial gene expression values $g(0)$, initial values of SNN parameters $P(0)$, and different connectivity;
2. Run each SNN model over a period of time T and record the LFP;
3. Calculate the spectral characteristics of the LFP using FFT;
4. Compare the spectral characteristics of SNN LFP to the characteristics of the target EEG signal. Evaluate the closeness of the LFP signal for each SNN to the target EEG signal characteristics. Proceed further according to the standard GA algorithm to find a SNN model that matches the EEG spectral characteristics better than previous solutions;
5. Repeat steps 1 to 4 until the desired GRN and SNN model behavior is obtained;
6. Analyze the GRN and the SNN parameters for significant gene patterns that cause the SNN model to manifest similar spectral characteristics as the real data.

The proposed CNGM modeling framework can be used to find patterns of gene regulation related to brain functions. In [4] some preliminary results of analysis performed on real human interictal EEG data are presented. The model performance and the real EEG data are compared for the following relevant to the problem sub-bands: delta (0.5–3.5 Hz), theta (3.5–7.5 Hz), alpha (7.5–12.5 Hz), beta 1 (12.5–18 Hz), beta 2 (18–30 Hz), gamma (above 30 Hz). This particular SNN had an evolved GRN with only 5 genes out of 16 (s100beta, GABRB, GABRA, mGLuR3, c-jun) and all other genes having constant expression values. A GRN is obtained that has a meaningful interpretation and can be used to model what will happen if a gene/protein is suppressed by administering a drug, for example.

In *evolving CNGM* new genes can be added to the GRN model at a certain time, in addition to the new spiking neurons and connections created incrementally, as it is in the *evolving SNN*. Developing new evolving CNGM to model brain functions and brain diseases, such as epilepsy, Alzheimer, Parkinson disease, Schizophrenia, mental retardation and others is a challenging problem for a future research [11, 4].

4.3 Open Questions

There were some questions that emerged from the first CNGM experiments:

- How many different GRNs would lead to similar LFPs and what do they have in common?
- What neuronal parameters to include in an ANN model and how to link them to activities of genes/proteins?
- What genes/proteins to include in the model and how to represent the gene interaction over time within each neuron?
- How to integrate in time the output activity of the ANN and the genes as it is known that neurons spike in millisecond intervals and the process of gene transcription and translation into proteins takes minutes?
- How to create and validate a CNGM in a situation of insufficient data?
- How to measure brain activity and the CNGM activity in order to validate the model?
- What useful information (knowledge) can be derived from a CNG model?
- How to adapt incrementally a CNGM model in a situation of new incoming data about brain functions and genes related to them?

Integrating principles from gene- and neuronal information processing in a single ANN model raises many other, more general, questions that need to be addressed in the future, for example:

- Is it possible to create a truly adequate CNGM of the whole brain? Would gene-brain maps help in this respect (see http://alleninstitute.org)?
- How can dynamic CNGM be used to trace over time and predict the progression of a brain diseases, such as epilepsy and Parkinson's?
- How to use CNGM to model gene mutation effects?
- How to use CNGM to predict drug effects?
- How CNGM can help understand better brain functions, such as memory and learning?
- What problems of CI can be efficiently solved with the use of a brain-gene inspired ANN?

5 Quantum Inspired CI

5.1 Quantum Level of Information Processing

At the quantum level, particles (e.g., atoms, electrons, ions, photons, etc.) are in a complex evolving state all the time. The atoms are the material that everything is made of. They can change their characteristics due to the frequency of external signals. Quantum computation is based upon physical principles from the theory of quantum mechanics [16].

One of the basic principles is the *linear superposition* of states. At a macroscopic or classical level a system exists only in a single basis state as energy, momentum, position, spin and so on. However, at microscopic or quantum level a quantum particle (e.g., atom, electron, positron, ion), or a quantum system, is in a superposition of all possible basis states. At the microscopic level any particle can assume different positions at the same time moment, can have different values of energy, can have different spins, and so on. This *superposition* principle is counterintuitive because in the classical physics one particle has only one position, energy, spin, etc.

If a quantum system interacts in any way with its environment, the superposition is assumed to be destroyed and the system *collapses* into one single real state as in the classical physics (Heisenberg). This process is governed by a probability amplitude. The square of the intensity for the probability amplitude is the quantum probability to observe the state.

Another quantum mechanics principle is the *entanglement* – two or more particles, regardless of their location, are in the same state with the same probability function. The two particles can be viewed as "correlated", undistinguishable, "synchronized", coherent. An example is a laser beam consisting of millions of photons having same characteristics and states.

Quantum systems are described by a probability density ψ that exists in a Hilbert space. The Hilbert space has a set of states $|\phi_i\rangle$ forming a basis. A system can exist in a certain quantum state $|\psi\rangle$ which is defined as

$$|\psi\rangle = \sum c_i|\phi_i\rangle, \ \sum |c_i|^2 = 1 \tag{14}$$

where the coefficients c_i may be complex. $|\psi\rangle$ is said to be in a superposition of the basis states $|\phi_i\rangle$. For example, the quantum inspired analogue of a single *bit* in classical computers can be represented as a *qu-bit* in a quantum computer:

$$|x\rangle = a|0\rangle + b|1\rangle \tag{15}$$

where $|0\rangle$ and $|1\rangle$ represent the states 0 and 1 and a and b their probability amplitudes respectively. The *qu-bit* is not a single value entity, but is a function of parameters which values are complex numbers. After the loss of coherence the *qu-bit* will "collapse" into one of the states $|0\rangle$ or $|1\rangle$ with the probability a^2 for the state $|0\rangle$ and probability b^2 for the state $|1\rangle$.

The state of a quantum particle (represented for example as a *qu-bit)* can be changed by an operator called *quantum gate.* A quantum gate is a reversible gate and can be represented as a unitary operator U acting on the *qu-bit* basis states. The defining property of an unitary matrix is that its conjugate transpose is equal to its inverse. There are several quantum gates already introduced, such as the NOT gate, controlled NOT gate, rotation gate, Hadamard gate, etc. (see [29, 31, 32, 33]).

5.2 Why Quantum Inspired CI?

Quantum mechanical computers and quantum algorithms try to exploit the massive quantum parallelism which is expressed in the principle of *superposition.* The principle of superposition can be applied to many existing methods of CI, where instead of a single state (e.g. a parameter value, or a finite automaton state, or a connection weight, etc.) a superposition of states will be used, described by a wave probability function, so that all these states will be computed in parallel, resulting in an increased speed of computation by many orders of magnitude [6, 24, 25, 29, 30, 31, 32, 33, 47, 48, 49, 50].

Quantum mechanical computers were proposed in the early 1980s and a description was formalized in the late 1980s. These computers, when implemented, are expected to be superior to classical computers in various specialized problems. Many efforts were undertaken to extend the principal ideas of quantum mechanics to other fields of interest. There are well known quantum algorithms such as Shor's quantum factoring algorithm [61] and Grover's database search algorithm [24, 31].

The advantage of quantum computing is that, while a system is *uncollapsed,* it can carry out more computing than a collapsed system, because, in a sense, it is computing in *many universes* at once. The above quantum principles have inspired research in both computational methods and brain study.

New theories (some of them, speculative at this stage) have been already formulated. For example, Penrose [48, 49] argues that solving the quantum measurement problem is pre-requisite for understanding the mind and that consciousness emerges as a macroscopic quantum state due to a coherence of quantum-level events within neurons.

5.3 Quantum Inspired Evolutionary Computation and Connectionist Models

Quantum inspired methods of evolutionary computation (QIEC) and other techniques have been already proposed and discussed in [25, 32] that include: genetic programming [63], particle swarm optimizers [43], finite automata and Turing machines, etc.

In QIEC, a population of n *qu-bit* individuals at time t can be represented as:

$$Q(t) = \{q_1^t, q_2^t, \ldots, q_n^t\} \tag{16}$$

where n is the size of the population.

Evolutionary computing with *qu-bit* representation has a better characteristic of population diversity than other representations, since it can represent linear superposition of states probabilistically. The *qu-bit* representation leads to a quantum parallelism of the system as it is possible to evaluate the fitness function on a superposition of possible inputs. The output obtained is also in the form of superposition which needs to be *collapsed* to get the actual solution.

Recent research activities focus on using quantum principles for ANN [15, 47, 65, 66, 68]. Considering quantum ANN seems to be important for at least two reasons. There is evidence for the role that quantum processes play in the living brain. Roger Penrose argued that a new physics binding quantum phenomena with general relativity can explain such mental abilities as *understanding, awareness* and *consciousness* [49]. The second motivation is the possibility that the field of classical ANN could be generalized to the promising new field of quantum computation [6]. Both considerations suggest a new understanding of mind and brain functions as well as new unprecedented abilities in information processing. Ezhov and Ventura are considering the quantum neural networks as the next natural step in the evolution of neurocomputing systems [15].

Several quantum inspired ANN models have been proposed and illustrated on small examples. In [68] QIEA is used to train a MLP ANN. Narayanan and Meneer simulated classical and quantum inspired ANN and compared their performance [47]. Their work suggests that there are indeed certain types of problems for which quantum neural networks will prove superior to classical ones.

Other relevant work includes quantum decision making, quantum learning models [42], quantum networks for signal recognition [66], quantum associative memory [65, 69]. There are also recent approaches to quantum competitive learning where the quantum system's potential for excellent performance is demonstrated on real-world data sets [70, 72].

6 Towards the Integration of Brain-, Gene-, and Quantum Information Processing Principles: A Conceptual Framework for a Future Research

6.1 Quantum Inspired SNN

In section 4 we presented a computational neuro-genetic model (CNGM) that integrated principles from neuronal information processing and gene information processing in the form of integrating SNN with GRN. Following some ideas from QI-ANN, we can expect that a *QI-SNN and QI-CNGM* would open new possibilities for modelling gene-neuron interactions related to brain functions and to new efficient AI applications.

The CNGM from section 4 linked principles of information processing in gene/protein molecules with neuronal spiking activity, and then – to the information processing of a neuronal ensemble, that is measured as local field potentials (LFP). How the quantum information processes in the atoms and particles (ions, electrons, etc), that make the large gene/protein molecules, relate to the spiking activity of a neuron and to the activity of a neuronal ensemble, is not known yet and it is a challenging question for the future.

What is known at present, is that the spiking activity of a neuron relates to the transmission of ions and neurotransmitter molecules across the synaptic clefts and to the emission of spikes. Spikes, as carriers of information, are electrical signals made of particles that are emitted in one neuron and transmitted along the nerves to many other neurons. These particles are characterised by their quantum properties. So, quantum properties may influence, under certain conditions, the spiking activity of neurons and of the whole brain, as brains obey the laws of quantum mechanics (as everything else in the material world does).

Similarly to a chemical effect of a drug to the protein and gene expression levels in the brain, that may affect the spiking activity and the functioning of the whole brain (modelling of these effects is subject of the computational neurogenetic modeling), external factors like radiation, light, high frequency signals etc. can influence the quantum properties of the particles in the brain through gate operators. According to Penrose [49] microtubules in the neurons are associated with quantum gates, even though what constitutes a quantum gate in the brain is still a highly speculative topic.

So, the question is: *Is it possible to create a SNN model and a CNGM that incorporate some quantum principles?*

A quantum inspired SNN (QI-SNN) can be developed as an extension of the concept of evolving SNN [71] using the superposition principle, where instead of many SNN maps, each representing one object (e.g. a face), there will be a single SNN, where both connections and the neurons are represented as particles, being in many states at the same time defined as probability wave function. When an input vector is presented to the QI-SNN, the network collapses in a single SNN defining the class of the recognised input vector.

6.2 A Conceptual Framework of a QI-CNGM

Here we extend the concept of CNGM (see eq. 9–11) introducing the level of quantum information processing. That results in a conceptual and hypothetical QI-CNGM, we intend to investigate and develop as a future research.

The following is a list of equations that include quantum particle states and functions (hypothetical at this stage) into the equations eq. 9–11 (equations

18–20), starting with a new equation (17) that is concerned only with the level of quantum particle states:

A future state Q' of a particle or a group of particles (e.g. ions, electrons, etc.) depends on the current state Q and on the frequency spectrum Eq of an external signal, according to the Max Planck constant:

$$Q' = \mathrm{Fq}\ (Q, \mathrm{Eq})), \tag{17}$$

A future state of a molecule M' or a group of molecules (e.g. genes, proteins) depends on its current state M, on the quantum state Q of the particles, and on an external signal Em:

$$M' = \mathrm{Fm}\ (Q, M, \mathrm{Em}), \tag{18}$$

A future state N' of a spiking neuron, or an ensemble of neurons will depend on its current state N, on the state of the molecules M, on the state of the particles Q and on external signals En:

$$N' = \mathrm{Fn}\ (N, M, Q, \mathrm{En}), \tag{19}$$

And finally, a future neuronal state C' of the brain will depend on its current state C and also on the neuronal N, on the molecular M, and on the quantum Q states of the brain:

$$C' = \mathrm{Fc}\ (C, N, M, Q, \mathrm{Ec}), \tag{20}$$

The above hypothetical model of integrated function representations is based on the following assumptions:

- A large amount of atoms are characterised by the same quantum properties, possibly related to the same gene/protein expression profile of a large amount of neurons characterised by spiking activity that can be represented as a function.
- A large neuronal ensemble can be represented by a single LFP function.
- A cognitive process can be represented, at an abstract level, as a function Fc that depends on all lower levels of neuronal, genetic and quantum activities.

6.3 Open Questions

Several reasons can be given in support to the research in integrating principles from quantum-, molecular-, and brain information processing into future CI models:

- This may lead to a better understanding of neuronal-,molecular-, and quantum information processes.
- This may lead to new computer devices – exponentially faster and more accurate than the current ones.
- At the nano-level of microelectronic devices, quantum processes would have a significant impact and new methods of computation would be needed anyway.

7 Conclusions and Directions for Further Research

This chapter presents some CI models inspired by principles from different levels of information processing in the brain – including neuronal level, gene/protein level, and quantum level, and argues that CI models that integrate principles from different levels of information processing would be useful tools for a better understanding of brain functions and for the creation of more powerful methods and systems of computational intelligence.

Many open questions need to be answered in the future, some of them are listed below:

- How quantum processes affect the functioning of a living system in general?
- How quantum processes affect cognitive and mental functions?
- Is it true that the brain is a quantum machine – working in a probabilistic space with many states (e.g. thoughts) being in a superposition all the time and only when we formulate our thought through speech or writing, then the brain "collapses" in a single state?
- Is fast pattern recognition in the brain, involving far away segments, a result of both parallel spike transmissions and particle entanglement?
- Is communication between people and between living organisms in general, a result of entanglement processes?
- How does the energy in the atoms relate to the energy of the proteins, the cells and the whole brain?
- Would it be beneficial to develop different quantum inspired (QI) computational intelligence techniques, such as: QI-SVM, QI-GA, QI-decision trees, QI-logistic regression, QI-cellular automata, QI-ALife?
- How do we implement the QI computational intelligence algorithms in order to benefit from their high speed and accuracy? Should we wait for the quantum computers to be realised many years from now, or we can implement them efficiently on specialised computing devices based on classical principles of physics?

Further directions in our research are:

- Building a brain–gene-quantum ontology system that integrates facts, information, knowledge and CI models of different levels of information processing in the brain and their interaction.
- Building novel brain-, gene-, and quantum inspired CI models, studying their characteristics and interpreting the results.
- Applying the new methods to solving complex CI problems in neuroinformatics and brain diseases, bioinformatics and cancer genetics, multimodal information processing and biometrics.

Acknowledgement

The paper is supported partially by a grant AUTX0201 funded by the Foundation of Research Science and Technology of New Zealand and also by the Knowledge Engineering and Discovery Research Institute KEDRI (http://www.kedri.info), Auckland University of Technology.

References

[1] Amari S, Kasabov N (1998) Brain-like Computing and Intelligent Information Systems. Springer Verlag, New York
[2] Arbib M (1987) Brains, Machines and Mathematics. Springer-Verlag, Berlin.
[3] Arbib M (eds) (2003) The Handbook of Brain Theory and Neural Networks, MIT Press, Cambridge, MA.
[4] Benuskova, L. and N. Kasabov (2007) Computational Neurogenetic Modelling, Springer, New York.
[5] Bishop C (1995) Neural networks for pattern recognition. Oxford University Press, Oxford, UK
[6] Brooks, M. (1999) Quantum computing and communications, Springer Verlag, Berlin/Heidelberg.
[7] Brown, C., Shreiber, M., Chapman, B., and Jacobs, G. (2000) Information science and bioinformatics, in: N.Kasabov, ed, Future directions of intelligent systems and information sciences, Physica Verlag (Springer Verlag), 251–287
[8] Carpenter G, Grossberg S, Markuzon N, Reynolds JH, Rosen DB (1991) Fuzzy ARTMAP: A neural network architecture for incremental supervised learning of analogue multi-dimensional maps. IEEE Trans of Neural Networks 3(5): 698–713.
[9] Carpenter G, Grossberg S (1991) Pattern recognition by self-organizing neural networks. Massachusetts. The MIT Press, Cambridge, MA, U.S.A.
[10] Chan, Z., N.Kasabov, Lesley Collins (2006) *A Two-Stage Methodology for Gene Regulatory Network Extraction from Time-Course Gene Expression Data*, Expert Systems with Applications: An International Journal (ISSN: 0957-4174), Special issue on Intelligent Bioinformatics Systems, 59–63.
[11] Chin, H., Moldin, S. (eds) (2001) Methods in Genomic Neuroscience, CRC Press.
[12] Collin P. Williams, Scott H. Clearwater (1998) *Explorations in Quantum Computing*, ISBN: 038794768X, Berlin, Germany: Springer-Verlag.
[13] Destexhe A (1998) Spike-and-wave oscillations based on the properties of GABA_B receptors. J Neurosci 18: 9099–9111.
[14] Dimitrov, D. S, I. Sidorov and N. Kasabov (2004) *Computational Biology,* in: M. Rieth and W. Sommers (eds) Handbook of Theoretical and Computational Nanotechnology, Vol. 6 American Scientific Publisher, Chapter 1.

[15] Ezhov, A. and D. Ventura (2000) Quantum neural networks, in: N. Kasabov (ed) Future Directions for Intelligent Systems and Information Sciences, Springer Verlag.
[16] Feynman, R. P., R. B. Leighton, and M. Sands (1965) The Feynman Lectures on Physics, Addison-Wesley Publishing Company, Massachusetts.
[17] Fogel DB (1995) Evolutionary Computation – Toward a New Philosophy of Machine Intelligence. IEEE Press, New York.
[18] Freeman J, Saad D (1997) On-line learning in radial basis function networks. Neural Computation 9 (7)
[19] Freeman W (2000) Neurodynamics. Springer-Verlag, London.
[20] Fritzke B (1995) A growing neural gas network learns topologies. Advances in Neural Information Processing Systems 7: 625–632.
[21] Gerstner W, Kistler WM (2002) Spiking Neuron Models. Cambridge Univ. Press, Cambridge, MA.
[22] Grossberg S (1969) On learning and energy – entropy dependence in recurrent and nonrecurrent signed networks. J Stat Phys 1: 319–350.
[23] Grossberg S (1982) Studies of Mind and Brain. Reidel, Boston.
[24] Grover, L. K. (1996) A fast quantum mechanical algorithm for database search, in STOC '96: Proceedings of the twenty-eighth annual ACM symposium on Theory of computing, New York, NY, USA, ACM Press, 212–219.
[25] Han, K.-H. and J.-H. Kim (2002) Quantum-inspired evolutionary algorithm for a class of combinatorial optimization, IEEE Transactions on Evolutionary Computation, 580–593.
[26] Haykin S (1994) Neural Networks – A Comprehensive Foundation. Prentice Hall, Engelwood Cliffs, NJ.
[27] Hebb D (1949) The Organization of Behavior. John Wiley and Sons, New York.
[28] Heskes TM, Kappen B (1993) On-line learning processes in artificial neural networks. In: (eds) Mathematic Foundations of Neural Networks, vol. Elsevier, Amsterdam, 199–233.
[29] Hey, T. (1999) Quantum computing: An introduction, *Computing & Control Engineering Journal*, Piscataway, NJ: IEEE Press, June, vol. 10, no. 3, 105–112.
[30] Hinton GE (1989) Connectionist learning procedures. Artificial Intelligence 40: 185–234.
[31] Hogg, T. and D. Portnov (2000), Quantum optimization, Information Sciences, 128, 181–197.
[32] Jang, J.-S., K.-H. Han, and J.-H. Kim (2003) Quantum-inspired evolutionary algorithm-based face verification, Lecture Notes in Computer Science, 2147–2156.
[33] Kak, S.C. Quantum neural computation, Research report, Louisiana State University, Dep. Electr. and Comp. Eng., Baton Rouge, LA 70803-5901, USA

[34] Kasabov N (1996) Foundations of Neural Networks, Fuzzy Systems and Knowledge Engineering. The MIT Press, CA, MA
[35] Kasabov N (1998) Evolving Fuzzy Neural Networks – Algorithms, Applications and Biological Motivation. In: Yamakawa T, Matsumoto G (eds) Methodologies for the conception, design and application of soft computing, World Scientific, 271–274.
[36] Kasabov N (2001) Evolving fuzzy neural networks for on-line supervised/unsupervised, knowledge–based learning. IEEE Trans. SMC – part B, Cybernetics 31(6): 902–918
[37] Kasabov N. (2007) Evolving connectionist systems: The Knowledge Engineering Approach, Springer Verlag, London, New York, Heidelberg, in print (first edition 2002)
[38] Kasabov N, Song Q (2002) DENFIS: Dynamic, evolving neural-fuzzy inference systems and its application for time-series prediction. IEEE Trans. on Fuzzy Systems 10:144–154
[39] Kasabov N., Chan, S. H., Jain, V., Sidirov, I., and Dimitrov S. D. (2004) Gene Regulatory Network Discovery from Time-Series Gene Expression Data – A Computational Intelligence Approach, Lecture Notes in Computer science (LNCS), Springer-Verlag, Vol. 3316, 1344–1353.
[40] Kasabov N. and L. Benuskova (2004)Computational Neurogenetics, International Journal of Theoretical and Computational Nanoscience, Vol. 1 (1) American Scientific Publisher, 2004, 47–61.
[41] Kohonen T (1997) Self-Organizing Maps. Springer, Verlag.
[42] Kouda, N., N. Matsui, H. Nishimura, and F. Peper (2005) Qu-bit neural network and its learning efficiency, Neural Comput. Appl., 14, 114–121.
[43] Liu, J., W. Xu, and J. Sun, Quantum-behaved particle swarm optimization with mutation operator, in 17th IEEE International Conference on Tools with Artificial Intelligence (ICTAI'05), 2005.
[44] Maass W, Bishop CM (eds) (1999) Pulsed Neural Networks, The MIT Press, Cambridge, MA.
[45] Marcus G (2004) The Birth of the Mind: How a Tiny Number of Genes Creates the Complexity of the Human Mind. Basic Books, New York.
[46] Moody J, Darken C (1989) Fast learning in networks of locally-tuned processing units Neural Computation 1: 281–294
[47] Narayanan, A. and T. Meneer, Quantum artificial neural network architectures and components, Information Sciences, (2000), 199–215.
[48] Penrose, R., Shadows of the mind. A search for the missing science of consciousness, Oxford University Press, 1994.
[49] Penrose, R., The Emperor's new mind, Oxford Univ.Press, Oxford, 1989
[50] Perkowski, M.A. Multiple-valued quantum circuits and research challenges for logic design and computational intelligence communities, IEEE Comp.Intelligence Society Magazine, November, 2005
[51] Platt, J (1991) A resource allocating network for function interpolation. Neural Computation 3: 213–225

[52] Poggio T (1994) Regularization theory, radial basis functions and networks. In: From Statistics to Neural Networks: Theory and Pattern Recognition Applications. NATO ASI Series, No.136, 83–104
[53] Pribram, K. (1993) Rethinking Neural Networks: Quantum Fields and Biological data. Proceeding of the first Appalachian Conference on behavioral Neurodynamics. Lawrence Erlbaum Associates Publishers, Hillsdate new Yersy
[54] Resconi, G. and A.J.van Der Wal (2000), A data model for the morphogenetic neuron, Int.J.General Systems,Vol.29(1), 141–149.
[55] Resconi, G., G.J.Klir, E.Pessa (1999), Conceptual Foundations of quantum mechanics the role of evidence theory, quantum sets and modal logic. International Journal of Modern Physics C. Vol.10 No.1, 29–62
[56] Rolls, E. and A.Treves (1998) Neural Networks and Brain Function, Oxford University Press
[57] Rosenblatt F (1962) Principles of Neurodynamics. Spartan Books, New York.
[58] Rumelhart DE, Hinton GE, Williams RJ (eds) (1986) Learning internal representations by error propagation. Parallel Distributed Processing: Explorations in the Microstructure of Cognition, MIT Press / Bradford Books, Cambridge, MA.
[59] Rummery GA, Niranjan M (1994) On-line Q-learning using connectionist system. Cambridge University Engineering Department, CUED/F-INENG/TR, pp 166
[60] Schaal S, Atkeson C (1998) Constructive incremental learning from only local information. Neural Computation 10: 2047–2084
[61] Shor, P. W. (1997) Polynomial-time algorithms for prime factorization and discrete logarithms on a quantum computer, SIAM J. Comput., 26, 1484–1509.
[62] Song Q, Kasabov N (2005) TWNFI – Transductive Neural-Fuzzy Inference System with Weighted Data Normalization and Its Application in Medicine. IEEE Tr Fuzzy Systems, December, Vol.13, 6, 799–808
[63] Spector, L. (2004) Automatic Quantum Computer Programming: A Genetic Programming Approach, Kluwer Academic Publishers, 2004.
[64] Taylor JG (1999) The Race for Consciousness. MIT Press, Cambridge, MA.
[65] Trugenberger, C. A. (2002) Quantum pattern recognition, Quantum Information Processing, 1, pp. 471–493.
[66] Tsai, X.-Y., H.-C. Huang, and S.-J. Chuang (2005) Quantum NN vs. NN in signal recognition, in ICITA'05: Proceedings of the Third International Conference on Information Technology and Applications (ICITA'05) Volume 2, Washington, DC, USA, IEEE Computer Society, pp. 308–312.
[67] Vapnik V (1998) Statistical Learning Theory. John Wiley & Sons Inc.
[68] Venayagamoorthy, G. K and Gaurav, S. (2006) Quantum-Inspired Evolutionary Algorithms and Binary Particle Swarm Optimization for Training

MLP and SRN Neural Networks, Journal of Theoretical and Computational Nanoscience, January.
[69] Ventura, D. and T. Martinez (2000) Quantum associative memory, Inf. Sci. Inf. Comput. Sci., 124, 273–296.
[70] Ventura, D.(1999) Implementing competitive learning in a quantum system, in Proceedings of the International Joint Conference of Neural Networks, IEEE Press.
[71] Wysoski, S., L. Benuskova and N. Kasabov (2006) On-line learning with structural adaptation in a network of spiking neurons for visual pattern recognition, Proc. ICANN 2006, LNCS, Springer, Part I, Vol.413, 61–70
[72] Xie, G. and Z. Zhuang (2003) A quantum competitive learning algorithm, Liangzi Dianzi Xuebao/Chinese Journal of Quantum Electronics (China), 20, 42–46.
[73] Yamakawa T, Kusanagi II, Uchino E, Miki T (1993) A new Effective Algorithm for Neo Fuzzy neuron Model, In: Proc. Fifth IFSA World Congress, IFSA, Seoul, Korea, pp 1017–1020.
[74] Yao X (1993) Evolutionary artificial neural networks. Intl J Neural Systems 4(3): 203–222.
[75] Zadeh LA (1965) Fuzzy Sets. Information and Control 8: 338–353.

The Science of Pattern Recognition. Achievements and Perspectives

Robert P.W. Duin[1] and Elżbieta Pękalska[2]

[1] ICT group, Faculty of Electr. Eng., Mathematics and Computer Science
Delft University of Technology, The Netherlands
r.duin@ieee.org

[2] School of Computer Science, University of Manchester, United Kingdom
pekalska@cs.man.ac.uk

Summary. Automatic pattern recognition is usually considered as an engineering area which focusses on the development and evaluation of systems that imitate or assist humans in their ability of recognizing patterns. It may, however, also be considered as a science that studies the faculty of human beings (and possibly other biological systems) to discover, distinguish, characterize patterns in their environment and accordingly identify new observations. The engineering approach to pattern recognition is in this view an attempt to build systems that simulate this phenomenon. By doing that, scientific understanding is gained of what is needed in order to recognize patterns, in general.

Like in any science understanding can be built from different, sometimes even opposite viewpoints. We will therefore introduce the main approaches to the science of pattern recognition as two dichotomies of complementary scenarios. They give rise to four different schools, roughly defined under the terms of expert systems, neural networks, structural pattern recognition and statistical pattern recognition. We will briefly describe what has been achieved by these schools, what is common and what is specific, which limitations are encountered and which perspectives arise for the future. Finally, we will focus on the challenges facing pattern recognition in the decennia to come. They mainly deal with weaker assumptions of the models to make the corresponding procedures for learning and recognition wider applicable. In addition, new formalisms need to be developed.

1 Introduction

We are very familiar with the human ability of pattern recognition. Since our early years we have been able to recognize voices, faces, animals, fruits or inanimate objects. Before the speaking faculty is developed, an object like a ball is recognized, even if it barely resembles the balls seen before. So, except for the memory, the skills of abstraction and generalization are essential to find our way in the world. In later years we are able to deal with much more

Robert P.W. Duin and Elżbieta Pękalska: *The Science of Pattern Recognition. Achievements and Perspectives*, Studies in Computational Intelligence (SCI) **63**, 221–259 (2007)
www.springerlink.com

complex patterns that may not directly be based on sensorial observations. For example, we can observe the underlying theme in a discussion or subtle patterns in human relations. The latter may become apparent, e.g. only by listening to somebody's complaints about his personal problems at work that again occur in a completely new job. Without a direct participation in the events, we are able to see both analogy and similarity in examples as complex as social interaction between people. Here, we learn to distinguish the pattern from just two examples.

The pattern recognition ability may also be found in other biological systems: the cat knows the way home, the dog recognizes his boss from the footsteps or the bee finds the delicious flower. In these examples a direct connection can be made to sensory experiences. Memory alone is insufficient; an important role is that of generalization from observations which are similar, although not identical to the previous ones. A scientific challenge is to find out how this may work.

Scientific questions may be approached by building models and, more explicitly, by creating simulators, i.e. artificial systems that roughly exhibit the same phenomenon as the object under study. Understanding will be gained while constructing such a system and evaluating it with respect to the real object. Such systems may be used to replace the original ones and may even improve some of their properties. On the other hand, they may also perform worse in other aspects. For instance, planes fly faster than birds but are far from being autonomous. We should realize, however, that what is studied in this case may not be the bird itself, but more importantly, the ability to fly. Much can be learned about flying in an attempt to imitate the bird, but also when differentiating from its exact behavior or appearance. By constructing fixed wings instead of freely movable ones, the insight in how to fly grows. Finally, there are engineering aspects that may gradually deviate from the original scientific question. These are concerned with how to fly for a long time, with heavy loads, or by making less noise, and slowly shift the point of attention to other domains of knowledge.

The above shows that a distinction can be made between the scientific study of pattern recognition as the ability to abstract and generalize from observations and the applied technical area of the design of artificial pattern recognition devices without neglecting the fact that they may highly profit from each other. Note that patterns can be distinguished on many levels, starting from simple characteristics of structural elements like strokes, through features of an individual towards a set of qualities in a group of individuals, to a composite of traits of concepts and their possible generalizations. A pattern may also denote a single individual as a representative for its population, model or concept. Pattern recognition deals, therefore, with patterns, regularities, characteristics or qualities that can be discussed on a low level of sensory measurements (such as pixels in an image) as well as on a high level of the derived and meaningful concepts (such as faces in images). In this work, we will focus on the scientific aspects, i.e. what we know about the way pattern recognition works and, especially, what can be learned from our attempts to build artificial recognition devices.

A number of authors have already discussed the science of pattern recognition based on their simulation and modeling attempts. One of the first, in the beginning of the sixties, was Sayre [64], who presented a philosophical study on perception, pattern recognition and classification. He made clear that classification is a task that can be fulfilled with some success, but recognition either happens or not. We can stimulate the recognition by focussing on some aspects of the question. Although we cannot set out to fully recognize an individual, we can at least start to classify objects on demand. The way Sayre distinguishes between recognition and classification is related to the two subfields discussed in traditional texts on pattern recognition, namely unsupervised and supervised learning. They fulfill two complementary tasks. They act as automatic tools in the hand of a scientist who sets out to find the regularities in nature.

Unsupervised learning (also related to exploratory analysis or cluster analysis) gives the scientist an automatic system to indicate the presence of yet unspecified patterns (regularities) in the observations. They have to be confirmed (verified) by him. Here, in the terms of Sayre, a pattern is recognized. **Supervised learning** is an automatic system that verifies (confirms) the patterns described by the scientist based on a representation defined by him. This is done by an automatic classification followed by an evaluation.

In spite of Sayre's discussion, the concepts of pattern recognition and classification are still frequently mixed up. In our discussion, classification is a significant component of the pattern recognition system, but unsupervised learning may also play a role there. Typically, such a system is first presented with a set of known objects, the training set, in some convenient representation. Learning relies on finding the data descriptions such that the system can correctly characterize, identify or classify novel examples. After appropriate preprocessing and adaptations, various mechanisms are employed to train the entire system well. Numerous models and techniques are used and their performances are evaluated and compared by suitable criteria. If the final goal is prediction, the findings are validated by applying the best model to unseen data. If the final goal is characterization, the findings may be validated by complexity of organization (relations between objects) as well as by interpretability of the results.

Fig. 1 shows the three main stages of pattern recognition systems: Representation, Generalization and Evaluation, and an intermediate stage of Adaptation [20]. The system is trained and evaluated by a set of examples, the Design Set. The components are:

- **Design Set.** It is used both for training and validating the system. Given the background knowledge, this set has to be chosen such that it is representative for the set of objects to be recognized by the trained system.

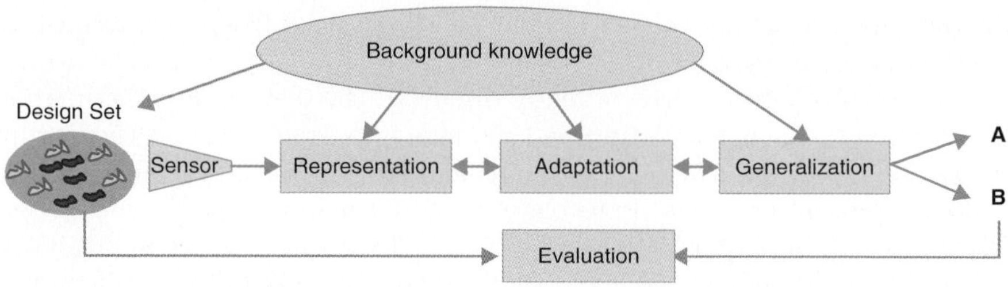

Fig. 1. Components of a pattern recognition system

There are various approaches how to split it into suitable subsets for training, validation and testing. See e.g. [22, 32, 62, 77] for details.

- **Representation.** Real world objects have to be represented in a formal way in order to be analyzed and compared by mechanical means such as a computer. Moreover, the observations derived from the sensors or other formal representations have to be integrated with the existing, explicitly formulated knowledge either on the objects themselves or on the class they may belong to. The issue of representation is an essential aspect of pattern recognition and is different from classification. It largely influences the success of the stages to come.
- **Adaptation.** It is an intermediate stage between Representation and Generalization, in which representations, learning methodology or problem statement are adapted or extended in order to enhance the final recognition. This step may be neglected as being transparent, but its role is essential. It may reduce or simplify the representation, or it may enrich it by emphasizing particular aspects, e.g. by a nonlinear transformation of features that simplifies the next stage. Background knowledge may appropriately be (re)formulated and incorporated into a representation. If needed, additional representations may be considered to reflect other aspects of the problem. Exploratory data analysis (unsupervised learning) may be used to guide the choice of suitable learning strategies.
- **Generalization or Inference.** In this stage we learn a concept from a training set, the set of known and appropriately represented examples, in such a way that predictions can be made on some unknown properties of new examples. We either generalize towards a concept or infer a set of general rules that describe the qualities of the training data. The most common property is the class or pattern it belongs to, which is the above mentioned classification task.
- **Evaluation.** In this stage we estimate how our system performs on known training and validation data while training the entire system. If the results are unsatisfactory, then the previous steps have to be reconsidered.

Different disciplines emphasize or just exclusively study different parts of this system. For instance, perception and computer vision deal mainly with the representation aspects [21], while books on artificial neural networks [62], machine learning [4, 53] and pattern classification [15] are usually restricted to generalization. It should be noted that these and other studies with the words "pattern" and "recognition" in the title often almost entirely neglect the issue of representation. We think, however, that *the main goal of the field of pattern recognition is to study generalization in relation to representation* [20].

In the context of representations, and especially images, generalization has been thoroughly studied by Grenander [36]. What is very specific and worthwhile is that he deals with infinite representations (say, unsampled images), thereby avoiding the frequently returning discussions on dimensionality and directly focussing on a high, abstract level of pattern learning. We like to mention two other scientists that present very general discussions on the pattern recognition system: Watanabe [75] and Goldfarb [31, 32]. They both emphasize the structural approach to pattern recognition that we will discuss later on. Here objects are represented in a form that focusses on their structure. A generalization over such structural representations is very difficult if one aims to learn the *concept*, i.e. the underlying, often implicit definition of a pattern class that is able to generate possible realizations. Goldfarb argues that traditionally used numeric representations are inadequate and that an entirely new, structural representation is necessary. We judge his research program as very ambitious, as he wants to learn the (generalized) structure of the concept from the structures of the examples. He thereby aims to make explicit what usually stays implicit. We admit that a way like his has to be followed if one ever wishes to reach more in concept learning than the ability to name the right class with a high probability, without having built a proper understanding.

In the next section we will discuss and relate well-known general scientific approaches to the specific field of pattern recognition. In particular, we like to point out how these approaches differ due to fundamental differences in the scientific points of view from which they arise. As a consequence, they are often studied in different traditions based on different paradigms. We will try to clarify the underlying cause for the pattern recognition field. In the following sections we sketch some perspectives for pattern recognition and define a number of specific challenges.

2 Four Approaches to Pattern Recognition

In science, new knowledge is phrased in terms of existing knowledge. The starting point of this process is set by generally accepted evident views, or observations and facts that cannot be explained further. These foundations, however, are not the same for all researchers. Different types of approaches may be distinguished originating from different starting positions. It is almost a type of taste from which perspective a particular researcher begins. As a consequence, different 'schools' may arise. *The point of view, however, determines what we see.* In other words, staying within a particular framework of thought we cannot achieve more than what is derived as a consequence of the corresponding assumptions and constraints. To create more complete and objective methods, we may try to integrate scientific results originating from different approaches into a single pattern recognition model. It is possible that confusion arises on how these results may be combined and where they essentially differ. But the combination of results of different approaches may also appear to be fruitful, not only for some applications, but also for the scientific understanding of the researcher that broadens the horizon of allowable starting points. This step towards a unified or integrated view is very important in science as only then a more complete understanding is gained or a whole theory is built.

Below we will describe four approaches to pattern recognition which arise from two different dichotomies of the starting points. Next, we will present some examples illustrating the difficulties of their possible interactions. This discussion is based on earlier publications [16, 17].

2.1 Platonic and Aristotelian Viewpoints

Two principally different approaches to almost any scientific field rely on the so-called Platonic and Aristotelian viewpoints. In a first attempt they may be understood as top-down and bottom-up ways of building knowledge. They are also related to deductive (or holistic) and inductive (or reductionistic) principles. These aspects will be discussed in Section 4.

The Platonic approach starts from generally accepted concepts and global ideas of the world. They constitute a coherent picture in which many details are undefined. The primary task of the Platonic researcher is to recognize in his[3] observations the underlying concepts and ideas that are already accepted by him. Many theories of the creation of the universe or the world rely on this scenario. An example is the drifts of the continents or the extinction of the mammoths. These theories do not result from a reasoning based on observations, but merely from a more or less convincing global theory (depending on the listener!) that seems to extrapolate far beyond the hard facts. For the Platonic researcher, however, it is not an extrapolation, but an adaptation of previous formulations of the theory to new facts. That is the way this approach works: existing ideas that have been used for a long time are gradually adapted to new incoming observations. The change does not rely on an essential paradigm shift in the concept, but on finding better, more appropriate relations with the observed world in definitions and explanations. The essence of the theory has been constant for a long time. So, in practise the

[3] For simplicity, we refer to researchers in a male form; we mean both women and men.

Platonic researcher starts from a theory which can be stratified into to a number of hypotheses that can be tested. Observations are collected to test these hypotheses and, finally, if the results are positive, the theory is confirmed.

The observations are of primary interest in the Aristotelian approach. Scientific reasoning stays as closely as possible to them. It is avoided to speculate on large, global theories that go beyond the facts. The observations are always the foundation on which the researcher builds his knowledge. Based on them, patterns and regularities are detected or discovered, which are used to formulate some tentative hypotheses. These are further explored in order to arrive at general conclusions or theories. As such, the theories are not global, nor do they constitute high level descriptions. A famous guideline here is the so-called Occam's razor principle that urges one to avoid theories that are more complex than strictly needed for explaining the observations. Arguments may arise, however, since the definition of complexity depends, e.g. on the mathematical formalism that is used.

The choice for a particular approach may be a matter of preference or determined by non-scientific grounds, such as upbringing. Nobody can judge what the basic truth is for somebody else. Against the Aristotelians may be held that they do not see the overall picture. The Platonic researchers, on the other hand, may be blamed for building castles in the air. Discussions between followers of these two approaches can be painful as well as fruitful. They may not see that their ground truths are different, leading to pointless debates. What is more important is the fact that they may become inspired by each other's views. One may finally see real world examples of his concepts, while the other may embrace a concept that summarizes, or constitutes an abstraction of his observations.

2.2 Internal and the External Observations

In the contemporary view science is *'the observation, identification, description, experimental investigation, and theoretical explanation of phenomena'*[4] or *'any system of knowledge that is concerned with the physical world and its phenomena and that entails unbiased observations and systematic experimentation.*[5] So, the aspect of observation that leads to a possible formation of a concept or theory is very important. Consequently, the research topic of the science of pattern recognition, which aims at the generalization from observations for knowledge building, is indeed scientific. Science is in the end a brief explanation summarizing the observations achieved through abstraction and their generalization.

Such an explanation may primarily be observed by the researcher in his own thinking. Pattern recognition research can thereby be performed by introspection. The researcher inspects himself how he generalizes from

[4] http://dictionary.reference.com/
[5] http://www.britannica.com/

observations. The basis of this generalization is constituted by the primary observations. This may be an entire object ('I just see that it is an apple') or its attributes ('it is an apple because of its color and shape'). We can also observe pattern recognition in action by observing other human beings (or animals) while they perform a pattern recognition task, e.g. when they recognize an apple. Now the researcher tries to find out by experiments and measurements how the subject decides for an apple on the basis of the stimuli presented to the senses. He thereby builds a model of the subject, from senses to decision making.

Both approaches result into a model. In the external approach, however, the senses may be included in the model. In the internal approach, this is either not possible or just very partially. We are usually not aware of what happens in our senses. Introspection thereby starts by what they offer to our thinking (and reasoning). As a consequence, models based on the internal approach have to be externally equipped with (artificial) senses, i.e. with sensors.

2.3 The Four Approaches

The following four approaches can be distinguished by combining the two dichotomies presented above:

(1) Introspection by a Platonic viewpoint: object modeling.
(2) Introspection by an Aristotelian viewpoint: generalization.
(3) Extrospection by an Aristotelian viewpoint: system modeling.
(4) Extrospection by a Platonic viewpoint: concept modeling.

These four approaches will now be discussed separately. We will identify some known procedures and techniques that may be related to these. See also Fig. 2.

Object modeling. This is based on introspection from a Platonic viewpoint. The researcher thereby starts from global ideas on how pattern recognition systems may work and tries to verify them in his own thinking and reasoning. He thereby may find, for instance, that particular color and shape descriptions of an object are sufficient for him to classify it as an apple. More generally, he may discover that he uses particular reasoning rules operating on a fixed set of possible observations. The so-called syntactic and structural approaches to pattern recognition [26] thereby belong to this area, as well as the case-based reasoning [3]. There are two important problems in this domain: how to constitute the general concept of a class from individual object descriptions and how to connect particular human qualitative observations such as 'sharp edge' or 'egg shaped' with physical sensor measurements.

Generalization. Let us leave the Platonic viewpoint and consider a researcher who starts from observations, but still relies on introspection. He wonders what he should do with just a set of observations without any framework. An important point is the nature of observations. Qualitative observations such as 'round', 'egg-shaped' or 'gold colored' can be judged as recognitions in themselves based on low-level outcomes of senses. It is difficult to

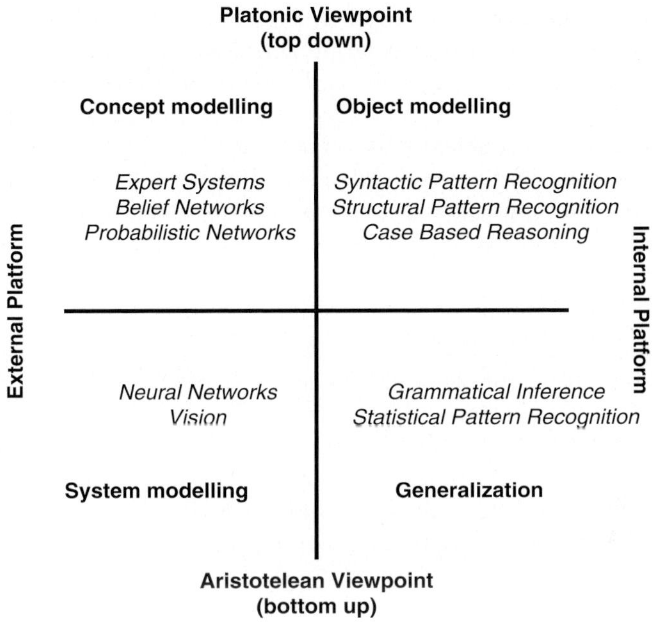

Fig. 2. Four approaches to Pattern Recognition

neglect them and to access the outcomes of senses directly. One possibility for him is to use artificial senses, i.e. of sensors, which will produce quantitative descriptions. The next problem, however, is how to generalize from such numerical outcomes. The physiological process is internally unaccessible. A researcher who wonders how he himself generalizes from low level observations given by numbers may rely on statistics. This approach thereby includes the area of statistical pattern recognition.

If we consider low-level inputs that are not numerical, but expressed in attributed observations as 'red, egg-shaped', then the generalization may be based on logical or grammatical inference. As soon, however, as the structure of objects or attributes is not generated from the observations, but derived (postulated) from a formal global description of the application knowledge, e.g. by using graph matching, the approach is effectively top-down and thereby starts from object or concept modeling.

System modeling. We now leave the internal platform and concentrate on research that is based on the external study of the pattern recognition abilities of humans and animals or their brains and senses. If this is done in a bottom-up way, the Aristotelian approach, then we are in the area of low-level modeling of senses, nerves and possibly brains. These models are based on the physical and physiological knowledge of cells and the proteins and minerals that constitute them. Senses themselves usually do not directly generalize from observations. They may be constructed, however, in such a way that this process is strongly favored on a higher level. For instance, the way the eye (and the retina, in particular) is constructed, is advantageous for the detection of edges and movements as well as for finding interesting details in a global, overall picture. The area of vision thereby profits from this approach. It is studied how nerves process the signals they receive from the senses on a level close to the brain. Somehow this is combined towards a generalization of what is observed by the senses. Models of systems of multiple nerves are called neural networks. They appeared to have a good generalization ability and are thereby also used in technical pattern recognition applications in which the physiological origin is not relevant [4, 62].

Concept modeling. In the external platform, the observations in the starting point are replaced by ideas and concepts. Here one still tries to externally model the given pattern recognition systems, but now in a top-down manner. An example is the field of expert systems: by interviewing experts in a particular pattern recognition task, it is attempted to investigate what rules they use and in what way they are using observations. Also belief networks and probabilistic networks belong to this area as far as they are defined by experts and not learned from observations. This approach can be distinguished from the above system modeling by the fact that it is in no way attempted to model a physical or physiological system in a realistic way. The building blocks are the ideas, concepts and rules, as they live in the mind of the researcher. They are adapted to the application by external inspection of an expert, e.g. by interviewing him. If this is done by the researcher internally by introspection, we have closed the circle and are back to what we have called object modeling, as the individual observations are our internal starting point. We admit that the difference between the two Platonic approaches is minor here (in contrast to the physiological level) as we can also try to interview ourselves to create an objective (!) model of our own concept definitions.

2.4 Examples of Interaction

The four presented approaches are four ways to study the science of pattern recognition. Resulting knowledge is valid for those who share the same starting point. If the results are used for building artificial pattern recognition devices, then there is, of course, no reason to restrict oneself to a particular approach. Any model that works well may be considered. There are, however, certain difficulties in combining different approaches. These may be caused by differences in culture, assumptions or targets. We will present two examples, one for each of the two dichotomies.

Artificial neural networks constitute an alternative technique to be used for generalization within the area of statistical pattern recognition. It has taken, however, almost ten years since their introduction around 1985 before neural networks were fully acknowledged in this field. In that period, the neural network community suffered from lack of knowledge on the competing classification procedures. One of the basic misunderstandings in the pattern recognition field was caused by its dominating paradigm stating that learning systems should never be larger than strictly necessary, following the Occam's razor principle. It could have not been understood how largely oversized systems such as neural networks would have ever been able to generalize without adapting to peculiarities in the data (the so-called overtraining). At the same time, it was evident in the neural network community that the larger neural network the larger its flexibility, following the analogy that a brain with many neurons would perform better in learning than a brain with a few ones. When this contradiction was finally solved (an example of Kuhn's paradigm shifts [48]), the area of statistical pattern recognition was enriched with a new set of tools. Moreover, some principles were formulated towards understanding of pattern recognition that otherwise would have only been found with great difficulties.

In general, it may be expected that the internal approach profits from the results in the external world. It is possible that thinking, the way we generalize from observations, changes after it is established how this works in nature. For instance, once we have learned how a specific expert solves his problems, this may be used more generally and thereby becomes a rule in structural pattern recognition. The external platform may thereby be used to enrich the internal one.

A direct formal fertilization between the Platonic and Aristotelian approaches is more difficult to achieve. Individual researchers may build some understanding from studying each other's insights, and thereby become mutually inspired. The Platonist may become aware of realizations of his ideas and concepts. The Aristotelian may see some possible generalizations of the observations he collected. It is, however, still one of the major challenges in science to formalize this process.

How should existing knowledge be formulated such that it can be enriched by new observations? Everybody who tries to do this directly encounters the problem that observations may be used to reduce uncertainty (e.g. by the parameter estimation in a model), but that it is very difficult to formalize uncertainty in existing knowledge. Here we encounter a fundamental 'paradox' for a researcher summarizing his findings after years of observations and studies: he has found some answers, but almost always he has also generated more new questions. Growing knowledge comes with more questions. In any formal system, however, in which we manage to incorporate uncertainty (which is already very difficult), this uncertainty will be reduced after having incorporating some observations. We need an automatic hypothesis generation in order to generate new questions. How should the most likely ones be determined? We need to look from different perspectives in order to stimulate the creative process and bring sufficient inspiration and novelty to hypothesis generation. This is necessary in order to make a step towards building a complete theory. This, however, results in the computational complexity mentioned in the literature [60] when the Platonic structural approach to pattern recognition has to be integrated with the Aristotelian statistical approach.

The same problem may also be phrased differently: how can we express the uncertainty in higher level knowledge in such a way that it may be changed (upgraded) by low level observations? Knowledge is very often structural and has thereby a qualitative nature. On the lowest level, however, observations are often treated as quantities, certainly in automatic systems equipped with physical sensors. And here the Platonic – Aristotelian polarity meets the internal – external polarity: by crossing the border between concepts and observations we also encounter the border between qualitative symbolic descriptions and quantitative measurements.

3 Achievements

In this section we will sketch in broad terms the state of the art in building systems for generalization and recognition. In practical applications it is not the primary goal to study the way of bridging the gap between observations and concepts in a scientific perspective. Still, we can learn a lot from the heuristic solutions that are created to assist the human analyst performing a recognition task. There are many systems that directly try to imitate the decision making process of a human expert, such as an operator guarding a chemical process, an inspector supervising the quality of industrial production or a medical doctor deriving a diagnosis from a series of medical tests. On the basis of systematic interviews the decision making can become explicit and imitated by a computer program: an **expert system** [54]. The possibility to improve such a system by learning from examples is usually very limited and restricted to logical inference that makes the rules as general as possible, and the estimation of the thresholds on observations. The latter is needed as the human expert is not always able to define exactly what he means, e.g. by 'an unusually high temperature'.

In order to relate knowledge to observations, which are measurements in automatic systems, it is often needed to relate **knowledge uncertainty** to imprecise, noisy, or generally invalid measurements. Several frameworks have been developed to this end, e.g. fuzzy systems [74], Bayesian belief networks [42] and reasoning under uncertainty [82]. Characteristic for these approaches is that the given knowledge is already structured and needs explicitly defined parameters of uncertainty. New observations may adapt these parameters by relating them to observational frequencies. The knowledge structure is not learned; it has to be given and is hard to modify. An essential problem is that the variability of the external observations may be probabilistic, but the uncertainty in knowledge is based on 'belief' or 'fuzzy' definitions. Combining them in a single mathematical framework is disputable [39].

In the above approaches either the general knowledge or the concept underlying a class of observations is directly modeled. In **structural pattern**

recognition [26, 65] the starting point is the description of the structure of a single object. This can be done in several ways, e.g. by strings, contour descriptions, time sequences or other order-dependent data. Grammars that are inferred from a collection of strings are the basis of a syntactical approach to pattern recognition [26]. The incorporation of probabilities, e.g. needed for modeling the measurement noise, is not straightforward. Another possibility is the use of graphs. This is in fact already a reduction since objects are decomposed into highlights or landmarks, possibly given by attributes and also their relations, which may be attributed as well. Inferring a language from graphs is already much more difficult than from strings. Consequently, the generalization from a set of objects to a class is usually done by finding typical examples, prototypes, followed by graph matching [5, 78] for classifying new objects.

Generalization in structural pattern recognition is not straightforward. It is often based on the comparison of object descriptions using the entire available training set (the nearest neighbor rule) or a selected subset (the nearest prototype rule). Application knowledge is needed for defining the representation (strings, graphs) as well as for the dissimilarity measure to perform graph matching [51, 7]. The generalization may rely on an analysis of the matrix of dissimilarities, used to determine prototypes. More advanced techniques using the dissimilarity matrix will be described later.

The **1-Nearest-Neighbor Rule** (1-NN) is the simplest and most natural classification rule. It should always be used as a reference. It has a good asymptotic performance for metric measures [10, 14], not worse than twice the Bayes error, i.e. the lowest error possible. It works well in practice for finite training sets. Fig. 3 shows how it performs on the Iris data set in comparison to the linear and quadratic classifiers based on the assumption of normal

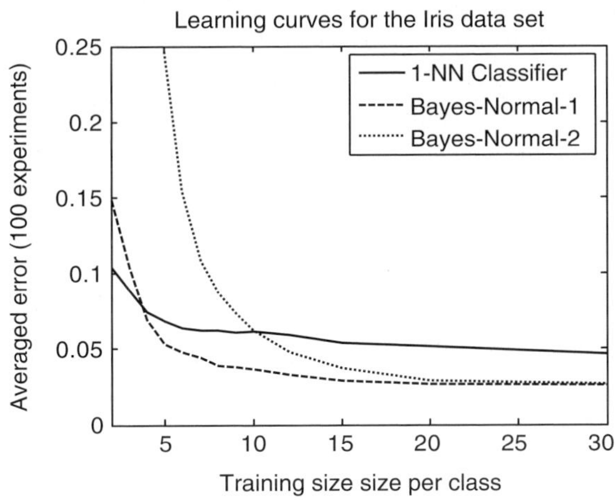

Fig. 3. Learning curves for Iris data

distributions [27]. The k-NN rule, based on a class majority vote over the k nearest neighbors in the training set, is, like the Parzen classifier, even Bayes consistent. These classifiers approximate the Bayes error for increasing training sets [14, 27].

However, such results heavily rely on the assumption that the training examples are identically and independently drawn (iid) from the same distribution as the future objects to be tested. This assumption of a fixed and stationary distribution is very strong, but it yields the best possible classifier. There are, however, other reasons, why it cannot be claimed that pattern recognition is solved by these statistical tools. The 1-NN and k-NN rules have to store the entire training set. The solution is thereby based on a comparison with all possible examples, including ones that are very similar, and asymptotically identical to the new objects to be recognized. By this, a class or a concept is not learned, as the decision relies on memorizing all possible instances. There is simply no generalization.

Other classification procedures, giving rise to two **learning curves** shown in Fig. 3, are based on specific model assumptions. The classifiers may perform well when the assumptions hold and may entirely fail, otherwise. An important observation is that models used in statistical learning procedures have almost necessarily a statistical formulation. Human knowledge, however, certainly in daily life, has almost nothing to do with statistics. Perhaps it is hidden in the human learning process, but it is not explicitly available in the context of human recognition. As a result, there is a need to look for effective model assumptions that are not phrased in statistical terms.

In Fig. 3 we can see that a more complex quadratic classifier performs initially worse than the other ones, but it behaves similarly to a simple linear classifier for large training sets. In general, complex problems may be better solved by complex procedures. This is illustrated in Fig. 4, in which the resulting error curves are shown as functions of complexity and training size. Like in Fig. 3, small training sets require simple classifiers. Larger training sets may be used to train more complex classifiers, but the error will increase, if pushed too far. This is a well-known and frequently studied phenomenon in

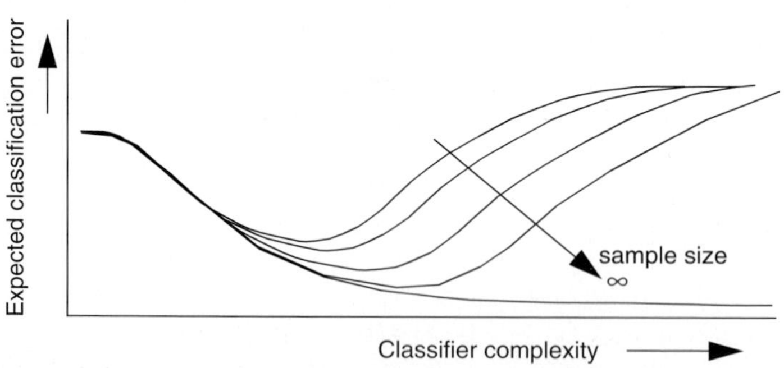

Fig. 4. Curse of dimensionality

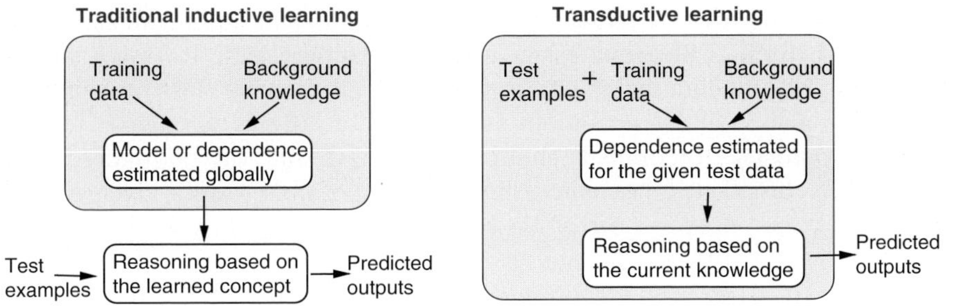

Fig. 5. Inductive (left) and transductive (right) learning paradigms; see also [8]. Background knowledge is here understood in terms of properties of the representations and the specified assumptions on a set of learning algorithms and related parameters

relation to the dimensionality (complexity) of the problem. Objects described by many features often rely on complex classifiers, which may thereby lead to worse results if the number of training examples is insufficient. This is the **curse of dimensionality**, also known as the Rao's paradox or the peaking phenomenon [44, 45]. It is caused by the fact that the classifiers badly generalize, due to a poor estimation of their parameters or their focus/adaptation to the noisy information or irrelevant details in the data. The same phenomenon can be observed while training complex neural networks without taking proper precautions. As a result, they will adapt to accidental data configurations, hence they will *overtrain*. This phenomenon is also well known outside the pattern recognition field. For instance, it is one of the reasons one has to be careful with extensive mass screening in health care: the more diseases and their relations are considered (the more complex the task), the more people will we be unnecessarily sent to hospitals for further examinations.

An important conclusion from this research is that the cardinality of the set of examples from which we want to infer a pattern concept bounds the complexity of the procedure used for generalization. Such a method should be simple if there are just a few examples. A somewhat complicated concept can only be learnt if sufficient prior knowledge is available and incorporated in such a way that the simple procedure is able to benefit from it.

An extreme consequence of the lack of prior knowledge is given by Watanabe as the **Ugly Duckling Theorem** [75]. Assume that objects are described by a set of atomic properties and we consider predicates consisting of all possible logic combinations of these properties in order to train a pattern recognition system. Then, all pairs of objects are equally similar in terms of the number of predicates they share. This is caused by the fact that all atomic properties, their presence as well as their absence, have initially equal weights. As a result, the training set is of no use. Summarized briefly, if we do not know anything about the problem we cannot learn (generalize and/or infer) from observations.

An entirely different reasoning pointing to the same phenomenon is the **No-Free-Lunch Theorem** formulated by Wolpert [81]. It states that all classifiers perform equally well if averaged over all possible classification problems. This also includes a random assignment of objects to classes. In order to understand this theorem it should be realized that the considered set of all possible classification problems includes all possible ways a given data set can be distributed over a set of classes. This again emphasizes that learning cannot be successful without any preference or knowledge.

In essence, it has been established that without prior or background knowledge, no learning, no generalization from examples is possible. Concerning specific applications based on strong models for the classes, it has been shown that additional observations may lower the specified gaps or solve uncertainties in these models. In addition, if these uncertainties are formulated in statistical terms, it will be well possible to diminish their influence by a statistical analysis of the training set. It is, however, unclear what the minimum prior knowledge is that is necessary to make the learning from examples possible. This is of interest if we want to uncover the roots of concept formation, such as learning of a class from examples. There exists one principle, formulated at the very beginning of the study of automatic pattern recognition, which may point to a promising direction. This is the principle of **compactness** [1], also phrased as a compactness hypothesis. It states that we can only learn from examples or phenomena if their representation is such that small variations in these examples cause small deviations in the representation. This demands that the representation is based on a continuous transformation of the real world objects or phenomena. Consequently, it is assumed that a sufficiently small variation in the original object will not cause the change of its class membership. It will still be a realization of the same concept. Consequently, we may learn the class of objects that belong to the same concept by studying the domain of their corresponding representations.

The Ugly Duckling Theorem deals with discrete logical representations. These cannot be solved by the compactness hypothesis unless some metric is assumed that replaces the similarity measured by counting differences in predicates. The No-Free-Lunch Theorem clearly violates the compactness assumption as it makes object representations with contradictory labelings equally probable. In practice, however, we encounter only specific types of problems.

Building proper **representations** has become an important issue in pattern recognition [20]. For a long time this idea has been restricted to the reduction of overly large feature sets to the sizes for which generalization procedures can produce significant results, given the cardinality of the training set. Several procedures have been studied based on feature selection as well as linear and nonlinear feature extraction [45]. A pessimistic result was found that about any hierarchical ordering of (sub)space separability that fulfills the necessary monotonicity constraints can be constructed by an example based on normal distributions only [11]. Very advanced procedures are needed to find such 'hidden' subspaces in which classes are well separable [61]. It has to be doubted, however, whether such problems arise in practice, and whether such feature selection procedures are really necessary in problems with finite sample sizes. This doubt is further supported by an insight that feature reduction procedures should rely on global and not very detailed criteria if their purpose is to reduce the high dimensionality to a size which is in agreement with the given training set.

Feed-forward neural networks are a very general tool that, among others, offer the possibility to train a single system built between sensor and classification [4, 41, 62]. They thereby cover the representation step in the input layer(s) and the generalization step in the output layer(s). These layers are simultaneously optimized. The number of neurons in the network should be sufficiently large to make the interesting optima tractable. This, however, brings the danger of overtraining. There exist several ways to prevent that by incorporating some regularization steps in the optimization process. This replaces the adaptation step in Fig. 1. A difficult point here, however, is that it is not sufficiently clear how to choose regularization of an appropriate strength. The other important application of neural networks is that the use of various regularization techniques enables one to control the nonlinearity of the resulting classifier. This gives also a possibility to use not only complex, but also moderately nonlinear functions. Neural networks are thereby one of the most general tools for building pattern recognition systems.

In statistical learning, Vapnik has rigorously studied the problem of adapting the complexity of the generalization procedure to a finite training set [72, 73]. The resulting **Vapnik-Chervonenkis** (VC) dimension, a complexity measure for a family of classification functions, gives a good insight into the mechanisms that determine the final performance (which depends on the training error and the VC dimension). The resulting error bounds, however, are too general for a direct use. One of the reasons is that, like in the No-Free-Lunch Theorem, the set of classification problems (positions and labeling of the data examples) is not restricted to the ones that obey the compactness assumption.

One of the insights gained by studying the complexity measures of polynomial functions is that they have to be as simple as possible in terms of the number of their free parameters to be optimized. This was already realized by Cover in 1965 [9]. Vapnik extended this finding around 1994 to arbitrary non-linear classifiers [73]. In that case, however, the number of free parameters is not necessarily indicative for the complexity of a given family of functions, but the VC dimension is. In Vapnik's terms, the VC dimension reflects the flexibility of a family of functions (such as polynomials or radial basis functions) to separate arbitrarily labeled and positioned n-element data in a vector space of a fixed dimension. This VC dimension should be finite and small to guarantee the good performance of the generalization function.

This idea was elegantly incorporated to the **Support Vector Machine** (SVM) [73], in which the number of parameters is as small as a suitably determined subset of the training objects (the support vectors) and into independent of the dimensionality of the vector space. One way to phrase this principle is that the structure of the classifier itself is simplified as far as possible (following the Occam's razor principle). So, after a detor along huge neural networks possibly having many more parameters than training examples, pattern recognition was back to the *small-is-beautiful* principle, but now better understood and elegantly formulated.

The use of **kernels** largely enriched the applicability of the SVM to nonlinear decision functions [66, 67, 73]. The kernel approach virtually generates nonlinear transformations of the combinations of the existing features. By using the representer theorem, a linear classifier in this nonlinear feature space can be constructed, because the kernel encodes generalized inner products of the original vectors only. Consequently, well-performing nonlinear classifiers built on training sets of almost any size in almost any feature space can be computed by using the SVM in combination with the 'kernel trick' [66].

This method has still a few limitations, however. It was originally designed for separable classes, hence it suffers when high overlap occurs. The use of slack variables, necessary for handling such an overlap, leads to a large number of support vectors and, consequently, to a large VC dimension. In such cases, other learning procedures have to be preferred. Another difficulty is that the class of admissible kernels is very narrow to guarantee the optimal solution. A kernel K has to be (conditionally) positive semidefinite (cpd) functions of two variables as only then it can be interpreted as a generalized inner product in reproducing kernel Hilbert space induced by K. Kernels were first considered as functions in Euclidean vector spaces, but they are now also designed to handle more general representations. **Special-purpose kernels** are defined in a number of applications such as text processing and shape recognition, in which good features are difficult to obtain. They use background knowledge from the application in which similarities between objects are defined in such a way that a proper kernel can be constructed. The difficulty is, again, the strong requirement of kernels as being cpd.

The next step is the so-called **dissimilarity representation** [56] in which general proximity measures between objects can be used for their representation. The measure itself may be arbitrary, provided that it is meaningful for the problem. Proximity plays a key role in the quest for an integrated structural and statistical learning model, since it is a natural bridge between these two approaches [6, 56]. Proximity is the basic quality to capture the characteristics of a set objects forming a group. It can be defined in various ways and contexts, based on sensory measurements, numerical descriptions, sequences, graphs, relations and other non-vectorial representations, as well as their combinations. A representation based on proximities is, therefore, universal.

Although some foundations are laid down [56], the ways for effective learning from general proximity representations are still to be developed. Since measures may not belong to the class of permissable kernels, the traditional SVM, as such, cannot be used. There exist alternative interpretations of indefinite kernels and their relation to pseudo-Euclidean and Krein spaces [38, 50, 55, 56, 58], in which learning is possible for **non-Euclidean representations**. In general, proximity representations are embedded into suitable vector spaces equipped with a generalized inner product or norm, in which numerical techniques can either be developed or adapted from the existing ones. It has been experimentally shown that many classification techniques may perform well for general dissimilarity representations.

4 Perspectives

Pattern recognition deals with discovering, distinguishing, detecting or characterizing patterns present in the surrounding world. It relies on extraction and representation of information from the observed data, such that after integration with background knowledge, it ultimately leads to a formulation of new knowledge and concepts. The result of learning is that the knowledge already captured in some formal terms is used to describe the present interdependencies such that the relations between patterns are better understood (interpreted) or used for generalization. The latter means that a concept, e.g. of a class of objects, is formalized such that it can be applied to unseen examples of the same domain, inducing new information, e.g. the class label of a new object. In this process new examples should obey the same deduction process as applied to the original examples.

In the next subsections we will first recapitulate the elements of logical reasoning that contribute to learning. Next, this will be related to the Platonic and Aristotelian scientific approaches discussed in Section 2. Finally, two novel pattern recognition paradigms are placed in this view.

4.1 Learning by Logical Reasoning

Learning from examples is an active process of concept formation that relies on abstraction (focus on important characteristics or reduction of detail) and analogy (comparison between different entities or relations focussing on some aspect of their similarity). Learning often requires dynamical, multi-level (seeing the details leading to unified concepts, which further build higher level concepts) and possibly multi-strategy actions (e.g. in order to support good predictive power as well as interpretability). A learning task is basically defined by input data (design set), background knowledge or problem context and a learning goal [52]. Many inferential strategies need to be synergetically integrated to be successful in reaching this goal. The most important ones are inductive, deductive and abductive principles, which are briefly presented next. More formal definitions can be sought in the literature on formal logic, philosophy or e.g. in [23, 40, 52, 83].

Inductive reasoning is the *synthetic* inference process of arriving at a conclusion or a general rule from a limited set of observations. This relies on a formation of a concept or a model, given the data. Although such a derived inductive conclusion cannot be proved, its reliability is supported by empirical observations. As along as the related deductions are not in contradiction with experiments, the inductive conclusion remains valid. If, however, future observations lead to contradiction, either an adaption or a new inference is necessary to find a better rule. To make it more formal, induction learns a general rule R (concerning A and B) from numerous examples of A and B. In practice, induction is often realized in a quantitative way. Its strength relies then on probability theory and the law of large numbers, in which given a large number of cases, one can describe their properties in the limit and the corresponding rate of convergence.

Deductive reasoning is the *analytic* inference process in which existing knowledge of known facts or agreed-upon rules is used to derive a conclusion. Such a conclusion does not yield 'new' knowledge since it is a logical consequence of what has already been known, but implicitly (it is not of a greater generality than the premises). Deduction, therefore, uses a logical argument to make explicit what has been hidden. It is also a valid form of proof provided that one starts from true premises. It has a predictive power, which makes it complementary to induction. In a pattern recognition system, both evaluation and prediction rely on deductive reasoning. To make it more formal, let us assume that A is a set of observations, B is a conclusion and R is a general rule. Let B be a logical consequence of A and R, i.e. $(A \wedge R) \models B$, where $\models$ denotes entailment. In a deductive reasoning, given A and using the rule R, the consequence B is derived.

Abductive reasoning is the *constructive* process of deriving the most likely or best explanations of known facts. This is a creative process, in which possible and feasible hypotheses are generated for a further evaluation. Since both abduction and induction deal with incomplete information, induction may be viewed in some aspects as abduction and vice versa, which leads to some confusion between these two [23, 52]. Here, we assume they are different. Concerning the entailment $(A \wedge R) \models B$, having observed the consequence B in the context of the rule R, A is derived to *explain* B.

In all learning paradigms there is an interplay between inductive, abductive and deductive principles. Both deduction and abduction make possible to conceptually understand a phenomenon, while induction verifies it. More precisely, abduction generates or reformulates new (feasible) ideas or hypotheses, induction justifies the validity of these hypothesis with observed data and deduction evaluates and tests them. Concerning pattern recognition systems, *abduction* explores data, transforms the representation and suggests feasible classifiers for the given problem. It also generates new classifiers or reformulates the old ones. Abduction is present in an initial exploratory step or in the Adaptation stage; see Fig. 1. Induction trains the classifier in the Generalization stage, while deduction predicts the final outcome (such as label) for the test data by applying the trained classifier in the Evaluation stage.

Since abduction is hardly emphasized in learning, we will give some more insights. In abduction, a peculiarity or an artifact is observed and a hypothesis is then created to explain it. Such a hypothesis is suggested based on existing knowledge or may extend it, e.g. by using analogy. So, the abductive process is creative and works towards new discovery. In data analysis, visualization facilitates the abductive process. In response to visual observations of irregularities or bizarre patterns, a researcher is inspired to look for clues that can be used to explain such an unexpected behavior. Mistakes and errors can therefore serve the purpose of discovery when strange results are inquired with a critical mind. Note, however, that this process is very hard to implement into automatic recognition systems as it would require to encode not only the detailed domain knowledge, but also techniques that are able to detect 'surprises' as well as strategies for their possible use. In fact, this requires a conscious interaction. Ultimately, only a human analyst can interactively respond in such cases, so abduction can be incorporated into semi-automatic systems well. In traditional pattern recognition systems, abduction is usually defined in the terms of data and works over pre-specified set of transformations, models or classifiers.

4.2 Logical Reasoning Related to Scientific Approaches

If pattern recognition (learning from examples) is merely understood as a process of concept formation from a set of observations, the inductive principle is the most appealing for this task. Indeed, it is the most widely emphasized in the literature, in which 'learning' is implicitly understood as 'inductive learning'. Such a reasoning leads to inferring new knowledge (rule or model) which is hopefully valid not only for the known examples, but also for novel, unseen objects. Various validation measures or adaptation steps are taken to support the applicability of the determined model. Additionally, care has to be taken that the unseen objects obey the same assumptions as the original objects used in training. If this does not hold, such an empirical generalization becomes invalid. One should therefore exercise in critical thinking while designing a complete learning system. It means that one has to be conscious which assumptions are made and be able to quantify their sensibility, usability and validity with the learning goal.

On the other hand, deductive reasoning plays a significant role in the Platonic approach. This top-down scenario starts from a set of rules derived from expert knowledge on problem domain or from a degree of belief in a hypothesis. The existing prior knowledge is first formulated in appropriate terms. These are further used to generate inductive inferences regarding the validity of the hypotheses in the presence of observed examples. So, deductive formalism (description of the object's structure) or deductive predictions (based on the Bayes rule) precede inductive principles. A simple example in the Bayesian inference is the well-known Expectation-Maximization (EM) algorithm used in problems with incomplete data [13]. The EM algorithm iterates between the

E-step and M-step until convergence. In the E-step, given a current (or initial) estimate of the unknown variable, a conditional expectation is found, which is maximized in the M-step and derives a new estimate. The E-step is based on deduction, while the M-step relies on induction. In the case of Bayesian nets, which model a set of concepts (provided by an expert) through a network of conditional dependencies, predictions (deductions) are made from the (initial) hypotheses (beliefs over conditional dependencies) using the Bayes theorem. Then, inductive inferences regarding the hypotheses are drawn from the data. Note also that if the existing prior knowledge is captured in some rules, learning may become a simplification of these rules such that their logical combinations describe the problem.

In the Aristotelian approach to pattern recognition, observation of particulars and their explanation are essential for deriving a concept. As we already know, abduction plays a role here, especially for data exploration and characterization to explain or suggest a modification of the representation or an adaptation of the given classifier. Aristotelian learning often relies on the Occam's razor principle which advocates to choose the simplest model or hypothesis among otherwise equivalent ones and can be implemented in a number of ways [8].

In summary, the Platonic scenario is dominantly inductive-deductive, while the Aristotelian scenario is dominantly inductive-abductive. Both frameworks have different merits and shortcomings. The strength of the Platonic approach lies in the proper formulation and use of subjective beliefs, expert knowledge and possibility to encode internal structural organization of objects. It is model-driven. In this way, however, the inductive generalization becomes limited, as there may be little freedom in the description to explore and discovery of new knowledge. The strength of the Aristotelian approach lies in a numerical induction and a well-developed mathematical theory of vector spaces in which the actual learning takes place. It is data-driven. The weakness, however, lies in the difficulty to incorporate the expert or background knowledge about the problem. Moreover, in many practical applications, it is known that the implicit assumptions of representative training sets, identical and identically distributed (iid) samples as well as stationary distributions do not hold.

4.3 Two New Pattern Recognition Paradigms

Two far-reaching novel paradigms have been proposed that deal with the drawbacks of the Platonic and Aristotelian approaches. In the Aristotelian scenario, Vapnik has introduced *transductive* learning [73], while in the Platonic scenario, Goldfarb has advocated a new structural learning paradigm [31, 32]. We think these are two major perspectives of the science of pattern recognition.

Vapnik [73] formulated the main learning principle as: *'If you posses a restricted amount of information for solving some problem, try to solve the problem directly and never solve a more general problem as an intermediate step.'* In the traditional Aristotelian scenario, the learning task is often transformed to the problem of function estimation, in which a decision function is determined *globally* for the entire domain (e.g. for all possible examples in a feature vector space). This is, however, a solution to a more general problem than necessary to arrive at a conclusion (output) for *specific* input data. Consequently, the application of this common-sense principle requires a reformulation of the learning problem such that novel (unlabeled) examples are considered in the context of the given training set. This leads to the transductive principle which aims at estimating the output for a *given* input only when required and may differ from an instance to instance. The training sample, considered either globally, or in the local neighborhoods of test examples, is actively used to determine the output. As a result, this leads to confidence measures of single predictions instead of globally estimated classifiers. It provides ways to overcome the difficulty of iid samples and stationary distributions. More formally, in a transductive reasoning, given an entailment $A \models (B \cup C)$, if the consequence B is observed as the result of A, then the consequence C becomes more likely.

The truly transductive principle requires an active synergy of inductive, deductive and abductive principles in a conscious decision process. We believe it is practised by people who analyze complex situations, deduce and validate possible solutions and make decisions in novel ways. Examples are medical doctors, financial advisers, strategy planners or leaders of large organizations. In the context of automatic learning, transduction has applications to learning from partially labeled sets and otherwise missing information, information retrieval, active learning and all types of diagnostics. Some proposals can be found e.g. in [34, 46, 47, 73]. Although many researchers recognize the importance of this principle, many remain also reluctant. This may be caused by unfamiliarity with this idea, few existing procedures, or by the accompanying computational costs as a complete decision process has to be constantly inferred anew.

In the Platonic scenario, Goldfarb and his colleagues have developed structural inductive learning, realized by the so-called evolving transformation systems (ETS) [31, 32]. Goldfarb first noticed the intrinsic and impossible to overcome inadequacy of vector spaces to truly learn from examples [30]. The reason is that such quantitative representations loose all information on object structure; there is no way an object can be generated given its numeric encoding. The second crucial observation was that all objects in the universe have a formative history. This led Goldfarb to the conclusion that an object representation should capture the object's formative evolution, i.e. the way the object is created through a sequence of suitable transformations in time. The creation process is only possible through structural operations. So, *'the resulting representation embodies temporal structural information in the form of a formative, or generative, history'* [31]. Consequently, objects are treated as evolving structural processes and a class is defined by structural processes, which are 'similar'. This is an inductive structural/symbolic class representation, the central concept in ETS. This representation is learnable from a (small) set of examples and has the capability to generate objects from the class.

The generative history of a class starts from a single progenitor and is encoded as a multi-level hierarchical system. On a given level, the basic structural elements are defined together with their structural transformations, such that both are used to constitute a new structural element on a higher level. This new element becomes meaningful on that level. Similarity plays an important role, as it is used as a basic quality for a class representation as a set of similar structural processes. Similarity measure is learned in training to induce the optimal finite set of weighted structural transformations that are necessary on the given level, such that the similarity of an object to the class representation is large. 'This mathematical structure allows one to capture dynamically, during the learning process, the compositional structure of objects/events within a given inductive, or evolutionary, environment' [31].

Goldfarb's ideas bear some similarity to the ones of Wolfram, presented in his book on 'a new kind of science' [80]. Wolfram considers computation as the primary concept in nature; all processes are the results of cellular-automata[6] type of computational processes, and thereby inherently numerical. He observes that repetitive use of simple computational transformations can cause very complex phenomena, especially if computational mechanisms are used at different levels. Goldfarb also discusses dynamical systems, in which complexity is built from simpler structures, through hierarchical folding up (or enrichment). The major difference is that he considers structure of primary interest, which leads to evolving temporal structural processes instead of computational ones.

In summary, Goldfarb proposes a revolutionary paradigm: an ontological model of a class representation in an epistemological context, as it is learnable from examples. This is a truly unique unification. We think it is the most complete and challenging approach to pattern recognition to this date, a breakthrough. By including the formative history of objects into their representation, Goldfarb attributes them some aspects of human consciousness. The far reaching consequence of his ideas is a generalized measurement process that will be one day present in sensors. Such sensors will be able to measure 'in structural units' instead of numerical units (say, meters) as it is currently done. The inductive process over a set of structural units lies at the foundation of new inductive informatics. The difficulty, however, is that the current formalism in mathematics and related fields is not yet prepared for adopting these far-reaching ideas. We, however, believe, they will pave the road and be found anew or rediscovered in the next decennia.

---

[6] Cellular automata are discrete dynamical systems that operate on a regular lattice in space and time, and are characterized by 'local' interactions.

5 Challenges

A lot of research effort is needed before the two novel and far-reaching paradigms are ready for practical applications. So, this section focuses on several challenges that naturally come in the current context and will be summarized for the design of automatic pattern recognition procedures. A number of fundamental problems, related to the various approaches, have already been identified in the previous sections and some will return here on a more technical level. Many of the points raised in this section have been more extensively discussed in [17]. We will emphasize these which have only been touched or are not treated in the standard books [15, 71, 76] or in the review by Jain et al. [45]. The issues to be described are just a selection of the many which are not yet entirely understood. Some of them may be solved in the future by the development of novel procedures or by gaining an additional understanding. Others may remain an issue of concern to be dealt with in each application separately. We will be systematically describe them, following the line of advancement of a pattern recognition system; see also Fig. 1:

- **Representation and background knowledge.** This is the way in which individual real world objects and phenomena are numerically described or encoded such that they can be related to each other in some meaningful mathematical framework. This framework has to allow the generalization to take place.
- **Design set.** This is the set of objects available or selected to develop the recognition system.
- **Adaptation.** This is usually a transformation of the representation such that it becomes more suitable for the generalization step.
- **Generalization.** This is the step in which objects of the design set are related such that classes of objects can be distinguished and new objects can be accurately classified.
- **Evaluation.** This is an estimate of the performance of a developed recognition system.

5.1 Representation and Background Knowledge

The problem of representation is a core issue for pattern recognition [18, 20]. Representation encodes the real world objects by some numerical description, handled by computers in such a way that the individual object representations can be interrelated. Based on that, later a generalization is achieved, establishing descriptions or discriminations between classes of objects. Originally, the issue of representation was almost neglected, as it was reduced to the demand of having discriminative features provided by some expert. Statistical learning is often believed to start in a given feature vector space. Indeed, many books on pattern recognition disregard the topic of representation, simply by assuming that objects are somehow already represented [4, 62]. A systematic study on representation [20, 56] is not easy, as it is application or domain-dependent (where the word *domain* refers to the nature or character of problems and the resulting type of data). For instance, the representations of a time signal, an image of an isolated 2D object, an image of a set of objects on some background, a 3D object reconstruction or the collected set of outcomes of a medical examination are entirely different observations that need individual approaches to find good representations. Anyway, if the starting point of a pattern recognition problem is not well defined, this cannot be improved later in the process of learning. It is, therefore, of crucial importance to study the representation issues seriously. Some of them are phrased in the subsequent sections.

The use of vector spaces. Traditionally, objects are represented by vectors in a feature vector space. This representation makes it feasible to perform some generalization (with respect to this linear space), e.g. by estimating density functions for classes of objects. The object structure is, however, lost in such a description. If objects contain an inherent, identifiable structure or organization, then relations between their elements, like relations between neighboring pixels in an image, are entirely neglected. This also holds for spatial properties encoded by Fourier coefficients or wavelets weights. These original structures may be partially rediscovered by deriving statistics over a set of vectors representing objects, but these are not included in the representation itself. One may wonder whether the representation of objects as vectors in a space is not oversimplified to be able to reflect the nature of objects in a proper way. Perhaps objects might be better represented by convex bodies, curves or by other structures in a metric vector space. The generalization over sets of vectors, however, is heavily studied and mathematically well developed. How to generalize over a set of other structures is still an open question.

The essential problem of the use of vector spaces for object representation is originally pointed out by Goldfarb [30, 33]. He prefers a structural representation in which the original object organization (connectedness of building structural elements) is preserved. However, as a generalization procedure for structural representations does not exist yet, Goldfarb starts from the evolving transformation systems [29] to develop a novel system [31]. As already indicated in Sec. 4.3 we see this as a possible direction for a future breakthrough.

Compactness. An important, but seldom explicitly identified property of representations is compactness [1]. In order to consider classes, which are bounded in their domains, the representation should be constrained: objects that are similar in reality should be close in their representations (where the closeness is captured by an appropriate relation, possibly a proximity measure). If this demand is not satisfied, objects may be described capriciously and, as a result, no generalization is possible. This compactness assumption puts some restriction on the possible probability density functions used to describe classes in a representation vector space. This, thereby, also narrows the set of possible classification problems. A formal description of the probability distribution of this set may be of interest to estimate the expected performance of classification procedures for an arbitrary problem.

In Sec. 3, we pointed out that the lack of a formal restriction of pattern recognition problems to those with a compact representation was the basis of pessimistic results like the No-Free-Lunch Theorem [81] and the classification error bounds resulting from the VC complexity measure [72, 73]. One of the main challenges for pattern recognition to find a formal description of compactness that can be used in error estimators the average over the set of possible pattern recognition problems.

Representation types. There exists numerous ways in which representations can be derived. The basic 'numerical' types are now distinguished as:

- *Features.* Objects are described by characteristic attributes. If these attributes are continuous, the representation is usually compact in the corresponding feature vector space. Nominal, categorical or ordinal attributes may cause problems. Since a description by features is a reduction of objects to vectors, different objects may have identical representations, which may lead to class overlap.
- *Pixels or other samples.* A complete representation of an object may be approximated by its sampling. For images, these are pixels, for time signals, these are time samples and for spectra, these are wavelengths. A pixel representation is a specific, boundary case of a feature representation, as it describes the object properties in each point of observation.
- *Probability models.* Object characteristics may be reflected by some probabilistic model. Such models may be based on expert knowledge or trained from examples. Mixtures of knowledge and probability estimates are difficult, especially for large models.
- *Dissimilarities, similarities or proximities.* Instead of an absolute description by features, objects are relatively described by their dissimilarities to a collection of specified objects. These may be carefully optimized prototypes or representatives for the problem, but also random subsets may work well [56]. The dissimilarities may be derived from raw data, such as images, spectra or time samples, from original feature representations or from structural representations such as strings or relational graphs. If the dissimilarity measure is nonnegative and zero only for two identical objects, always belonging to the same class, the class overlap may be avoided by dissimilarity representations.
- *Conceptual representations.* Objects may be related to classes in various ways, e.g. by a set of classifiers, each based on a different representation, training set or model. The combined set of these initial classifications or clusterings constitute a new representation [56]. This is used in the area of combining clusterings [24, 25] or combining classifiers [49].

In the structural approaches, objects are represented in qualitative ways. The most important are strings or sequences, graphs and their collections and hierarchical representations in the form of ontological trees or semantic nets.

Vectorial object descriptions and proximity representations provide a good way for generalization in some appropriately determined spaces. It is, however, difficult to integrate them with the detailed prior or background knowledge that one has on the problem. On the other hand, probabilistic models and, especially, structural models are well suited for such an integration. The later, however, constitute a weak basis for training general classification schemes. Usually, they are limited to assigning objects to the class model that fits best, e.g. by the nearest neighbor rule. Other statistical learning techniques are applied to these if given an appropriate proximity measure or a vectorial representation space found by graph embeddings [79].

It is a challenge to find representations that constitute a good basis for modeling object structure and which can also be used for generalizing from examples. The next step is to find representations not only based on background knowledge or given by the expert, but to learn or optimize them from examples.

5.2 Design Set

A pattern recognition problem is not only specified by a representation, but also by the set of examples given for training and evaluating a classifier in various stages. The selection of this set and its usage strongly influence the overall performance of the final system. We will discuss some related issues.

Multiple use of the training set. The entire design set or its parts are used in several stages during the development of a recognition system. Usually, one starts from some exploration, which may lead to the removal of wrongly represented or erroneously labeled objects. After gaining some insights into the problem, the analyst may select a classification procedure based on the observations. Next, the set of objects may go through some preprocessing and normalization. Additionally, the representation has to be optimized, e.g. by a feature/object selection or extraction. Then, a series of classifiers has to be trained and the best ones need to be selected or combined. An overall evaluation may result in a re-iteration of some steps and different choices.

In this complete process the same objects may be used a number of times for the estimation, training, validation, selection and evaluation. Usually, an expected error estimation is obtained by a cross-validation or hold-out method [32, 77]. It is well known that the multiple use of objects should be avoided as it biases the results and decisions. Re-using objects, however, is almost unavoidable in practice. A general theory does not exist yet, that predicts how much a training set is 'worn-out' by its repetitive use and which suggests corrections that can diminish such effects.

Representativeness of the training set. Training sets should be representative for the objects to be classified by the final system. It is common to take a randomly selected subset of the latter for training. Intuitively, it seems to be useless to collect many objects represented in the regions where classes do not overlap. On the contrary, in the proximity of the decision boundary, a higher sampling rate seems to be advantageous. This depends on the complexity of the decision function and the expected class overlap, and is, of course, inherently related to the chosen procedure.

Another problem are the unstable, unknown or undetermined class distributions. Examples are the impossibility to characterize the class of non-faces in the face detection problem, or in machine diagnostics, the probability distribution of all casual events if the machine is used for undetermined production purposes. A training set that is representative for the class distributions cannot be found in such cases. An alternative may be to sample the *domain* of the classes such that all possible object occurrences are approximately covered. This means that for any object that could be encountered in practice there exists a sufficiently similar object in the training set, defined in relation to the specified class differences. Moreover, as class density estimates cannot be derived for such a training set, class posterior probabilities cannot be estimated. For this reason such a type of domain based sampling is only appropriate for non-overlapping classes. In particular, this problem is of interest for non-overlapping (dis)similarity based representations [18].

Consequently, we wonder whether it is possible to use a more general type of sampling than the classical iid sampling, namely the domain sampling. If so, the open questions refer to the verification of dense samplings and types of new classifiers that are explicitly built on such domains.

5.3 Adaptation

Once a recognition problem has been formulated by a set of example objects in a convenient representation, the generalization over this set may be considered, finally leading to a recognition system. The selection of a proper generalization procedure may not be evident, or several disagreements may exist between the realized and preferred procedures. This occurs e.g. when the chosen representation needs a non-linear classifier and only linear decision functions are computationally feasible, or when the space dimensionality is high with respect to the size of the training set, or the representation cannot be perfectly embedded in a Euclidean space, while most classifiers demand that. For reasons like these, various adaptations of the representation may be considered. When class differences are explicitly preserved or emphasized, such an adaptation may be considered as a part of the generalization procedure. Some adaptation issues that are less connected to classification are discussed below.

Problem complexity. In order to determine which classification procedures might be beneficial for a given problem, Ho and Basu [43] proposed to investigate its complexity. This is an ill-defined concept. Some of its aspects include data organization, sampling, irreducibility (or redundancy) and the interplay between the local and global character of the representation and/or of the classifier. Perhaps several other attributes are needed to define complexity such that it can be used to indicate a suitable pattern recognition solution to a given problem; see also [2].

Selection or combining. Representations may be complex, e.g. if objects are represented by a large amount of features or if they are related to a large set of prototypes. A collection of classifiers can be designed to make use of this fact and later combined. Additionally, also a number of representations may be considered simultaneously. In all these situations, the question arises on which should be preferred: a selection from the various sources of information or some type of combination. A selection may be random or based on a systematic search for which many strategies and criteria are possible [49]. Combinations may sometimes be fixed, e.g. by taking an average, or a type of a parameterized combination like a weighted linear combination as a principal component analysis; see also [12, 56, 59].

The choice favoring either a selection or combining procedure may also be dictated by economical arguments, or by minimizing the amount of necessary measurements, or computation. If this is unimportant, the decision has to be made according to the accuracy arguments. Selection neglects some information, while combination tries to use everything. The latter, however, may suffer from overtraining as weights or other parameters have to be estimated and may be adapted to the noise or irrelevant details in the data. The sparse solutions offered by support vector machines [67] and sparse linear programming approaches [28, 35] constitute a way of compromise. How to optimize them efficiently is still a question.

Nonlinear transformations and kernels. If a representation demands or allows for a complicated, nonlinear solution, a way to proceed is to transform the representation appropriately such that linear aspects are emphasized. A simple (e.g. linear) classifier may then perform well. The use of kernels, see Sec. 3, is a general possibility. In some applications, indefinite kernels are proposed as being consistent with the background knowledge. They may result in non-Euclidean dissimilarity representations, which are challenging to handle; see [57] for a discussion.

5.4 Generalization

The generalization over sets of vectors leading to class descriptions or discriminants was extensively studied in pattern recognition in the 60's and 70's of the previous century. Many classifiers were designed, based on the assumption of normal distributions, kernels or potential functions, nearest neighbor rules, multi-layer perceptrons, and so on [15, 45, 62, 76]. These types of studies were later extended by the fields of multivariate statistics, artificial neural networks and machine learning. However, in the pattern recognition community, there is still a high interest in the classification problem, especially in relation to practical questions concerning issues of combining classifiers, novelty detection or the handling of ill-sampled classes.

Handling multiple solutions. Classifier selection or classifier combination. Almost any more complicated pattern recognition problem can be solved in multiple ways. Various choices can be made for the representation, the adaptation and the classification. Such solutions usually do not only differ in the total classification performance, they may also make different errors. Some type of combining classifiers will thereby be advantageous [49]. It is to be expected that in the future most pattern recognition systems for real world problems are constituted of a set of classifiers. In spite of the fact that this area is heavily studied, a general approach on how to select, train and combine solutions is still not available. As training sets have to be used for optimizing several subsystems, the problem how to design complex systems is strongly related to the above issue of multiple use of the training set.

Classifier typology. Any classification procedure has its own explicit or built-in assumptions with respect to data inherent characteristics and the class distributions. This implies that a procedure will lead to relatively good performance if a problem fulfils its exact assumptions. Consequently, any classification approach has its problem for which it is the best. In some cases such a problem might be far from practical application. The construction of such problems may reveal which typical characteristics of a particular procedure are. Moreover, when new proposals are to be evaluated, it may be demanded that some examples of its corresponding typical classification problem are published, making clear what the area of application may be; see [19].

Generalization principles. The two basic generalization principles, see Section 4, are probabilistic inference, using the Bayes-rule [63] and the minimum description length principle that determines the most simple model in agreement with the observations (based on Occam's razor) [37]. These two principles are essentially different[7]. The first one is sensitive to multiple copies of an existing object in the training set, while the second one is not. Consequently, the latter is not based on densities, but just on object differences or distances. An important issue is to find in which situations each of these principle should be recommended and whether the choice should be made in the beginning, in the selection of the design set and the way of building a representation, or it should be postpone until a later stage.

The use of unlabeled objects and active learning. The above mentioned principles are examples of statistical inductive learning, where a classifier is

[7] Note that Bayesian inference is also believed to implement the Occam's razor [8] in which preference for simpler models is encoded by encouraging particular prior distributions. This is, however, not the primary point as it is in the minimum description length principle.

induced based on the design set and it is later applied to unknown objects. The disadvantage of such approach is that a decision function is in fact designed for all possible representations, whether valid or not. Transductive learning, see Section 4.3, is an appealing alternative as it determines the class membership only for the objects in question, while relying on the collected design set or its suitable subset [73]. The use of unlabeled objects, not just the one to be classified, is a general principle that may be applied in many situations. It may improve a classifier based on just a labeled training set. If this is understood properly, the classification of an entire test set may yield better results than the classification of individuals.

Classification or class detection. Two-class problems constitute the traditional basic line in pattern recognition, which reduces to finding a discriminant or a binary decision function. Multi-class problems can be formulated as a series of two-class problems. This can be done in various ways, none of them is entirely satisfactory. An entirely different approach is the description of individual classes by so-called one-class classifiers [69, 70]. In this way the focuss is given to class description instead of to class separation. This brings us to the issue of the structure of a class.

Traditionally classes are defined by a distribution in the representation space. However, the better such a representation, the higher its dimensionality, the more difficult it is to estimate a probability density function. Moreover, as we have seen above, it is for some applications questionable whether such a distribution exist. A class is then a part of a possible non-linear manifold in a high-dimensional space. It has a structure instead of a density distribution. It is a challenge to use this approach for building entire pattern recognition systems.

5.5 Evaluation

Two questions are always apparent in the development of recognition systems. The first refers to the overall performance of a particular system once it is trained, and has sometimes a definite answer. The second question is more open and asks which good recognition procedures are in general.

Recognition system performance. Suitable criteria should be used to evaluate the overall performance of the entire system. Different measures with different characteristics can be applied, however, usually, only a single criterion is used. The basic ones are the average accuracy computed over all validation objects or the accuracy determined by the worst-case scenario. In the first case, we again assume that the set of objects to be recognized is well defined (in terms of distributions). Then, it can be sampled and the accuracy of the entire system is estimated based on the evaluation set. In this case, however, we neglect the issue that after having used this evaluation set together with the training set, a better system could have been found. A more interesting point is how to judge the performance of a system if the distribution of objects is ill-defined or if a domain based classification system is used as discussed above. Now, the largest mistake that is made becomes a crucial factor for this type of judgements. One needs to be careful, however, as this may refer to an unimportant outlier (resulting e.g. from invalid measurements).

Practice shows that a single criterion, like the final accuracy, is insufficient to judge the overall performance of the whole system. As a result, multiple performance measures should be taken into account, possibly at each stage. These measures should not only reflect the correctness of the system, but also its flexibility to cope with unusual situations in which e.g. specific examples should be rejected or misclassification costs incorporated.

Prior probability of problems. As argued above, any procedure has a problem for which it performs well. So, we may wonder how large the class of such problems is. We cannot state that any classifier is better than any other classifier, unless the distribution of problems to which these classifiers will be applied is defined. Such distributions are hardly studied. What is done at most is that classifiers are compared over a collection of benchmark problems. Such sets are usually defined ad hoc and just serve as an illustration. The collection of problems to which a classification procedure will be applied is not defined. As argued in Section 3, it may be as large as all problems with a compact representation, but preferably not larger.

6 Discussion and Conclusions

Recognition of patterns and inference skills lie at the core of human learning. It is a human activity that we try to imitate by mechanical means. There are no physical laws that assign observations to classes. It is the human consciousness that groups observations together. Although their connections and interrelations are often hidden, some understanding may be gained in the attempt of imitating this process. The human process of learning patterns from examples may follow along the lines of trial and error. By freeing our minds of fixed beliefs and petty details we may not only understand single observations but also induce principles and formulate concepts that lie behind the observed facts. New ideas can be born then. These processes of abstraction and concept formation are necessary for development and survival. In practice, (semi-)automatic learning systems are built by imitating such abilities in order to gain understanding of the problem, explain the underlying phenomena and develop good predictive models.

It has, however, to be strongly doubted whether statistics play an important role in the human learning process. Estimation of probabilities, especially in multivariate situations is not very intuitive for majority of people. Moreover, the amount of examples needed to build a reliable classifier by statistical means is much larger than it is available for humans. In human recognition, proximity based on relations between objects seems to come before features are searched and may be, thereby, more fundamental. For this reason and the above observation, we think that the study of proximities, distances and domain based classifiers are of great interest. This is further encouraged by the fact that such representations offer a bridge between the possibilities of learning in vector spaces and the structural description of objects that preserve relations between objects inherent structure. We think that the use of proximities for representation, generalization and evaluation constitute the most intriguing issues in pattern recognition.

The existing gap between structural and statistical pattern recognition partially coincides with the gap between knowledge and observations. Prior knowledge and observations are both needed in a subtle interplay to gain new knowledge. The existing knowledge is needed to guide the deduction process and to generate the models and possible hypotheses needed by induction, transduction and abduction. But, above all, it is needed to select relevant examples and a proper representation. If and only if the prior knowledge is made sufficiently explicit to set this environment, new observations can be processed to gain new knowledge. If this is not properly done, some results may be obtained in purely statistical terms, but these cannot be integrated with what was already known and have thereby to stay in the domain of observations. The study of automatic pattern recognition systems makes perfectly clear that learning is possible, only if the Platonic and Aristotelian scientific approaches cooperate closely. This is what we aim for.

References

[1] A.G. Arkadev and E.M. Braverman. *Computers and Pattern Recognition.* Thompson, Washington, DC, 1966.

[2] M. Basu and T.K. Ho, editors. *Data Complexity in Pattern Recognition.* Springer, 2006.

[3] R. Bergmann. *Developing Industrial Case-Based Reasoning Applications.* Springer, 2004.

[4] C.M. Bishop. *Neural Networks for Pattern Recognition.* Oxford University Press, 1995.

[5] H. Bunke. Recent developments in graph matching. In *International Conference on Pattern Recognition*, volume 2, pages 117–124, 2000.

[6] H. Bunke, S. Günter, and X. Jiang. Towards bridging the gap between statistical and structural pattern recognition: Two new concepts in graph matching. In *International Conference on Advances in Pattern Recognition*, pages 1–11, 2001.

[7] H. Bunke and K. Shearer. A graph distance metric based on the maximal common subgraph. *Pattern Recognition Letters*, 19(3-4):255–259, 1998.

[8] V.S. Cherkassky and F. Mulier. *Learning from data: Concepts, Theory and Methods.* John Wiley & Sons, Inc., New York, NY, USA, 1998.

[9] T.M. Cover. Geomerical and statistical properties of systems of linear inequalities with applications in pattern recognition. *IEEE Transactions on Electronic Computers*, EC-14:326–334, 1965.

[10] T.M. Cover and P.E. Hart. Nearest Neighbor Pattern Classification. *IEEE Transactions on Information Theory*, 13(1):21–27, 1967.

[11] T.M. Cover and J.M. van Campenhout. On the possible orderings in the measurement selection problem. *IEEE Transactions on Systems, Man, and Cybernetics*, SMC-7(9):657–661, 1977.

[12] I.M. de Diego, J.M. Moguerza, and A. Muñoz. Combining kernel information for support vector classification. In *Multiple Classifier Systems*, pages 102–111. Springer-Verlag, 2004.

[13] A. Dempster, N. Laird, and D. Rubin. Maximum likelihood from incomplete data via the EM algorithm. *Journal of the Royal Statistical Society*, Series B, 39(1):1–38, 1977.

[14] L. Devroye, L. Györfi, and G. Lugosi. *A Probabilistic Theory of Pattern Recognition.* Springer-Verlag, 1996.

[15] R.O. Duda, P.E. Hart, and D.G. Stork. *Pattern Classification.* John Wiley & Sons, Inc., 2nd edition, 2001.

[16] R.P.W. Duin. Four scientific approaches to pattern recognition. In *Fourth Quinquennial Review 1996-2001. Dutch Society for Pattern Recognition and Image Processing*, pages 331–337. NVPHBV, Delft, 2001.

[17] R.P.W. Duin and E. Pękalska. Open issues in pattern recognition. In *Computer Recognition Systems*, pages 27–42. Springer, Berlin, 2005.

[18] R.P.W. Duin, E. Pękalska, P. Paclík, and D.M.J. Tax. The dissimilarity representation, a basis for domain based pattern recognition? In L. Goldfarb, editor, *Pattern representation and the future of pattern recognition, ICPR 2004 Workshop Proceedings*, pages 43–56, Cambridge, United Kingdom, 2004.

[19] R.P.W. Duin, E. Pękalska, and D.M.J. Tax. The characterization of classification problems by classifier disagreements. In *International Conference on Pattern Recognition*, volume 2, pages 140–143, Cambridge, United Kingdom, 2004.

[20] R.P.W. Duin, F. Roli, and D. de Ridder. A note on core research issues for statistical pattern recognition. *Pattern Recognition Letters*, 23(4):493–499, 2002.

[21] S. Edelman. *Representation and Recognition in Vision.* MIT Press, Cambridge, 1999.

[22] B. Efron and R.J. Tibshirani. *An Introduction to the Bootstrap.* Chapman & Hall, London, 1993.

[23] P. Flach and A. Kakas, editors. *Abduction and Induction: essays on their relation and integration.* Kluwer Academic Publishers, 2000.

[24] A. Fred and A.K. Jain. Data clustering using evidence accumulation. In *International Conference on Pattern Recognition*, pages 276–280, Quebec City, Canada, 2002.

[25] A. Fred and A.K. Jain. Robust data clustering. In *Conf. on Computer Vision and Pattern Recognition*, pages 442 –451, Madison - Wisconsin, USA, 2002.

[26] K.S. Fu. *Syntactic Pattern Recognition and Applications.* Prentice-Hall, 1982.

[27] K. Fukunaga. *Introduction to Statistical Pattern Recognition.* Academic Press, 1990.

[28] G.M. Fung and O.L. Mangasarian. A Feature Selection Newton Method for Support Vector Machine Classification. *Computational Optimization and Aplications*, 28(2):185–202, 2004.

[29] L. Goldfarb. On the foundations of intelligent processes – I. An evolving model for pattern recognition. *Pattern Recognition*, 23(6):595–616, 1990.

[30] L. Goldfarb, J. Abela, V.C. Bhavsar, and V.N. Kamat. Can a vector space based learning model discover inductive class generalization in a symbolic environment? *Pattern Recognition Letters*, 16(7):719–726, 1995.

[31] L. Goldfarb and D. Gay. What is a structural representation? Fifth variation. Technical Report TR05-175, University of New Brunswick, Fredericton, Canada, 2005.

[32] L. Goldfarb and O. Golubitsky. What is a structural measurement process? Technical Report TR01-147, University of New Brunswick, Fredericton, Canada, 2001.

[33] L. Goldfarb and J. Hook. Why classical models for pattern recognition are not pattern recognition models. In *International Conference on Advances in Pattern Recognition*, pages 405–414, 1998.

[34] T. Graepel, R. Herbrich, and K. Obermayer. Bayesian transduction. In *Advances in Neural Information System Processing*, pages 456–462, 2000.

[35] T. Graepel, R. Herbrich, B. Schölkopf, A. Smola, P. Bartlett, K.-R. Müller, K. Obermayer, and R. Williamson. Classification on proximity data with LP-machines. In *International Conference on Artificial Neural Networks*, pages 304–309, 1999.

[36] U. Grenander. *Abstract Inference.* John Wiley & Sons, Inc., 1981.

[37] P. Grünwald, I.J. Myung, and Pitt M., editors. *Advances in Minimum Description Length: Theory and Applications.* MIT Press, 2005.

[38] B. Haasdonk. Feature space interpretation of SVMs with indefinite kernels. *IEEE Transactions on Pattern Analysis and Machine Intelligence*, 25(5):482–492, 2005.

[39] I. Hacking. *The emergence of probability.* Cambridge University Press, 1974.

[40] G. Harman and S. Kulkarni. *Reliable Reasoning: Induction and Statistical Learning Theory.* MIT Press, to appear.

[41] S. Haykin. *Neural Networks, a Comprehensive Foundation, second edition.* Prentice-Hall, 1999.

[42] D. Heckerman. A tutorial on learning with Bayesian networks. In M. Jordan, editor, *Learning in Graphical Models*, pages 301–354. MIT Press, Cambridge, MA, 1999.

[43] T.K. Ho and M. Basu. Complexity measures of supervised classification problems. *IEEE Transactions on Pattern Analysis and Machine Intelligence*, 24(3):289–300, 2002.
[44] A. K. Jain and B. Chandrasekaran. Dimensionality and sample size considerations in pattern recognition practice. In P. R. Krishnaiah and L. N. Kanal, editors, *Handbook of Statistics*, volume 2, pages 835–855. North-Holland, Amsterdam, 1987.
[45] A.K. Jain, R.P.W. Duin, and J. Mao. Statistical pattern recognition: A review. *IEEE Transactions on Pattern Analysis and Machine Intelligence*, 22(1):4–37, 2000.
[46] T. Joachims. Transductive inference for text classification using support vector machines. In I. Bratko and S. Dzeroski, editors, *International Conference on Machine Learning*, pages 200–209, 1999.
[47] T. Joachims. Transductive learning via spectral graph partitioning. In *International Conference on Machine Learning*, 2003.
[48] T.S. Kuhn. *The Structure of Scientific Revolutions.* University of Chicago Press, 1970.
[49] L.I. Kuncheva. *Combining Pattern Classifiers. Methods and Algorithms.* Wiley, 2004.
[50] J. Laub and K.-R. Müller. Feature discovery in non-metric pairwise data. *Journal of Machine Learning Research*, pages 801–818, 2004.
[51] A. Marzal and E. Vidal. Computation of normalized edit distance and applications. *IEEE Transactions on Pattern Analysis and Machine Intelligence*, 15(9):926–932, 1993.
[52] R.S. Michalski. Inferential theory of learning as a conceptual basis for multistrategy learning. *Machine Learning*, 11:111–151, 1993.
[53] T. Mitchell. *Machine Learning.* McGraw Hill, 1997.
[54] Richard E. Neapolitan. *Probabilistic reasoning in expert systems: theory and algorithms.* John Wiley & Sons, Inc., New York, NY, USA, 1990.
[55] C.S. Ong, S. Mary, X.and Canu, and Smola A.J. Learning with non-positive kernels. In *International Conference on Machine Learning*, pages 639–646, 2004.
[56] E. Pękalska and R.P.W. Duin. *The Dissimilarity Representation for Pattern Recognition. Foundations and Applications.* World Scientific, Singapore, 2005.
[57] E. Pękalska, R.P.W. Duin, S. Günter, and H. Bunke. On not making dissimilarities Euclidean. In *Joint IAPR International Workshops on SSPR and SPR*, pages 1145–1154. Springer-Verlag, 2004.
[58] E. Pękalska, P. Paclík, and R.P.W. Duin. A Generalized Kernel Approach to Dissimilarity Based Classification. *Journal of Machine Learning Research*, 2:175–211, 2002.
[59] E. Pękalska, M. Skurichina, and R.P.W. Duin. Combining Dissimilarity Representations in One-class Classifier Problems. In *Multiple Classifier Systems*, pages 122–133. Springer-Verlag, 2004.

[60] L.I. Perlovsky. Conundrum of combinatorial complexity. *IEEE Transactions on Pattern Analysis and Machine Intelligence*, 20(6):666–670, 1998.
[61] P. Pudil, J. Novovićova, and J. Kittler. Floating search methods in feature selection. *Pattern Recognition Letters*, 15(11):1119–1125, 1994.
[62] B. Ripley. *Pattern Recognition and Neural Networks.* Cambridge University Press, Cambridge, 1996.
[63] C.P. Robert. *The Bayesian Choice.* Springer-Verlag, New York, 2001.
[64] K.M. Sayre. *Recognition, a study in the philosophy of artificial intelligence.* University of Notre Dame Press, 1965.
[65] M.I. Schlesinger and Hlavác. *Ten Lectures on Statistical and Structural Pattern Recognition.* Kluwer Academic Publishers, 2002.
[66] B. Schölkopf and A.J. Smola. *Learning with Kernels.* MIT Press, Cambridge, 2002.
[67] J. Shawe-Taylor and N. Cristianini. *Kernel methods for pattern analysis.* Cambridge University Press, UK, 2004.
[32] M. Stone. Cross-validation: A review. *Mathematics, Operations and Statistics*, (9):127–140, 1978.
[69] D.M.J. Tax. *One-class classification. Concept-learning in the absence of counter-examples.* PhD thesis, Delft University of Technology, The Netherlands, 2001.
[70] D.M.J. Tax and R.P.W. Duin. Support vector data description. *Machine Learning*, 54(1):45–56, 2004.
[71] F. van der Heiden, R.P.W. Duin, D. de Ridder, and D.M.J. Tax. *Classification, Parameter Estimation, State Estimation: An Engineering Approach Using MatLab.* Wiley, New York, 2004.
[72] V. Vapnik. *Estimation of Dependences based on Empirical Data.* Springer Verlag, 1982.
[73] V. Vapnik. *Statistical Learning Theory.* John Wiley & Sons, Inc., 1998.
[74] L.-X. Wang and J.M. Mendel. Generating fuzzy rules by learning from examples. *IEEE Transactions on Systems, Man, and Cybernetics*, 22(6):1414–1427, 1992.
[75] S. Watanabe. *Pattern Recognition, Human and Mechanical.* John Wiley & Sons, 1985.
[76] A. Webb. *Statistical Pattern Recognition.* John Wiley & Sons, Ltd., 2002.
[77] S.M. Weiss and C.A. Kulikowski. *Computer Systems That Learn.* Morgan Kaufmann, 1991.
[78] R.C. Wilson and E.R. Hancock. Structural matching by discrete relaxation. *IEEE Transactions on Pattern Analysis and Machine Intelligence*, 19(6):634–648, 1997.
[79] R.C. Wilson, B. Luo, and E.R. Hancock. Pattern vectors from algebraic graph theory. *IEEE Transactions on Pattern Analysis and Machine Intelligence*, 27(7):1112–1124, 2005.
[80] S. Wolfram. *A new kind of science.* Wolfram Media, 2002.
[81] D.H. Wolpert. *The Mathematics of Generalization.* Addison-Wesley, 1995.

[82] R.R. Yager, M. Fedrizzi, and J. (Eds) Kacprzyk. *Advances in the Dempster-Shafer Theory of Evidence.* Wesley, 1994.

[83] C.H. Yu. Quantitative methodology in the perspectives of abduction, deduction, and induction. In *Annual Meeting of American Educational Research Association*, San Francisco, CA, 2006.

Towards Comprehensive Foundations of Computational Intelligence

Włodzisław Duch

[1] Department of Informatics, Nicolaus Copernicus University, Grudziądzka 5, Toruń, Poland,
[2] School of Computer Engineering, Nanyang Technological University, Singapore.

Summary. Although computational intelligence (CI) covers a vast variety of different methods it still lacks an integrative theory. Several proposals for CI foundations are discussed: computing and cognition as compression, meta-learning as search in the space of data models, (dis)similarity based methods providing a framework for such meta-learning, and a more general approach based on chains of transformations. Many useful transformations that extract information from features are discussed. Heterogeneous adaptive systems are presented as particular example of transformation-based systems, and the goal of learning is redefined to facilitate creation of simpler data models. The need to understand data structures leads to techniques for logical and prototype-based rule extraction, and to generation of multiple alternative models, while the need to increase predictive power of adaptive models leads to committees of competent models. Learning from partial observations is a natural extension towards reasoning based on perceptions, and an approach to intuitive solving of such problems is presented. Throughout the paper neurocognitive inspirations are frequently used and are especially important in modeling of the higher cognitive functions. Promising directions such as liquid and laminar computing are identified and many open problems presented.

1 Introduction

Computational intelligence emerged from interactions of several research communities with overlapping interests, inspired by observations of natural information processing. H. Spencer in his "Principles of Psychology" published in 1855 drew neural network diagrams and made a conjecture that all intelligence may be interpreted in terms of successful associations between psychological states driven by the strength of connections between the internal states (see the early history of connectionism in [174]).

Research in artificial neural networks (ANNs) grew out from attempts to drastically simplify biophysical neural models, and for a long time was focused on logical and graded response (sigmoidal) neurons [5] used for classification, approximation, association, and vector quantization approaches used

Włodzisław Duch: *Towards Comprehensive Foundations of Computational Intelligence*, Studies in Computational Intelligence (SCI) **63**, 261–316 (2007)
www.springerlink.com

for clusterization and self-organization [108]. Later any kind of basis function expansion, used since a long time in approximation theory [145] and quite common in pattern recognition, became "a neural network" [4], with radial basis function (RBF) networks [142] becoming a major alternative to multilayer perceptron networks (MLPs). However, as pointed out by Minsky and Papert [133] there are some problems that such networks cannot solve. Although these authors were wrong about the XOR (or the parity) problem which is easily solved by adding hidden neurons to the network, they were right about the topological invariants of patterns, in particular about the problem of connectedness (determining if the pattern is connected or disconnected). Such problems can only be solved with a different type of neurons that include at least one phase-sensitive parameter [113], or with spiking neurons [175]. Computational neuroscience is based on spiking neurons [69], and although mathematical characterization of their power has been described [121] their practical applications are still limited. On the other hand feedforward artificial neural networks found wide applications in data analysis [146] and knowledge extraction [62]. Better understanding of mathematical foundations brought extensions of neural techniques towards statistical pattern recognition models, such as the Support Vector Machines (SVMs) [157] for supervised learning and Independent Component Analysis [92] and similar techniques for unsupervised learning.

Most networks are composed of elements that perform very simple functions, such as squashed weighted summation of their inputs, or some distance-based function [57, 58]. Connectionist modeling in psychology [153] introduced nodes representing whole concepts, or states of network subconfigurations, although their exact relations to neural processes were never elucidated. It was natural from this point of view to extend connectionist networks to all kinds of graphical models [101], including Bayesian belief networks and abstract network models for parallel distributed processing. Concept nodes, representing information derived from perception or from some measurements, are not too precise and thus may be represented in terms of fuzzy membership functions. Fuzzy models may be formally derived as generalization of multivalued logics, but the field has also rather obvious cognitive inspirations related to models of intelligent behavior at a higher, psychological level, rather than elementary neural level. Sets of fuzzy rules have a natural graphical representation [118], and are deeply connected to neural networks [37]. Fuzzy rules organized in a network form may be tuned by adaptive techniques used in neural networks, therefore they are called neurofuzzy systems [135, 138]. Thus fuzzy and neural systems are at the core of computational intelligence.

Brains and sensory organs have been structured and optimized by the evolutionary processes to perform specific functions. This inspiration led to introduction of evolutionary programming [66, 73], and later also other biologically-inspired optimization approaches, such as ant, swarm, and immunological system algorithms [16, 107, 30], that can be used for optimization of adaptive parameters in neural and neurofuzzy systems. Although the algorithms based on these diverse biological inspirations are used for similar applications the time-scales of evolutionary, behavioral and immunological processes involved are very different, and the type of intelligence they are related to is also quite different.

The three main branches of computational intelligence are thus inspired by evolutionary processes that structured brains and intelligence, low-level brain processes that enable perception and sensorimotor reactions (primary sensory and motor cortices), and intermediate level fuzzy concepts and associative reasoning (higher sensory areas in temporal and parietal lobes). To cover all phenomena related to intelligence in a computational framework representation and reasoning based on complex knowledge structures is needed. Traditionally artificial intelligence (AI) has been concerned with high level cognition, using the symbolic knowledge modeling to solve problems that require sequential reasoning, planning and understanding of language, but ignoring learning and associative memories. High-level cognition requires different approach than perception-action sequences at the lower cognitive level, where artificial neural networks, pattern recognition and control techniques are used. Knowledge used in reasoning and understanding language is based on a large number of concepts with complex structure, huge amount of diverse information that may be combined in an infinite number of ways. Making inferences from partial observations requires systematic search techniques and may draw inspirations from decision-making processes in which prefrontal cortex of the brain is involved. One of the big challenges facing CI community is integration of the good-old fashioned artificial intelligence (GOFAI). Although some attempts in this direction have been made [90] typical textbooks on artificial intelligence [181, 154] include very little information on neural networks or fuzzy systems, with learning reduced to probabilistic, Bayesian models, maximum likelihood approaches, and symbolic machine learning methods. Overlap between the two communities in terms of conference topics, or journal publications has always been minimal. Each branch of CI has its natural areas of application requiring methods that may not overlap significantly. Even neural networks and pattern recognition communities, despite a considerable overlap in applications, tend to be separated.

This situation may change in near future. Development of artificial pets or other autonomous systems that should survive in hostile environment is a great challenge for signal analysis to model perception, control systems for behavioral modeling, and perception-based reasoning including attention. Autonomous agents, such as robots, need to reason using both abstract knowledge and information based on perceptions, information coming from sensors, categorized into information granules that are easier to handle. Efforts in cognitive robotics require combination of high behavioral competence with human-level higher cognitive competencies. Autonomous agents should be based on cognitive architectures that integrate low and high-level cognitive functions. This area is slowly gaining popularity and will be a natural meeting ground for all branches of computational intelligence. Development of chatterbots that involve people in interesting conversation is based now on natural language processing and knowledge representation techniques, but it remains to be seen how far one can go in this direction without including real perception. Another driving force that should encourage the use of search techniques that form the basis for AI problem solving is the "crises of the richness" that afflicts computational intelligence. Recent component-based data mining packages[1] contain hundreds of learning methods, input transformations, pre- and post-processing components that may be combined in thousands of ways. Some form of meta-learning that should automatically discover interesting models is urgently needed and it has to be based on search in the model space. Heuristic search techniques have been developed to solve complex combinatorial problems and CI has reached the stage now where they should be used.

In this paper an attempt is made to outline foundations for a large part of computational intelligence research, identify open problems and promising directions, show how to solve this "crises of the richness" and how to go beyond pattern recognition, towards problems that are of interest in artificial intelligence, such as learning from partial observations or perceptions, and systematic reasoning based on perceptions. The emphasis here is on architectures and capabilities of models, rather then learning techniques, therefore evolutionary and other optimization approaches are not discussed. In the second section foundations of computational intelligence are discussed. Computing and cognition seen from the perspective of compression of information is analyzed, introducing a new measure of syntactic and semantic information content, heterogeneous systems that may discover specific bias in the data, and meta-learning scheme to discover good data models in an automatic way. In the third section a general CI theory based on composition of transformations is outlined, showing how new information is generated, extracted from the data, how to use heterogeneous learning systems, redefine the goal of learning in case of difficult problems, understand data in the similarity-based framework and use many individual models in meta-learning schemes. Section four shows how to go beyond pattern recognition using intuitive computing and correlation machines. Neurocognitive inspirations are very important at every step and are discussed in section five. A summary of open problems closes this paper.

2 Searching for Computational Intelligence Foundations

A quick glance on some books with "computational intelligence" title [109, 134, 138] shows that the field still lacks a coherent framework. Many branches of CI are presented one after another, with distant biological inspirations as a common root, although in case of such statistical techniques as kernel

[1] See: http://en.wikipedia.org/wiki/Data_mining

methods or SVMs such inspirations cannot be provided. Different experts define computational intelligence as a collection of computational techniques that are glued together only for historical or even personal reasons. Modern pattern recognition textbooks [63, 176, 166] start from the Bayesian probabilistic foundations that may be used to justify both discriminat as well as the nearest neighbor type of learning methods. Supervised and unsupervised pre-processing techniques, classification, rule extraction, approximation and data modeling methods, cost functions and adaptive parameter optimization algorithms are used as components that may be combined in thousands of ways to create adaptive systems.

Can there be a common foundation for most computational intelligence methods guiding the creation of adaptive systems? Computational learning theory [104] is a rigorous mathematical approach to learning, but it covers only the theoretical aspects of learning and is rarely used in practice. In brain science it has been commonly assumed (although only recently tested [162]) that sensory systems and neurons in the primary sensory cortex are well adapted through the evolutionary and developmental processes to the statistical properties of the visual, auditory and other types of signals they process. Neurons in the higher sensory areas react to progressively more complex aspects of signal structure. Difficult, ill-determined problems (including perception and natural language processing) may be solved only by using extensive background knowledge. Despite relatively rigid architecture of primary sensory areas neurons can quickly adapt to changes in the sensory signal statistics, for example variance, contrast, orientation and spatial scale. Thus a series of transformations is made before sufficient amount of information is derived from the signal to perform object identification, and then information about the object is used for associations and further reasoning. Inspirations from brain sciences serve below to discuss some challenges facing CI, and will be explored in more details later in this paper.

2.1 Some Challenges for CI

Intelligent systems should have goals, select appropriate data, extract information from data, create percepts and reason with them to find new knowledge. Goal setting may be a hierarchical process, with many subgoals forming a plan of action or solution to a problem. Humans are very flexible in finding alternative ways to solve a given problem, and a single-objective solutions are rarely sufficient. Brains have sufficient resources to search for alternative solutions to the problem, recruiting many specialized modules in this process. An important challenge for computational intelligence is thus to create flexible systems that can use their modules to explore various ways to solve the same problem, proposing multiple solutions that may have different advantages. This idea will be explored below using a meta-learning search process in the space of all possible models that may be composed from available transformations. The great advantage of Lisp programming is that the program may modify itself. There are no examples of CI programs that could adjust themselves in a deeper way, beyond parameter optimization, to the problem analyzed. Such adjustment has been partially implemented in the similarity-based meta-learning scheme [35, 53], and is also in accord with evolving programs and connectionist systems [103] that to a limited degree change their structure.

Most CI algorithms have very limited goals, such as prediction (using approximation method) or diagnosis (classification) based on data with some fixed structure. Such algorithms are essential building blocks of general intelligent systems, although their goals and information flow is determined by the user who tries to find a method that works for a given data. For example, running neural network software the user has to make many decisions, designing the network, selecting the training method, setting parameters, preparing the data, evaluating results and repeating the whole cycle. In effect the user acts as an external controller for the system, while the brain has parts controlling other parts [151, 152]. With sufficiently large library of different procedures for data preprocessing, feature selection and transformation, creation of data models, optimization of these models and postprocessing of results (already hundreds of components are available in such packages as Weka [182], Yale [131] and others) the control problem becomes quite difficult and the number of possible variants of such approach guarantees a constant supply of conference papers for many years to come.

Most efforts in the computational intelligence field goes into the improvement of individual algorithms. For example, model selection [166] in neural networks is usually restricted to architectures (number of nodes, each performing the same type of functions) and improvements of the training schemes; in the decision tree the focus is on the node splitting criteria and pruning strategies. The current focus on accuracy improvements of individual models, dominating in the academic journal and conference papers, is hard to justify both from the practical and theoretical point of view. The 'no free lunch' theorem [63, 176] shows that there is no single learning algorithm that is inherently superior to all other algorithms. In real world applications there may be many additional considerations, different methods may offer different advantages, for example presenting results in comprehensible way or using features of the problem that can be obtained with lower costs or effort. These considerations are almost never addressed in the literature on learning systems. In practice "Experience with a broad range of techniques is the best insurance for solving arbitrary new classification problems (Chapter 9.2.1, [63]). Moreover, one should find all algorithms that work sufficiently well for a given data, but offer different advantages. Although data mining packages include now many algorithms still some of the best algorithms used in the *StatLog* project [130] are not implemented in research or commercial software. It is doubtful that new algorithms are going to be always significantly better. Most programs have many parameters and it is impossible to master them all.

The first challenge for CI is thus to create flexible systems that can configure themselves finding several interesting solutions for a given tasks. Instead of a single learning algorithm priorities may be set to define what will be an interesting solution. A system that automatically creates algorithms on demand should search for configurations of computational modules in the space of all models restricted by the user priorities. For example, if the goal is to understand or make a comprehensible model of the data, methods that extract rules from data or that find interesting prototypes in the data should be preferred, although methods that provide interesting visualizations may also be considered. A lot of knowledge about reliability of data samples, possible outliers, suspected cases, relative costs of features or their redundancies is usually ignored as there is no way to pass it to and to use it in CI programs. Models that are not working well on all data may work fine on some subsets of data and be still useful. In practical applications validation and verification of solutions may be of great importance.

These challenges have been only partially addressed so far by CI community. Systematic generation of interesting models has been the topic of meta-learning research. In the simplest version meta-learning may be reduced to a search for good models among those available in the data mining packages, a search in the model/parameter space. In the machine learning field the multistrategy learning has been introduced by Michalski [64]. Learning of a single model may be sufficiently difficult, therefore to be feasible search in the space of many possible models should be heuristically guided. The Metal project [72] tried to collect information about data characteristics and correlate it with the methods that performed well on a given data. A system recommending classification methods for a given data has been built using this principle, but it works well only in a rather simple cases. Not much is known about the use of heuristic knowledge to guide the search for interesting models. One problem with most meta-learning approaches is that the granularity of the existing models is too large and thus only a small subspace of all possible models is explored.

Although computational intelligence covers a vast variety of different methods it lacks integrative theory. Bayesian approaches to classification and approximation in pattern recognition [63, 176, 166] covers mostly statistical methods, leaving many neural and fuzzy approaches in CI largely outside of their scope. There is an abundance of specialized, ad hoc algorithms for information selection, clusterization, classification and approximation tasks. An integrative theory is needed, providing a good view of interrelations of various approaches, and a good start for meta-learning, that could automatically find an appropriate tool for a given task. Creating such a theory is a great challenge. Some guiding principles that address it are described below.

2.2 Computing and Cognition as Compression

Neural information processing in perception and cognition is based on the principles of economy, or information compression [22]. In computing these ideas have been captured by such concepts as the minimum (message) length encoding, minimum description length, or general algorithmic complexity [119]. An approach to information compression by multiple alignment, unification and search has been proposed as a unifying principle in computing, analysis and production of natural language, fuzzy pattern recognition, probabilistic reasoning and unsupervised inductive learning [183, 184], but so far only models for sequential data have been considered. The difficulty in applying such principles to real problems are due to the importance of underlying structures for handling information and knowledge representation. Multiple alignment is sufficient for sequential information that is stored in strings but not for structured information that is stored in the brain subnetworks in a way that is not yet fully understood.

Information compression and encoding of new information in terms of old has been used to define the measure of syntactic and semantic information introduced in [56]. This information is based on the size of the minimal graph representing a given data structure or knowledge–base specification, thus it goes beyond alignment of sequences. A chunk of information has different value for someone who does not have any associations with it and treats it as a fact to be ignored or remembered, and a very different value for an expert who may have to restructure many existing associations to accommodate it. Semantic information measure, introduced in [56], is proportional to the change of algorithmic (Chaitin-Kolmogorov) information [20] that is needed to describe the whole system, and therefore measures relative complexity, depending on the knowledge already accumulated. Algorithmic information or the relative complexity of an object y in respect to a given object x is defined as the minimal length of the program p for obtaining y from x. Algorithmic information captures some intuitive features of information: a binary string obtained by truly random process cannot be compressed and carries the amount of information equal to the number of its digits. An apparently random string may, however, be easily computable from some short algorithm, for example a square root or the value of π. Because of the Gödel and related theorems it is in general not possible to decide whether the string is random or simple to compute. Although algorithmic information has correct intuitive features it is usually very hard to compute, relying on a concept of universal computer.

Suppose that information is encoded in n-element binary strings. There are 2^n possible strings and if they are chosen in a completely random way only a little compression will be possible. Each new string will contribute about n bits of information and strings will be represented in form of a binary tree. Encoding new information in terms of the old is possible in various ways if some parts of the new bit strings are already present in the previously analyzed old bit strings. For example, if the new string B_n differs from an old

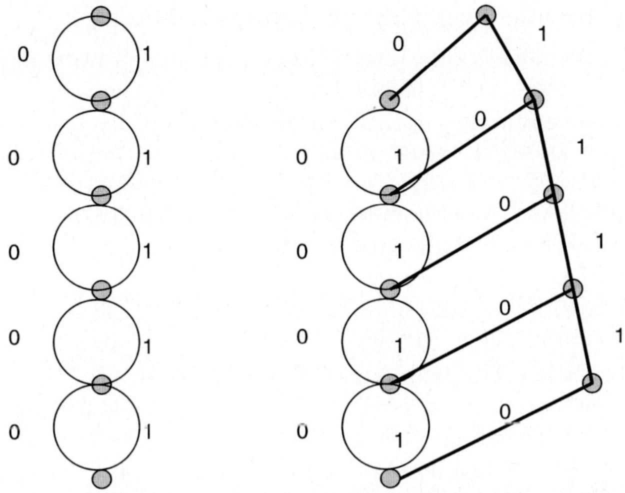

Fig. 1. Left: minimal graph representing a set of all 32 binary 5-bit strings. Right: minimal graph for a set of all 5-bit binary strings, without 11111 string. Such graphs are created by folding binary trees to minimal graphs

string B_o only by the last bit b_n the whole B_n string contributes only one bit of information relatively to B_o. If many bit strings are received the whole information contained in them may be presented in a binary tree and folded into a minimal graph. If all 2^n strings are present the minimal graph may be represented in a very compact form (Fig. 1, left). However, if all but the last string 11111 are received the minimal graph (created by folding binary tree with just one edge removed) is rather complicated (Fig. 1, right). Adding the last string carries – from the point of view of the whole system that builds internal representation of the structure of the incoming information – a large amount of information, reducing the 5-bit graph by 7 edges and 4 vertices. The number of edges plus vertices changed in the minimal graph representing all data is thus a useful measure of the structural information that is gained by receiving new information. If the strings repeat themselves no new structural information is gained, although frequency of repeated information chunks may be useful for other purposes.

Semantic contents or meaning of information may only be defined in a context of cognitive system, for example an expert system based on a set of rules stored in a knowledge base, or a semantic network. Knowledge base, together with the rules of inference, defines a universum of facts, representing knowledge or some model of the world. Information gain relatively to the knowledge base is defined as the change of the size of the minimal knowledge base that can accommodate this fact and its consequences. When a new fact is learned by a human it may take from several seconds to days or even years before this fact is truly accommodated and the meaning is fully understood.

Years of repetition of basic facts in mathematics and natural sciences are required to really digest this information: once it has been integrated into the "knowledge base" the information contained in the school book can be quickly reviewed and new information encoded in terms of the old. Adding new rule to the knowledge base requires accommodation of this knowledge, and thus modification of existing knowledge. For example, learning that some fruits are edible may require a short sequence of symbols to transmit, but will have large influence on the knowledge base, creating new goals and modifying behavior of humans and animals.

Perfect minimal encoding is probably never accomplished in real brains, but even an imperfect approximation of this process gives a useful measure of semantic information. For example, cyclomatic complexity [127] of a software module is calculated from a connected graph showing topology of control flow within the program as $CC = E - N + p$, where E is the number of edges of the graph, N the number of nodes of the graph and p the number of connected components. Software with cyclomatic complexity over 50 is considered untestable and very high risk. Adding new module to such software leads to a large change in the amount of information that the system gains, making it hard to predict all the consequences.

The use of algorithmic information measures in computational intelligence is still rare. However, CI systems that encode new information in terms of the known information are certainly not new. They include constructive neural networks that add new nodes only if the current approximation is not sufficient [87, 50], similarity-based systems that accept new reference vector checking first if it is not redundant, decision trees that are pruned to increase their generalization and ignore data that are already correctly handled, information selection methods that increase the pool of features or their combinations only when new data justifies it, and many other approaches.

2.3 Meta-learning via Search in the Model Space

Meta-learning requires detailed characterization of the space of possible models, the ability to create and modify CI models in a systematic way. This should not be done in a random way, as the space of all possible models is too large to explore. Some framework that covers large number of different methods is needed. For feedforward neural networks such framework involves possible architectures, from simplest to more complex, and a taxonomy of different types of transfer functions [57, 58], allowing for systematic exploration of different network models. Although the focus of neural network community has been on learning algorithms and network architectures, it is clear that selection of transfer functions is decisive for the speed of convergence in approximation and classification problems. As already shown in [57] (see also [11, 12]) some problems may require $O(n^2)$ parameters using localized functions and only $O(n)$ parameters when non-local functions are used. The n-parity problem may be trivially solved using a periodic function with a single parameter [40] while the multilayer perceptron (MLP) networks need $O(n^2)$ parameters and learn it only with great difficulty. It is hard to create basis functions expansion that will not have the universal approximator property, yet the fact that MLPs and radial basis function (RBF) networks are universal approximators has somehow blinded the researches who ignored quite obvious fact that it is the speed of convergence (especially for multidimensional data) that is most important. In principle neural networks may learn any mappings, but the ability to learn quickly and accurately requires flexible "brain modules", or transfer functions that are appropriate for the problem to be solved. This problem will not disappear thanks to better learning procedures or architectures, the main research topics in neural computing.

Meta-learning methods should help to build the final model from components performing different transformations on available data. Almost all adaptive systems are homogenous, i.e. they are built from many processing elements of the same type. MLP neural networks and decision trees use nodes that partition the input space by hyperplanes. Networks based on localized functions (RBF, Gaussian classifiers) frequently use nodes that provide spherical or ellipsoidal decision borders. This cannot be the best inductive bias for all data, frequently requiring large number of processing elements even in cases when simple solutions exist. Neurocognitive inspirations that go beyond simple neurons may point the way here. A single cortical column in the brain provides many types of microcircuits that respond in a qualitatively different way to the incoming signals [123]. Other cortical columns may combine these responses in a perceptron-like fashion to enable complex discriminations. At the level of higher cognition brains do not recognize all objects in the same feature space. Even within the same sensory modality a small subset of complex features is selected, allowing to distinguish one class of objects from another (for example, in case of vision a simple sketch is sufficient). Object recognition or category assignment by the brain is probably based on evaluation of similarity to memorized prototypes of objects using a few characteristic features.

In contrast to human categorization most pattern recognition systems implicitly assume that classification is done using the same features in all regions of the input space. Memory-based techniques use single distance (or similarity) function to distinguish all objects, statistical or neural methods provide hyperplanes (MLPs) or Gaussian functions (RBF networks) for discrimination, but rarely both. Decision trees are usually univariate, employing a decision rule for the threshold value of a single feature, partitioning the input space into hyperrectangles. Multivariate decision trees provide several hyperplanes at high computational cost. Support Vector Machines use one kernel globally optimized for a given dataset [27]. All these systems may be called "homogenous" since they search for a solution providing the same type of elements, the same type of decision borders in the whole feature space. Committees of the homogenous systems are frequently used to improve and stabilize results [17]. Combining systems of different types in a committee is a step towards heterogeneous systems that use different types of decision borders, but such models may become quite complex and difficult to understand.

A rather obvious extension of traditional approach is to use class-specific features that may be quite distinct for different classes. This approach is used in an implicit way in feedforward neural networks with strong regularization that leads to creation of hidden nodes strongly linked only to selected features and used for specific classes. This type of solutions may also emerge in hierarchical systems, such as decision trees, where each class may end in a different branch using different features, although at least one feature is always shared among all classes. Using feature selection for separate classifiers $C_k(\mathbf{X})$ that distinguish a single class from the rest may lead to completely distinct sets of features. In the K-class problem a set of $i = 1..K$ selector transformations $\mathbf{Z}_i = \mathcal{T}_i(\mathbf{X}_i)$ may be defined by feature selection techniques, and these classifiers are restricted to their own $C_k(\mathbf{Z}_k)$ subspaces. The final decision depends on the relative confidence of individual classifiers that may assign the same vector to the class they specialize in. Even though individual classifiers are trained on different feature subsets their confidence may be scaled by a second transformation, such as additional scaling or a linear mixture of their predictions. Alternatively, new adaptive transformations may be trained on the K-dimensional vectors ${}^1\mathbf{X} = C_k(\mathbf{X})$ obtained as predictions for all training data. In this case the classifiers are used as data transformations and their number may be larger than K. Binary classifiers (such as decision trees) usually give only one answer $C_k(\mathbf{X}) = 1$, but some other classifiers may actually create K probabilities, so the final dimensionality after this transformation may reach at least K^2.

A more sophisticated approach to class-specific use of features has been presented by Baggenstoss [8]. It is based on estimation of probability density functions (PDFs) in the reduced low-dimensional feature space selected separately for each class, and mapping these PDFs back to the original input space, where Bayesian classification is performed. To constrain the inverse mapping (it is obviously not unique) a reference hypothesis is used for which $\mathcal{P}(\mathbf{X}|H_0)$ and $\mathcal{P}(\mathbf{Z}|H_0)$ are both known, and likelihood rations are preserved, that is:

$$\mathcal{P}(\mathbf{X}) = \mathcal{P}(\mathbf{Z})\mathcal{P}(\mathbf{X}|H_0)/\mathcal{P}(\mathbf{Z}|H_0) \tag{1}$$

The "correction factor" to the PDF calculated in $\mathbf{Z}$ space is simply the ratio of the two reference hypothesis PDFs. The PDF projection theorem opens many possibilities worth exploration, although the choice of the reference hypothesis may sometimes be non-trivial.

2.4 Similarity-based Framework for Meta-learning

Similarity (or dissimilarity, measured by some distance) is a very fundamental concept that can be used as a basis for computational intelligence methods

[140]. For additive similarity measures models based on similarity to prototypes are equivalent to models based on fuzzy rules and membership functions [48]. Similarity functions may be related to distance functions by many transformations, for example exponential transformation $S(\mathbf{X},\mathbf{Y}) = \exp(-D(\mathbf{X},\mathbf{Y}))$. Additive distance functions $D(\mathbf{X},\mathbf{Y})$ are then converted to the multiplicative similarity factors (membership functions). For example, Euclidean distance function $D_2(\mathbf{X},\mathbf{Y})^2 = \sum_i W_i(X_i - Y_i)^2$ is equivalent to a multivariate Gaussian similarity function $S_2(\mathbf{X},\mathbf{Y}) = \exp(-D_2(\mathbf{X},\mathbf{Y})^2)$ centered at $\mathbf{Y}$ with ellipsoidal contours of constant values $D_2(\mathbf{X},\mathbf{Y}) = \text{const}$, equal to the product of univariate Gaussian membership functions $S_2(\mathbf{X},\mathbf{Y}) = \prod_i G(X_i,Y_i) = \prod_i \exp[-W_i(X_i-Y_i)^2]$. Using such transformations fuzzy rules (F-rules) with product norms may always be replaced by prototype-based rules (P-rules) with appropriate similarity functions. On the other hand all additive distance functions may be replaced by product T-norms with membership functions given by exponential one-dimensional distance factors. For example, the Manhattan distance function $D_1(\mathbf{X},\mathbf{P}) = \sum_{i=1} |X_i - P_i|$ leads to a product of $\exp(-|X_i - P_i|)$ membership functions. However, non-additive distance functions (for example the Mahalanobis distance) are difficult to approximate using products or combinations of one-dimensional fuzzy membership functions, unless explicit correlations between fuzzy features are taken into account.

Prototype-based rules (P-rules), although rarely used, are in many cases easier to comprehend than fuzzy rules (F-rules) and create new possibilities for data understanding. Relations between these two types of rules have so far not been explored in details. Two types of prototype-based rules may be distinguished: minimum distance rules (discriminative approach), and the threshold rules (similarity-based approach). Minimum distance rule selects the prototype that is closer:
IF $P = \arg\min_{\mathbf{P}'} D(\mathbf{X},\mathbf{P}')$ THAN $\text{Class}(\mathbf{X}) = \text{Class}(\mathbf{P})$,
while threshold rules select the prototype that is sufficiently similar, or closer than some threshold distance:
IF $D(\mathbf{X},\mathbf{P}) \leq d_P$ THEN C.

The minimum distance rule for two prototypes defines a hyperplane bisecting the line connecting these prototypes. There are many methods to find optimal hyperplanes [63, 87] and they can be used to select optimal position of prototypes. On the other hand variants of LVQ methods [108] lead to prototypes and thus hyperplanes that do not seem to correspond to any discriminant analysis algorithms. In particular the idea of maximizing margins, not only minimizing errors, used in SVM algorithms based on solutions to regularized least square problem in the primial or dual space [21], has not yet been used for prototype optimization or selection [97, 81]. Any hyperplane defined by its bias and normal vector $(W_0, \mathbf{W})$ is equivalent to a minimal distance rule for two prototypes $\mathbf{P}, \mathbf{P}'$ such that $\mathbf{W}/||\mathbf{W}|| = (\mathbf{P}-\mathbf{P}')/||\mathbf{P}-\mathbf{P}'||$, and $W_0 = \frac{1}{2}||\mathbf{P} - \mathbf{P}'||$. Thus discrimination hyperplanes do not specify by themselves interesting prototypes, they can move in the subspace parallel to the **W** plane, and be placed in important positions, for example close to cluster centers.

Decision boundaries of the threshold rules depend on the type of the distance function $D(\mathbf{X},\mathbf{P})$. They frequently provide localized decision regions and may not be able to label vectors in some areas of feature space, unless a default (ELSE) rule is used. Distance function $D_{\mathbf{W}}(\mathbf{X},\mathbf{P})$ between prototype **P** and point **X** that has constant value for all points lying on a plane perpendicular to **W** is calculated by:

$$D_{\mathbf{W}}(\mathbf{X},\mathbf{P}) = |\sum_{i}^{N} s_i(\mathbf{X}_i - \mathbf{P}_i)|; \quad s_i = \mathbf{W}_i/||\mathbf{W}|| \tag{2}$$

The threshold rule IF $D(\mathbf{X},\mathbf{P}) \leq d_P$ THEN C, with $d_P = \frac{1}{2}||\mathbf{P}-\mathbf{P}'||$ is equivalent to the minimum distance rule for propotypes $\mathbf{P},\mathbf{P}'$. Relations between various discriminant analysis techniques on the one hand, and optimization of prototypes on the other hand, have just started to be explored. Relations between similarity based methods and fuzzy rule-based systems have also not yet been analyzed in depth. More attention has been devoted to relations between RBF networks and fuzzy logic models [105].

RBF and MLP networks may be viewed as a particular implementation of hierarchical sets of fuzzy threshold logic rules based on sigmoidal membership functions, equivalent to crisp logic networks applied to the input data with uncertainty [37]. Leaving uncertainty (fuzziness) on the input side makes the networks or the logical rule-based systems easier to understand, and achieves similar results as the Type-2 fuzzy systems [129]. Moreover, it shows deep connections between neural and fuzzy systems, with different types of input uncertainty equivalent to crisp input values with specific transfer functions. Many natural assumptions about uncertainty of input variable x lead to probability that rule $\Pr(x > \theta)$ is true given by the membership functions of sigmoidal shape, for example semi-linear functions for uncertainties that are constant in some interval or to erf functions (almost identical to logistic functions) for Gaussian uncertainties.

The radial basis functions became a synonym for all basis function expansions, although in approximation theory already in the classical book of Achieser published in 1956 [2] many such expansions were considered. RBF networks are equivalent to the fuzzy systems only in special cases, for example when the Gaussian membership functions are used [105], but in the literature RBF is frequently used as a synonym for Gaussian node networks, although any functions that depends only on Euclidean distance, or a radial variable $\phi(r) = \phi(||\mathbf{X}-\mathbf{R}||)$, is suitable for RBF expansion. However, it is not clear that a radial dependence is always the best assumption for a given data. Another useful category of basis set expansion methods uses separable basis functions (SBF), where each node implements a product of one-dimensional functions $\phi(\mathbf{X}) = \prod_i f_i(X_i)$. Approximation abilities of SBF networks are similar to those of RBF networks, and the separable function realized by their nodes may be interpreted in a natural way as the product of fuzzy membership functions. They may also form a natural generalization of the Naive Bayes (NB) approach, with each network node implementing a local NB model, and the whole network functioning as a committee of such models, aggregating evidence for each class and using some voting procedure or final decision [112]. It is not clear why so much research has been devoted to the RBF networks while neural networks based on separable functions are virtually unknown: the Feature Space Mapping (FSM) network [50, 60, 3, 44] seems to be the only existing implementation of the Separable Basis Function networks so far.

A general framework for similarity-based methods (SBMs) has been formulated using the concept of similarity [43, 35]. This framework includes typical feedforward neural network models (MLP, RBF, SBF), some novel networks (Distance-Based Multilayer Perceptrons (D-MLPs, [41]) and the nearest neighbor or minimum-distance networks [33, 43]), as well as many variants of the nearest neighbor methods, improving upon the traditional approach by providing more flexible decision borders. This framework has been designed to enable meta-learning based on a search in the space of all possible models that may be systematically constructed. New algorithms are generated by applying admissible extensions to the existing algorithms and the most promising are retained and extended further. Training is performed using parameter optimization techniques [52, 53]. Symbolic values used with probabilistic distance functions allow to avoid ad hoc procedure to replace them with numerical values. To understand the structure of the data prototype-based interpretation of the results is used, simplifying predictive models and providing prototype rules (P-rules) that may be converted to fuzzy rules (F-rules) [48].

In the SBM approach objects (samples, cases, states) $\{\mathbf{O}^i\}$, $i = 1 \ldots n$ may be represented by a set of numerical or symbolic features $X^i_j = X_j(\mathbf{O}^i), j = 1...N$ characterizing these objects, or by a procedure that evaluates directly similarity $D(\mathbf{O}^i, \mathbf{O}^k)$ between objects, allowing for comparison of objects with complex structure. For classification problems a function or a procedure to estimate $p(C_i|\mathbf{X}; M), i = 1..K$, the posterior probability of assigning vector $\mathbf{X}$ to a class C_i, is defined, while for approximation problems a function $Y(\mathbf{X}; M)$ is defined. In both cases the model M includes various procedures, parameters and optimization methods. A general similarity-based model for classification problems is constructed from the following components:

- input pre-processing transformation, either providing directly dissimilarities of objects $D(\mathbf{O}^i, \mathbf{O}^k)$ or mapping them to symbolic/numerical descriptions $\mathbf{X}(\mathbf{O})$ that define the feature space; in both cases a data matrix is obtained;
- a procedure to select relevant features or their combinations;
- function $d_j(X_j; Y_j)$ to calculate similarity of X_j, Y_j feature values, $j = 1..N$;

- function $D(\mathbf{X}, \mathbf{Y}) = D(\{d_j(X_j; Y_j)\})$ that combines similarities defined for each attribute to compute similarities of vectors;
- specification of the neighborhoods, such as the number of reference vectors k taken into account around of $\mathbf{X}$, or the number of hidden nodes in feedforward neural networks based on functions with localized support in the feature space;
- the weighting function $G(D) = G(D(\mathbf{X}, \mathbf{R}))$ estimating contribution of each reference vector $\mathbf{R}$;
- a set of prototype (reference) vectors $\{\mathbf{R}\}$ selected from the set of training vectors $\{\mathbf{X}^i\}$ and optimized using some procedures;
- a set of class probabilities or membership values $p_i(\mathbf{R}), i = 1 \dots K$ for each reference vector;
- a misclassification risk matrix $\mathcal{R}(C_i, C_j), i, j = 1 \dots K$;
- a scaling function $K(\cdot)$ estimating the influence of the error, for a given training example, on the total cost function;
- a function (or a matrix) $\mathcal{S}(\cdot, \cdot)$ evaluating similarity (or more frequently dissimilarity) of the classes; if class labels are soft, or if they are given by a vector of probabilities $p_i(\mathbf{X})$, classification task is in fact a mapping;
- a total cost function $E[\mathrm{d}_T; M]$ that is minimized at the training stage directly, in a crossvalidation, bootstrap or other procedure;
- procedures to optimize parameters of the model at each stage.

This framework may be used to automatically generate a sequence of models with growing complexity. It has been used in a meta-learning scheme to solve classification problems [53], starting from the simplest k-NN model parameterized by $M = \{k, D(\cdot, \cdot), \{\mathbf{X}\}\})$, i.e. the whole training dataset used as the reference set, k nearest prototypes included with the same weight, using a typical distance function, such as the Euclidean or the Manhattan distance. Probabilities are computed as $p(C_i|\mathbf{X}; M) = N_i/k$, where N_i is the number of nearest vectors that belong to class C_i. The initially model has thus only one parameter k for optimization. If such model is not sufficiently accurate new procedures/parameters are added to create a more complex model. The search tree in the space of all models is generated by extending current model in the simplest way, using a measure that penalizes for increased model complexity and rewards for increased accuracy. Various model selection criteria may be applied to control this process [176]. Several good models are maintained in a beam search procedure, and the search stops when additional complexity of possible extensions of existing models does not justify increased accuracy.

Scalar products or cosine distances are a particular way to measure similarity, quite useful especially when vectors have significantly different norms (as in the evaluation of text similarity [125]). Similarity may be evaluated in a more sophisticated way by learning [114] and designing various kernels that evaluate similarity between complex objects [27, 157]. Although kernels are usually designed for SVM methods, for example in text classification [120] or bioinformatics [173, 7, 171] applications, they may be directly used in the SBM

framework, because kernels are specific (dis)similarity functions. In particular positive semidefinitive kernels used in SVM approach correspond to some Euclidean dissimilarity matrices [140]. SVM is just one particular method of a single hyperplane optimization in a space where kernel $K(\mathbf{X}, \mathbf{Y})$ serves as a scalar product. However, for data models requiring different resolutions in different areas of feature spaces SVM may not be the best approach, and a combination of kernels for feature space expansion, easily accommodated in the SBM framework, should be used. Meta-learning is not a single method, but an approach that leads to a tailor-made methods created on demand, therefore it may to a large degree avoid the "no-free-lunch" theorem restrictions [176]. SVM models are certainly an important family among a large number of models that may be explored during metalearning.

3 Transformation-based CI Theory

Similarity-based methods using only direct similarity comparisons of some vector representations are restricted to a single transformation (not counting pre-processing), $\mathbf{Y} = \mathcal{T}(\mathbf{X}; \mathbf{R})$. Although in principle this is sufficient for universal approximation [85] in practice it may slow down the convergence and make a discovery of simple data models very difficult. SBM framework is generalized here to include multiple transformation steps. In the stacking approach [185, 163, 159] one classifier produces outputs that are used as inputs for another classifier. Wolpert showed [185] that biases of stacked classifiers may be deducted step by step, improving generalization of the final system. The same is true if a series of transformations is composed together to produce a data model. Transformation-based approach fits very well to modern component-based data mining and may be presented in from of graphical models, although quite different than probabilistic models presented in [101].

General similarity transformations may act on objects and produce vectors, either by analysis of object properties or object similarity to some reference objects. In both cases feature-based vector description $\mathbf{X}$ of the object $\mathbf{O}$ is produced, although the size of this vector may be different. In the first case the number of features N is independent of the number of objects, while in the second case all training objects may be used as a reference and $\mathbf{X}_i = K(\mathbf{O}, \mathbf{O}_i)$ feature values calculated using kernel $K(\cdot, \cdot)$ function (or procedure) to provide n-dimensional vector.

CI calculations may be presented as a series of transformations, divided into several stages. Starting from the raw input data ${}^0\mathbf{X} = \mathbf{X}$ that defines initial feature space, first transformation $\mathcal{T}_1$ scales individual features, filters them, combining pairs or small subsets of features. This leads to a new dataset ${}^1\mathbf{X} = \mathcal{T}_1({}^0\mathbf{X})$ with vectors based on a new set of features that may have different dimensionality than the original data. The second transformation ${}^2\mathbf{X} = \mathcal{T}_2({}^1\mathbf{X})$ extracts multidimensional information from pre-processed features ${}^1\mathbf{X}$. This is further analyzed by subsequent transformations that either

aim at separation of the data or at mapping to a specific structures that can be easily recognized by the final transformation. The final transformation provides desired information. These transformations can in most cases be presented in a layered, graphical form.

Chains of transformations created on demand should lead to optimal CI methods for a given data. This requires characterization of elementary transformations. The goal here is to describe all CI methods as a series of transformations but at a higher level than the psuedocode. Several types of vector transformations should be considered: component, selector, linear combinations and non-linear functions. Component transformations work on each vector component separately, for example shifting and scaling component values when calculating distances or scalar products. Selector transformations define subsets of vectors or subsets of features using various criteria for information selection, or similarity to the known cases (nearest neighbors), or distribution of feature values and class labels. Non-linear functions may serve as kernels or as neural transfer functions [57]. Transformations composed from these elementary types may always be presented in a network form.

3.1 Extracting Information from Single or Small Subsets of Features

New information from available features may be extracted using various types of network nodes in several ways. First, by providing diverse basis functions or receptive fields for sampling the data separately in each dimension, although two or higher-dimensional receptive fields may also be used in some applications (as is the case for image or signal processing filters, such as wavelets). Fuzzy and neurofuzzy systems usually include a "fuzzification step", defining for each feature several membership functions $\mu_k(X_i)$ that act as large receptive fields. Projecting each feature value X_i on these receptive fields μ_k increases the dimensionality of the original data. This may lead to some improvement of results as even a linear transformation in the extended space has a greater chance to separate data better.

Second, information may be extracted by scaling the features using logarithmic, sigmoidal, exponential, polynomial and other simple functions; such transformations help to make the density of points in one dimension more uniform, circumventing some problems that would require multiresolution algorithms. Although they are rarely mentioned as a part of the learning algorithms adaptive preprocessing at this stage may have a critical influence on the final data model.

Third, information is extracted by creating new features using linear combinations, tensor products, ratios, periodic functions or using similarity measures on subsets of input variables. Non-linear feature transformations, such as tensor products of features, are particularly useful, as Pao has already noted introducing functional link networks [139, 1]. Rational function neural networks [87] in signal processing [117] and other applications use ratios of

polynomial combinations of features; a linear dependence on a ratio $y = x_1/x_2$ is not easy to approximate if the two features x_1, x_2 are used directly. Groups of several strongly correlated features may be replaced by a single combination performing principal component analysis (PCA) restricted to small subspaces. To decide which groups should be combined standardized Pearson's linear correlation is calculated:

$$r_{ij} = 1 - \frac{|C_{ij}|}{\sigma_i \sigma_j} \in [-1, +1] \tag{3}$$

where the covariance matrix is:

$$C_{ij} = \frac{1}{n-1} \sum_{k=1}^{n} \left(X_i^{(k)} - \bar{X}_i \right) \left(X_j^{(k)} - \bar{X}_j \right); \quad i,j = 1 \cdots d \tag{4}$$

These coefficient may be clustered, for example by using dendrogram techniques. Depending on the clustering thresholds they provide reduced number of features, but also features at different scales, from a combination of a few features to a global PCA combinations of all features. This approach may help to discover hierarchical sets of features that are useful in problems requiring multiscale analysis. Another way to obtain features for multiscale problems is to do clusterization in the data space and make local PCA within the clusters to find features that are most useful in various areas of space.

Linear combinations derived from interesting projection directions may provide low number of interesting features, but in some applications non-linear processing is essential. The number of possible transformations at this stage is very large. Feature selection techniques [84], and in particular filter methods wrapped around algorithms that search for interesting feature transformations (called "frappers" in [39]), may be used to quickly evaluate the usefulness of proposed transformations. The challenge is to provide a single framework for systematic selection and creation of interesting transformations in a meta-learning scheme. Evolutionary techniques may prove to be quite useful in this area.

3.2 Extracting Information from All Features

After transforming individual features or small groups of features to create $^1\mathbf{X}$ space additional transformations that involve all features are considered. Frequently these transformations are used to reduce dimensionality of the data. Srivastava and Liu [164] point out that the choice of optimal transformation depends on the application and the data set. They have presented an elegant geometrical formulation using Stiefel and Grassmann manifolds, providing a family of algorithms for finding orthogonal linear transformations of features that are optimal for specific tasks and specific datasets. They find PCA to be optimal transformation for image reconstruction under mean-squared error, Fisher discriminant for classification using linear discrimination, ICA for

signal extraction from a mixture using independence, optimal linear transformation of distances for the nearest neighbor rule in appearance-based recognition of objects, transformations for optimal generalization (maximization of margin), sparse representations of natural images and retrieval of images from a large database. In all these applications optimal transformations are different and may be found by defining appropriate cost functions and optimizing them using stochastic gradient techniques. Some of their cost functions may be difficult to optimize and it is not yet clear that sophisticated techniques based on differential geometry, advocated in [164], offer significant practical advantages, although they certainly provide an interesting insight into the problem. Simpler learning algorithms based on numerical gradient techniques and systematic search algorithms give surprisingly good results and can be applied to optimization of difficult functions [110], but a detailed comparison of such methods has not yet been made.

In some cases instead of reduction of dimensionality expansion of the feature space may be useful. Random linear projection of input vectors into a high-dimensional space ${}^2\mathbf{X} = \mathbf{L}({}^1\mathbf{X})$ is the simplest expansion, with the random matrix $\mathbf{L}$ that has more rows than columns (see neurobiological justification of such projections in [123]). If highly nonlinear low-dimensional decision borders are needed large number of neurons should be used in the hidden layer, providing linear projection into high-dimensional space followed by filtering through neural transfer functions to normalize the output from this transformation. Enlarging the data dimensionality increases the chance to make the data separable, and this is frequently the goal of this transformation, ${}^2\mathbf{X} = \mathcal{T}_2({}^1\mathbf{X}; {}^1\mathbf{W})$. If (near) separability can be achieved this way the final transformation may be linear $\mathbf{Y} = \mathcal{T}_3({}^2\mathbf{X}; {}^2\mathbf{W}) = \mathbf{W}_2 \cdot {}^2\mathbf{X}$. A combination of random projection and linear model may work well [91] – this is basically achieved by random initialization of feedforward neural networks and a linear discriminant (LDA) solution for the output model, a method used to start a two-phase RBF learning [158]. However, one should always check whether such transformation is really justified from the point of view of model complexity, because linear discrimination may work also quite well for many datasets in the original feature space, and many non-random ways to create interesting features may give better results. It may also be worthwhile to add pre-processed ${}^1\mathbf{X} = \mathcal{T}_1(\mathbf{X})$ features to the new features generated by the second transformation ${}^2\mathbf{X} = ({}^1\mathbf{X}, \mathcal{T}_2({}^1\mathbf{X}; {}^1\mathbf{W}))$, because they are easier to interpret and frequently contain useful information – in case of linear transformations this simply adds additional diagonal part to the weight matrix.

In general the higher the dimensionality of the transformed space the greater the chance that the data may be separated by a hyperplane [87]. One popular way of creating highly-dimensional representations without increasing computational costs is by using the kernel trick [157]. Although this problem is usually presented in the dual space the solution in the primal space is conceptually simpler [115, 21]. Regularized linear discriminant (LDA) solution is found in the new feature space ${}^2X = \mathbf{K}(\mathbf{X}) = K({}^1\mathbf{X}, \mathbf{X})$, mapping $\mathbf{X}$

using kernel functions for each training vector. Feature selection techniques may be used to leave only components corresponding to "support vectors that provide essential support for classification, for example only those close to the decision borders or those close to the centers of cluster, depending on the type of the problem. Any CI method may be used in the kernel-based feature space $K(\mathbf{X})$, although if the dimensionality is large data overfitting is a big danger, therefore only the simplest and most robust models should be used. SVM solution to use LDA with margin maximization is certainly a good strategy.

Consider for example a two-class case. In m-dimensional space the expected maximum number of separable vectors randomly assigned to one of the classes is $2m$ [24, 87]. For k-bit strings there are $n = 2^k$ vectors and 2^n Boolean functions that may be separated in the space with $n/2$ dimensions with high probability. In case of k-bit Boolean problems localized kernels are not useful as the number of vectors n grows exponentially fast with k, but separation of all vectors is possible in the space generated by polynomial kernel of degree k, providing new features based on $n-1$ monomials $x_a, x_a x_b, x_a x_b x_c, \ldots, x_1 \ldots x_k$. Indeed SVM algorithms with such kernel are capable of learning all Boolean functions although they do not generalize well, treating each vector as a separate case. For example, SVM with polynomial kernel of degree k (or with a Gaussian kernel) may solve the k-bit parity problem. However, removing a single string from the training set in the leave-one-out test will lead to perfect learning of the training sets, but always wrong predictions of the test vector, thus achieving 0% accuracy! Unfortunately only for parity the answer will always be wrong (each vector is surrounded by the nearest neighbors from the opposite class), for other Boolean functions one cannot count on it. This problem may be solved in a simple way in the original feature space if the goal of learning is redefined (see Sect. 3.5 below).

If the final transformation is linear $\mathbf{Y} = {}^3\mathbf{X} = \mathcal{T}_3({}^2\mathbf{X}; {}^2\mathbf{W})$ parameters ${}^2\mathbf{W}$ are either determined in an iterative procedure simultaneously with parameters ${}^1\mathbf{W}$ from previous transformations (as in the backpropagation algorithms [87]), or they may be sequentially determined by calculating the pseudoinverse transformation, as is frequently practiced in the two-phase RBF learning [158], although in experiments on more demanding data simultaneous adaptation of all parameters (in RBF networks they include centers, scaling parameters, and output layer weights) gives better results. The initial random transformation may use arbitrary basis functions, although for localized functions simple clustering instead of a random projection is recommended. Most basis function networks provide receptive fields in the subspace of the original feature space or on the pre-processed input data. Transformations of this kind may be presented as a layer of network nodes that perform vector mapping $\mathcal{T}_2({}^1\mathbf{X}; {}^1\mathbf{W})$ based on some specific criterion. Many interesting mappings are linear and define transformations equivalent to those provided by the Exploratory Projection Pursuit Networks (EPPNs) [100, 67]. Quadratic cost functions used for optimization of linear transformations may lead to formulation of the problem in terms of linear equations, but most cost functions or optimization criteria

are non-linear even for linear transformations. A few such transformations are listed below:

- Statistical tests for dependency between class and feature value distributions, such as Pearson's correlation coefficient, χ^2 and other measures that may be used to determine best orthogonal combination of features in each node.
- Principal Component Analysis (PCA) in its many variants, with each node computing principal component [63, 176, 166]).
- Linear Discriminatory Analysis (LDA), with each node computing LDA direction (using one of the numerous LDA algorithms [63, 176, 166]).
- Fisher Discriminatory Analysis (FDA), with each node computing canonical component using one of many FDA algorithms [176, 167].
- Independent Component Analysis, with each node computing one independent component [92, 23].
- Linear factor analysis, computing common and unique factors from data [77].
- Canonical correlation analysis [71].
- KL, or Kullback-Leibler networks with orthogonal or non-orthogonal components; networks maximizing mutual information [170] are a special case here, with product vs. joint distribution of classes/feature values.
- Classical scaling, or linear transformation embedding input vectors in a space where distances are preserved [140].
- Linear approximations to multidimensional scaling [140].
- Separability criterion used on orthogonalized data [78].

Non-linearities may be introduced in transformations in several ways: either by adding non-linear functions to linear combinations of features, or using distance functions, or transforming components and combining results [57, 58]. Linear transformations in kernel space are equivalent to non-linear transformations in the original feature space. A few non-linear transformations are listed below:

- Kernel versions of linear transformations, including radial and other basis set expansion methods [157].
- Weighted distance-based transformations, a special case of general kernel transformations, that use (optimized) reference vectors [43].
- Perceptron nodes based on sigmoidal functions with scalar product or distance-based activations [42, 41], as in layers of MLP networks, but with targets specified by some criterion (any criterion used for linear transformations is sufficient).
- Heterogeneous transformations using several types of kernels to capture details at different resolution [57].
- Heterogeneous nodes based or several type of non-linear functions to achieve multiresolution transformations [57].

- KL, or Kullback-Leibler networks with orthogonal or non-orthogonal components; networks maximizing mutual information [170] are a special case here, with product vs. joint distribution of classes/feature values.
- χ^2 and other statistical tests for dependency to determine best combination of features.
- Factor analysis, computing common and unique factors, reducing noise in the data.
- Nodes implementing fuzzy separable functions, or other fuzzy functions [50].

Many other transformations of this sort are known and may be used at this stage in transformation-based systems. A necessary step for meta-learning is to create taxonomy, similarities and relations among such transformations to enable systematic search in the space of possible models, but this has not yet been done. An obvious division is between fixed transformations that are based on local criterion, with well-defined targets, and adaptive transformations that are based on criterions optimizing several steps simultaneously (as in backpropagation), where the targets are defined only for composition of transformations. Fixed $\mathcal{T}_2$ transformations have coefficients calculated directly from the input data or data after $\mathcal{T}_1$ transformation. Activity of the network nodes has then clear interpretation, and the number of nodes may be determined from estimation of such criteria as the information gain compared to the increased complexity. A general way to calculate fixed transformation coefficients is to create a single component (for example, one LDA hyperplane), and then orthogonalize the input vectors to this component, repeating the process in an iterative way. General projection pursuit transformations [100, 67] may provide a framework for various criteria used in fixed transformations.

Transformations may also have adaptive coefficients, determined either by an optimization procedure for the whole system (for example, global optimization or backpropagation of errors), or by certain targets set for this mapping (see Sect. 3.5 below). The interpretation of node functions is not so clear as for the fixed targets for individual transformations, but the final results may be better. Fixed transformations may be very useful for initialization of adaptive transformations or may be useful to find better solutions of more complex fixed transformations. For example, multidimensional scaling requires very difficult minimization and seems most of the time to converge to a better solution if PCA transformations are performed first. Nonlinear Mapping Pursuit Networks (MPN), similar to EPPNs, may be defined and used as a fixed transformation layer, followed by linear model in the same way as it is done in the functional link networks [139].

Adding more transformation layers with distance-based conditions, that is using similarity in the space created by evaluation of similarity in the original input space, leads to higher-order nearest neighbor methods, rather unexplored area. There are many other possibilities. For example, consider a parity-like problem with vectors from the same class that are spread far apart and surrounded by vectors from other classes [40]. The number of nodes covering such data using localized functions will be proportional to the number of vectors. Such transformations will not be able to generalize. Kernel methods based on localized or polynomial kernels will also not be able to generalize. MLP networks may solve the problem but need architectures specially designed for each problem of this type and are hard to train. Linear projections may provide interesting views on such data, but the number of directions $\mathbf{W}$ that should be considered to find good projections grows exponentially fast with the number of input dimensions (bits). In case of n-bit parity projection $Y = \mathbf{W} \cdot \mathbf{X}$ counts the number of 1 bits, producing odd and even numbers for the two parity classes. A periodic functions (such as cosine) is sufficient to solve the parity problem, but is not useful to handle other logical problems. Interesting transformations should find directions $\mathbf{W}$ that project a large number of training vectors $\mathbf{X}$ into localized groups. A window function $G(||\mathbf{W} \cdot \mathbf{X} - Y||)$ may capture a region where a delocalized large cluster of vectors from a single class is projected. A constructive network that adds new nodes to capture all interesting projections should be able to solve the problem. The linear output transformations will simply add outputs of nodes from all clusters that belong to a given class.

This type of geometric thinking leads to transformations that will be very useful in metalearning systems, facilitating learning of arbitrary Boole'an problems.

3.3 Heterogeneous Adaptive Systems

Many transformations may lead to the same goal because transformations with non-polynomial transfer functions are usually universal approximators [116]. The speed of convergence and the complexity of networks needed to solve a given problem is more interesting. Approximations of complex decision borders or approximations of multidimensional mappings by neural networks require flexibility that may be provided only by networks with sufficiently large number of parameters. This leads to the bias-variance dilemma [14, 63, 176], since large number of parameters contribute to a large variance of neural models and small number of parameters increase their bias.

Regularization techniques may help to avoid overparameterization and reduce variance, but training large networks has to be expensive. In MLP models regularization methods decrease the weights, forcing the network function to be more smooth. This is a good bias for approximation of continuous functions, but it is not appropriate for data generated using logical rules, where sharp, rectangular decision borders are needed. Two orthogonal hyperplanes cannot be represented accurately with soft sigmoidal functions used in MLPs or with Gaussian functions used by RBF networks. The inadequacy of fundamental model limits the accuracy of neural solutions, and no improvement of the learning rule or network architecture will change this. Transformations based on scalar products (hyperplanes, delocalized decision borders) solve some problems with $O(N)$ parameters, while the use of localized functions (for example Gaussians) requires $O(N^2)$ parameters; on other problems this situation is reversed [57]. Therefore discovering the proper bias for a given data is very important. Some real world examples showing the differences between RBF and MLP networks that are mainly due to the transfer functions used were presented in [57] and [46, 47].

The simplest transformation that has the chance to discover appropriate bias for complex data may require several different types of elementary functions. Heterogeneous adaptive systems (HAS) introduced in [51] provide different types of decision borders at each stage of building data model, enabling discovery of the most appropriate bias for the data. Neural [59, 96, 45], decision tree [51, 80] and similarity-based systems [53, 179, 180] of this sort have been described, finding for some data simplest and most accurate models known so far.

Heterogeneous neural algorithms that use several transfer functions within one network may be introduced in several ways. A constructive method that selects the most promising function from a pool of candidates adding new node to the transformation has been introduced in [45, 96, 59]. Other constructive algorithms, such as the cascade correlation [65], may also be used for this purpose. Each candidate node using different transfer function should be trained and the most useful candidate added to the network.

The second approach starts from transformation that uses many types of functions using information selection or regularization techniques to reduce the number of functions [96]. Initially the network may be too complex but at the end only the functions that are best suited to solve the problem are left. In the ontogenic approach neurons are removed and added during the learning process [96].

The third approach starts from flexible transfer functions that are parameterized in a way that makes them universal. Linear activation based on a scalar product $\mathbf{W} \cdot \mathbf{X}$ is combined with higher order terms used in distance measures to create functions that for some parameters are localized, and for other parameters non-localized. Several functions of such kind have been proposed in [57]. In particular bicentral functions are very useful and flexible, with decision regions of convex shapes, suitable for classification. These functions are products of N pairs of sigmoidal functions (Fig. 2):

$$
\begin{aligned}
Bi2s(\mathbf{X}; \mathbf{t}, \mathbf{B}, \mathbf{s}) &= \prod_{i=1}^{N} \sigma(A2_i^+)\ (1 - \sigma(A2_i^-)) \qquad (5)\\
&= \prod_{i=1}^{N} \sigma(e^{s_i} \cdot (x_i - t_i + e^{b_i}))\ (1 - \sigma(e^{s'_i} \cdot (x_i - t_i - e^{b_i})))
\end{aligned}
$$

The first sigmoidal factor in the product is growing for increasing input x_i while the second is decreasing, localizing the function around t_i. Shape

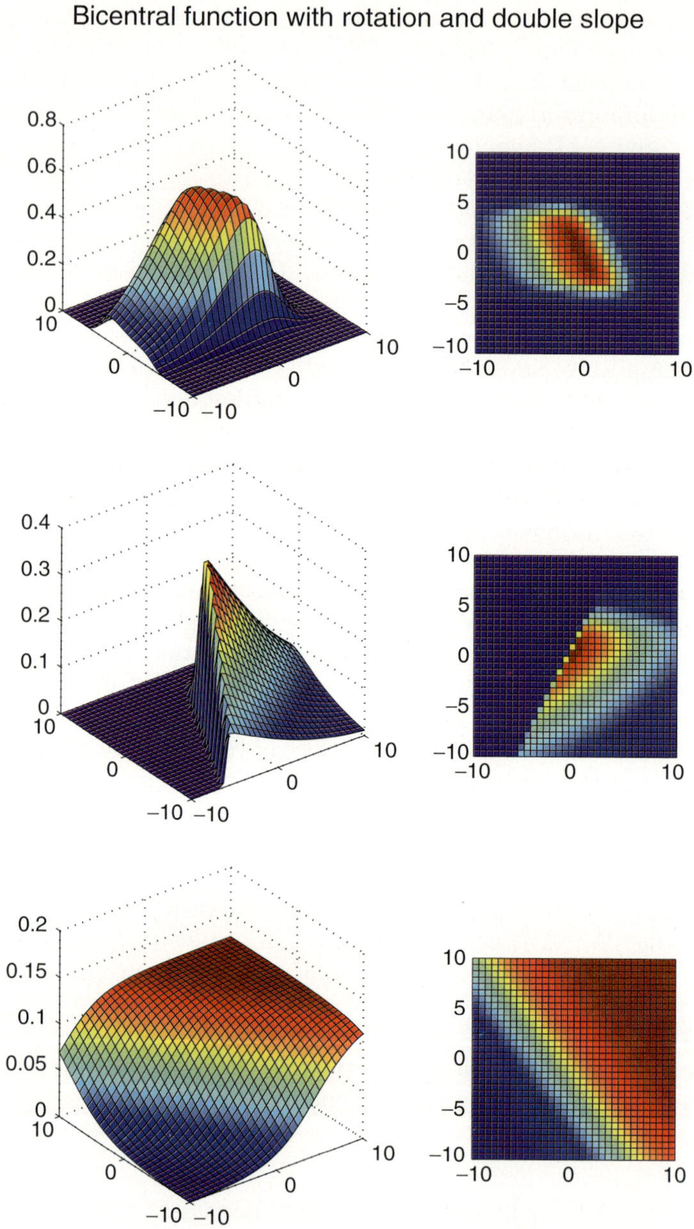

Fig. 2. A few shapes of general bicentral functions (Eq. 5)

adaptation of the density $Bi2s(\mathbf{X};\mathbf{t},\mathbf{B},\mathbf{s})$ is possible by shifting centers $\mathbf{t}$, rescaling $\mathbf{B}$ and $\mathbf{s}$. Product form leads to well-localized convex contours of bicentral functions. Exponentials e^{s_i} and e^{b_i} are used instead of s_i and b_i

parameters to prevent oscillations during the learning procedure (learning becomes more stable). Using small slope s_i and/or s'_i the bicentral function may delocalize or stretch to *left* and/or *right* in any dimension. This allows creation of such contours of transfer functions as half-infinite channel, half-hyper ellipsoidal, soft triangular, etc.

Although the costs of using this function is a bit higher than using the sigmoidal function (each function requires $4N$ parameters) more flexible decision borders are produced. Rotations of these contours require additional N parameters. An important advantage of the bicentral functions comes from their separability, enabling analysis of each dimension or a subspace of the input data independently: one can forget some of the input features and work in the remaining subspace. This is very important in classification when some of the features are missing and allows to implement associative memories using feedforward networks [50, 3]. Bicentral functions with rotations (as well as multivariate Gaussian functions with rotation) have been implemented so far only in two neural network models, the Feature Space Mapping [50, 3] and the IncNet [102, 99, 98].

Very little experience with optimization of transfer functions in heterogenous systems has been gathered so far. Neural networks using different transfer functions should use lower number of nodes, and thus the function performed by the network may be more transparent. For example, one hyperplane may be used to divide the input space into two classes and one additional Gaussian function to account for local anomaly. Analysis of the mapping performed by an MLP network trained on the same data will not be so simple. More algorithms to create such models are needed.

3.4 Geometrical Perspective

Composition of transformations may also be seen from geometrical perspective. Informational geometry [93] is aimed at characterization of the space of all possible probability distributions, and thus works in the space of model parameters. Geometry of heteroassociative vector transformations, from the input feature space to the output space, is also interesting. It is clear that different sensory signals are recognized in different feature subspaces, but even in a single sensory modality different objects are recognized paying attention to different features. This shows that vector space model for characterization of objects is too simple to account for object recognition in perception.

At each point of the input space relative importance of features may change. One way to implement this idea [35] is to create local non-symmetric similarity function $D(\mathbf{X} - \mathbf{Y}; \mathbf{X})$, smoothly changing between different regions of the input space. For example this may be a Minkovsky function $D(\mathbf{X} - \mathbf{Y}; \mathbf{X}) = \sum_i s_i(\mathbf{X})|X_j - Y_i)|$ with the scaling factor that depend on the point $\mathbf{X}$ of the input space, in particular many of them may be zero. Such scaling factors may be calculated for each training vector using local PCA, and interpolated between the vectors. Local Linear Embedding (LLE) is a popular method of this sort [150] and many other manifold learning methods have been developed. Alternatively a smooth mapping may be generated training MLP or other neural networks to approximate desired scaling factors.

Prototype rules for data understanding and transformation may be created using geometrical learning techniques that construct a convex hull encompassing the data, for example an enclosing polytope, cylinder, a set of ellipsoids or some other surface enclosing the data points. Although geometrical algorithms may be different than neural or SVM algorithms, the decision surfaces they provide are similar to those offered by feedforward networks. A covering may be generated by a set of balls or ellipsoids following principal curve, for example using the piecewise linear skeletonization approximation to principal curves [106]. An algorithm of this type creating a "hypersausage" decision regions has been published recently [161]. More algorithms of this type should be developed, and their relations with neural algorithms investigated.

From geometrical perspective kernel transformations are capable of smoothing or flatting decision borders. Using the vectors $\mathbf{R}^{(i)}$ that are close to the decision border as support vectors for kernel (distance) calculation creates new features, placing support vectors on a hyperplane (distance for all $\mathbf{R}^{(i)}$ is zero). Therefore a single hyperplane after such transformation is frequently sufficient to achieve good separation of data. However, if the data has complex structure, disjoint clusters from the same class, or requires special transformation for extraction of information this may not be an optimal approach.

After the second transformation (or a series of transformations) all data is converted to the second internal representation ${}^{2}\mathbf{X}$, and the final transformation is added to extract simple structure from multidimensional data.

3.5 Redefining the Goal of Learning

The first two transformations should discover interesting structures in data or increase the chance of data separability, as in the case of kernel transformations. More transformations may be applied, either at the pre-processing stage (normalization, whitening, calculation of Fourier, Hadamard or wavelet coefficients etc), or at the information extraction stage. The role of the final transformations is to compress this information, find interesting views on the data from the point of view of certain goals. These transformations usually involves a drastic reduction of dimensionality. The number of outputs in the approximation problems is equal to the number of approximated components, or the problem is broken into several single-output functions. In the K-class classification problems the number of outputs is usually $K-1$, with zero output for the default class. In the Error Correcting Output Codes (ECOC) approach [32] learning targets that are easier to distinguish are defined, setting a number of binary targets that define a prototype "class signature" vectors. The final transformation compares then the distance from the actual output to these class prototypes.

The learning targets used in most CI methods for classification are aimed at linear separability. The final linear transformation provides a hyperplane that divides the data, transformed by a series of mappings $\mathcal{T}_k(..\mathcal{T}_2(\mathcal{T}_1(\mathbf{X})\ldots))$, into two halfspaces. Linear transformation is the simplest and quite natural if the kernel transformation increases the dimensionality and flattens the decision borders, making the data linearly separable. However, for difficult problems, such as learning of Boolean functions, this will not work. Instead of thinking about the decision hyperplane it is better to focus on interesting projections or more general transformations of data. For linearly separable data $\mathbf{W}\cdot\mathbf{X}$ projection creates two distinct clusters. For non-separable data an easier target is to obtain several well separated clusters. For example, in k-bit parity problem projection on $\mathbf{W} = [1, 1..1]$ shows $k+1$ clusters clearly providing a satisfactory solution to the problem. Thus instead of aiming at linear separation using the final transformation based on hyperplane, the goal of learning may be redefined by assuming another well-defined transformation. In particular using the interval-based transformation as the final step easily "disarms" the remaining non-linearity in data, and greatly simplifies the task of all previous transformations.

The dataset $\mathbf{X}_i$ of points belonging to two classes is called k-separable if a direction $\mathbf{W}$ exist such that points $y_i = \mathbf{W}\cdot\mathbf{X}_i$ are clustered in k intervals, each containing vectors from a single class only. A dataset that is k-separable may also be $k+m$ separable, until $k+m = n$ is reached. Although usually the minimal k is of interest sometimes higher k's may be preferred if the margins between projected clusters are larger, or if among k clusters some have very small number of elements. Problems that may be solved by linear projection on no less than k-clusters belonging to alternating classes are called k-separable problems [40]. This concept allows for characterization of the space of all non-separable problems. The difficulty of learning grows quickly with the minimum k required to solve a given problem. Linear (2-separable) problems are quite simple and may be solved with linear SVM or any other variant of LDA model. Kernel transformations may convert some problems into linearly separable in higher dimensional space. Problems requiring $k = 3$, for example the XOR problem, are already slightly more difficult for non-local transformations (for example MLPs), and problems with high k quickly become intractable for general classification algorithms.

Among all Boolean classification problems linear separability is quite rare. For 3 bits there are 8 vertices in the cube and $2^8 = 256$ possible Boolean functions. Two functions are constant (always 0 or 1), 102 are linearly separable, 126 are 3-separable and 26 are 4-separable functions. For more than half of the 3-bit Boolean functions there is no linear projection that will separate the data. Almost half (126) of all the functions give at least 3 alternating clusters. For the 4-bit problem there are 16 hypercube vertices, with Boolean functions corresponding to 16-bit numbers, from 0 to 65535 (64K functions). The number of linearly separable functions is 1880, or less than 3% of all functions, with about 22%, 45% and 29% being 3 to 5-separable. About 188 functions were found that seem to be either 4 or 5-separable, but in fact contain projection of at least two hypercube vertices with different labels on the same point. Although the percentage of linearly separable functions rapidly decreases relatively low k-separability indices resolve most of the Boolean functions.

Changing the goal of learning may thus help to discover much simpler data models than those provided by kernel methods. The final transformation separating the classes on a line with k-intervals has only $k-1$ parameters. Periodic or quasi-periodic separability in one dimension is worth considering to avoid high number of intervals. Other simplified transformations that handle different types of non-linearities may be defined in two or more dimensions. Mapping on the chessboard targets, or on localized Voronoi cells defined by prototypes localized on a regular multidimensional grid, may handle directly quite difficult non-linearities.

New targets require a different approach to learning because the vector labels Y_X do not specify to which interval a given vector $\mathbf{X}$ belongs. Gradient training is still possible if soft windows based on combinations of sigmoidal functions are used. Thus for 3-separable problems the final transformation is from 3 intervals: $[-\infty, a], [a, b], [b, +\infty]$ to $-1, +1, -1$ values. For the middle interval a soft window functions may be set $S(x; a, b) = \tanh(x-a) - \tanh(x-b) - 1 \in [-1, +1]$. Quadratic error function suitable for learning is:

$$E(a, b, \mathbf{W}) = \sum_{\mathbf{X}} (S(\mathbf{W} \cdot \mathbf{X}; a, b) - Y_X)^2 \tag{6}$$

Starting from small random weights the center y_0 of projected data $y = \mathbf{W} \cdot \mathbf{X}$, the range $[y_{\min}, y_{\max}]$ is estimated, and a, b values are set to $y_0 \pm |y_{\max} - y_{\min}|/4$. The weights and the $[a, b]$ interval are trained using gradient method. It is also possible to implement 3-separable backpropagation learning in purely neural architecture based on a single linear neuron or perceptron for projection plus a combination of two neurons creating a "soft trapezoidal window" function $S(x; a, b)$ that passes only the output in the $[a, b]$ interval [47]. The two additional neurons (Fig. 3) have fixed weights ($+1$ and -1) and biases a, b, adding only two adaptive parameters. An additional parameter determining the slope of the window shoulders may be introduced to scale

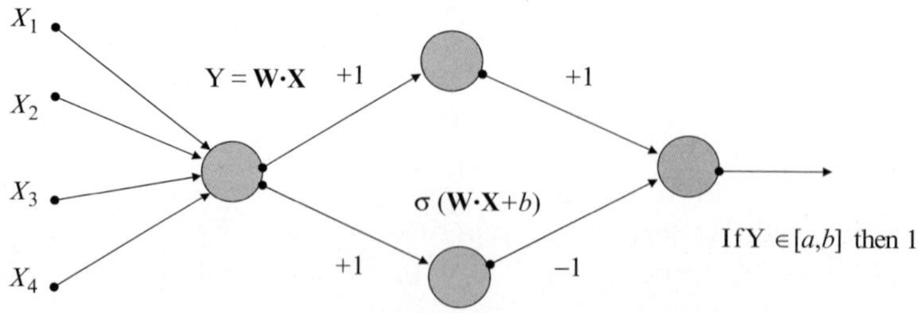

Fig. 3. MLP solution to the 3-separable case

the $\mathbf{W} \cdot \mathbf{X}$ values as the weights grow. The input layer may of course be replaced by hidden layers that implement additional mappings, for example kernel mappings, thus making this at least as powerful as SVM methods.

This network architecture has $n+2$ parameters and is able to separate a single class bordered by vectors from other classes. For n-dimensional 3-separable problems standard MLP architecture requires at least two hidden neurons connected to an output neuron with $2(n+1)+3$ parameters. For k-separability this architecture will simply add one additional neuron for each new interval, with one bias parameter. n-bit parity problems require only n neurons (one linear perceptron and $n-1$ neurons with adaptive biases for intervals), while in the standard approach $O(n^2)$ parameters are needed [94]. Tests of such architectures showed (W. Duch, R. Adamczak, M. Grochowski, in preparation) that indeed one may learn dificult Boolean functions this way. In fact we are training here a single bi-central function with rotations (Eq. 5), creating simplest possible model of the data.

Algorithms of this type, projecting data on many disjoined pure clusters, may have biological justification. Neurons in association cortex form strongly connected microcircuits found in cortical columns, resonating with different frequencies when an incoming signal $X(t)$ appears. This essentially projects the signal into high-dimensional space. A perceptron neuron observing the activity of a column containing many microcircuits learns to react to signals in an interval around particular frequency in a supervised way based on Hebbian principles. It is sufficient to combine outputs from selected microcircuits correlated with the category that is being learned. In case of signals microcircuits may be treated as resonators specializing in discovering interesting signal structures, for example Gabor filters in vision. A parallel array of one-bit threshold quantizers with sums of inputs is a crude approximation to such model. It achieves not only optimal signal detection, but even for suprathreshold input signals it improves its performance when additional noise is added, a phenomenon called "suprathreshold stochastic resonance" [149]. In case of abstract reasoning combination of disjoint projections on the $\mathbf{W} \cdot \mathbf{X}$ line is more useful than simple quantizers.

3.6 Prototype-based Rules for Data Understanding

Most attempts to understand the structure of data in machine learning is focussed on extraction of logical rules [62, 47]. Relations between fuzzy logic systems and basis set networks are fairly obvious and have been described in details [105, 95]. The use of Gaussian functions in the Radial Basis Function (RBF) networks is equivalent to the use of sets of fuzzy rules with Gaussian membership functions. Although it is an interesting fact in itself, it has not lead to any new development, in most applications simple Gaussian classifiers are created. To optimize fuzzy systems neural adaptation techniques may be used, leading to neurofuzzy systems [50, 135, 138].

Fuzzy set $\mathcal{F}$ is defined by the universe $\mathcal{X}$ and the membership functions $\chi_F(X)$, specifying the degree to which elements of this universe belong to the set $\mathcal{F}$. This degree should not be interpreted as probability [111] and in fact at least four major interpretations of the meaning of membership functions may be distinguished [13]. One natural interpretation is based on the degree to which all elements $X \in \mathcal{X}$ are similar to the typical elements (that is those with $\chi_F(X) \approx 1$) of $\mathcal{F}$. From this point of view fuzzy modeling seems to be a special case of similarity modeling, field that have not yet been fully developed. On the other hand fuzzy models are quite successful and may contribute to new similarity measures. Relations between fuzzy rules and similarity to prototypes are worth more detailed exploration.

An analogy with human object recognition is quite fruitful. Perceiving and recognizing an object requires attention to be paid to its most characteristic features. First feature values X_i are measured by our sensory systems with relatively high precision (deriving, for example, physical parameters of sound), and then primary and secondary sensory cortex transforms these input values using higher-order receptive fields that integrate spatio-temporal patterns facilitating recognition of complex patterns (for example, phonemes). In fuzzy modeling each feature X_i of an object $\mathbf{X}$ is filtered through a large receptive field $\mathcal{F}_{ij}$, defined by a membership function $\mu_{F_j}(X_i)$. Simple MFs, such as triangular, trapezoidal or Gaussian, are used to model the degree to which some value X_i belongs to the receptive field $\mathcal{F}_{ij}$. Comparing to the sophisticated processing of sensory signals by the brain this is a very crude approach in which larger receptive fields are obtained directly from individual features using membership functions, instead of non-linear combinations of several features. Brain-like information processing may of course be more accurately modeled using hierarchical fuzzy systems.

Selection of prototypes and features together with similarity measures offers new, so far unexplored alternative to neurofuzzy methods [49, 180, 15]. Duality between similarity measures and membership functions allows for generation of propositional rules based on individual membership functions, but there are significant differences. Fuzzy rules first apply large receptive fields (membership functions) to these individual features, combining them in conditions of rules later. P-rules in their natural form first create a combination of features (via similarity functions) and then apply various membership functions to this combination. Neurofuzzy systems generate fuzzy rules and optimize membership functions [135, 138] using input vectors defined in fixed-dimensional feature spaces. Similarity may be evaluated between objects with complex structures that are not easy to describe using a common sets of features. In particular the use of probabilistic, data dependent distance functions allows for definition of membership functions for symbolic data (such as the sequential DNA or protein strings) that may be difficult to derive in other way.

Experiments in cognitive psychology show that logical rules are rarely used to define natural categories, human categorization is based on memorization of exemplars and prototypes [148]. Similarity functions may be used to model the importance of different features in evaluating similarity between the new case in relation to stored prototypes. Multiplicative similarity factors may easily be converted to additive distance factors and vice versa. Rule-based classifiers are useful only if the rules they use are reliable, accurate, stable and sufficiently simple to be understood [47]. Prototype-based rules are useful addition to the traditional ways of data exploration based on crisp or fuzzy logical rules. They may be helpful in cases when logical rules are too complex or difficult to obtain. A small number of prototype-based rules with specific similarity functions associated with each prototype may provide complex decision borders that are hard to approximate using logical systems, but are still sufficiently simple to understand them. For example, such simple rules have been generated for some medical datasets using heterogeneous decision tree [80]. A single P-rule for breast cancer data classifying as malignant cancer all cases that are closer to prototype case (taken as one of the training cases) than a certain threshold, achieves 97.4% accuracy (sensitivity 98.8% and specificity 96.8%). The accuracy in this case is at least as good as that of any alternative system tried on this data.

Combining various feature selection and prototype selection methods with similarity functions leads to many interesting algorithms. An interesting possibility is to use the prototype-based rules to describe exceptions in the crisp or fuzzy logic systems. Systematic investigation of various membership functions, T-norms and co-norms, and their relation to distance functions is certainly worth pursuing. The algorithms for generation of P-rules should be competitive to the existing neurofuzzy algorithms and will become an important addition to the methods of computational intelligence. Although similarity measures provide great flexibility in creating various decision borders this may turn to be a disadvantage if the primary goal is to understand the data rather than make most accurate predictions (neurofuzzy approaches have of course the same problem). Optimized similarity measures may not agree with human intuition and in some cases larger number of prototypes with simpler similarity measures may offer more acceptable solution.

3.7 Multiple Models for Meta-learning

Multi-objective optimization problems do not have a single best solution. Usually data mining systems return just a single best model but finding a set of Pareto optimal models if several criteria are optimized is much more ambitious goal. For example, accuracy should be maximized, but variance should be minimized, or sensitivity should be maximized while the false alarm rate should be minimal. The search process for optimal models in meta-learning should explore many different models. Models that are close to the Pareto front [132] should be retained and evaluated by domain experts.

A forest of decision trees [79] and heterogeneous trees [80] is an example of a simple meta-search in a model space restricted to decision trees.

Heterogeneous trees use different types of rule premises, splitting the branches not only using individual features, but also using tests based on distances from the training data vectors. These trees work in fact in a kernel space, but the optimization process is quite different than in the SVM case. In case when linear discrimination works well standard decision trees may give poor results, but adding distance-based conditions with optimal support vectors far from decision borders provides flat spherical borders that work as well as hyperplanes. The beam search maintains at each stage k decision trees (search states), ordering them by their accuracy estimated using cross-validation on the training data [80]. This algorithm has found some of the simplest and most accurate decision rules that gave different tradeoffs between sensitivity and specificity.

The metalearning search procedure creates many individual models and it would be wasteful not to use them, unless only models of specific types are of interest (for example, models that are easily comprehensible). Metalearning usually leads to several interesting models, as different types of optimization channels are enabled by the search procedure. If a committee of models is desired diversification of individual models that should perform well in different regions of input space may be necessary, especially for learning of difficult tasks. The mixture of models allows to approximate complicated probability distributions quite accurately improving stability of individual models. Individual models are frequently unstable [17], i.e. quite different models are created as a result of repeated training (if learning algorithms contains stochastic elements) or if the training set is slightly perturbed [6].

Although brains are massively parallel computing devices attention mechanisms are used to inhibit parts of the neocortex that are not competent in analysis of a given type of signal. All sensory inputs (except olfactory) travel through the thalamus where their importance and rough category is estimated. Thalamic nuclei activate only those brain areas that may contribute useful information to the analysis of a given type of signals [168]. This observation may serve as an inspiration for construction of better algorithms for data analysis. In the metasearch process all models that handle sufficiently many cases mistreated by other models should be maintained.

A committee based on competent models, with various factors determining regions of competence (or incompetence) may be used to integrate decisions of individual models [54, 55]. The competence factor should reach $F(\mathbf{X}; M_l) \approx 1$ in all areas where the model M_l has worked well and $F(\mathbf{X}; M_l) \approx 0$ near the training vectors where errors were made. A number of functions may be used for that purpose: a Gaussian function $F(||\mathbf{X} - \mathbf{R}_i||; M_l) = 1 - G(||\mathbf{X} - \mathbf{R}_i||^a; \sigma_i)$, where $a \geq 1$ coefficient is used to flatten the function, a simpler $1/\left(1 + ||\mathbf{X} - \mathbf{R}_i||^{-a}\right)$ inverse function, or a logistic function $1-\sigma(a(||\mathbf{X}-\mathbf{R}_i||-b))$, where a defines its steepness and b the radius where the value drops to 1/2. Because many factors are multiplied in the incompetence function of the model each factor should quickly reach 1 outside of the incompetence area.

This is achieved by using steep functions or defining a threshold values above which exactly 1 is taken.

Results of $l = 1 \dots m$ models providing estimation of probabilities $\mathcal{P}(C_i|\mathbf{X}; M_l)$ for $i = 1 \dots K$ classes may be combined in many different ways [112]: using majority voting, averaging results of all models, selecting a single model that shows highest confidence (i.e. gives the largest probability), selecting a subset of models with confidence above some threshold, or using simple linear combination. For class C_i coefficients of linear combination are determined from the least-mean square solution of:

$$\mathcal{P}(C_i|\mathbf{X}; M) = \sum_{l=1}^{m} \sum_{m} W_{i,l} F(\mathbf{X}; M_l) \mathcal{P}(C_i|\mathbf{X}; M_l) \tag{7}$$

The incompetence factors simply modify probabilities $F(\mathbf{X}; M_l)$ $\mathcal{P}(C_i|\mathbf{X}; M_l)$ that are used to set linear equations for all training vectors $\mathbf{X}$, therefore the solution is done in the same way as before. After renormalization $\mathcal{P}(C_i|\mathbf{X}; M)/\sum_j \mathcal{P}(C_j|\mathbf{X}; M)$ give final probabilities for classification. In contrast to AdaBoost and similar procedures [10] explicit information about competence, or quality of classifier performance in different feature space areas, is used here. Many variants of committee or boosting algorithms with competence are possible [112], focusing on generation of diversified models, Bayesian framework for dynamic selection of most competent classifier [70], regional boosting [124], confidence-rated boosting predictions [156], task clustering and gating approach [9], or stacked generalization [185].

A committee may be build as a network of networks, or a network where each element has been replaced by a very complex processing element made from individual network. This idea fits well to the transformation-based learning. Incompetence factors may be used to create virtual subnetworks, with different effective path of information flow. Modulation of the activity of modules is effective only if the information about the current state is distributed to all modules simultaneously. In the brain this role may be played by the working memory. Here it can be replaced by networks of modules adjusting their internal states (local knowledge that each module has learned) and their interactions (modulations of weights) to the requirements of the information flow through this system.

4 Beyond Pattern Recognition

The transformation-based approach described here is quite general and may be used for all kinds of pattern recognition problems, classification, approximation, pattern completion, association and unsupervised learning, extending what has already been done in the similarity-based approach [35, 43]. Computational intelligence should go beyond that, using partial observations (perceptions) to reason and learn from them.

Biological systems may be viewed as associative machines, but associative memories do not capture essential features of this process. Real brains constantly learn to pay attention to relevant features and use correlations between selected feature values and correct decisions. People may learn to act appropriately without explicitly realizing the rules behind their actions, showing that this process may proceed intuitively, without conscious control. Associative machines should learn from observations correlation of subsets of features and apply many such correlations in decision making process. Problems solved by symbolic artificial intelligence are of this sort. Bits and pieces of knowledge should be combined in a search for a final solution, and this leads in all interesting cases to a combinatorial explosion.

Not much progress has been made along this line of research in the last decades, although already in the PDP Books [153] several articles addressed combinatorial problems that are beyond pattern recognition. Boltzmann machines and harmony theory have been used to answer questions about complex systems from partial observations [153], but they proved to be very inefficient because the stochastic training algorithm needs time that grows exponentially with the size of the problem. Helmholtz machines [28], and recently introduced multi-layer restricted Boltzmann machines and deep belief networks [89] have been used only for pattern recognition problems so far. These models are based on stochastic algorithms and binary representations and thus are rather restricted.

Inferences about complex behavior from partial observations require systematic search. Suppose that a number of relations between small subsets of all variables characterizing complex system or situation are known *a priori* or from observations. For example, representing discrete changes of 3 variables ($\Delta A = +$ for increase, $\Delta A = -$ for decrease and $\Delta A = 0$ for no change) $3^3 = 27$ possibilities are distinguished, from all three variables decreasing, $(\Delta A, \Delta B, \Delta C) = (-, -, -)$ to all three increasing $(\Delta A, \Delta B, \Delta C) = (+, +, +)$. If these variables are constrained by additive $A = B + C$, multiplicative $A = B \cdot C$ or inverse additive $A^{-1} = B^{-1} + C^{-1}$ relations $A = f(B, C)$ then for all these relations 14 of the 27 possibilities cannot occur, for example $\Delta A = 0$ is impossible if both $\Delta B = \Delta C = -$ or both are $+$). If many such relations are applicable for N variables out of 3^N possible solutions only a few will be in agreement with all constraints. Assuming specific relations: $f(A_1, A_2) = A_3; f(A_2, A_3) = A_4; \ldots f(A_{N-2}, A_{N-1}) = A_N$ leaves only $4N+1$ possible solutions. For a large N it is a negligible fraction of all 3^N possibilities. Relations among 3 variables – representing partial observations – are stored in "knowledge atoms", or nodes arranged in one-dimensional array, connected to relevant input features. If the values of any two variables A_i, A_{i+1} or A_i, A_{i+2} are known then one of such nodes will provide the value of the third variable. In at most $N - 2$ steps, in each step selecting nodes that have only one unknown input, all values of variables are determined. Suppose now that only a single variable in two nodes has specific value, for example A_1 and A_4. An assumption about the value of A_2 should be made, starting 3 branches of a search tree with $A_2 = -, 0, +$. In the first step this leads to an inference of A_3 value, and in the second step $f(A_2, A_3) = A_4$ is checked, leaving only those branches for which both relations are true.

This is obviously a very fortuitous set of relations, but in most real situations very small search trees, sometimes reduced to a single branch, are still sufficient. An application to the analysis of a simple electric circuit (7 variables, currents I, voltages V and resistances R) [153] using network that keeps in its nodes relations between I, V, R (Ohm's law) and Kirchoff laws ($V = V_1 + V_2$ and $R = R_1 + R_2$) has been reported [50], showing how *a priori* knowledge enables solutions to problems involving qualitative changes of currents, voltages and resistances. The feature space representation works as a powerful heuristics in the reasoning process. In the seven variable problem considered here there are $3^7 = 2187$ different strings of 7 variables, but only 111 may actually be realized. In more complicated cases, when the number of variables involved is larger and the number of values these variable may take is also larger, the percentage of situations that fulfills all the constraints is vanishingly small. Network nodes may also implement functions that determine numerical values of unknown variables (such as $V = I \cdot R$).

Such network of knowledge atoms may solve mathematical problems without explicit transformation of equations, and when qualitative reasoning is sufficient it serves as a model of intuitive computing. For more complex situations hierarchical decomposition of the problem is necessary, depending on the questions asked. For example, changes of parameters of a single or a few elements of a complex electrical circuit may be decomposed into blocks, and there is no need to assign values to all variables. People in such cases analyze graphical structure of connections and nodes representing the problem, starting from elements mentioned in the problem statement.

Network nodes may also estimate probabilities of different observations, and then the algorithm may use them in sequential Markov-chain reasoning or in other variants of such algorithms approximating joint distribution of variables in various ways. The confabulation architecture [88] is an example of such algorithm that uses products of a few (usually 4) conditional probabilities in a specific way
$\arg\max_j(\min[p(i_1|j)p(i_2|j)p(i_3|j)p(i_4|j)]$,
where $p(i_1|j)$ is the probability that word i_1 precedes word j in a sequence. This algorithm trained on a large corpus of English stories produces plausible words j, although it cannot capture the meaning of the story or learn a strategy of games played, due to the lack of long-term dependencies and structural representation of the concepts.

Similar approaches may be useful in many other fields. In intelligent control random actions may be correlated with probability distributions of different results, creating several scenarios [169]. Problems of this type are somewhere between pattern recognition and typical artificial intelligence problems. Neural networks (a core CI technology) may be used as heuristics to constrain search (a core AI technology) in problem solving. Robots, including autonomous vehicles, need to combine reasoning with pattern recogntion in a real time. It would be very worthwhile to collect data for real problems of this kind, encouraging the development of algorithms for their solutions.

The inference process in this approach resembles human reasoning that either proceeds sequentially, or temporarily assumes one of the possibilities, checking if it is consistent with all knowledge assumed to be true at a given stage. The interplay between left and right hemisphere representations leads to generalization of constraints that help to reason at the meta-level [38]. These ideas may form a basis for an associative machine that could reason using both perceptions (observations) and *a priori* knowledge. In this way pattern recognition (lower level cognitive functions) may be combined in a natural way with reasoning (higher level cognitive functions).

A very interesting approach to representation of objects as evolving structural entities/processes has been developed by Goldfarb and his collaborators [76, 74, 75]. Structure of objects is a result of temporal evolution, a series of transformations describing the formative history of these objects. This is more ambitious than the syntactic approach in pattern recognition [68], where objects are composed of atoms, or basic structures, using specific rules that belong to some grammar. In the evolving transformation system (ETS) object structure is a temporal recording of structured events, making syntax and semantics inseparable. ETS formalism leads to a new concept of class which is represented by similar structural processes. Classes defined by decision borders do not capture the meaning of objects. Inductive learning process should discover class structures as a series of transformations that change the primitive elements to their observed structure. In this way a generative model is produced that may generate an infinite set of examples of objects from a given class.

This line of thinking is in agreement with the modified goal of learning presented in Sect. 3.5. Various initial transformations should discover interesting representations of different aspects of objects, learn how to measure their (dis)similarities introducing kernel functions. These transformations may be viewed as receptive fields of sensors observing the data, or as selection of operations that compare objects in some specific way. For comparison of strings, to take the simplest example, various substring operations may be used. ETS is also in agreement with the idea of computing and learning as compression, as evolving transformations compress the information. Application of ETS to structural representation of molecules has been presented [75], and structural representation of spoken language has been analyzed from this point of view [83].

5 Neurocognitive Inspirations, Computing, Cognition and Compression

How to scale up our present models to perform more interesting cognitive functions? Neurocognitive inspirations lead to modular networks that should process information in a hierarchical way that roughly should correspond to functions of various brain areas, and these networks become modules that are used to build next-level supernetworks, functional equivalents of larger brain areas. The principles on which models should be based at each level are similar [61]: networks of interacting modules should adjust to the flow of information (learn) changing their internal knowledge and their interactions with other modules. Efficient algorithms for learning are known only at the lowest level, when very simple interactions and local knowledge of processing elements are assumed. The process of learning leads to emergence of novel, complex behaviors and competencies. Maximization of system information processing capacity may be one guiding principle in building such systems: if the supernetwork is not able to model all relations in the environment then it should recruit additional members that will specialize in learning facts, relations or behaviors that have been missing.

Very complex supernetworks, such as the individual brains, may be further treated as units that cooperate to create higher-level structures, such as groups of experts, institutions, think-tanks or universities, commanding huge amounts of knowledge that is required to solve problems facing the whole society. Brain-storming is an example of interaction that may bring ideas up that are further evaluated and analyzed in a logical way by groups of experts. The difficult part is to create ideas. Creativity requires novel combination, generalization of knowledge that each unit has, applying it in novel ways. This process may not fundamentally differ from generalization in neural networks, although it takes place at much higher level of complexity. The difficult part is to create a system that has sufficiently rich, dense representation of useful knowledge to be able to solve the problem by combining or adding new concepts/elements [38].

The brain has much more computing power than our current machinery and thus may solve problems in a different way. Nevertheless brain resources are limited and the mechanism of encoding the new information using old memory structures is quite natural, leading to a great conservation of resources and enabling associative recall. This is also a source of serious difficulty in defining the meaning of symbols, encoded by activity of neural microcircuits that is constantly changing, spreading activation to other concepts. As a result relations between concepts change depending on the context, making the invariant meaning of concepts only a rough approximation. In experimental psychology this process, known as semantic priming, is one of the most popular subjects of investigation [128].

How can this priming process be approximated? An attractor network model has been created to explain results of psychological experiments [25]. However, such dynamical models are rather complex and do not scale well with the size of the network. Processing sequential information by simpler mechanisms, such as spreading activation in appropriately structured networks, is more suitable for information retrieval [26]. The challenge here is to create large semantic networks with overall structure and weighted links that facilitate associations and reproduce priming effects. Wordnet (http://wordnet.princeton.edu) and many other dictionaries provide some useful relations that may be used in network construction, although these relations capture a small fraction of knowledge and associations that humans have about each concept. Medical applications of spreading activation networks are easier to create because huge lexical resources are available [126].

Activation of semantic networks may be seen as dynamical construction of a relevant feature space, or non-zero subspace for vector models in information retrieval. New information is projected into this space, expanding it by adding new dimensions with non-zero components. Although the use of semantic networks is quite natural geometrical description of this process is still interesting. In geometrical model activation of the network concept node corresponds to a change of a metric around this concept (active concept associates with other active concepts), and this changes similarity relations at a given moment. Concepts in some specific meanings attract other concepts that become closer, while other concepts increase their distance, facilitating disambiguation. Static vector space models do not take that into account such dynamical changes, therefore spreading activation models should have an important advantages here.

Projecting new information on the semantic network creates strong associations with existing network nodes, encoding partially new knowledge in terms of the old one. Basic perceptual and conceptual objects of mind are created early in the developmental process, therefore perceptual information will be predominantly encoded in terms of the old knowledge. Seeing or hearing new objects will be remembered by adding new nodes that bind together (symbolize) specific configuration of node activations, responsible for recognition of natural categories. These nodes in turn are linked to new nodes coding abstract concepts that do not activate directly any perceptual nodes, allowing for higher and higher levels of abstraction.

Computing and cognition may be seen as information compression, although there are clearly exceptions. In the brain signals may be projected into higher-dimensional space, for example information from retina is projected on visual cortex with order of magnitude expansion in the number of cells involved. In computing we have seen that kernel methods make implicit projection into a highly-dimensional space to achieve separability before final reduction of information is made. Although the idea of cognition as compression is worth exploring it has been so far developed only for sequential, one-dimensional systems [184], and to some extent in our approach to concept disambiguation in medical domain [126]. Multiple alignment may be applied to sequential data, while computational intelligence methods work with many other types of data, including signals, vectors, spatial structures (like chemical molecules), or multimedia data. To account for preliminary processing of sensory and motor signals a more general approach to data transformation is needed. Instead of a sequence alignment calculus based on "small world" active subnetworks in the huge semantic network encoding relations between mind objects is required. New knowledge activates old similar structures, extends them in new directions, and thus is encoded in a compressed way. This process may be approximated using graphs representing active network nodes that represent current mind state. Associations between different network configurations are determined by transition probabilities between network states.

Several interesting developments emerged from neurocognitive inspirations, the two most important theories being the liquid computing [123] and laminar computing [82, 147]. Liquid computing concentrates on microcircuits and columnar architecture, and laminar computing on the layered architecture of neocortex. The neocortex microcircuits composed of densely connected neurons within a diameter of 500μm are heterogeneous and differ across brain regions. Many properties of these microcircuits are stereotypical, therefore a concept of generic microcircuit template may be a useful abstraction allowing for understanding of dynamical properties of the cortex. Maass and Markram [122] argue that online computing is indistinguishable from learning, because temporal integration of information in a stream of data constantly changes the system. Learning is thus an integral part of microcircuit dynamics. Boolean functions and approximations to Boolean functions may be computed by feedforward networks with stereotypical basis functions, including sigmoidal or any other nonlinear functions. Computing such functions is thus relatively easy, there are many biophysical mechanisms that can influence neuronal states and thus can be interpreted as performing computations and learning. A useful approximation to microcircuit dynamics may be provided by finite state automata [18, 29, 34]. The Liquid State Machine (LSM) model aims at better approximation at the microscopic level, based on "liquid" high dimensional states of neural microcircuits that change in real time. In [34] another approximation relating neurodynamical states to psychological space and mental events has been advocated to bridge neuroscience and psychology.

LSM treats complex dynamical systems as non-linear filters transforming input signals $x(t)$ into activations of microcircuit (hidden) neurons $h(t) = L(x(t))$. The output stream $y(t) = M(h(t))$ is then provided using "read-out neurons" that learn to select and recombine activations of the hidden layer. In fact this is a simple transformation model that projects signals into high-dimensional spaces $h(t)$ created by microcircuits in form of temporal filters or basis functions, where separation is relatively straightforward. Oscillators based on neurons with diverse time constants are needed to create good projections. Linear or simple perceptron readout neurons are sufficient to approximate any time-varying signal. A single LSM may be used for various tasks, depending on the training of these output neurons. It has a fading memory for signals, depending on the time constants of the neurons. In fact any larger reservoir of neurons with some memory will be sufficient to create projections of signals to high-dimensional space. The Echo State Networks, or more general Reservoir Computing, use untrained recurrent neural networks as "reservoirs of activity" to implement this projection. These approaches are equivalent to the Liquid State Machines [86] and are being applied to time series prediction, dynamical system identification and speech recognition with great success [86].

The question how to use neural elements to compute Boolean functions has many solutions [122]. A complementary question that is also worth asking is: how can neural circuits discover approximate but rather complicated logic (Boolean function) in data, going beyond trivial associations? How are different activations of the hidden layer recombined to enable this? This is what humans and animals evidently do all the time. LSM has so far been used for problems where simple associations are sufficient. k-separability and other more general goals of learning may be responsible for the ability of cortex to learn complex approximate logic using liquid states. One good algorithm for k-separability combines projections with discovery of local clusters. This is essentially what is needed for object-centered representations in vision [31] and has been used to model the outputs of parietal cortex neurons [143, 144]. Any continuous sensory-motor transformation may be approximated in this way [155]. Although precise neural implementation of such basis functions is not clear they may result from the activity of microcircuits. Generalization of k-separability to time-dependent transformations done by liquid state machines is rather straightforward. Therefore it is quite likely that k-separability is an interesting abstraction of a basic biological mechanism.

Laminar computing is based on inspirations derived from organization of neocortex into six layers with specific architecture, and information flow in the brain on macroscopic scales. The bottom-up interactions provide signals from the senses, while the top-down interactions provide expectations or prior knowledge, that helps to solve ill-defined problems. Finally the horizontal interactions enable competition and specialization. Laminar computing has been studied initially in the visual cortex [82] but seems to have captured more general principles of cortical computations [147]. It explains how distributed data is grouped into coherent object representations, how attention selects important events and how cortex develops and learns to express environmental constraints. The laminar computing models have been used to explain many neurophysiological and psychophysical results about visual system, but it has also been used to develop new algorithms for image processing. In practical applications a version of Adaptive Resonant Theory [19] called LAMINART [147], has been used. So far this is the most advanced approach to perception that will certainly play a very important role in the growing field of autonomous mental development and cognitive robotics.

6 Summary of Open Problems

Many open problems have been identified in the preceding chapters. Below is a summary of some of them:

- Solid foundations of computational intelligence that go beyond probabilistic pattern recognition approaches are needed.

The current state is clearly not satisfactory. Bayesian probabilistic framework forms the foundation of many pattern recognition methods [63, 176, 166] and graphical networks [101], but it does not cover most branches of computational intelligence. Good foundations should help to create methods that adjust themselves to the data, finding the simplest possible model that accurately describes the data. The no-free-lunch theorem shows that a single approach is not sufficient [63] and automatic selection of methods included in larger data mining packages is still too difficult as these methods are not presented from the point of view of common foundations.

Earlier suggestions to base CI foundations on similarity concept [35, 43, 141] led to a framework in which meta-learning could be implemented [53] using search techniques in the model space. It has been found that similarity-based rule systems are equivalent to fuzzy systems [49], providing an alternative way to understand data, and that neural algorithms can also be presented in this framework. Heterogeneous constructive systems of this type are especially useful and have already discovered some of the simplest descriptions of data [80, 51]. A framework based on transformations, presented in this paper for the first time, is even more general, as it includes all kinds of pre-processing and unsupervised methods for initial data transformations, and looks at learning as a series of data transformations, defining new goals of learning. The work on hyperkernels also goes in similar direction [137].

Although this framework is still in the initial stage of its development it has a chance to provide foundations for computational intelligence models. Adaptation of these models requires some optimization techniques, processes that operate on admissible transformations. A new generation of data mining software, capable of implementing arbitrary sequences of transformations for meta-learning, is in development (K. Grąbczewski and N. Jankowski, in preparation).

- Existing methods cannot learn difficult problems.

CI algorithms are rarely addressing real, difficult problems, and many researchers are convinced that universal approximators, such as neural networks, are good tools to learn the structure of any data. In fact off-the-shelf systems work well only for problems that are "not-too-far" from the linearly separable problems. Fortunately many interesting datasets in machine learning domain are of this kind. Problems with inherent approximate Boolean logic become quickly intractable with growing complexity of the logical functions. Growing complexity of such problems may be characterized using the notion of k-separability [40], or setting more general goal for learning. We are probably not aware how many problems in bioinformatics or text processing are intractable and therefore are ignoring them. Datasets for such difficult problems are needed to increase awareness of the need for such methods, but first we should be able to solve at least some of these problems.

- Problems requiring reasoning based on perceptions should be explored.

The basic function of simple brains is to link perception with action. Agents with more complex brains have internal states and goals, and need to perceive and reason before making actions. Agent-based approach in artificial intelligence [154] is usually based on symbolic reasoning and logical rules. There has been some work on hybrid neural-symbolic architectures [178], but not much effort devoted to neural architectures capable of representations of predicate relations. The connectionist model Shruti [160, 177] seems to be an exception, although it has not gained wider popularity. Although spreading activation networks allow for some reasoning (mostly disambiguation and coherent interpretation of concepts [126]) this is not sufficient to solve problems requiring combinatorial search, compositionality, problems arising in sentence parsing or sequential reasoning. How to control spreading activation to account for systematic thinking process? New mathematical techniques for representation of objects and relations by active subgraphs in large semantic networks seem to be required. Search processes may be constrained by "intuitive computing" using neural heuristics [34]. Reinforcement learning [165], reservoir learning [86], laminar computing [82, 147] and chunking [136] should also be used as general mechanisms for sequence learning and divide-and-conquer sub-problem parsing. No cognitive architectures have captured so far all these processes, and combining them together is an interesting challenge.

Neurocognitive inspirations for understanding language and general cognitive processes lead to distributed connectionist systems that can be approximated by networks with local activations, and that in turn may be partially understood in terms of symbolic processes and probabilistic or finite automata [178]. Investigation of relations between approximations of brain functions at different level of complexity is quite fruitful, leading to new models of mental processes based on psychological spaces [34, 61]. At the simplest level perceptrons are found, with a single internal parameter and synaptic interactions based on fixed weight connections. Enhanced perceptrons sensitive to phase synchronization [113] are able to solve the famous connectedness and other problems posed by Minsky and Papert [133]. Networks with spiking neurons also have this capability [175], but it is not clear if they have additional powers – characterization of different complexity classes seems to be incomplete here. At a higher level complex processing elements modeling whole microcircuits are found, and even higher networks of networks and societies of minds. Although new biological mechanisms at the level of synaptic learning are certainly very important, simple abstractions of complex functions such as self-organization proved to be quite interesting. At the level of approximating microcuircuit and minicolumn functions simpler mechanisms, implementing for example some form of k-separability learning, may also be useful. Perhaps correlation-based learning, as in the Alopex algorithm [172], would be sufficient for biological implementation? Such approximations may also be viewed as information compression [184].

- Methodology of evaluation and development of CI methods is urgently needed.

Every year hundreds of methods are introduced, some of them rediscovered many times, others being minor variations on the known themes. It is well known that a data on which a given method works well may always be found [63]. There is no simple way to evaluate *a priori* how a new method will perform on a real data, but at least it is possible to understand what type of decision borders it is using and thus what type of problems it may solve. Efforts to provide standard benchmarks have largely failed. For example, the Data for Evaluating Learning in Valid Experiments (Delve) project[2] has been quite promising, presenting problems of growing complexity, but the project has unfortunately been abandoned.

Without standard methodology of evaluating new approaches no progress will be possible. There is a tendency to create algorithms and test them on new datasets, ignoring good reference results that are hard to compete with. Even though quite interesting datasets are available from some competitions many papers end with one or two trivial examples. Benchmarks that go beyond pattern recognition are particularly hard to find. It is up to the editors and reviewers of journals to enforce comparison with the simplest methods that may solve similar problems, and require solutions to problems of increasing complexity. For example, in extraction of logical rules comparison with rules extracted by popular decision trees should be required [62], or in classification problems comparison with linear discrimination and the nearest neighbor model. Collecting raw data, results of data analysis and software implementing different methods should be encouraged by data repositories and competitions at conferences.

Comparison of accuracy does not address problems in real applications. There may be many performance measures, different costs involved, tradeoffs between model simplicity and accuracy, rejection rate and confidence, or costs of obtaining different features and making different types of errors. Verification and validation of CI models is of great importance in industrial applications where software should always perform up to the intended specifications. Testing the system on data that is similar to the training data may not be sufficient. The standard neural network testing is not able to validate applications used to assess safety of nuclear reactors, chemical plants, military equipment or medical life support systems. These issues are largely ignored by the CI academic community and mostly explored by researchers working in NASA, IBM and other research oriented companies. Visualization of data transformations performed by CI systems, including analysis of perturbation data, are very useful tools [36], although still rarely used.

Cognitive robotics may be an ultimate challenge for computational intelligence. Various robotic platforms that could be used for testing new ideas in

[2] http://www.cs.utoronto.ca/delve/

semi-realistic situations would be very useful. They have to combine perception, object recognition, reasoning, planning and control in real-time environment. However, computational intelligence has even more general ambitions, looking for solutions to all hard problems for which effective algorithms are not known.

Acknowledgments

I am grateful for the support by the Polish Committee for Scientific Research, research grant 2005-2007.

References

[1] F. Corbacho A. Sierra, J.A. Macias. Evolution of functional link networks. *IEEE Transactions on Evolutionary Computation*, 5:54–65, 2001.
[2] N.I. Achieser. *Theory of Approximation.* Frederick Ungar, New York, 1956. Reprinted: Dover Publications, New York 1992.
[3] R. Adamczak, W. Duch, and N. Jankowski. New developments in the feature space mapping model. In *Third Conference on Neural Networks and Their Applications*, pages 65–70, Kule, Poland, Oct 1997.
[4] J.A. Anderson, A. Pellionisz, and E. Rosenfeld. *Neurocomputing 2.* MIT Press, Cambridge, MA, 1990.
[5] J.A. Anderson and E. Rosenfeld. *Neurocomputing - foundations of research.* MIT Press, Cambridge, MA, 1988.
[6] R. Avnimelech and N. Intrator. Boosted mixture of experts: An ensemble learning scheme. *Neural Computation*, 11:483–497, 1999.
[7] F.R. Bach and M.I. Jordan. Kernel independent component analysis. *Journal of Machine Learning Research*, 3:1–48, 2002.
[8] P.M. Baggenstoss. The pdf projection theorem and the class-specific method. *IEEE Transactions on Signal Processing*, 51:672–668, 2003.
[9] B. Bakker and T. Heskes. Task clustering and gating for bayesian multitask learning. *Journal of Machine Learning Research*, 4:83–99, 2003.
[10] E. Bauer and R. Kohavi. An empirical comparison of voting classification algorithms: bagging, boosting and variants. *Machine learning*, 36:105–142, 1999.
[11] Y. Bengio, O. Delalleau, and N. Le Roux. The curse of highly variable functions for local kernel machines. *Advances in Neural Information Processing Systems*, 18:107–114, 2006.
[12] Y. Bengio, M. Monperrus, and H. Larochelle. Non-local estimation of manifold structure. *Neural Computation*, 18:2509–2528, 2006.
[13] T. Bilgiç and I.B. Türkşen. Measurements of membership functions: Theoretical and empirical work. In D. Dubois and H. Prade, editors,

Fundamentals of Fuzzy Sets, Vol. 1, pages 195–232. Kluver, Boston, 2000.
[14] C.M. Bishop. *Neural Networks for Pattern Recognition.* Oxford University Press, 1995.
[15] M. Blachnik, W. Duch, and T. Wieczorek. Selection of prototypes rules – context searching via clustering. *Lecture Notes in Artificial Intelligence*, 4029:573–582, 2006.
[16] E. Bonabeau, M. Dorigo, and G. Theraulaz. *Swarm Intelligence: From Natural to Artificial Systems.* Oxford University Press, 1999.
[17] L. Breiman. Bias-variance, regularization, instability and stabilization. In C.M. Bishop, editor, *Neural Networks and Machine Learning*, pages 27–56. Springer-Verlag, 1998.
[18] Y. Burnod. *An Adaptive Neural Network. The Cerebral Cortex.* Prentice-Hall, London, 1990.
[19] G.A. Carpenter and S. Grossberg. Adaptive resonance theory. In M.A. Arbib, editor, *The Handbook of Brain Theory and Neural Networks, 2nd ed*, pages 87–90. MIT Press, Cambridge, MA, 2003.
[20] G. Chaitin. *Algorithmic Information Theory.* Cambridge University Press, 1987.
[21] O. Chapelle. Training a support vector machine in the primal. *Neural Computation*, 19:1155–1178, 2007.
[22] N. Chater. The search for simplicity: A fundamental cognitive principle? *Quarterly Journal of Experimental Psychology*, 52A:273–302, 1999.
[23] A. Cichocki and S. Amari. *Adaptive Blind Signal and Image Processing. Learning Algorithms and Applications.* J. Wiley & Sons, New York, 2002.
[24] T.M. Cover. Geometrical and statistical properties of systems of linear inequalities with applications in pattern recognition. *IEEE Transactions on Electronic Computers*, 14:326–334, 1965.
[25] G.S. Cree, K. McRae, and C. McNorgan. An attractor model of lexical conceptual processing: Simulating semantic priming. *Cognitive Science*, 23(3):371–414, 1999.
[26] F. Crestani. Application of spreading activation techniques in information retrieval. *Artifical Intelligence Review*, 11(6):453–482, 1997.
[27] N. Cristianini and J. Shawe-Taylor. *An Introduction to Support Vector Machines and other Kernel-Based Learning Methods.* Cambridge University Press, 2000.
[28] P. Dayan and G.E. Hinton. Varieties of helmholtz machines. *Neural Networks*, 9:1385–1403, 1996.
[29] A.M. de Callataÿ. *Natural and Artificial Intelligence. Misconceptions about Brains and Neural Networks.* Elsevier, Amsterdam, 1992.
[30] L.N. de Castro and J.I. Timmis. *Artificial Immune Systems: A New Computational Intelligence Approach.* Springer, 2002.
[31] S. Deneve and A. Pouget. Basis functions for object-centered representations. *Neuron*, 37:347–359, 2003.

[32] T.G. Dietterich and G. Bakiri. Solving multiclass learning problems via error-correcting output codes. *Journal Of Artificial Intelligence Research*, 2:263–286, 1995.

[33] W. Duch. Neural minimal distance methods. In *Proceedings 3-rd Conference on Neural Networks and Their Applications*, pages 183–188, Kule, Poland, Oct 1997.

[34] W. Duch. Platonic model of mind as an approximation to neurodynamics. In S i. Amari and N. Kasabov, editors, *Brain-like computing and intelligent information systems*, pages 491–512. Springer, 1997.

[35] W. Duch. Similarity based methods: a general framework for classification, approximation and association. *Control and Cybernetics*, 29:937–968, 2000.

[36] W. Duch. Coloring black boxes: visualization of neural network decisions. In *Int. Joint Conf. on Neural Networks, Portland, Oregon*, volume I, pages 1735–1740. IEEE Press, 2003.

[37] W. Duch. Uncertainty of data, fuzzy membership functions, and multilayer perceptrons. *IEEE Transactions on Neural Networks*, 16:10–23, 2005.

[38] W. Duch. Computational creativity. In *World Congress on Computational Intelligence, Vancouver, Canada*, pages 1162–1169. IEEE Press, 2006.

[39] W. Duch. Filter methods. In I. Guyon, S. Gunn, M. Nikravesh, and L. Zadeh, editors, *Feature extraction, foundations and applications*, pages 89–118. Physica Verlag, Springer, Berlin, Heidelberg, New York, 2006.

[40] W. Duch. *k*-separability. *Lecture Notes in Computer Science*, 4131:188–197, 2006.

[41] W. Duch, R. Adamczak, and G.H.F. Diercksen. Distance-based multilayer perceptrons. In M. Mohammadian, editor, *International Conference on Computational Intelligence for Modelling Control and Automation*, pages 75–80, Amsterdam, The Netherlands, 1999. IOS Press.

[42] W. Duch, R. Adamczak, and G.H.F. Diercksen. Neural networks in non-euclidean spaces. *Neural Processing Letters*, 10:201–210, 1999.

[43] W. Duch, R. Adamczak, and G.H.F. Diercksen. Classification, association and pattern completion using neural similarity based methods. *Applied Mathemathics and Computer Science*, 10:101–120, 2000.

[44] W. Duch, R. Adamczak, and G.H.F. Diercksen. Feature space mapping neural network applied to structure-activity relationship problems. In Soo-Young Lee, editor, *7th International Conference on Neural Information Processing (ICONIP'2000)*, pages 270–274, Dae-jong, Korea, 2000.

[45] W. Duch, R. Adamczak, and G.H.F. Diercksen. Constructive density estimation network based on several different separable transfer functions. In *9th European Symposium on Artificial Neural Networks*, pages 107–112, Bruges, Belgium, Apr 2001.

[46] W. Duch, R. Adamczak, and K. Grąbczewski. Extraction of logical rules from backpropagation networks. *Neural Processing Letters*, 7:1–9, 1998.
[47] W. Duch, R. Adamczak, and K. Grąbczewski. A new methodology of extraction, optimization and application of crisp and fuzzy logical rules. *IEEE Transactions on Neural Networks*, 12:277–306, 2001.
[48] W. Duch and M. Blachnik. Fuzzy rule-based systems derived from similarity to prototypes. *Lecture Notes in Computer Science*, 3316:912–917, 2004.
[49] W. Duch and M. Blachnik. Fuzzy rule-based systems derived from similarity to prototypes. In N.R. Pal, N. Kasabov, R.K. Mudi, S. Pal, and S.K. Parui, editors, *Lecture Notes in Computer Science*, volume 3316, pages 912–917. Physica Verlag, Springer, New York, 2004.
[50] W. Duch and G.H.F. Diercksen. Feature space mapping as a universal adaptive system. *Computer Physics Communications*, 87:341–371, 1995.
[51] W. Duch and K. Grąbczewski. Heterogeneous adaptive systems. In *IEEE World Congress on Computational Intelligence*, pages 524–529. IEEE Press, Honolulu, May 2002.
[52] W. Duch and K. Grudziński. Search and global minimization in similarity-based methods. In *International Joint Conference on Neural Networks*, page Paper 742, Washington D.C., 1999. IEEE Press.
[53] W. Duch and K. Grudziński. Meta-learning via search combined with parameter optimization. In L. Rutkowski and J. Kacprzyk, editors, *Advances in Soft Computing*, pages 13–22. Physica Verlag, Springer, New York, 2002.
[54] W. Duch and L. Itert. Competent undemocratic committees. In L. Rutkowski and J. Kacprzyk, editors, *Neural Networks and Soft Computing*, pages 412–417. Physica Verlag, Springer, 2002.
[55] W. Duch and L. Itert. Committees of undemocratic competent models. In L. Rutkowski and J. Kacprzyk, editors, *Proc. of Int. Conf. on Artificial Neural Networks (ICANN), Istanbul*, pages 33–36, 2003.
[56] W. Duch and N. Jankowski. Complex systems, information theory and neural networks. In *First Polish Conference on Neural Networks and Their Applications*, pages 224–230, Kule, Poland, Apr 1994.
[57] W. Duch and N. Jankowski. Survey of neural transfer functions. *Neural Computing Surveys*, 2:163–213, 1999.
[58] W. Duch and N. Jankowski. Taxonomy of neural transfer functions. In *International Joint Conference on Neural Networks*, volume III, pages 477–484, Como, Italy, 2000. IEEE Press.
[59] W. Duch and N. Jankowski. Transfer functions: hidden possibilities for better neural networks. In *9th European Symposium on Artificial Neural Networks*, pages 81–94, Brusells, Belgium, 2001. De-facto publications.
[60] W. Duch, N. Jankowski, A. Naud, and R. Adamczak. Feature space mapping: a neurofuzzy network for system identification. In *Proceedings of the European Symposium on Artificial Neural Networks*, pages 221–224, Helsinki, Aug 1995.

[61] W. Duch and J. Mandziuk. Quo vadis computational intelligence? In P. Sincak, J. Vascak, and K. Hirota, editors, *Machine Intelligence: Quo Vadis?*, volume 21, pages 3–28. World Scientific, Advances in Fuzzy Systems – Applications and Theory, 2004.
[62] W. Duch, R. Setiono, and J. Zurada. Computational intelligence methods for understanding of data. *Proceedings of the IEEE*, 92(5):771–805, 2004.
[63] R.O. Duda, P.E. Hart, and D.G. Stork. *Patter Classification.* J. Wiley & Sons, New York, 2001.
[64] R.S. Michalski (ed). *Multistrategy Learning.* Kluwer Academic Publishers, 1993.
[65] S.E. Fahlman and C. Lebiere. The cascade-correlation learning architecture. In D.S. Touretzky, editor, *Advances in Neural Information Processing Systems 2*, pages 524–532. Morgan Kaufmann, 1990.
[66] L.J. Fogel, A.J. Owens, and M.J. Walsh, editors. *Artificial Intelligence through Simulated Evolution.* Wiley & Sons, 1966.
[67] J.H. Friedman. Exploratory projection pursuit. *Journal of the American Statistical Association*, 82:249–266, 1987.
[68] K.S. Fu. *Syntactic Pattern Recognition and Applications.* Prentice-Hall, New York, 1982.
[69] W. Gerstner and W.M. Kistler. *Spiking Neuron Models. Single Neurons, Populations, Plasticity.* Cambridge University Press, 2002.
[70] G. Giacinto and F. Roli. Dynamic classifier selection based on multiple classifier behaviour. *Pattern Recognition*, 34:179–181, 2001.
[71] A. Gifi. *Nonlinear Multivariate Analysis.* Wiley, Boston, 1990.
[72] Ch. Giraud-Carrier, R. Vilalta, and P. Brazdil. Introduction to the special issue on meta-learning. *Machine Learning*, 54:197–194, 2004.
[73] D. Goldberg. *Genetic Algorithms in Optimization and Machine Learning.* Addison-Wesley, 1989.
[74] L. Goldfarb and D. Gay. What is a structural representation? fifth variation. Technical Report Technical Report TR05-175, Faculty of Computer Science, University of New Brunswick, Canada, 2005.
[75] L. Goldfarb, D. Gay, O. Golubitsky, and D. Korkin. What is a structural representation? a proposal for a representational formalism. *Pattern Recognition*, (submitted), 2006.
[76] L. Goldfarb and S. Nigam. The unified learning paradigm: A foundation for ai. In V. Honovar and L. Uhr, editors, *Artificial Intelligence and Neural Networks: Steps Toward Principled Integration*, pages 533–559. Academic Press, Boston, 1994.
[77] R.L. Gorsuch. *Factor Analysis.* Erlbaum, Hillsdale, NJ, 1983.
[78] K. Grąbczewski and W. Duch. The separability of split value criterion. In *Proceedings of the 5th Conf. on Neural Networks and Soft Computing*, pages 201–208, Zakopane, Poland, 2000. Polish Neural Network Society.
[79] K. Grąbczewski and W. Duch. Forests of decision trees. *Neural Networks and Soft Computing, Advances in Soft Computing*, pages 602–607, 2002.

[80] K. Grąbczewski and W. Duch. Heterogenous forests of decision trees. *Springer Lecture Notes in Computer Science*, 2415:504–509, 2002.

[81] M. Grochowski and N. Jankowski. Comparison of instance selection algorithms. II. results and comments. *Lecture Notes in Computer Science*, 3070:580–585, 2004.

[82] S. Grossberg. How does the cerebral cortex work? development, learning, attention, and 3d vision by laminar circuits of visual cortex. *Behavioral and Cognitive Neuroscience Reviews*, 2:47–76, 2003.

[83] A. Gutkin. Towards formal structural representation of spoken language: An evolving transformation system (ETS) approach. PhD Thesis, School of Informatics, University of Edinburgh, UK, 2005.

[84] I. Guyon, S. Gunn, M. Nikravesh, and L. Zadeh. *Feature extraction, foundations and applications.* Physica Verlag, Springer, Berlin, Heidelberg, New York, 2006.

[85] L. Györfi, M. Kohler, A. Krzyżak, and H. Walk. *A Distribution-Free Theory of Nonparametric Regression.* Springer Series in Statistics, Springer-Verlag, New York, 2002.

[86] H. Haas H. Jaeger. Harnessing nonlinearity: Predicting chaotic systems and saving energy in wireless communication. *Science*, 304:78–80, 2004.

[87] S. Haykin. *Neural Networks - A Comprehensive Foundation.* Maxwell MacMillian Int., New York, 1994.

[88] R. Hecht-Nielsen. Cogent confabulation. *Neural Networks*, 18:111–115, 2005.

[89] G.E. Hinton, S. Osindero, and Y. Teh. A fast learning algorithm for deep belief nets. *Neural Computation*, 18:381–414, 2006.

[90] V. Honavar and L. Uhr, editors. *Artificial Intelligence and Neural Networks: Steps Toward Principled Integration.* Academic Press, Boston, 1994.

[91] G.-B. Huang, L. Chen, and C.-K. Siew. Universal approximation using incremental constructive feedforward networks with random hidden nodes. *IEEE Transactions on Neural Networks*, 17:879–892, 2006.

[92] A. Hyvärinen, J. Karhunen, and E. Oja. *Independent Component Analysis.* Wiley & Sons, New York, NY, 2001.

[93] S-i. Amari and H. Nagaoka. *Methods of information geometry.* American Mathematical Society, 2000.

[94] E.M. Iyoda, H. Nobuhara, and K. Hirota. A solution for the n-bit parity problem using a single translated multiplicative neuron. *Neural Processing Letters*, 18(3):233–238, 2003.

[95] J.S.R. Jang and C.T. Sun. Functional equivalence between radial basis function neural networks and fuzzy inference systems. *IEEE Transactions on Neural Networks*, 4:156–158, 1993.

[96] N. Jankowski and W. Duch. Optimal transfer function neural networks. In *9th European Symposium on Artificial Neural Networks*, pages 101–106, Bruges, Belgium, 2001. De-facto publications.

[97] N. Jankowski and M. Grochowski. Comparison of instance selection algorithms. I. algorithms survey. *Lecture Notes in Computer Science*, 3070:598–603, 2004.

[98] N. Jankowski and V. Kadirkamanathan. Statistical control of growing and pruning in RBF-like neural networks. In *Third Conference on Neural Networks and Their Applications*, pages 663–670, Kule, Poland, October 1997.

[99] N. Jankowski and V. Kadirkamanathan. Statistical control of RBF-like networks for classification. In *7th International Conference on Artificial Neural Networks*, pages 385–390, Lausanne, Switzerland, October 1997. Springer-Verlag.

[100] C. Jones and R. Sibson. What is projection pursuit. *Journal of the Royal Statistical Society A*, 150:1–36, 1987.

[101] M. Jordan and Eds. T.J. Sejnowski. *Graphical Models. Foundations of Neural Computation.* MIT Press, 2001.

[102] V. Kadirkamanathan. A statistical inference based growth criterion for the RBF networks. In Vlontzos, editor, *Proceedings of the IEEE. Workshop on Neural Networks for Signal Processing*, pages 12–21, New York, 1994.

[103] N. Kasabov. *Evolving Connectionist Systems - Methods and Applications in Bioinformatics, Brain Study and Intelligent Machines.* Springer, Perspectives in Neurocomputing, 2002.

[104] M.J. Kearns and U.V. Vazirani. *An Introduction to Computational Learning Theory.* MIT Press, Cambridge, MA, 1994.

[105] V. Kecman. *Learning and Soft Computing.* MIT Press, Cambridge, MA, 2001.

[106] B. Kégl and A. Krzyzak. Piecewise linear skeletonization using principal curves. *IEEE Transactions on Pattern Analysis and Machine Intelligence*, 24:59–74, 2002.

[107] J. Kennedy, R.C. Eberhart, and Y. Shi. *Swarm Intelligence.* Morgan Kaufmann, 2001.

[108] T. Kohonen. *Self-organizing maps.* Springer-Verlag, Heidelberg Berlin, 1995.

[109] A. Konar. *Computational Intelligence. Principles, Techniques and Applications.* Springer, New York, 2005.

[110] M. Kordos and W. Duch. Variable step search mlp training method. *International Journal of Information Technology and Intelligent Computing*, 1:45–56, 2006.

[111] B. Kosko. *Neural Networks and Fuzzy Systems.* Prentice Hall International, 1992.

[112] L.I. Kuncheva. *Combining Pattern Classifiers. Methods and Algorithms.* J. Wiley & Sons, New York, 2004.

[113] N. Kunstman, C. Hillermeier, B. Rabus, and P. Tavan. An associative memory that can form hypotheses: a phase-coded neural network. *Biological Cybernetics*, 72:119–132, 1994.

[114] G.R.G. Lanckriet, N. Cristianini, P. Bartlett, L. El Ghaoui, and M.I. Jordan. Learning the kernel matrix with semidefinite programming. *Journal of Machine Learning Research*, 5:27–72, 2004.

[115] Y.J. Lee and O.L. Mangasarian. Ssvm: A smooth support vector machine for classification. *Computational Optimization and Applications*, 20:5–22, 2001.

[116] M. Leshno, V.Y. Lin, Pinkus, and S. Schocken. Multilayer feedforward networks with a nonpolynomial activation function can approximate any function. *Neural Networks*, 6:861–867, 1993.

[117] H. Leung and S. Haykin. Detection and estimation using an adaptive rational function filters. *IEEE Transactions on Signal Processing*, 12:3365–3376, 1994.

[118] H. Li, C.L.P. Chen, and H.P. Huang. *Fuzzy Neural Intelligent Systems: Mathematical Foundation and the Applications in Engineering.* CRC Press, 2000.

[119] M. Li and P. Vitányi. *An Introduction to Kolmogorov Complexity and its Applications.* Springer-Verlag, 1997 (2nd ed).

[120] H. Lodhi, C. Saunders, J. Shawe-Taylor, N. Cristianini, and C. Watkins. Text classification using string kernels. *Journal of Machine Learning Research*, 2:419–444, 2002.

[121] W. Maass and Eds. C.M. Bishop, editors. *Pulsed Neural Networks.* MIT Press, Cambridge, MA, 1998.

[122] W. Maass and H. Markram. Theory of the computational function of microcircuit dynamics. In S. Grillner and A. M. Graybiel, editors, *Microcircuits. The Interface between Neurons and Global Brain Function*, pages 371–392. MIT Press, 2006.

[123] W. Maass, T. Natschläger, and H. Markram. Real-time computing without stable states: A new framework for neural computation based on perturbations. *Neural Computation*, 14:2531–2560, 2002.

[124] R. Maclin. Boosting classifiers regionally. In *Proc. 15th National Conference on Artificial Intelligence, Madison, WI.*, pages 700–705, 1998.

[125] C.D. Manning and H. Schütze. *Foundations of Statistical Natural Language Processing.* MIT Press, Cambridge, MA, 1999.

[126] P. Matykiewicz, W. Duch, and J. Pestian. Nonambiguous concept mapping in medical domain. *Lecture Notes in Artificial Intelligence*, 4029:941–950, 2006.

[127] T.J. McCabe and C.W. Butler. Design complexity measurement and testing. *Communications of the ACM*, 32:1415–1425, 1989.

[128] T.P. McNamara. *Semantic Priming. Perspectives from Memory and Word Recognition.* Psychology Press, 2005.

[129] J.M. Mendel. *Uncertain Rule-Based Fuzzy Logic Systems: Introduction and New Directions.* Prentice-Hall, 2000.

[130] D. Michie, D. J. Spiegelhalter, and C. C. Taylor. *Machine learning, neural and statistical classification.* Elis Horwood, London, 1994.

[131] I. Mierswa, M. Wurst, R. Klinkenberg, M. Scholz, and T. Euler. Yale: Rapid prototyping for complex data mining tasks. In *Proc. 12th ACM SIGKDD Int. Conf. on Knowledge Discovery and Data Mining (KDD 2006)*, pages 935–940, 2006.
[132] K. Miettinen. *Nonlinear Multiobjective Optimization.* Kluwer Academic Publishers, 1999.
[133] M. Minsky and S. Papert. *Perceptrons: An Introduction to Computational Geometry.* MIT Press, 1969.
[134] S. Mitra and T. Acharya. *Data Mining: Multimedia, Soft Computing, and Bioinformatics.* J. Wiley & Sons, New York, 2003.
[135] D. Nauck, F. Klawonn, R. Kruse, and F. Klawonn. *Foundations of Neuro-Fuzzy Systems.* John Wiley & Sons, New York, 1997.
[136] A. Newell. *Unified theories of cognition.* Harvard Univ. Press, Cambridge, MA, 1990.
[137] C.S. Ong, A. Smola, and B. Williamson. Learning the kernel with hyperkernels. *Journal of Machine Learning Research*, 6:1045–1071, 2005.
[138] S.K. Pal and S. Mitra. *Neuro-fuzzy Pattern Recognition: Methods in Soft Computing Paradigm.* J. Wiley & Sons, New York, 1999.
[139] Y.H. Pao. *Adaptive Pattern Recognition and Neural Networks.* Addison-Wesley, Reading, MA, 1989.
[140] E. Pękalska and R.P.W. Duin, editors. *The dissimilarity representation for pattern recognition: foundations and applications.* New Jersey; London: World Scientific, 2005.
[141] E. Pękalska, P. Paclik, and R.P.W. Duin. A generalized kernel approach to dissimilarity-based classification. *Journal of Machine Learning Research*, 2:175–211, 2001.
[142] T. Poggio and F. Girosi. Network for approximation and learning. *Proceedings of the IEEE*, 78:1481–1497, 1990.
[143] A. Pouget and T.J. Sejnowski. Spatial transformation in the parietal cortex using basis functions. *Journal of Cognitive Neuroscience*, 9:222–237, 1997.
[144] A. Pouget and T.J. Sejnowski. Simulating a lesion in a basis function model of spatial representations: comparison with hemineglect. *Psychological Review*, 108:653–673, 2001.
[145] M.J.D. Powell. Radial basis functions for multivariable interpolation: A review. In J.C. Mason and M.G. Cox, editors, *Algorithms for Approximation of Functions and Data*, pages 143–167, Oxford, 1987. Oxford University Press.
[146] J.R. Rabunal and J. Dorado, editors. *Artificial Neural Networks in Real-life Applications.* Hershey, PA, Idea Group Pub, 2005.
[147] R. Raizada and S. Grossberg. Towards a theory of the laminar architecture of cerebral cortex: Computational clues from the visual system. *Cerebral Cortex*, 13:100–113, 2003.
[148] I. Roth and V. Bruce. *Perception and Representation.* Open University Press, 1995. 2nd ed.

[149] D. Rousseau and F. Chapeau-Blondeau. Constructive role of noise in signal detection from parallel arrays of quantizers. *Signal Processing*, 85:571–580, 2005.

[150] S. Roweis and L. Saul. Nonlinear dimensionality reduction by locally linear embedding. *Science*, 290(5500):2323–2326, 2000.

[151] A. Roy. Artificial neural networks - a science in trouble. *SIGKDD Explorations*, 1:33–38, 2000.

[152] A. Roy. A theory of the brain: There are parts of the brain that control other parts. In *Proc. of the Int. Joint Conf. on Neural Networks (IJCNN 2000)*, volume 2, pages 81–86. IEEE Computer Society Press, 2000.

[153] D.E. Rumelhart and J.L. McClelland (eds). *Parallel Distributed Processing, Vol. 1: Foundations.* MIT Press, Cambridge, MA, 1986.

[154] S.J. Russell and P. Norvig. *Artificial Intelligence. A Modern Approach.* Prentice-Hall, Englewood Cliffs, NJ, 1995.

[155] E. Salinas and T.J. Sejnowski. Gain modulation in the central nervous system: where behavior, neurophysiology, and computation meet. *Neuroscientist*, 7:430–440, 2001.

[156] R.E. Schapire and Y. Singer. Improved boosting algorithms using confidence-rated predictions. *Machine Learning*, 37:297–336, 1999.

[157] B. Schölkopf and A.J. Smola. *Learning with Kernels. Support Vector Machines, Regularization, Optimization, and Beyond.* MIT Press, Cambridge, MA, 2001.

[158] F. Schwenker, H.A. Kestler, and G. Palm. Three learning phases for radial-basis-function networks. *Neural Networks*, 14:439–458, 2001.

[159] A.K. Seewald. Exploring the parameter state space of stacking. In *Proceedings of the 2002 IEEE International Conference on Data Mining (ICDM 2002)*, pages 685–688, 2002.

[160] L. Shastri. Advances in Shruti - a neurally motivated model of relational knowledge representation and rapid inference using temporal synchrony. *Applied Intelligence*, 11:79–108, 1999.

[161] W. Shoujue and L. Jiangliang. Geometrical learning, descriptive geometry, and biomimetic pattern recognition. *Neurocomputing*, 67:9–28, 2005.

[162] E. Simoncelli and B.A. Olshausen. Natural image statistics and neural representation. *Annual Review of Neuroscience*, 24:1193–1216, 2001.

[163] P. Smyth and D. Wolpert. Linearly combining density estimators via stacking. *Machine Learning*, 36:59–83, 1999.

[164] Anuj Srivastava and Xiuwen Liu. Tools for application-driven linear dimension reduction. *Neurocomputing*, 67:136–160, 2005.

[165] R.S. Sutton and A.G. Barto. *Reinforcement Learning: An Introduction.* MIT Press, Cambridge, MA, 1998.

[166] R. Tibshirani, T. Hastie and J. Friedman. *The Elements of Statistical Learning.* Springer-Verlag, 2001.

[167] J.D. Tebbens and P. Schlesinger, Improving Implementation of Linear Discriminant Analysis for the High Dimension/Small Sample Size Problem, Computational Statistics and Data Analysis (in press).

[168] R.F. Thompson. *The Brain. The Neuroscience Primer.* W.H. Freeman and Co, New York, 1993.

[169] S. Thrun, W. Burgard, and D. Fox. *Probabilistic Robotics.* MIT Press, 2005.

[170] K. Torkkola. Feature extraction by non-parametric mutual information maximization. *Journal of Machine Learning Research*, 3:1415–1438, 2003.

[171] K. Tsuda and W.S. Noble. Learning kernels from biological networks by maximizing entropy. *Bioinformatics*, 20:i326–i333, 2004.

[172] K.P. Unnikrishnan and K.P. Venugopal. Alopex: a correlation-based learning algorithm for feeedforward and recurrent neural networks. *Neural Computation*, 6:469–490, 1994.

[173] J.-P. Vert. A tree kernel to analyze phylogenetic profiles. *Bioinformatics*, 18:S276–S284, 2002.

[174] S.F. Walker. A brief history of connectionism and its psychological implications. In A. Clark and R. Lutz, editors, *Connectionism in Context*, pages 123–144. Springer-Verlag, Berlin, 1992.

[175] D.L. Wang. On connectedness: a solution based on oscillatory correlation. *Neural Computation*, 12:131–139, 2000.

[176] A.R. Webb. *Statistical Pattern Recognition.* J. Wiley & Sons, 2002.

[177] C. Wendelken and L. Shastri. Multiple instantiation and rule mediation in shruti. *Connection Science*, 16:211–217, 2004.

[178] S. Wermter and R. Sun. *Hybrid Neural Systems.* Springer, 2000.

[179] T. Wieczorek, M. Blachnik, and W. Duch. Influence of probability estimation parameters on stability of accuracy in prototype rules using heterogeneous distance functions. *Artificial Intelligence Studies*, 2:71–78, 2005.

[180] T. Wieczorek, M. Blachnik, and W. Duch. Heterogeneous distance functions for prototype rules: influence of parameters on probability estimation. *Artificial Intelligence Studies* (in print).

[181] P.H. Winston. *Artificial Intelligence.* Addison-Wesley, Reading, MA, third edition, 1992.

[182] I.H. Witten and E. Frank. *Data Mining: Practical machine learning tools and techniques.* Morgan Kaufmann, 2nd Ed, 2005.

[183] J.G. Wolff. Information compression by multiple alignment, unification and search as a unifying principle in computing and cognition. *Artificial Intelligence Review*, 19:193–230, 2003.

[184] J.G. Wolff. *Unifying Computing and Cognition. The SP Theory and its Applications.* CognitionResearch.org.uk (Ebook edition), 2006. http://www.cognitionresearch.org.uk.

[185] D. Wolpert. Stacked generalization. *Neural Networks*, 5:241–259, 1992.

Knowledge-Based Clustering in Computational Intelligence

Witold Pedrycz

Department of Electrical & Computer Engineering University of Alberta
Edmonton T6R 2G7 Canada
and Systems Research Institute, Polish Academy of Sciences
Warsaw, Poland

Summary. Clustering is commonly regarded as a synonym of unsupervised learning aimed at the discovery of structure in highly dimensional data. With the evident plethora of existing algorithms, the area offers an outstanding diversity of possible approaches along with their underlying features and potential applications. With the inclusion of fuzzy sets, fuzzy clustering became an integral component of Computational Intelligence (CI) and is now broadly exploited in fuzzy modeling, fuzzy control, pattern recognition, and exploratory data analysis. A lot of pursuits of CI are human-centric in the sense they are either initiated or driven by some domain knowledge or the results generated by the CI constructs are made easily interpretable. In this sense, to follow the tendency of human-centricity so profoundly visible in the CI domain, the very concept of fuzzy clustering needs to be carefully revisited. We propose a certain paradigm shift that brings us to the idea of *knowledge*-based clustering in which the development of information granules – fuzzy sets is governed by the use of data as well as domain knowledge supplied through an interaction with the developers, users and experts. In this study, we elaborate on the concepts and algorithms of knowledge-based clustering by considering the well known scheme of Fuzzy C-Means (FCM) and viewing it as an operational model using which a number of essential developments could be easily explained. The fundamental concepts discussed here involve clustering with domain knowledge articulated through partial supervision and proximity-based knowledge hints. Furthermore we exploit the concepts of collaborative as well as consensus driven clustering. Interesting and useful linkages between information granularity and privacy and security of data are also discussed.

1 Introductory Comments

The human-centric facet of Computational Intelligence (CI) becomes profoundly visible in a significant number of developments. One could mention here system modeling, pattern recognition, and decision-making. In data analysis tasks completed in the setting of the CI, the phenomenon of human-centricity manifests quite vividly in several ways and the needs there are well

Witold Pedrycz: *Knowledge-Based Clustering in Computational Intelligence*, Studies in Computational Intelligence (SCI) **63**, 317–341 (2007)
www.springerlink.com

articulated. First, the results are presented at some suitable level of abstraction secured by the use of information granules. Likewise the semantics of the information granules that are used to organize findings about data is conveyed in the language of fuzzy sets whose interpretation is quite intuitive. In this sense, we envision that the available mechanisms of presentation of results to the end-user are quite effective. Second, the communication with the human at the entry point when the data sets become analyzed is not that well developed. Domain knowledge available there is crucial to the build up of models (say, fuzzy models) and the establishment of their transparency and readability. It is worth stressing that the transparency and accuracy are the two dominant requirements of fuzzy models we are interested in satisfying to the highest possible extent.

The effective two-way communication is a key to the success of CI constructs, especially if we are concerned with the ways how all computing becomes navigated. For instance, the mechanisms of relevance feedback that become more visible in various interactive systems hinge upon the well-established and effective human-centric schemes of processing in which we effectively accept user hints and directives and release results in a highly comprehensible format.

Given the existing algorithms of clustering that are pivotal to the design of information granules (and as such playing an important role in the CI constructs), we become cognizant that the principles guiding processing realized by them need to be augmented. The main quest is to assure that the fuzzy clustering operates not only on data (its data-driven optimization underpinnings are well known) but takes full advantage of various sources of knowledge available when dealing with the problem at hand. In particular, we anticipate that any guidance available from the user could be incorporated as a part of the optimization environment. This point of view as to the unified treatment of data and knowledge in clustering augments the existing principle of data analysis and gives rise to the concept of knowledge-based clustering. The ultimate objective of this study is to introduce and explore various scenarios where knowledge could be seamlessly included into the algorithmic architecture of fuzzy clustering. We discuss several fundamental concepts such as clustering with partial supervision and proximity knowledge hints, collaborative clustering and a consensus mode of clustering.

The organization of the material reflects the main concepts discussed in the study. For the sake of completeness, in Section 3, we study with a brief note on the FCM algorithm by highlighting the main features that make its role visible in the CI domain. Section 3 is devoted to the formulation of the main challenges and spells out a number of open questions. In Section 4, we cover the mechanisms of human-oriented guidance such as partial supervision and proximity-based clustering. Distributed data mining in the unsupervised mode is discussed in Section 5. Collaborative fuzzy clustering is presented in Section 6 where we formulate the problem, discuss privacy aspects linked with information granularity, and present the underlying principles. The vertical mode of collaboration is presented along with the detailed design phases (Section 7). Further we elaborate on consensus based clustering. Concluding comments are covered in Section 9.

2 Fuzzy C-Means (FCM) as an Example of the CI Algorithm of Data Analysis

To make a consistent exposure of the overall material and assure linkages with the ensuing optimization developments, we confine ourselves to one of the objective function based fuzzy clustering. More specifically, we consider a Fuzzy C-Means (FCM) [4] governed by the following objective function

$$Q = \sum_{i=1}^{c} \sum_{k=1}^{N} u_{ik}^{m} \|\mathbf{x}_k - \mathbf{v}_i\|^2 \tag{1}$$

where $\mathbf{x}_k$ denotes an multidimensional data point (pattern), $\mathbf{v}_i$ is an i-th prototype and $U = [u_{ik}]$, $i = 1, 2, \ldots, c$; $k = 1, 2, \ldots, N$ is a partition matrix. $\|.\|$ denotes a certain distance function and "m" stands for a fuzzification coefficient; $m > 1.0$. The minimization of (1) is realized with respect to the partition matrix and the prototypes. The optimization scheme and all specific features of the minimization of Q are well reported in the literature, refer for instance to [1, 20]. What is of interest to us here is an observation that fuzzy clustering is inherently associated with the granularity of information. In a nutshell fuzzy clustering leads to the abstraction of data into a format of information granules. Two essential and somewhat orthogonal dimensions of the granulation process are envisioned: (a) numeric realization of the granulation through a collection of the prototypes, and (b) a collection of information granules – fuzzy sets represented by successive rows of the partition matrix. Interestingly enough, there is a direct correspondence between these two representations. Given a collection of prototypes we can determine the entries of the partition matrix. And vice versa, a given partition matrix along with the data gives rise to the prototypes. The interpretability of the results of the FCM is its significant and highly valuable feature of the algorithm. As a collection of fuzzy sets (described by the corresponding rows of the generated partition matrix) offer a holistic view at the structure of data, this feature of the FCM emphasizes its linkages with the main thrusts of Computational Intelligence.

3 Challenges and Open Issues

Indisputably, fuzzy clustering (including FCM) is one of the well-established conceptual and algorithmic avenues that has become an integral, highly visible construct present in numerous modeling pursuits encountered in fuzzy systems, neurofuzzy systems, and Computational Intelligence, in general. Given all of those, arises an obvious question as to the further developments that could support some open issues and anticipated or already existing challenges. They could eventually contribute to the changes of the landscape of this area in the years to come.

While any projection in the rapidly developing areas could be quire risky, there are several challenges which could be quite visible and influential in the buildup and progression of the area in the future.

Knowledge-based orientation of fuzzy clustering. A heavy and visible reliance on numeric data is an evident feature of fuzzy clustering as it could be seen today. There are, however, other important factors one has to take into account when discovering the structure in data. Various sources of knowledge are available from experts, data analysts, users and they come in various formats. The fundamental challenge concerns efficient ways of their incorporation into the clustering schemes, both as a concept and the algorithmic enhancement. This is not a straightforward task given the fact that clustering has to reconcile numeric aspects (data) and knowledge component (human factors). In essence, the knowledge-based orientation of clustering is in line of human-centricity of Computational Intelligence and the development of interaction schemes.

Distributed character of processing This challenge has emerged because of the inherently distributed nature of data. Those tend to be distributed at individual locations (say, sensor networks) and this poses an interesting quest as to the distributed clustering. The centralized mode that is predominant today in fuzzy clustering requires a careful revision. The clustering techniques available nowadays that predominantly revolve around a single, huge and centrally available dataset do require a careful re-visiting and reformulation.

Communication, collaboration and consensus building All of those aspects are associated in one way or another with the distributed nature of data sets. Given the distributed character of data, it is also very likely that they cannot be shared because of the privacy and security restrictions. On the other hand, some collaboration and interaction would be highly desirable given the fact that the structure in some datasets could be quite similar and sharing the knowledge about the discovery of clusters within one dataset with other sites could be beneficial. How to facilitate collaboration and consensus building in data analysis while respecting security requirements becomes an evident challenge.

Each of these challenges comes with a suite of their own quite specific problems that do require a very careful attention both at the conceptual as well as algorithmic level. We have highlighted the list of challenges and in the remainder of this study present some of the possible formulations of the associated problems and look at their solutions. It is needless to say that our proposal points at some direction that deems to be of relevance however does not pretend to offer a complete solution to the problem. Some algorithmic pursuits are also presented as an illustration of some possibilities emerging there.

4 Data and Knowledge in Clustering: Forming a Human-Centric Perspective of Computational Intelligence in Data Analysis

In fuzzy clustering, we are ultimately faced with the problem of optimization driven by data. This clearly emphasizes the role of data in the processes of revealing the structure. While this is the evident and dominant tendency, a shift of this data-oriented paradigm is contemplated in light of the fact that not only the data are essential but any domain knowledge available from users, designers has to play a pivotal role. Considering such domain knowledge as an important and indispensable component of data analysis, it becomes clear that it cast data analysis in some human-centric perspective. To be more descriptive, we may refer to pursuits carried out in this way as a knowledge-based clustering. There are two fundamental issues that need to be addressed in the setting of the knowledge-based clustering: (a) what type of knowledge-based hints could be envisioned, and (b) how they could be incorporated as a part of the optimization (more specifically, what needs to be done with regard to the possible augmentation of the objective function and how the ensuing optimization scheme has to be augmented to efficiently cope with the modified objective function).

5 Fuzzy Clustering and Mechanisms of Human-Oriented Guidance

In this section, we highlight several commonly encountered alternatives that emerge when dealing with domain knowledge and building formal mechanisms which reformulate the underlying objective function. We focus on two formats of domain knowledge being available in this setting that is labeling of some selected data points and assessments of proximity of some pairs of data.

5.1 Mechanisms of Partial Supervision

The effect of partial supervision involves a subset of labeled data, which come with their class membership. These knowledge-based hints have to be included into the objective function and reflect that some patterns have been labeled. In the optimization, we expect that the structure to be discovered conforms to the membership grades already provided for these selected patterns. More descriptively, we can treat the labeled patterns to form a grid of "anchor" points using which we attempt to discover the entire structure in the data set.

Put it differently, such labeled data should help us navigate a process of revealing the structure. The generic objective function shown in the form (1) has to be revisited and expanded so that the structural information (labeled data points) is taken into consideration. While there could be different alternatives possible with this regard, we consider the following additive expansion of the objective function, [20, 21, 22]

$$Q = \sum_{i=1}^{c}\sum_{k=1}^{N} u_{ik}^2 \|\mathbf{x}_k - \mathbf{v}_i\|^2 + \alpha \sum_{i=1}^{c}\sum_{k=1}^{N} (u_{ik} - f_{ik} b_k)^2 \|\mathbf{x}_k - \mathbf{v}_i\|^2 \tag{2}$$

The first term is aimed at the discovery of the structure in data and is the same as in the standard FCM. The second term (weighted by some positive scaling factor α) addresses the effect of partial supervision. It requires careful attention because of the way in which it has been introduced into the objective function and the role it plays during its optimization. There are two essential data structures containing information about the initial labeling process (labeled data points)

- the vector of labels, denoted by $\mathbf{b} = [b_1 b_2 \ldots b_N]^T$. Each pattern $\mathbf{x}_k$ comes with a Boolean indicator: we assign b_k equal to1 if the pattern has been already labeled and $b_k = 0$ otherwise.
- The partition matrix $F = [f_{ik}]$, $i = 1, 2, \ldots, c$; $k = 1, 2, \ldots N$ which contains membership grades assigned to the selected patterns (already identified by the nonzero values of b). If $b_k = 1$ then the corresponding column shows the provided membership grades. If $b_k = 0$ then the entries of the corresponding k-th column of F do not matter; technically we could set them up to zero.

The nonnegative weight factor (α) helps set up a suitable balance between the supervised and unsupervised mode of learning. Apparently when $\alpha = 0$ then we end up with the standard FCM. Likewise if there are no labeled patterns ($\mathbf{b} = 0$) then the objective function reads as

$$Q = (1+\alpha) \sum_{i=1}^{c}\sum_{k=1}^{N} u_{ik}^2 d_{ik}^2 \tag{3}$$

and becomes nothing but a scaled version of the standard objective function encountered in the FCM optimization process. If the values of α increase significantly, we start discounting any structural aspect of optimization (where properly developed clusters tend to minimize) and rely primarily on the information contained in the labels of the patterns. Subsequently, any departure from the required membership values in F would lead to the significant increase in the values of the objective function.

One could consider a slightly modified version of the objective function

$$Q = \sum_{i=1}^{c}\sum_{k=1}^{N} u_{ik}^2 d_{ik}^2 + \alpha \sum_{i=1}^{c}\sum_{k=1}^{N} (u_{ik} - f_{ik})^2 b_k\, d_{ik}^2 \tag{4}$$

where the labeling vector **b** shows up in a slightly different format. In essence, this function captures the essence of partial supervision. For some slight variations on the issue of partial supervision, the reader may refer to the work by [3, 1, 15, 17, 28].

Once the objective function (2) has been optimized, the resulting entries of the partition matrix U assume the form

$$u_{ik} = \frac{1}{1+\alpha}\left[\frac{1+\alpha\left(1-b_k\sum\limits_{i=1}^{c} f_{ik}\right)}{\sum\limits_{j=1}^{c}\left(\frac{d_{ik}}{d_{jk}}\right)^2} + \alpha f_{ik} b_k\right] \tag{5}$$

For $\alpha = 0$, the formula returns the result produced by the "standard" FCM. Moving on to the computations of the prototypes, the necessary condition for the minimum of Q with respect to the prototypes comes in the form $\frac{\partial Q}{\partial v_{st}} = 0$, $s = 1, 2, .., c$; $t = 1, 2, \ldots, n$. Calculating the respective partial derivatives one obtains

$$\begin{aligned}\frac{\partial Q}{\partial v_{st}} &= \frac{\partial}{\partial v_{st}}\left[\sum_{i=1}^{c}\sum_{k=1}^{N} u_{ik}^2 \sum_{j=1}^{n}(x_{kj}-v_{ij})^2\right.\\ &\quad \left. + \alpha\sum_{i=1}^{c}\sum_{k=1}^{N}(u_{ik}-f_{ik}b_k)^2\sum_{j=1}^{n}(x_{kj}-v_{ij})^2\right]\\ &= \frac{\partial}{\partial v_{st}}\left[\sum_{i=1}^{c}\sum_{k=1}^{N}\left[u_{ik}^2+(u_{ik}-f_{ik}b_k)^2\right]\sum_{j=1}^{n}(x_{kj}-v_{ij})^2\right]\end{aligned} \tag{6}$$

Let us introduce the following shorthand notation

$$\Psi_{ik} = u_{ik}^2 + (u_{ik} - f_{ik}b_k)^2 \tag{7}$$

This leads to the optimality condition of the form

$$\frac{\partial Q}{\partial v_{st}} = 2\sum_{k=1}^{N}\Psi_{sk}(x_{kt} - v_{st}) = 0 \tag{8}$$

and finally we derive the prototypes in the following form

$$\mathbf{v}_s = \frac{\sum\limits_{k=1}^{N}\Psi_{sk}\mathbf{x}_k}{\sum\limits_{k=1}^{N}\Psi_{sk}} \tag{9}$$

5.2 Clustering with Proximity Hints

The concept of proximity is one of the fundamental notions when assessing the mutual dependency between membership occurring two patterns. Consider two patterns with their corresponding columns in the partition matrix denoted by "k" and "l", that is $\mathbf{u}_k$ and $\mathbf{u}_l$, respectively. The proximity between them, $\text{Prox}(\mathbf{u}_k, \mathbf{u}_l)$, is defined in the following form [23, 25]

$$\text{Prox}(\mathbf{u}_k, \mathbf{u}_l) = \sum_{i=1}^{c} \min(u_{ik}, u_{il}) \tag{10}$$

Note that the proximity function is symmetric and returns 1 for the same pattern (k = 1); however this relationship is not transitive. In virtue of the properties of any partition matrix we immediately obtain

$$\text{Prox}(\mathbf{u}_k, \mathbf{u}_1) = \sum_{i=1}^{c} \min(u_{ik}, u_{il}) = \text{Prox}(\mathbf{u}_1, \mathbf{u}_k) \tag{11}$$

$$\text{Prox}(\mathbf{u}_k, \mathbf{u}_k) = \sum_{i=1}^{c} \min(u_{ik}, u_{ik}) = 1$$

Let us illustrate the concept of proximity for c = 2. In this case $u_{1k} = 1 - u_{2k}$ so that we can confine ourselves to a single argument. The resulting plot (with the first coordinates of the patterns, u_{1k} and u_{11}) is included in Figure 1.

The incorporation of the proximity-based knowledge hints leads to the two optimization processes. The first one is the same as captured by the original objective function. In the second one we reconcile the proximity hints

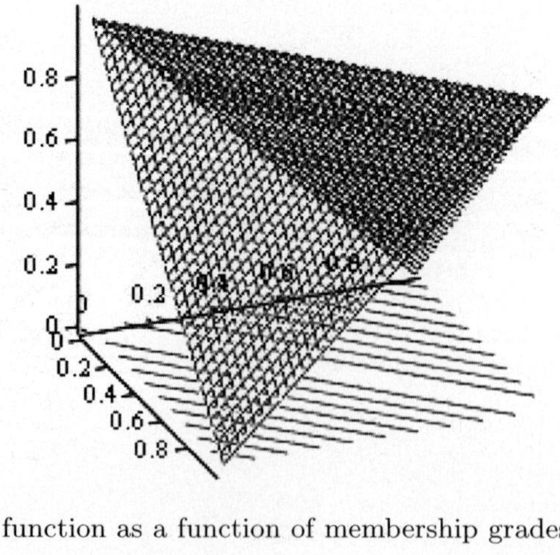

Fig. 1. Proximity function as a function of membership grades encountered in the partition matrix

with the proximity values induced by the partition matrix generated by the generic FCM. Denote the proximity values delivered by the user as $\mathrm{Prox}[k_1, k_2]$ where k_1 and k_2 are the indexes of the data points for which the proximity value is provided. Obviously these hints are given for some pairs of data so to emphasize that we introduce a Boolean predicate $B[k_1, k_2]$ defined in the following manner

$$B[k_1, k_2] = \begin{cases} 1, \text{if the value of } \mathrm{Prox}[k_1, k_2] \text{ has} \\ \quad \text{been specified for the pair } (k_1, k_2) \\ 0, \text{otherwise} \end{cases} \tag{12}$$

Note that for any pair of data, the corresponding induced level of proximity that is associated with the partition matrix produced by the FCM is computed as given by (10). We request that the proximity knowledge-based hints offered by the designer coincide with the induced proximity values implied by the structure revealed by the FCM on the basis of numeric data. Computationally, we express this requirement by computing the expression (which is a sum of distances between the corresponding values of the proximity values)

$$\sum_{k_1}\sum_{k_2} \|\mathrm{Prox}[k_1, k_2] - \sum_{i=1}^{c} \min(u_{ik_1}, u_{ik_2})\|^2 B[k_1, k_2] \tag{13}$$

By making changes to the entries of the partition matrix U, we minimize the value of the expression given above thus arriving at some agreement between the data and the domain knowledge. The optimization activities are then organized into two processes exchanging results as outlined in Figure 2. There are two optimization activities. The first one, being driven by data

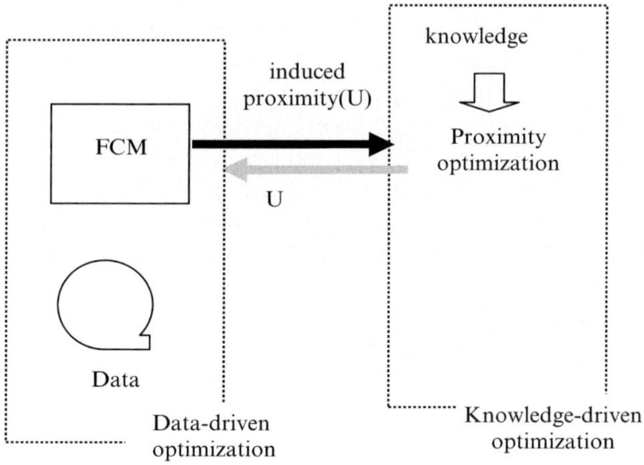

Fig. 2. The optimization data – and knowledge-driven processes of proximity-based fuzzy clustering

produces some partition matrix. The values of this matrix are communicated to the second optimization process driven by the proximity-based knowledge hints. At this stage, the proximity values induced by the partition matrix are compared with the proximities coming as knowledge hints and (13) is minimized giving rise to the new values of the partition matrix U which in turn is communicated to the data driven optimization phase. At this point, this "revised" partition matrix is used to further minimize the objective function following the iterative scheme of the FCM.

6 Distributed Data Mining

Quite commonly we encounter situations where databases are distributed rather than centralized [10, 19, 29]. There are different outlets of the same company and each of them operates independently and collects data about customers populating their independent databases. The data are not available to others. In banking, each branch may run its own database and such databases could be geographically remote from each other. In health institutions, there could be separate datasets with a very limited communication between the individual institutions. In sensor networks (which become quite popular given the nature of various initiatives such as intelligent houses, information highway, etc.), we encounter local databases that operate independently from each other and are inherently distributed. They are also subject to numerous technical constraints (e.g., a fairly limited communication bandwidth, limited power supply, etc) which significantly reduce possible interaction between the datasets. Under these circumstances, the "standard" data mining activities are faced now new challenges that need to be addressed. It becomes apparent that processing all data in a centralized manner cannot be exercised. On the other hand, data mining of each of the individual databases could benefit from availability of findings coming from others. The technical constraints and privacy issues dictate a certain level of interaction. There are two general modes of interaction that is collaborative clustering and consensus clustering both of which are aimed at the data mining realized in the distributed environment. The main difference lies in the level of interaction. The collaborative clustering is positioned at the more active side where the structures are revealed in a more collective manner through some ongoing interaction. The consensus driven clustering is focused on the reconciliation of the findings while there is no active involvement at the stage of constructing clusters.

7 Collaborative Clustering

Given the distributed character of data residing at separate databases, we are ultimately faced with the need for some collaborative activities of data mining. With the distributed character of available data come various issues of privacy, security, limited communication capabilities that have to be carefully investigated. We show that the notion of information granularity that is at heart of fuzzy sets plays a pivotal role in this setting.

7.1 Privacy and Security of Computing Versus Levels of Information Granularity

While the direct access to the numeric data is not allowed because of the privacy constraints [2, 5, 6, 7, 8, 9, 10, 11, 12, 13, 14, 16, 18, 30, 32, 33] all possible interaction could be realized through some interaction occurring at the higher level of abstraction delivered by information granules. In objective function based fuzzy clustering, there are two important facets of information granulation conveyed by (a) partition matrices and (b) prototypes. Partition matrices are, in essence, a collection of fuzzy sets which reflect the nature of the data. They do not reveal detailed numeric information. In this sense, there is no breach of privacy and partition matrices could be communicated not revealing details about individual data points. Likewise prototypes are reflective of the structure of data and form a summarization of data. Given a prototype, detailed numeric data are hidden behind them and cannot be reconstructed back to the original form of the individual data points. In either case, no numeric data are directly made available.

The level of information granularity [34] is linked with the level of detail and in this sense when changing the level of granularity possible leakage of privacy could occur. For instance, in limit when the number of clusters becomes equal to the number of data points, each prototype is just the data point and not privacy is retained. Obviously, this scenario is quite unrealistic as the structure (the number of clusters) is kept quite condensed when contrasted with all data. The schematic view of privacy offered through information granulation resulting within the process of clustering is illustrated in Figure 3. We note here that the granular constructs (either prototypes or partition matrices) build some granular interfaces.

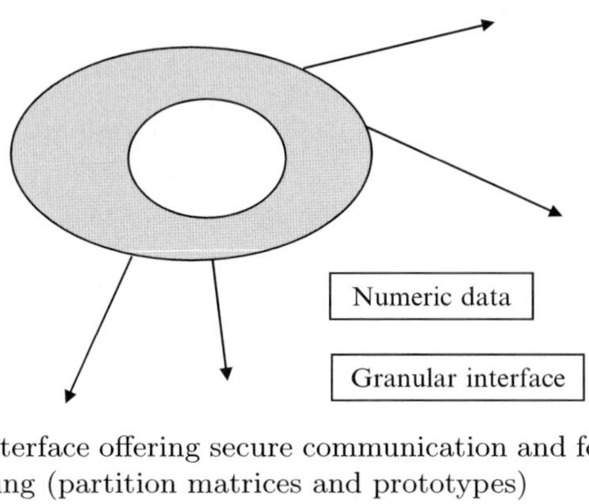

Fig. 3. Granular interface offering secure communication and formed by the results of the fuzzy clustering (partition matrices and prototypes)

7.2 The Underlying Principle of Collaborative Clustering

When dealing with distributed databases we are often interested in a collaborative style of discovery of relationships [24, 25] that could be common to all of the databases. There are a lot of scenarios where such collaborative pursuits could be deemed highly beneficial. We could envision a situation where the databases are located in quite remote locations and given some privacy requirements as well as possible technical constraints we are not allowed to collect (transfer) all data into a single location and run any centralized algorithm of data mining, say clustering. On the other hand, at the level of each database each administrator/analyst involved in its collection, maintenance and other activities could easily appreciate the need for some joint activities of data mining. Schematically, we can envision the overall situation as schematically visualized in Figure 4.

While the collaboration can assume a variety of detailed schemes, the two of them are the most essential. We refer to them as horizontal and vertical modes of collaboration or briefly horizontal and vertical clustering. More descriptively, given are "P" data sets **X**[1], **X**[2], .. **X**[p] where **X**[ii] stands for the ii-th dataset (we adhere to the consistent notation of using square brackets to identify a certain data set) in *horizontal* clustering we have the same objects that are described in *different* feature spaces. In other words, these could be the same collection of patients coming with their records built within each medical institution. The schematic illustration of this mode of clustering portrayed in Figure 4 underlines the fact that any possible collaboration occurs at the structural level viz. through the information granules (clusters) built over the data; the clusters are shown in the form of auxiliary interface layer surrounding the data. The net of directed links shows how the collaboration between different data sets takes place. The width of the links emphasizes the fact that an intensity of collaboration could be different depending upon

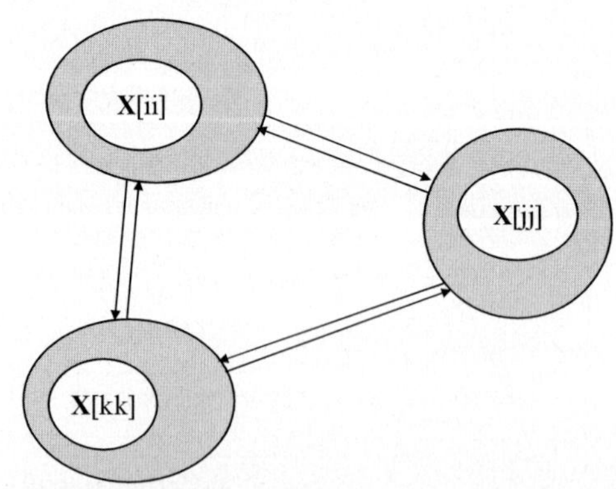

Fig. 4. A scheme of collaborative clustering involving several datasets and interacting at the level of granular interfaces

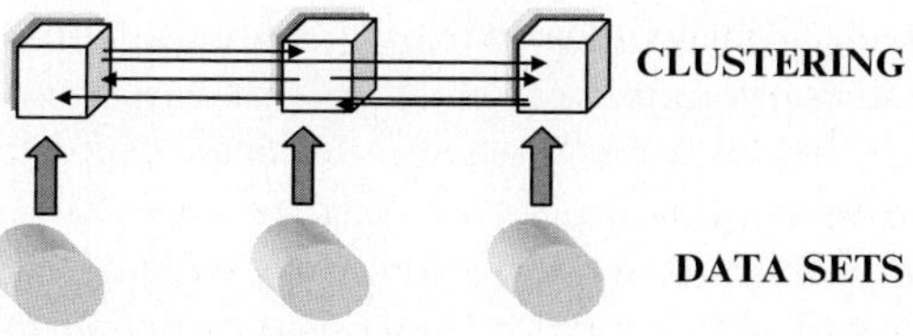

Fig. 5. A general scheme of horizontal clustering; all communication is realized through some granular interface

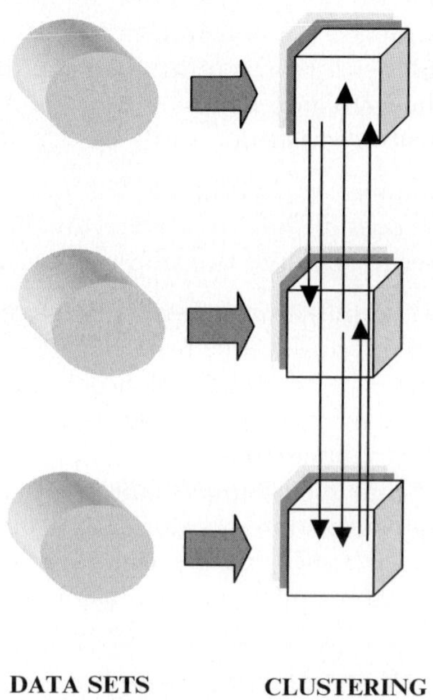

Fig. 6. A general scheme of vertical clustering; note a "stack" of data sets communicating through some layer of granular communication

the dataset being involved and the intension of the collaboration say, a willingness of some organization to accept findings from external sources).

The mode of *vertical* clustering, Figure 6, is complementary to the one already presented. Here the data sets are described in the same feature space but deal with *different* patterns. In other words, we consider that $\mathbf{X}[1]$, $\mathbf{X}[2]$, $\ldots$, $\mathbf{X}[P]$ are defined in the same feature space while each of them consists of different patterns, $\dim(\mathbf{X}[1]) = \dim(\mathbf{X}[2]) = \ldots \dim(\mathbf{X}[P])$ while $\mathbf{X}[ii]$ $\mathbf{X}[jj]$. We can show the data sets as being stack on each other (hence the name of this clustering mode).

Collaboration happens through some mechanisms of interaction. While the algorithmic details are presented in the subsequent section, it is instructive to underline the nature of the possible collaboration.

- in horizontal clustering we deal with the same patterns and different feature spaces. The communication platform one can establish is through the partition matrix. As we have the same objects, this type of collaboration makes sense. The confidentiality of data has not been breached: we do not operate on individual patterns but the resulting information granules (fuzzy relations, that is partition matrices). As this number is far lower than the number of data, the low granularity of these constructs moves us quite far from the original data
- in vertical clustering we are concerned with different patterns but the same feature space. Hence the communication at the level of the prototypes (which are high level representatives of the data) becomes feasible. Again, because of the aggregate nature of the prototypes, the confidentiality requirement has been satisfied.

There are also a number of hybrid models of collaboration where we encounter data sets with possible links of vertical and horizontal collaboration. The collaborative clustering exhibits two important features:

- The databases are distributed and there is no sharing of their content in terms of the individual records. This restriction is caused by some privacy and security concerns. The communication between the databases can be realized at the higher level of abstraction (which prevents us from any sharing of the detailed numeric data).
- Given the existing communication mechanisms, the clustering realized for the individual datasets takes into account the results about the structures of other datasets and *actively* engages them in the determination of the clusters; hence the term of collaborative clustering.

Depending upon the nature of the data located at each database and their mutual characteristics, we distinguish between two main fundamental modes of clustering such as horizontal and vertical clustering.

8 The Vertical Mode of Collaboration – The Main Flow of Processing

Let us start with setting up all necessary notation which will be subsequently used in the main phases of the development scheme. Let consider "P" databases $\mathbf{X}_1$, $\mathbf{X}_2, \ldots, \mathbf{X}_P$ whose elements (data points, patterns) are defined in the same feature space however each of these datasets consists of different data. Schematically, we can portray it in Figure 6. Given the privacy concerns, it becomes evident that sharing the data becomes impossible however as all data points are defined in the same space, communicating at the level of the prototypes becomes feasible. By noting that, we follow the same notation as included in Figure 6. The collections of the prototypes formed at the individual datasets are denoted by $\mathbf{v}_1[\text{ii}]$, $\mathbf{v}_2[\text{ii}]$, $\ldots$, $\mathbf{v}_c[\text{ii}]$ (the index in the square brackets pertains to the ii-th dataset).

The mode of *vertical* clustering, refer to Figure 6, is complementary to the one already presented. Here the data sets are described in the same feature space but deal with *different* patterns (data points). In other words, we consider that $\mathbf{X}[1], \mathbf{X}[2], \ldots, \mathbf{X}[P]$ are defined in the same feature space while each of them consists of different patterns, $\dim(\mathbf{X}[1]) = \dim(\mathbf{X}[2]) = \ldots \dim(\mathbf{X}[P])$ while $\mathbf{X}[ii]$ $\mathbf{X}[jj]$. We can show the data sets as being stack on each other (hence the name of this clustering mode).

In the discussion, we make a fundamental assumption about the same number of clusters. Whether this assumption is realistic or not, it still deserves more discussion. Later on we show how to relax this constraint and how this could be handled in an efficient manner.

8.1 The Development of Collaboration

The collaboration in the clustering process deserves a careful treatment. We do not know in advance if the structures emerging (or being discovered) at the level of the individual datasets are somewhat compatible and in this manner supportive of some collaborative activities. It could well be that in some cases the inherent structures of datasets are very different thus preventing from any effective collaboration to occur. The fundamental decision is whether we allow some datasets to collaborate or they should be eliminated from the collaboration from the very beginning. This important decision needs to be made upfront. One of the feasible possibilities would be to exercise some mechanisms of evaluating consistency of the clusters (structure) at site "ii" and some other dataset "jj". Consider that the fuzzy clustering has been completed separately for each dataset. The resulting structures represented by the prototypes are denoted by $\sim\mathbf{v}_1[ii], \sim\mathbf{v}_2[ii], \ldots, \sim\mathbf{v}_c[ii]$ for the ii-the dataset and $\sim\mathbf{v}_1[jj], \sim\mathbf{v}_2[jj], \ldots, \sim\mathbf{v}_c[jj]$. Consider the ii-th data set. The equivalent representation of the structure comes in the form of the partition matrix. For the ii-th dataset, the partition matrix is denoted by $\sim U[ii]$ whose elements are computed on the basis of the prototypes when using the dataset $\mathbf{X}[ii]$.

$$\sim u_{ik}[ii] = \frac{1}{\sum_{j=1}^{c} \left(\frac{\|\mathbf{x}_k - \sim\mathbf{v}_i[ii]\|}{|\mathbf{x}_k - \sim\mathbf{v}_j[ii]\|} \right)^{2/(m-1)}} \tag{14}$$

$\mathbf{x}_k \in \mathbf{X}[ii]$. The prototypes of the jj-th dataset being available for collaborative purposes when presented to $\mathbf{X}[ii]$ give rise to the partition matrix $\sim U[ii|jj]$ formed for the elements of $\mathbf{X}[ii]$ in the standard manner

$$\sim u_{ik}[ii|jj] = \frac{1}{\sum_{j=1}^{c} \left(\frac{\|\mathbf{x}_k - \sim\mathbf{v}_i[jj]\|}{|\mathbf{x}_k - \sim\mathbf{v}_j[jj]\|} \right)^{2/(m-1)}} \tag{15}$$

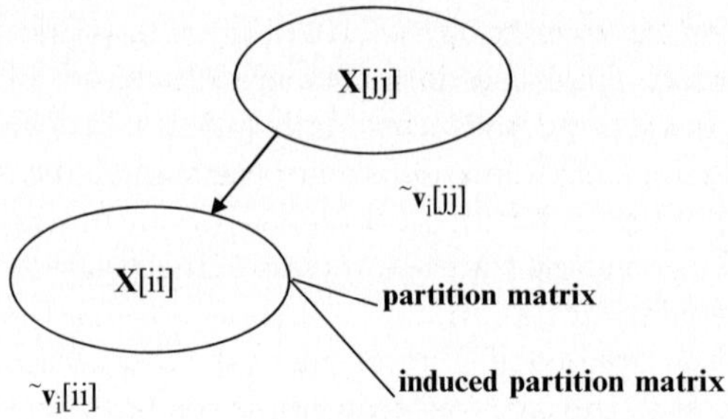

Fig. 7. Statistical verification of possibility of collaboration between datasets "ii" and "jj"

Again the calculations concern the data points of **X**[ii]. Refer to Figure 7 that highlights the essence of the interaction.

Given the partition matrix ~U[ii] and ~U[ii|jj] (induced partition matrix) we can check whether they are "compatible" meaning that the collaboration between these two datasets could be meaningful. We can test whether the histograms of membership grades of ~U[ii] and ~U[ii|jj] are statistically different (that is there is a statistically significant difference). This could be done using e.g., a standard nonparametric test such as χ^2. If the hypothesis of significant statistical difference between the partition matrices (that is corresponding structures) is not rejected, then we consider that the ii-th dataset can collaborate with the jj-th one. Noticeably, the relationship is not reciprocal so the issue of collaboration of the jj-th dataset with the ii-th needs to be investigated separately.

8.2 The Augmented Objective Function

The "standard" objective function minimized at the level of the ii-th dataset comes in the well-known form of the double sum, $\sum_{i=1}^{c[ii]} \sum_{k=1}^{N[ii]} u_{ik}^m[ii] \|\mathbf{x}_k - \mathbf{v}[ii]\|^2$. Given that the we admit collaboration with the jj-th dataset, in the search for the structure we take advantage of the knowledge of the prototypes representing the jj-th dataset and attempt to make the prototypes $\mathbf{v}_1[ii]$, $\mathbf{v}_2[ii]$, ..., $\mathbf{v}_c[ii]$ to be positioned closer to the corresponding prototypes $\mathbf{v}_1[jj]$, $\mathbf{v}_2[jj[$, ..., $\mathbf{v}_c[jj]$. This request is reflected in the form of the augmented objective function to come in the following format

$$Q[ii] = \sum_{k=1}^{N[ii]} \sum_{i=1}^{c} u_{ik}^2[ii] d_{ik}^2[ii] + \sum_{\substack{jj=1 \\ jj \neq ii}}^{P} \beta[ii, jj] \sum_{i=1}^{c} \sum_{k=1}^{N[ii]} u_{ik}^2[ii] \|\mathbf{v}_i[ii] - \mathbf{v}_i[jj]\|^2 \quad (16)$$

The first component is the same as the one guiding the clustering at the dataset **X**[ii] while the second part reflects the guidance coming from all other datasets that we identified as potential collaborators (which is done using the χ^2 test described in the previous section). The level of collaboration (which is asymmetric) is guided by the value collaboration coefficient. Its value is chosen on a basis of potential benefits of collaboration. This will be discussed in more detail in the next section. More specifically, β[ii,jj] is a collaboration coefficient supporting an impact coming from the jj-th dataset and affecting the structure to be determined in the ii-th data set. The number of patterns in the ii-th dataset is denoted by N[ii]. We use different letter to distinguish between the horizontal and vertical collaboration. The interpretation of (20) is quite obvious: the first term is the objective function directed towards the search of structure the ii-th dataset while the second articulates the differences between the prototypes (weighted by the partition matrix of the ii-th data set) which have to be made smaller through the refinement of the partition matrix (or effectively the moves of the prototypes in the feature space).

The optimization of Q[ii] involves the determination of the partition matrix U[ii] and the prototypes $\mathbf{v}_i$[ii]. As before we solve the problem for each dataset separately and allow the results interact so that this forms collaboration between the sets. The minimization of the objective function with respect to the partition matrix requires the use of the technique of Lagrange multipliers because of the existence of the standard constraints imposed on the partition matrix. We form an augmented objective function V incorporating the Lagrange multiplier λ and deal with each individual pattern (where $t = 1, 2, \ldots, N[ii]$),

$$V = \sum_{i=1}^{c} u_{it}^2[ii] d_{it}^2[ii] + \sum_{\substack{jj=1 \\ jj \neq ii}}^{P} \beta[ii, jj] \sum_{i=1}^{c} u_{it}^2[ii] \|\mathbf{v}_i[ii] - \mathbf{v}_i[jj]\|^2 - \lambda \left(\sum_{i=1}^{c} u_{it} - 1 \right) \tag{17}$$

Taking the derivative of V with respect to $u_{st}[ii]$ and making it zero, we have

$$\frac{\partial V}{\partial u_{st}} = 2u_{st}[ii] d_{st}^2[ii] + 2 \sum_{\substack{jj=1 \\ jj \neq ii}}^{P} \beta[ii, jj] u_{st}[ii] \|\mathbf{v}_i[ii] - \mathbf{v}_i[jj]\| - \lambda \tag{18}$$

For notational convenience, let us introduce the shorthand expression

$$D_{ii,jj} = \|\mathbf{v}_i[ii] - \mathbf{v}_i[jj]\|^2 \tag{19}$$

From (18) we derive

$$u_{st}[ii] = \frac{\lambda}{2 \left(d_{st}^2[ii] + \sum_{\substack{jj=1 \\ jj \neq ii}}^{P} \beta[ii, jj] D_{ii,jj} \right)} \tag{20}$$

In virtue of the standard normalization condition $\sum_{j=1}^{c} u_{jt}[ii] = 1$ one has

$$\frac{\lambda}{2} = \frac{1}{\sum_{j=1}^{c} \frac{1}{d_{jt}^{2}[ii] + \sum_{\substack{jj=1 \\ jj \neq ii}}^{P} \beta[ii,jj] D_{ii,jj}}} \tag{21}$$

With the following abbreviated notation

$$\varphi[ii] = \sum_{jj \neq ii}^{P} \beta[ii,jj]\, D_{ii,jj} \tag{22}$$

the partition matrix

$$u_{st}[ii] = \frac{1}{\sum_{j=1}^{c} \frac{d_{st}^{2}[ii] + \varphi[ii]}{d_{jt}^{2}[ii] + \varphi[ii]}} \tag{23}$$

For the prototypes, we complete calculations of the gradient of Q with respect to the coordinates of the prototype $\mathbf{v}[ii]$ and the solve the following system of equations

$$\frac{\partial Q[ii]}{\partial v_{st}[ii]} = 0, \quad s = 1, 2, .., c; t = 1, 2, ..n \tag{24}$$

We obtain

$$\frac{\partial Q[ii]}{\partial v_{st}[ii]} = 2\sum_{k=1}^{N} u_{sk}^{2}[ii](x_{kt} - v_{st}[ii]) + 2\sum_{jj \neq ii}^{P} \beta[ii,jj] \sum_{k=1}^{N} u_{sk}^{2}[ii](v_{st}[ii] - v_{st}[jj]) = 0 \tag{25}$$

Next

$$v_{st}[ii]\left(\sum_{jj \neq ii}^{P} \beta[ii,jj] \sum_{k=1}^{N[ii]} u_{sk}^{2}[ii] - \sum_{k=1}^{N[ii]} u_{sk}^{2}[ii]\right) = \sum_{jj \neq ii}^{P} \beta[ii,jj] \sum_{k=1}^{N[ii]} u_{sk}^{2}[ii] v_{st}[jj] - \sum_{k=1}^{N[ii]} u_{sk}^{2}[ii] x_{kt} \tag{26}$$

Finally we get

$$v_{st}[ii] = \frac{\sum_{jj \neq ii}^{P} \beta[ii,jj] \sum_{k=1}^{N[ii]} u_{sk}^{2}[ii] v_{st}[jj] - 2\sum_{k=1}^{N[ii]} u_{sk}^{2}[ii] x_{kt}}{\sum_{jj \neq ii}^{P} \beta[ii,jj] \sum_{k=1}^{N[ii]} u_{sk}^{2}[ii] - \sum_{k=1}^{N[ii]} u_{sk}^{2}[ii])} \tag{27}$$

An interesting application of vertical clustering occurs when dealing with huge data sets. Instead of clustering them in a single pass, we split them into individual data sets, cluster each of them separately and actively reconcile the results through the collaborative exchange of prototypes.

8.3 The Assessment of the Strength of Collaboration

The choice of a suitable level of collaboration realized between the datasets through clustering denoted by $\beta[ii,jj]$ deserves attention. Too high values of collaboration coefficient may lead to some instability of collaboration. Too low values of this coefficient may produce a very limited effect of collaboration that could be eventually made almost nonexistent in this manner. Generally speaking, the values of the collaboration coefficient could be asymmetric that is $\beta[ii, jj] \neq \beta[jj, ii]$. This is not surprising: we might have a case where at the level of dataset "ii" we are eager to collaborate and quite seriously accept findings coming from what has been discovered at dataset "jj" while the opposite might not be true. As the values of the collaboration coefficients could be different for any pair of datasets, the optimization of their values could be quite demanding and computationally intensive. To alleviate these shortcomings, let us express the coefficient $\beta[ii, jj]$ as the following product $\beta[ii, jj] = \omega\, f(ii, jj)$ meaning that we view it as a function of the specific datasets under collaboration calibrated by some constant ω (>0) whose value does not depend upon the indexes of the datasets. The choice of the function f(ii, jj) can be done in several different ways. In general, we can envision the following intuitive requirement: if the structure revealed at the site of the jj-th dataset is quite different from the one present at the ii-th data set, the level of collaboration could be set up quite low. If there is a high level of agreement between the structure revealed at the jj-th set with what has been found so far at the ii-th dataset, then f(ii, jj) should assume high values. Given these guidelines, we propose the following form of f(ii, jj)

$$f(ii, jj) = 1 - \frac{Q[ii|\,jj]}{Q[ii] + Q[ii|\,jj]} \tag{28}$$

Here Q[ii] denotes a value of the objective function obtained for clustering without any collaboration (viz. the partition matrix and the prototypes are formed on the basis of optimization realized for **X**[ii] only). Q[ii|jj] denotes the value of the objective function computed for the prototypes obtained for **X**[jj] (without any collaboration) and used for data in **X**[ii]; refer to Figure 8.

In essence, the values of Q[ii] and Q[ii|jj] reflect the compatibility of the structures in the corresponding data sets and in this manner tell us about a possible level of successful collaboration. The prototypes obtained for the dataset "jj" being used to determine the value of the objective function for the ii-th dataset could lead to quite comparable values of the objective function if the structure in **X**[jj] resembles the structure of **X**[ii]. In this case we envision

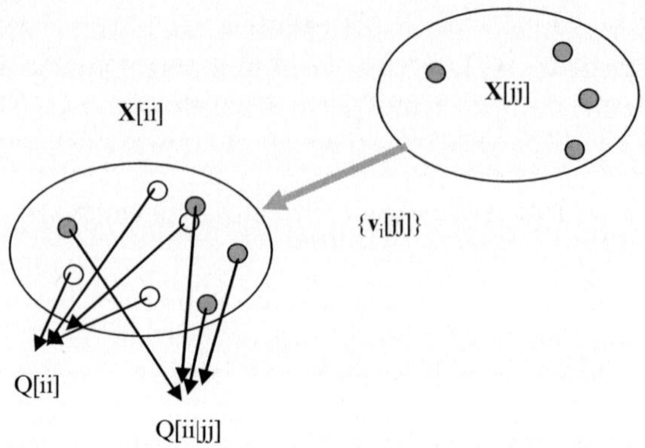

Fig. 8. Computing the values of Q[ii|jj] realized on a basis of the prototypes computed for X[jj]

Q[ii] < Q[ii|jj] yet Q[ii] $\approx$ Q[ii|jj]. On the other hand, if the structure in **X**[jj] is very different meaning that Q[ii|jj] >> Q[ii], the collaborative impact from what has been established for **X**[jj] could not be very advantageous. If Q[ii|jj] is close to Q[ii], f(ii, jj) approaches 1/2. In the second case, Q[ii|jj] >> Q[ii], the values of f(ii, jj) are close to zero.

Following the process described above, we are left now with a single coefficient (ω) controlling all collaborative activities for all datasets. This is far more practical yet its value needs to be properly selected. Here several alternatives could be sought:

(a) One could monitor the values of the overall objective function (1) during the course of optimization (minimization). The plot of the minimized objective function could be helpful here. The oscillations and a lack of convergence in the successive values of the objective function might strongly suggest that the values of ω are too high (too tight and intensive collaboration) and need to be reduced to assure smooth interaction between the datasets.

(b) We could also look at the differences between the results obtained without collaboration and with collaboration. For instance, a difference between the proximity matrices formed on a basis of the partition matrices constructed for the same dataset **X**[ii] without collaboration and with collaboration could serve as an indicator of the differences between the results. Such differences could be constrained by allowing only for some limited changes caused by the collaboration.

8.4 Dealing with Different Level of Granularity in the Collaboration Process

So far, we have made a strong assumption about the same number of clusters being formed at each individual dataset. This conjuncture could well be valid in many cases (as we consider collaboration realized at the same level of information granularity). It could be also quite inappropriate to made in some other cases. To cope with this problem, we need to move the optimization activities at the higher conceptual level by comparing results of clustering at the level of the proximity matrices. As indicated, when operating at this level of abstraction we are relieved from making any assumption about the uniform level of granularity occurring across all constructs.

9 Consensus–Based Fuzzy Clustering

In contrast to the collaborative clustering in which there is an ongoing active involvement of all datasets and the clustering algorithms running for individual datasets are impacted by the results developed at other sites, consensus-based clustering focuses mainly on the reconciliation of the individually developed structures. In this sense, building consensus is concerned with the formation of structure on the basis of the individual results of clustering developed separately (without any interaction) at the time of running the clustering algorithm. In this section, we are concerned with a collection of clustering methods being run on the same dataset. Hence U[ii], U[jj] stand here for the partition matrices produced by the corresponding clustering method. The essential step is concerned with the determination of some correspondence between the prototypes (partition matrices) formed for by each clustering method. Since there has not been any interaction when building clusters, there are no linkages between them once the clustering has been completed. The determination of this correspondence is an NP complete problem and this limits the feasibility of finding an optimal solution. One way of alleviating this problem is to develop consensus at the level of the partition matrix and the proximity matrices being induced by the partition matrices associated with other data. The use of the proximity matrices helps eliminate the need to identify correspondence between the clusters and handle the cases where there are different numbers of clusters used when running the specific clustering method.

The overall development process of consensus forming is accomplished in the following manner. Given the partition matrix U[ii], U[jj], etc. being developed individually, let us focus on the building consensus focused on U[ii]. Given the information about the structure coming in the form of U[ii] and other partition matrices U[jj], jj $\neq$ ii, the implied consensus-driven partition matrix ~U[ii] comes as a result of forming a sound agreement between the original partition matrix U[ii]. In other words, we would like to make ~U[ii] to be as close as possible to U[ii]. The minimization of the distance of the form $||\text{U[ii]} - {\sim}\text{U[ii]}||^2$ could be a viable optimization alternative. There are some other sources of structural information, see Figure 9. Here, however, we cannot establish a direct relationship between U[ii] (and ~U[ii]) and U[jj] given the reasons outlined before. The difficulties of this nature could be alleviated by

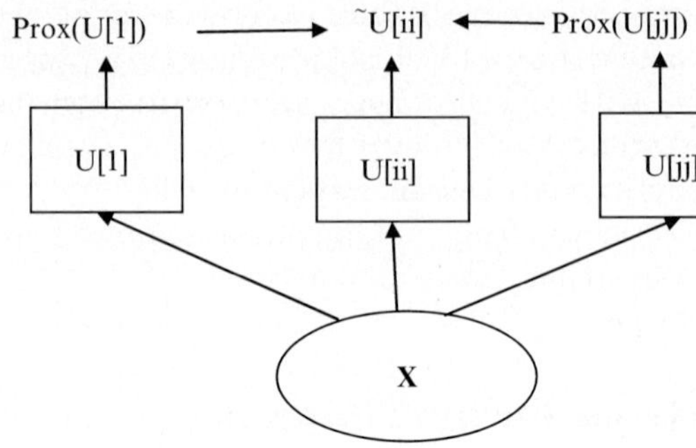

Fig. 9. A development of consensus-based clustering; the consensus building is focused on the partition matrix generated by the ii-th clustering method, U[ii]; here Prox(U[ii]) denotes a matrix of proximity values

considering the corresponding induced proximity matrices, say Prox(U[jj]). It is worth noting that a way in which the proximity matrix has been formed relieves us from the correspondence between the rows of the partition matrices (fuzzy clusters) and the number of clusters. In this sense, we may compare Prox (~U[ii]) and Prox (U[jj]) and searching for consensus by minimizing the distance $\|\text{Prox}(\sim U[ii])\text{-Prox}(U[jj])\|^2$. Considering all sources of structural information, the consensus building can be translated into the minimization of the following optimization problem

$$\|U[ii] - {\sim}U[ii]\|^2 + \gamma \sum_{jj \neq ii}^{P} \| \text{Prox}(U[jj]) - \text{Prox}({\sim}U[ii])\|^2 \tag{29}$$

The two components are reflective of the two essential sources of information about the structure. The positive weight factor (γ) is aimed at striking a sound compromise between the partition matrix U[ii] associated with the ii-th dataset. The result of the consensus reached for the ii-th method is the fuzzy partition matrix ~U[ii] minimizing the above performance index (29).

10 Concluding Notes

In this study, we emphasized the need for a revision of the paradigm of fuzzy clustering by augmenting it by the mechanisms of domain knowledge into its algorithmic layer. We have presented and discussed the key open issues and associate those to some evident challenges lying ahead in the progression of the discipline. Likewise we showed some pertinent links and outlined some promising avenues of algorithmic developments that might support the design of required conceptual platforms and specific detailed solutions.

We offered a number of algorithmic developments including clustering with partial supervision, collaborative clustering and clustering aimed at building consensus. In all of these we emphasized the role of human-centricity of the clustering framework and a distributed character of the available data. It has been shown how the issues of data privacy and security are alleviated through the use of granular information being inherently associated with the format of results generated in the process of clustering.

Acknowledgments

Support from the Canada Research Chair (CRC) program and the Natural Sciences and Engineering Research Council (NSERC) is gratefully acknowledged.

References

[1] Abonyi, J. and Szeifert, F. (2003). Supervised fuzzy clustering for the identification of fuzzy classifiers, *Pattern Recognition Letters*, 24, 14, 2195–2207.
[2] Agarwal, R. and Srikant, R. (2000). Privacy-preserving data mining. In: *Proc. of the ACM SIGMOD Conference on Management of Data*. ACM Press, New York, May 2000, 439–450.
[3] Bensaid, A. M., Hall, L. O., Bezdek, J. C. and Clarke L. P. (1996). Partially supervised clustering for image segmentation, *Pattern Recognition*, 29, 5, 859–871.
[4] Bezdek, J. C. (1981). *Pattern Recognition with Fuzzy Objective Function Algorithms*, Plenum Press, NY.
[5] Claerhout, B. and DeMoor, G.J.E. (2005). Privacy protection for clinical and genomic data: The use of privacy-enhancing techniques in medicine, *Int. Journal of Medical Informatics*, 74, 2–4, 257–265.
[6] Clifton, C. (2000). Using sample size to limit exposure to data mining, *Journal of Computer Security* 8,4, 281–307.
[7] Clifton, C. and Marks, D. (1996). Security and privacy implications of data mining. In: *Workshop on Data Mining and Knowledge Discovery*, Montreal, Canada, 15–19.
[8] Clifton, C. and Thuraisingham, B. (2001). Emerging standards for data mining, *Computer Standards & Interfaces*, 23, 3, 187–193.
[9] Coppi, R. and D'Urso, P. (2003). Three-way fuzzy clustering models for LR fuzzy time trajectories, *Computational Statistics & Data Analysis*, 43, 2, 149–177.
[10] Da Silva, J. C., Giannella, C., Bhargava, R., Kargupta, H. and Klusch, M. (2005). Distributed data mining and agents, *Engineering Applications of Artificial Intelligence*, 18, 7, 791–807.

[11] Du, W., Zhan, Z. (2002). Building decision tree classifier on private data. In: Clifton, C., Estivill-Castro, V. (Eds.), *IEEE ICDM Workshop on Privacy, Security and Data Mining, Conferences in Research and Practice in Information Technology*, vol. 14, Maebashi City, Japan, ACS, pp. 1–8.
[12] Evfimievski, A., Srikant, R., Agrawal, R. and Gehrke, J. (2004). Privacy preserving mining of association rules, *Information Systems*, 29, 4, 343–364.
[13] Johnsten, T. and Raghavan V.V. (2002). A methodology for hiding knowledge in databases. In: Clifton, C., Estivill-Castro, C. (Eds.), *IEEE ICDM Workshop on Privacy, Security and Data Mining, Conferences in Research and Practice in Information Technology*, vol. 14. Maebashi City, Japan, ACS, pp. 9–17.
[14] Kargupta, H., Kun, L., Datta, S., Ryan, J. and Sivakumar, K. (2003). Homeland security and privacy sensitive data mining from multi-party distributed resources, *Proc.* 12th *IEEE International Conference on Fuzzy Systems, FUZZ '03,*. Volume 2, 25–28 May 2003, vol.2, 1257–1260.
[15] Kersten, P.R. (1996). Including auxiliary information in fuzzy clustering, Proc. *1996 Biennial Conference of the North American Fuzzy Information Processing Society, NAFIPS*, 19–22 June 1996, 221 –224.
[16] Lindell, Y. and Pinkas, B. (2000). Privacy preserving data mining. In: *Lecture Notes in Computer Science*, vol. 1880, 36–54.
[17] Liu, H. and Huang, S.T. (2003). Evolutionary semi-supervised fuzzy clustering, *Pattern Recognition Letters*, 24, 16, 3105–3113.
[18] Merugu, S and Ghosh, J. (2005).A privacy-sensitive approach to distributed clustering, *Pattern Recognition Letters,* 26, 4, 399–410.
[19] Park, B. and Kargupta, H. (2003). Distributed data mining: algorithms, systems, and applications. In: Ye, N. (Ed.), *The Handbook of Data Mining.* Lawrence Erlbaum Associates, N. York, 341–358.
[20] Pedrycz, W. (1985). Algorithms of fuzzy clustering with partial supervision, *Pattern Recognition Letters*, 3, 1985, 13–20.
[21] Pedrycz, W. and Waletzky, J. (1997). Fuzzy clustering with partial supervision, *IEEE Trans. on Systems, Man, and Cybernetics*, 5, 787–795.
[22] Pedrycz, W. and Waletzky, J. (1997). Neural network front-ends in unsupervised learning, *IEEE Trans. on Neural Networks*, 8, 390–401.
[23] Pedrycz, W., Loia, V. and Senatore, S. (2004). P-FCM: A proximity-based clustering, *Fuzzy Sets & Systems*, 148, 2004, 21–41.
[24] Pedrycz, W. (2002). Collaborative fuzzy clustering, *Pattern Recognition Letters*, 23, 14, 1675–1686.
[25] Pedrycz, W. (2005). *Knowledge-Based Clustering: From Data to Information Granules*, J. Wiley, N. York.
[26] Pinkas, B. (2002). Cryptographic techniques for privacy-preserving data mining. *ACM SIGKDD Explorations Newsletter* 4, 2, 12–19.
[27] Strehl, A. and Ghosh, J. (2002). Cluster ensembles—a knowledge reuse framework for combining multiple partitions. *Journal of Machine Learning Research*, 3, 583–617.

[28] Timm, H., Klawonn, F. and Kruse, R. (2002). An extension of partially supervised fuzzy cluster analysis, *Proc. Annual Meeting of the North American Fuzzy Information Processing Society, NAFIPS 2002*, 27–29 June 2002, 63–68.

[29] Tsoumakas, G., Angelis, L. and Vlahavas, I. (2004). Clustering classifiers for knowledge discovery from physically distributed databases, *Data & Knowledge Engineering*, 49, 3, 223–242.

[30] Verykios, V.S., Bertino, E., Fovino, I.N., Provenza, L.P., Saygin, Y. and Theodoridis Y. (2004). State-of-the-art in privacy preserving data mining. *SIGMOD Record* 33, 1, 50–57.

[31] Wang K., Yu, P.S. and Chakraborty, S. (2004). Bottom-up generalization: a data mining solution to privacy protection, *Proc. 4th IEEE International Conference on Data Mining, ICDM 2004*, 1–4 Nov. 2004, 249–256

[32] Wang, S.L. and Jafari, A. (2005). Using unknowns for hiding sensitive predictive association rules, *Proc. 2005 IEEE International Conference on Information Reuse and Integration*, 223–228.

[33] Wang, E.T., Lee, G. and Lin, Y. T. (2005). A novel method for protecting sensitive knowledge in association rules mining, *Proc. 29th Annual International Computer Software and Applications Conference (COMPSAC 2005)*, vol. 2, 511–516.

[34] Zadeh, L. A. (2005). Toward a generalized theory of uncertainty (GTU) – an outline, *Information Sciences*, 172, 1–2, 1–40.

Generalization in Learning from Examples

Věra Kůrková

Institute of Computer Science, Academy of Sciences of the Czech Republic
vera@cs.cas.cz

Summary. Capability of generalization in learning from examples can be modeled using regularization, which has been developed as a tool for improving stability of solutions of inverse problems. Theory of inverse problems has been developed to solve various tasks in applied science such as acoustics, geophysics and computerized tomography. Such problems are typically described by integral operators. It is shown that learning from examples can be reformulated as an inverse problem defined by an evaluation operator. This reformulation allows one to characterize optimal solutions of learning tasks and design learning algorithms based on numerical solutions of systems of linear equations.

1 Learning of Artificial Neural Networks

Metaphorically, we can call machines *"artificial muscles"* as they extend capabilities of our biological muscles. Similarly, microscopes, telescopes, x-rays, electrocardiograms, tomography, as well as many other modern imaging systems play roles of *"artificial senses"* as they highly increase capabilities of our biological senses. Using such devices we can obtain huge amounts of empirical data from many areas of interest. But mostly such data in its raw form is not comprehensible to our brains. We are like King Midas, whose foolish wish to transform everything he touched into gold lead to his starvation. Just as Midas' body could not digest gold, our brains cannot digest machine-made data. But, in contrast to gold, some data can be transformed into comprehensible patterns.

In the 1940s in the early days of cybernetics, ideas of building *artificial neurons* and *networks* composed from them emerged. Inspired by simplified models of biological neural networks, artificial neural networks transform *inputs* of measured or preprocessed data into *outputs* in the form of codes or parameters of patterns. Such networks are beginning to be able to help us cope with the overload of data produced by our sensing devices.

The invention of the first "artificial muscles" required an understanding of the elementary laws of forces (which were already known to Archimedes in

Věra Kůrková: *Generalization in Learning from Examples*, Studies in Computational Intelligence (SCI) **63**, 343–363 (2007)
www.springerlink.com

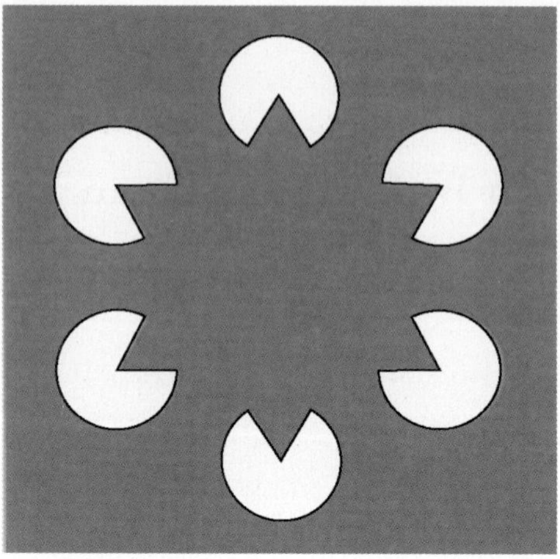

Fig. 1.

the 3rd century B.C. [36]) and the invention of optical "artificial senses", microscope and telescope, was possible because of an understanding of the laws of optics, which were discovered in the 17th century by developing ideas from Euclid's Optics [36]. Development of artificial neural networks also required some theoretical understanding of data analysis.

Although Popper's [34] explicit claim that *no patterns can be derived solely from empirical data* was stated in the 1960s, since the early stages of development of modern science, some *hypotheses* about patterns, among which the one fitting best to the measured data is searched for, have been assumed. For example, Kepler discovered that planetary orbits are elliptic assuming that planets move on the "most perfect curves" fitting to the data collected by Tycho Brahe.

Our visual perception often seems to work in this way as well. It tries to complete drawings like the one in Fig. 1 using familiar patterns. In Fig. 1 we imagine seeing a dark regular star blocking a pattern of light disks. Visual perception may also impose an interference patterns on repetitive drawings such as the one in Fig. 2. Human intelligence can even be measured by the capability to accomplish tasks of completing numerical or alphabetical patterns from partial information (such as finding a next element in an arithmetical sequence) in IQ tests.

Searching for curves fitting to astronomical data collected from observations of comets, Gauss and Legendre developed the *least square method.* In 1805, Legendre praised this method: "of all the principles that can be proposed, I think that there is none more general, more exact and more easy to apply than that consisting of minimizing the sum of the squares of errors" [5]. Many researchers of later generations shared Legendre's enthusiasm and used the least square method in statistical inference, pattern recognition, function approximation, curve or surface fitting, etc.

Fig. 2.

Fig. 3.

Neurocomputing brought a new terminology to data analysis: artificial neural networks *learn* functional relationships between their inputs and outputs from *training sets* of correctly processed *input-output pairs* (with inputs being measured or preprocessed data and outputs codes or parameters of patterns). Many neural-network learning algorithms are based on the least square method – for example, the back-propagation algorithm developed by Werbos in 1970s (see [42]) and reinvented by Rummelhart, Hinton and Williams [35]. Such algorithms iteratively adjust network parameters decreasing values of a functional computing the average of the squares of errors, which the network input-output function (determined by actual network parameters) makes on a set of examples. The functional is called *empirical error* because it is determined by a sample of empirical data chosen as a training set of examples as in Fig. 3.

For a *sample of input-output pairs of data* $z = \{(u_i, v_i) \in \Omega \times \mathbb{R}, i = 1, \dots, m\}$, where $\mathbb{R}$ denotes the set of real numbers and $\Omega \subset \mathbb{R}^d$, the *empirical error* functional $\mathcal{E}_z$ is defined at a function $f : \Omega \to \mathbb{R}$ as

$$\mathcal{E}_z(f) = \frac{1}{m} \sum_{i=1}^{m} (f(u_i) - v_i)^2. \tag{1}$$

The utilization of the least square method in neurocomputing differs from its traditional applications, where only optimization of *coefficients of a linear combination* of a fixed family of functions is performed. Neural network learning algorithms are searching an input-output function fitting to data in *nonlinear* function spaces. They optimize both *inner parameters* of the functions corresponding to computational units and *coefficients* of their linear combination.

Originally the units, called perceptrons, modeled some simplified properties of neurons, but later also other types of units (radial and kernel units) with suitable mathematical properties for function approximation became popular.

Optimization of the inner parameters of network units allows more flexibility, which potentially enables better efficiency in processing *high-dimensional data* [23], which are commonly produced through modern technology. This is one of the main advantages of artificial neural networks in contrast to linear methods. The theory of linear approximation shows that linear methods are not suitable for approximation of functions of a large number of variables since the model complexity of such methods grows *exponentially* with the *number of variables* [31].

2 Generalization

Various learning algorithms can adjust network parameters so that a network correctly processes all data from the training set. But it is desirable that the network also processes satisfactorily new data that have not been used for learning, i.e., it *generalizes* well.

The concept of generalization has been studied in philosophy in early 20th century. Husserl considered *eidetic generalization*, by which he meant the process of imagination of possible cases rather than observation of actual ones, while by *essences* or *eidos* he meant properties, kinds or types of ideal species that entities may exemplify, and by *eidetic variation* he meant possible changes an individual can undergo while remaining an instance of a given type of an essence [38]. As Husserl pointed out, one grasps the essence in its eidetic variation by grasping essential generalities about individuals of a given type or essence. Similarly, neural networks learn to recognize patterns by being trained on examples of correctly processed data. But Husserl also warned that apprehension of essences from their eidetic variation is usually incomplete.

Can these philosophical concepts be modeled mathematically and can such models inspire learning algorithms with the capability of generalization?

In algorithms based on minimization of the empirical error functional $\mathcal{E}_z$, network parameters are adjusted so that the network input-output function fits well to the training set. The smaller the value of the empirical error functional, the better the approximation of the input-output function to the training set. Often, there are many input-output functions approximating well the training set, some of them even interpolating it exactly as in Fig. 3. Although such interpolating input-output functions perfectly process all examples from the training set, they may fail on new data confirming Popper's claim that empirical data are not sufficient for obtaining any pattern. Thus, in addition to empirical data, one needs some *conceptual data* expressing prior knowledge about properties of a desired function.

In 1990s, Poggio and Girosi [32] proposed modifying the empirical error functional by adding to it a term to *penalize undesired properties* of the network input-output function. They replaced minimization of the empirical error $\mathcal{E}_z$ with minimization of

$$\mathcal{E}_z + \gamma\Psi,$$

where Ψ is a functional expressing some *global property* (such as smoothness) of the function to be minimized, and γ is a parameter controlling the trade-off between fitting to data and penalizing lack of the desired property. Their idea of such a modification of empirical error functional was inspired by *regularization*, which has been developed in the 1970s as a method of improving *stability* of certain integral equations. So Poggio and Girosi proposed modeling generalization mathematically in terms of *stability* guaranteeing that small changes or errors in empirical data cause merely minor changes in patterns.

In regularization, the functional Ψ is called the *stabilizer* and γ the *regularization* parameter. This parameter represents a compromise between good approximation of data and stability of the solution.

Girosi, Jones and Poggio [18] considered stabilizers penalizing functions with high frequencies in their Fourier representations. They used stabilizers of the form

$$\Psi(f) = \frac{1}{(2\pi)^{d/2}} \int_{\mathbb{R}^d} \frac{\tilde{f}(s)^2}{\tilde{k}(s)} \, ds, \tag{2}$$

where $\frac{1}{\tilde{k}}$ is a high-frequency filter (k is a function with positive Fourier transform such as the Gaussian $k(s) = e^{-\|s\|^2}$).

3 Inverse Problems

Integral equations, for the solution of which regularization was invented, belong to a wider class of *inverse problems*, in which unknown *causes* (such as shapes of functions, forces or distributions) are searched for from known *consequences* (empirical data). Inverse problems are fundamental in various domains of applied science such as astronomy, medical imaging, geophysics, heat conduction, seismology and meteorological forecasting. We shall show that learning can also be expressed as an inverse problem.

Often a dependence of consequences on causes is modeled mathematically by a *linear operator*. For such operator $A : \mathcal{X} \to \mathcal{Y}$ between two Hilbert spaces

$(\mathcal{X}, \|.\|_\mathcal{X})$, $(\mathcal{Y}, \|.\|_\mathcal{Y})$, an *inverse problem* determined by A is a task of finding for $g \in \mathcal{Y}$ (called *data*) some $f \in \mathcal{X}$ (called *solution*) such that

$$A(f) = g.$$

When $\mathcal{X}$ and $\mathcal{Y}$ are finite dimensional, linear operators can be represented by matrices. In infinite dimensional case, typical operators are integral ones. For example, Fredholm integral equations of the first and second kind

$$(I - \lambda L_K)(f) = g$$

$$L_K(f) = g,$$

resp., are inverse problems defined by integral operators $I - \lambda L_K$ and L_K, where

$$L_K(f)(x) = \int_a^b f(y)K(x,y)dy.$$

Another example of an integral operator defining an inverse problem is the Laplace transform

$$L(f)(p) = \int_0^{+\infty} e^{-pt} f(t)dt.$$

It describes the relaxation kinetics that occurs, for example, in nuclear magnetic resonance, photon correlation spectroscopy, and fluorescence or sedimentation equilibrium.

In 1902, Hadamard [20] introduced a concept of a *well-posed* problem in solving differential equations. Formally, well-posed inverse problems were defined by Courant [8] in 1962 as problems, where for all data there *exists a solution* which is *unique* and depends on the data *continuously*. So for a well-posed inverse problem, there exists a unique inverse operator $A^{-1} : \mathcal{Y} \to \mathcal{X}$. When A is continuous, then by the Banach open map theorem [16, p. 141] A^{-1} is continuous, too, and so the operator A is a homeomorphism of $\mathcal{X}$ onto $\mathcal{Y}$. When for some data, either there is no solution or there are multiple solutions or solutions do not depend on data continuously, the problem is called *ill-posed*.

Hadamard showed that some classical problems have unique solutions and he argued that all problems motivated by physical reality are well-posed. But the physics which Hadamard considered was the physics of the 19th century, when ill-posed problems were supposed to be anomalies. Similarly as in the mathematics of that time, continuous functions without derivatives, the Cantor's discontinuum, and the devil's staircase were supposed to be pathologies, but in the 20th century, they became central concepts of fractal geometry.

Also ill-posed problems have moved from the periphery of applied physics to being near its center. Many useful applications have been developed, the most famous one among them is the *computerized tomography*, for which the Nobel prize for physiology or medicine was awarded to Cormack and Hounsfield in 1979 [13]. Tomography is based on the idea that enough x-ray projections taken at different angles can yield a complete description of some internal structure. Mathematically, it requires reconstruction of a function of two variables from knowledge of its line integrals. Such a function models a space distribution of x-ray absorption and its line integrals correspond to measured total attenuation coefficients along the ray path lying between the x-ray tube and the detector. The attenuation coefficient is defined as the relative intensity loss, which varies for different body tissues (it has the highest value for bones).

Mathematically such a reconstruction can be described as an inverse problem with an operator called the *Radon transform.* This transform is defined for a function f of two variables as

$$R(f)(e,b) = \int_{H_{e,b}} f(x)dedb,$$

where $H_{e,b}$ denotes the line determined by its normal vector e and its translation from the origin b, i.e., $H_{e,b} = \{x \mid e \cdot x + b = 0\}$. In 1917, Radon used the transform carrying his name for solving gravitational field equations. Computerized tomography is a special case of Radon transform inversion, which is an ill-posed problem (its solutions do not exist for all data and the dependence of solutions on data is not continuous).

4 Pseudosolutions of Inverse Problems

For finite-dimensional inverse problems, in 1920 Moore proposed a method of *generalized inversion* based on a search for *pseudosolutions*, also called *least-square solutions*, for data, for which no solutions exist. His idea, published as an abstract [28], has not received too much attention until it was rediscovered by Penrose [30] in 1955. So it is called *Moore-Penrose pseudoinversion.* In the 1970s, it has been extended to the infinite-dimensional case, where similar properties as the ones of Moore-Penrose pseudoinverses of matrices hold for pseudoinverses of *continuous linear operators* between Hilbert spaces [19].

For an operator $A : \mathcal{X} \to \mathcal{Y}$, let $R(A) = \{g \in \mathcal{Y} \mid (\exists f \in \mathcal{X})(A(f) = g)\}$ denotes its *range* and $\pi_{clR(A)} : \mathcal{Y} \to clR(A)$ the *projection* of $\mathcal{Y}$ onto the closure of $R(A)$ in $(\mathcal{Y}, \|.\|_{\mathcal{Y}})$. Recall that every continuous operator A between two Hilbert spaces has an *adjoint* A^* satisfying for all $f \in \mathcal{X}$ and all $g \in \mathcal{Y}$,

$$\langle f, A^* g\rangle_{\mathcal{X}} = \langle Af, g\rangle_{\mathcal{Y}}.$$

Denoting $S(g) = \text{argmin}(\mathcal{X}, \|A(.) - g\|_{\mathcal{Y}})$, we can summarize properties of the pseudoinverse operator as follows (see, e.g., [19, pp. 37–46] and [4, pp. 56–60]):

If the range of A is closed, then there exists a unique continuous linear pseudoinverse operator $A^+ : \mathcal{Y} \to \mathcal{X}$ such that for every $g \in \mathcal{Y}$,

$$A^+(g) \in S(g)$$

$$\|A^+(g)\|_{\mathcal{X}} = \min_{f^o \in S(g)} \|f^o\|_{\mathcal{X}}$$

and for every $g \in \mathcal{Y}$, $AA^+(g) = \pi_{clR}(g)$ and

$$A^+ = (A^*A)^+A^* = A^*(AA^*)^+. \tag{3}$$

If the range is not closed, then A^+ is only defined for those $g \in \mathcal{Y}$, for which $\pi_{clR(A)}(g) \in R(A)$.

Continuous dependence of pseudosolutions on data cannot prevent their small variations to have large effects on forms of solutions. Stability of dependence of solutions on data can be measured by the *condition number* of the operator A defined as

$$cond(A) = \|A\| \, \|A^+\|.$$

When this number is large, pseudosolutions can be too sensitive to data errors. In such cases, the inverse problems are called *ill-conditioned.* Note that the concept of ill-conditioning is rather vague. More information about stability can be obtained from an analysis of behavior of the singular values of the operator A (see, e.g., [21]).

5 Regularization

A method of improving stability of solutions of ill-conditioned inverse problems, called *regularization*, was developed in 1960s. The basic idea in the treatment of ill-conditioned problems is to use some *a priori knowledge* about solutions to disqualify meaningless ones. In physically motivated inverse problems, such knowledge can be, for example, some *regularity condition* on the solution expressed in terms of existence of derivatives up to a certain order with bounds on the magnitudes of these derivatives or some *localization condition* such as a bound on the support of the solution or its behavior at infinity.

Among several types of regularization, the most popular one penalizes undesired solutions by adding a term called a stabilizer. It is called *Tikhonov's regularization* due to Tikhonov's unifying formulation [40]. Tikhonov's regularization replaces the problem of minimization of the functional

$$\|A(.) - g\|^2_{\mathcal{Y}}$$

with minimization of

$$\|A(.) - g\|^2_{\mathcal{Y}} + \gamma \Psi,$$

where Ψ is a functional called *stabilizer* and the *regularization parameter* γ plays the role of a trade-off between the least square solution and the penalization expressed by Ψ.

Typical choice of a stabilizer is the *square of the norm on* $\mathcal{X}$, for which the original problem is replaced with a minimization of the functional

$$\|A(.) - g\|_{\mathcal{Y}}^2 + \gamma\|.\|_{\mathcal{X}}^2.$$

For the stabilizer $\|.\|_{\mathcal{X}}^2$, regularized solutions always exist (in contrast to pseudosolutions, which in the infinite dimensional case do not exist for those data g, for which $\pi_{clR(A)}(g) \notin R(A)$). For every continuous operator $A : \mathcal{X} \to \mathcal{Y}$ between two Hilbert spaces and for every $\gamma > 0$, there exists a unique operator

$$A^\gamma : \mathcal{Y} \to \mathcal{X}$$

such that for every $g \in \mathcal{Y}$,

$$\{A^\gamma(g)\} = \operatorname{argmin}(\mathcal{X}, \|A(.) - g\|_{\mathcal{Y}}^2 + \gamma\|.\|_{\mathcal{X}}^2)$$

and

$$A^\gamma = (A^*A + \gamma I_{\mathcal{X}})^{-1}A^* = A^*(AA^* + \gamma I_{\mathcal{Y}})^{-1}, \tag{4}$$

where $I_{\mathcal{X}}, I_{\mathcal{Y}}$ denote the identity operators. Moreover for every $g \in \mathcal{Y}$, for which $A^+(g)$ exists,

$$\lim_{\gamma\to 0} A^\gamma(g) = A^+(g)$$

(see, e.g., [4, pp.68-70] and [19, pp.74-76]).

Even when the original inverse problem does not have a unique solution (and so it is ill-posed), for every $\gamma > 0$ the regularized problem has a *unique solution*. This is due to the uniform convexity of the functional $\|.\|_{\mathcal{Y}}^2$ (see, e.g., [26]). With γ decreasing to zero, the solutions $A^\gamma(g)$ of the regularized problems converge to the normal pseudosolution $A^+(g)$.

So for stabilizers equal to the squares of the norms $\|.\|_{\mathcal{X}}^2$ on Hilbert spaces of solutions, regularization is theoretically well understood. However to apply this theory, one needs to choose Hilbert spaces of solutions properly so that their norms model undesired properties of solutions. To make such a choice, some *a priori knowledge about meaningful solutions* is needed. For example, if such knowledge is in terms of *localization*, weighted $\mathcal{L}^2$-norms are suitable [11]. Stabilizers of the form

$$\|f\|_{\mathcal{L}_w^2}^2 = \int \frac{f(x)^2}{w(x)^2} dx$$

can force solutions to be localized where the weighting functions w are. When an a priori knowledge concerns *smoothness* expressed mathematically in terms of an existence and bounds on derivatives, suitable stabilizers can be the squares of weighted Sobolev norms

$$\|f\|^2_{d,2,w} = \left(\sum_{|\alpha| \leq d} \|D^\alpha f\|_{\mathcal{L}^2_w} \right)^2 ,$$

where d is the number of variables and for a multi-index $\alpha = (\alpha_1, \ldots, \alpha_d)$ with nonnegative integer components, $|\alpha| = \alpha_1 + \cdots + \alpha_d$ and $D^\alpha = (\partial/\partial x_1)^{\alpha_1} \ldots (\partial/\partial x_d)^{\alpha_d}$.

6 Learning from Data as an Inverse Problem

Learning of neural networks from examples is also an inverse problem – it requires to find for a *given training set* an *unknown input-output function* of a network of a given type. But the operator describing how such function determines data is of a different nature than operators modeling physical processes. The operator performs the *evaluations* of an input-output function at the input data from the training set: it assigns to a function a vector of its values at $u_1, \ldots, u_m$.

Let $(\mathbb{R}^m, \|.\|_{2,m})$ denote the m-dimensional Euclidean space with the weighted Euclidean norm

$$\|y\|_{2,m} = \sqrt{\frac{1}{m} \sum_{i=1}^m y_i^2}.$$

For an input data vector $u = (u_1, \ldots, u_m) \in \Omega^m$ and a Hilbert space $(\mathcal{X}, \|.\|_\mathcal{X})$ of functions on Ω, define an *evaluation operator*

$$L_u : (\mathcal{X}, \|.\|_\mathcal{X}) \to (\mathbb{R}^m, \|.\|_{2,m})$$

by

$$L_u(f) = (f(u_1), \ldots, f(u_m)) . \tag{5}$$

It is easy to check that the empirical error functional $\mathcal{E}_z$ with $z = \{(u_i, v_i) \in \Omega \times \mathbb{R}, i = 1, \ldots, m\}$, can be represented as

$$\mathcal{E}_z(f) = \frac{1}{m} \sum_{i=1}^m (f(u_i) - v_i)^2 = \|L_u(f) - v\|^2_{2,m}. \tag{6}$$

This representation proposed in [24, 25] and [12] allows one to express minimization of empirical error functional as an inverse problem

$$L_u(f) = v$$

of finding for a given *output data vector* $v = (v_1, \ldots, v_m) \in \mathbb{R}^m$ a solution of the form of an *input-output function* f fitting to the sample $z = \{(u_i, v_i) \in \Omega \times \mathbb{R}, i = 1, \ldots, m\}$. Finding a pseudosolution of this inverse problem is equivalent to the minimization of the empirical error functional $\mathcal{E}_z$ over $\mathcal{X}$.

But to take advantage of characterizations of pseudosolutions (4) and regularized solutions (3) from theory of inverse problems, solutions of the inverse problem defined by the operator L_u should be searched for in suitable Hilbert spaces, on which

- all evaluation operators of the form (5) are *continuous*
- norms can express some *undesired properties* of input-output functions.

Continuity of L_u guarantees that the formulas (3) and (4) hold and so descriptions of the pseudoinverse operator L_u^+ and the regularized operators L_u^γ can be easily obtained.

Neither the space $\mathcal{C}(\Omega)$ of all continuous functions on Ω nor the Lebesgue space $\mathcal{L}^2_{\rho_X}(\Omega)$ are suitable as the solution spaces: the first one is not a Hilbert space and the second one is not formed by pointwise defined functions. Moreover, the evaluation operator L_u is not continuous on any subspace of $\mathcal{L}^2_\lambda(\mathbb{R}^d)$ (where λ denotes the Lebesgue measure) containing the sequence $\{n\chi_n\}$, where χ_n denotes the characteristic function of $[0, \frac{1}{n}] \times [0,1]^{d-1}$ (see Fig. 4) or the sequence of scaled Gaussian functions $\left\{n^d e^{-(\frac{\|x\|}{n})^2}\right\}$. Indeed, all elements of these sequences have $\mathcal{L}^2_\lambda$-norms equal to 1, but the evaluation functional at zero maps them to unbounded sequences of real numbers and thus L_0 is not continuous (for a linear functional, continuity is equivalent to its boundedness).

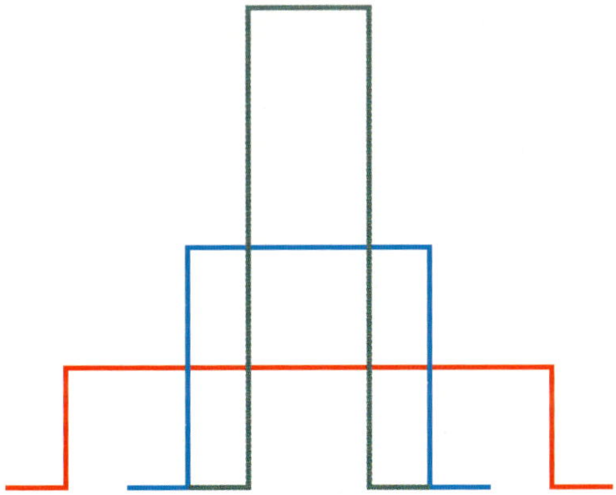

Fig. 4.

7 Reproducing Kernel Hilbert Spaces

Fortunately, there exists a large class of Hilbert spaces, on which *all evaluation functionals are continuous* and moreover, *norms* on such spaces can play *roles of measures of various types of oscillations* of input-output functions.

Girosi [17] showed that stabilizers

$$\Psi(f) = \frac{1}{(2\pi)^{d/2}} \int_{\mathbb{R}^d} \frac{\tilde{f}(s)^2}{\tilde{k}(s)} \, ds, \tag{7}$$

penalizing high-frequency oscillations (which were used in [18] to model generalization) can be represented as norms on certain Hilbert spaces.

Such spaces are called reproducing kernel Hilbert spaces. They were formally defined by Aronszajn [2] in 1950, but their theory includes many classical results on positive definite functions, matrices and integral operators with kernels. Aronszajn defined a *reproducing kernel Hilbert space (RKHS)* as a Hilbert space of pointwise defined real-valued functions on a nonempty set Ω such that all evaluation functionals are continuous, i.e., for every $x \in \Omega$, the evaluation functional $\mathcal{F}_x$, defined for any $f \in \mathcal{X}$ as

$$\mathcal{F}_x(f) = f(x),$$

is continuous (bounded).

The name of this class suggests that these spaces are related to kernels. Indeed, every RKHS is uniquely determined by a *symmetric positive semi-definite kernel* $K : \Omega \times \Omega \to \mathbb{R}$, i.e., a symmetric function of two variables satisfying for all m, all $(w_1, \ldots, w_m) \in \mathbb{R}^m$, and all $(x_1, \ldots, x_m) \in \Omega^m$,

$$\sum_{i,j=1}^{m} w_i \, w_j \, K(x_i, x_j) \geq 0.$$

A stronger property is *positive definiteness* requiring that if

$$\sum_{i,j=1}^{m} w_i w_j K(x_i, x_j) = 0$$

with $x_1, \ldots, x_m$ distinct, then for all $i = 1, \ldots, m$, $w_i = 0$.

By the Riesz Representation Theorem [16, p. 200], every evaluation functional $\mathcal{F}_x$ on a RKHS can be represented as an inner product with a function $K_x \in \mathcal{X}$, called the *representer* of x. It follows from properties of an inner product that the function $K : X \times X$ defined for all $x, y \in \Omega$ as

$$K(x, y) = \langle K_x, K_y \rangle,$$

where $\langle ., . \rangle$ denotes the inner product on $\mathcal{X}$, is symmetric and positive semi-definite. For a positive definite kernel, the set of representers $\{K_x \,|\, x \in X\}$ is linearly independent.

On the other hand, every positive semidefinite kernel $K : \Omega \times \Omega \to \mathbb{R}$ generates an RKHS denoted by

$$\mathcal{H}_K(\Omega).$$

It is formed by linear combinations of functions from $\{K_x \,|\, x \in \Omega\}$ and pointwise limits of all Cauchy sequences of these linear combinations with respect to the norm $\|.\|_K$ induced by the inner product $\langle .,. \rangle_K$ defined on representers as

$$\langle K_x, K_y \rangle_K = K(x, y)$$

and then extended to the whole space.

The equation

$$\mathcal{F}_x(f) = f(x) = \langle f, K_x \rangle_K$$

is called the *reproducing property.* So a representer K_x behaves like the Dirac delta distribution δ_x centered at x, for which

$$\delta_x(f) = f(x) = \langle f, \delta_x \rangle$$

holds [39, p.5]. But in contrast to the Dirac delta distribution, representers are *pointwise defined functions.* The reproducing property of pointwise defined functions is possible in RKHSs because they contain less functions than $\mathcal{L}^2$-spaces – only those satisfying certain restrictions on high-frequency oscillations of certain types defined by kernels.

A paradigmatic example of a kernel is the *Gaussian kernel*

$$K(x, y) = e^{-\|x-y\|^2}$$

on $\mathbb{R}^d \times \mathbb{R}^d$. Scaling and normalizing this kernel, one can construct a sequence converging to the Dirac delta distribution, namely $\left\{n^d e^{-(\frac{\|x\|}{n})^2}\right\}$. For every $b > 0$, the space $\mathcal{H}_{K_b}(\mathbb{R}^d)$ with scaled Gaussian kernel

$$K_b(x, y) = e^{-b\|x-y\|^2}$$

contains all functions computable by radial-basis function networks with a fixed width equal to b.

Other examples of positive semidefinite kernels are the *Laplace kernel*

$$K(x, y) = e^{-\|x-y\|},$$

homogeneous polynomial of degree p

$$K(x, y) = \langle x, y \rangle^p,$$

where $\langle \cdot, \cdot \rangle$ is any inner product on $\mathbb{R}^d$, *inhomogeneous polynomial* of degree p

$$K(x,y) = (1 + \langle x, y \rangle)^p,$$

and

$$K(x,y) = (a^2 + \|x - y\|^2)^{-\alpha}$$

with $\alpha > 0$ [10, p. 38].

As on every RKHS $\mathcal{H}_K(\Omega)$, all evaluation functionals are continuous, for every sample of input data $u = (u_1, \ldots, u_m)$, the operator

$$L_u : (\mathcal{H}_K(\Omega), \|.\|_K) \to (\mathbb{R}^m, \|.\|_{2,m})$$

is *continuous*. Moreover, its range is *closed* because it is finite dimensional. So one can apply results from theory of inverse problems [19] to show that for all data vectors $v = (v_1, \ldots, v_m)$, the normal pseudosolution $L_u^+(v)$ exists. Using the formula (3) we can describe its form as

$$f^+ = L_u^+(v) = \sum_{i=1}^{m} c_i K_{u_i}, \tag{8}$$

where

$$c = \mathcal{K}[u]^+ v, \tag{9}$$

and $\mathcal{K}[u]$ is the *Gram matrix of the kernel* K *with respect to the vector* u defined as

$$\mathcal{K}[u]_{i,j} = K(u_i, u_j).$$

The minimum of the empirical error $\mathcal{E}_z$ over $\mathcal{H}_K(\Omega)$ is achieved at the function f^+, which is a linear combination of the representers $K_{u_1}, \ldots, K_{u_m}$ determined by the input data $u_1, \ldots, u_m$. Thus f^+ can be interpreted as an *input-output function of a neural network with one hidden layer of kernel units and a single linear output unit.* For example, for the Gaussian kernel, f^+ is an input-output function of the Gaussian radial-basis function network with units centered at the input data $u_1, \ldots, u_m$. The coefficients $c = (c_1, \ldots, c_m)$ of the linear combination (corresponding to network output weights) can be computed by solving the system of linear equations (9).

Similarly as the function $f^+ = \sum_{i=1}^m c_i K_{u_i}$ minimizing the empirical error is a linear combination of the functions $K_{u_1}, \ldots, K_{u_m}$ defined by the input data $u_1, \ldots, u_m$, the solution f^γ of the regularized problem is of the form

$$f^\gamma = \sum_{i=1}^{m} c_i^\gamma K_{u_i}, \tag{10}$$

where

$$c^\gamma = (\mathcal{K}[u] + \gamma m \mathcal{I})^{-1} v. \tag{11}$$

This form follows from the formula (4).

Comparing the equations (8) and (9) with (10) and (11), one sees that both the functions f^+ and f^γ minimizing $\mathcal{E}_z$ and $\mathcal{E}_z + \|.\|_K^2$, resp., are linear combinations of representers $K_{u_1}, \ldots, K_{u_m}$ of input data $u_1, \ldots, u_m$, but the coefficients of the two linear combinations are different. For the pseudosolution, $c = (c_1, \ldots, c_m)$ is the image of the output vector $v = (v_1, \ldots, v_m)$ under Moore-Penrose pseudoinverse of the matrix $\mathcal{K}[u]$, while for the regularized solution, $c^\gamma = (c_1^\gamma, \ldots, c_m^\gamma)$ is the image of v under the inverse operator $(\mathcal{K}[u] + \gamma m \mathcal{I})^{-1}$.

This clearly shows the role of regularization – it makes *dependence of coefficients on output data more stable.* Growth of the regularization parameter γ leads from "under smoothing" to "over smoothing". However, the size of γ is constrained by the requirement of fitting f^γ to the sample of empirical data z, so γ cannot be too large.

The effect of regularization of the evaluation operator L_u depends on behavior of the eigenvalues of the Gram matrix $\mathcal{K}[u]$. Stability analysis of the regularized problem $\|L_u(.) - v\|_{2,m} + \gamma \|.\|_K^2$ can be investigated in terms of the finite dimensional problem of regularization of the Gram matrix $\mathcal{K}[u]$ with the parameter $\gamma' = \gamma m$, because the coefficient vector c^γ satisfies $c^\gamma = (\mathcal{K}[u] + \gamma m \mathcal{I})^{-1} v$. For K positive definite, the row vectors of the matrix $\mathcal{K}[u]$ are linearly independent. But when the distances between the data $u_1, \ldots, u_m$ are small, the row vectors might be nearly parallel and the small eigenvalues of $\mathcal{K}[u]$ might cluster near zero. In such a case, small changes of v can cause large changes of f^+. Various types of ill-posedness of $\mathcal{K}[u]$ can occur: the matrix can be *rank-deficient* when it has a cluster of small eigenvalues and a gap between large and small eigenvalues, or the matrix can represent a *discrete ill-posed problem*, when its eigenvalues gradually decay to zero without any gap in its spectrum [21].

Practical applications of learning algorithms based on theory of inverse problems are limited to such samples of data and kernels, for which iterative methods for solving systems of linear equations $c = \mathcal{K}[u]^+ v$ and $c^\gamma = (\mathcal{K}[u] + \gamma m \mathcal{I})^{-1} v$ are computationally efficient (see [33] for references to such applications).

In typical neural-network algorithms, networks with the number of hidden units n much smaller than the size m of the training set are used [22]. In [26] and [27], estimates of the speed of convergence of infima of empirical error over sets of functions computable by such networks to the global minimum $\mathcal{E}_z(f^\gamma)$ were derived. For reasonable data sets, such convergence is rather fast.

8 Three Reasons for Using Kernels in Machine Learning

In 1960s, applications of kernels to data analysis were introduced by Parzen [29] and Wahba (see [41]), who applied them to data smoothing by splines.

However, kernels were independently applied to learning theory, under the name *potential functions*. In 1964, Aizerman, Braverman and Rozonoer [1] proposed an algorithm solving classification tasks by transforming the geometry of input spaces by embedding them into higher dimensional inner product spaces [1]. Boser, Guyon and Vapnik [6] and Cortes and Vapnik [7] further developed this method of classification into the concept of the *support vector machine* (SVM), which is a one-hidden-layer network with kernel units in the hidden layer and one threshold output unit.

Linear separation simplifies classification. In some cases, even data which are not linearly separable can be transformed into linearly separable ones. A simple example is the body-mass index (BMI), which is equal to $\frac{w}{h^2}$, where w denotes the weight and h the height. The nonlinear mapping $(w, h) \to \frac{w}{h^2}$ allows to set a threshold defining obesity. Another example is the set of points belonging to two classes in Fig. 5. Because these two classes can be separated by an ellipse described by the nonlinear equation $\frac{x^2}{a} + \frac{y^2}{b} = 1$, after a proper nonlinear transformation they also can be separated linearly.

More sophisticated separations can be accomplished by embedding data into infinite dimensional RKHSs by the mapping $u \to K_u$ [37, 9]. So kernels can define transformations of geometries of data spaces to geometries of infinite dimensional Hilbert spaces.

A second reason for using kernels was found by Girosi [17] in 1998, when he realized that stabilizers of the form

$$\Psi(f) = \frac{1}{(2\,\pi)^{d/2}} \int_{\mathbb{R}^d} \frac{\tilde{f}(s)^2}{\tilde{k}(s)}\, ds \tag{12}$$

are special cases of squares of norms on RKHSs generated by *convolution kernels*, i.e., kernels

$$K(x, y) = k(x - y)$$

defined as translations of a function $k : \mathbb{R}^d \to \mathbb{R}$, for which the Fourier transform $\tilde{k}$ is positive. For such kernels, the value of the stabilizer $\|.\|_K^2$ at any

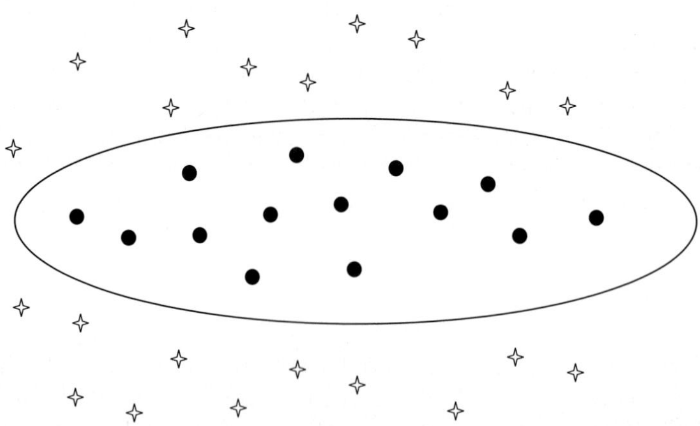

Fig. 5.

$f \in \mathcal{H}_K(\Omega)$ can be expressed as

$$\|f\|_K^2 = \frac{1}{(2\pi)^{d/2}} \int_{\mathbb{R}^d} \frac{\tilde{f}(\omega)^2}{\tilde{k}(\omega)} \, d\omega \tag{13}$$

[17], [10, p. 97]. A typical example of a convolution kernel with a positive Fourier transform is the Gaussian kernel.

The role of the squares of norms on RKHS can be illustrated by *Mercer kernels*, i.e., *continuous, symmetric, positive semidefinite functions* $K : \Omega \times \Omega \to \mathbb{R}$, where $\Omega \subset \mathbb{R}^d$ is *compact*. For a Mercer kernel K, $\|f\|_K^2$ can be expressed using eigenvectors and eigenvalues of the compact linear operator

$$L_K : \mathcal{L}^2_\mu(\Omega) \to \mathcal{L}^2_\mu(\Omega),$$

(where μ is a nondegenerate probability measure on Ω) defined for every $f \in \mathcal{L}^2_\mu(\Omega)$ as

$$L_K(f)(x) = \int_X f(y)\, K(x,y)\, dy.$$

By the Mercer Theorem (see, e.g., [10, p. 34])

$$\|f\|_K^2 = \sum_{i=1}^{\infty} \frac{a_i^2}{\lambda_i},$$

where the λ_i's are the eigenvalues of L_K and the a_i's are the coefficients of the representation $f = \sum_{i=1}^\infty a_i \phi_i$, where $\{\phi_i\}$ is the orthonormal basis of $\mathcal{H}_K(\Omega)$ formed by the eigenvectors of L_K. Note that the sequence $\{\lambda_i\}$ is either finite or convergent to zero (for K smooth enough, the convergence to zero is rather fast [14, p. 1119]). Thus, the stabilizer $\|.\|_K^2$ penalizes functions, for which the sequences of coefficients $\{a_i\}$ do not converge to zero sufficiently quickly, and so it constrains high-frequencies. The sequence $\{1/\lambda_i\}$ plays a role analogous to that of the function $1/\tilde{k}$ in the case of a convolution kernel.

A third reason for using kernels in learning theory follows from the above described reformulation of minimization of the empirical error functional as an inverse problem by Kůrková [24, 25] and De Vito et al. [12]. Application of tools from theory of inverse problems requires continuity of evaluation functionals. By the definition, RKHSs are the class of Hilbert spaces, where all evaluation functionals are continuous [2]. Characterization (10) of the minimum of the regularized task can be also obtained using functional derivatives as in [10, 33], but theory of inverse problems puts learning into a more general framework of applied mathematics.

9 Conclusion

Computational intelligence should be able to perform some of the tasks posed in IQ tests: to complete numerical, alphabetical or graphical patterns from their parts. Some of such tasks can be accomplished by artificial neural networks trained on examples of correctly completed patterns. The capability of networks to satisfactorily process also new data that were not used for learning is called *generalization*.

The concept of generalization became a subject of philosophical analysis in the early 20th century. Husserl considered *eidetic generalization* as a process of grasping an essence from *eidetic variation* of entities which exemplify it.

A mathematical model of generalization proposed by Poggio and Girosi in the 1990s expresses generalization as a *stability* of network input-output functions with respect to small variations in data. It was inspired by methods of improving stability of solutions of differential equations developed in the 1960s under the name *regularization*. This method can be applied to tasks, called *inverse problems*, where unknown causes (such as shapes of functions or distributions) are searched from *known consequences* (measured data).

But inverse problems of finding patterns fitting to data have a different nature than inverse problems from physics, which are typically defined by integral operators. Learning can be modeled in terms of *evaluation operators*.

The theory of inverse problems can be applied to this type of operators when solutions of learning tasks belong to a class of function spaces defined by *kernels*. Kernel models (corresponding to networks with kernel units) are suitable for data analysis because: (1) they can define mappings of input spaces into feature spaces, where data to be classified can be separated linearly, (2) norms on kernel spaces can play roles of undesirable attributes of network functions, the penalization of which improves generalization, and as (3) all evaluation functionals on function spaces defined by kernels are continuous, the theory of inverse problems can be applied to describe pseudosolutions and regularized solutions of learning tasks formulated as minimization of error functionals over kernel spaces.

Mathematical methods which work for tomography also apply to computational intelligence. Applications of results from theory of inverse problems combined with understanding of properties of kernels lead to description of optimal network functions and designs of learning algorithms. Their practical applications involve a variety of complexity problems related to computational efficiency of solutions of systems of linear equations and their stability. But the most difficult task is a proper choice of kernels, which requires some a priori knowledge of the class of patterns to be recognized.

Acknowledgement

This work was partially supported by the project 1ET100300517 of the program "Information Society" of the National Research Program of the Czech Republic and the Institutional research Plan AV0Z10300504. The author thanks P. C. Kainen from Georgetown University for fruitful comments and discussions.

References

[1] Aizerman M A, Braverman E M, Rozonoer L I (1964) Theoretical foundations of potential function method in pattern recognition learning. Automation and Remote Control 28:821–837

[2] Aronszajn N (1950) Theory of reproducing kernels. Transactions of AMS 68:33–404

[3] Berg C, Christensen J P R, Ressel P (1984) Harmonic Analysis on Semigroups. New York, Springer-Verlag

[4] Bertero M (1989) Linear inverse and ill-posed problems. Advances in Electronics and Electron Physics 75:1–120

[5] Bjorck A (1996) Numerical methods for least squares problem. SIAM

[6] Boser B, Guyon I, Vapnik V N (1992) A training algorithm for optimal margin clasifiers. In Proceedings of the 5th Annual ACM Workshop on Computational Learning Theory (Ed. Haussler D), pp. 144–152. ACM Press

[7] Cortes C, Vapnik V N (1995) Support-vector networks. Machine Learning 20:273–297

[8] Courant R, Hilbert D (1962) Methods of Mathematical Physics, vol. 2. New York, Wiley

[9] Cristianini N, Shawe-Taylor J (2000) An Introduction to Support Vector Machines. Cambridge, Cambridge University Press

[10] Cucker F, Smale S (2002) On the mathematical foundations of learning. Bulletin of the AMS 39:1–49

[11] De Mol C (1992) A critical survey of regularized inversion method. In Inverse Problems in Scattering and Imaging (Eds. Bertero M, Pike E R), pp. 346–370. Bristol, Adam Hilger

[12] De Vito E, Rosasco L, Caponnetto A, De Giovannini U, Odone F (2005) Learning from examples as an inverse problem. Journal of Machine Learning Research 6:883–904

[13] Di Chiro G, Brooks R A (1979) The 1979 Nobel prize in physiology or medicine. Science 206:1060–1062

[14] Dunford N, Schwartz J T (1963) Linear Operators. Part II: Spectral Theory. New York, Interscience Publishers

[15] Engl H W, Hanke M, Neubauer A (2000) Regularization of Inverse Problems. Dordrecht, Kluwer

[16] Friedman A (1982) Modern Analysis. New York, Dover

[17] Girosi F (1998) An equivalence between sparse approximation and support vector machines. Neural Computation 10:1455–1480 (AI Memo No 1606, MIT)

[18] Girosi F, Jones M, Poggio T (1995) Regularization theory and neural network architectures. Neural Computation 7:219–269

[19] Groetch C W (1977) Generalized Inverses of Linear Operators. New York, Dekker

[20] Hadamard J (1902) Sur les problèmes aux dérivées partielles et leur signification physique. Bulletin of University of Princeton 13:49
[21] Hansen P C (1998) Rank-Deficient and Discrete Ill-Posed Problems. Philadelphia, SIAM
[22] Kecman V (2001) Learning and Soft Computing. Cambridge, MIT Press.
[23] Kůrková V (2003) High-dimensional approximation by neural networks. Chapter 4 in Advances in Learning Theory: Methods, Models and Applications (Eds. Suykens J et al.), pp. 69–88. Amsterdam, IOS Press
[24] Kůrková V (2004) Learning from data as an inverse problem. In: COMPSTAT 2004 – Proceedings on Computational Statistics (Ed. Antoch J), pp. 1377–1384. Heidelberg, Physica-Verlag/Springer
[25] Kůrková V (2005) Neural network learning as an inverse problem. Logic Journal of IGPL 13:551–559
[26] Kůrková V, Sanguineti M (2005) Error estimates for approximate optimization by the extended Ritz method. SIAM Journal on Optimization 15:461–487
[27] Kůrková V, Sanguineti M (2005) Learning with generalization capability by kernel methods with bounded complexity. Journal of Complexity 21:350–367
[28] Moore E H (1920) Abstract. Bull. Amer. Math. Soc. 26:394–395
[29] Parzen E (1966) An approach to time series analysis. Annals of Math. Statistics 32:951–989
[30] Penrose R (1955) A generalized inverse for matrices. Proc. Cambridge Philos. Soc. 51:406–413
[31] Pinkus A (1985) n-width in Approximation Theory. Berlin, Springer-Verlag
[32] Poggio T, Girosi F (1990) Networks for approximation and learning. Proceedings IEEE 78:1481–1497
[33] Poggio T, Smale S (2003) The mathematics of learning: dealing with data. Notices of the AMS 50:536–544
[34] Popper K (1968) The Logic of Scientific Discovery. New York, Harper Torch Book
[35] Rummelhart D E, Hinton G E, Williams R J (1986) Learning internal representations by error propagation. In Parallel Distributed Processing (Eds. Rummelhart D E, McClelland J L), pp. 318–362. Cambridge, MIT Press
[36] Russo L (2004) The Forgotten Revolution. Berlin, Springer-Verlag
[37] Schölkopf B, Smola A J (2002) Learning with Kernels – Support Vector Machines, Regularization, Optimization and Beyond. Cambridge, MIT Press
[38] Smith D W, McIntyre R (1982) Husserl and Intentionality: A Study of Mind, Meaning, and Language. Dordrecht and Boston, D. Reidel Publishing Co.
[39] Strichartz R S (2003) A Quide to Distribution Theory and Fourier Transforms. Singapore, World Scientific

[40] Tikhonov A N, Arsenin V Y (1977) Solutions of Ill-posed Problems. Washington, D.C., W.H. Winston

[41] Wahba G (1990) Splines Models for Observational Data. Philadelphia, SIAM

[42] Werbos P J (1995) Backpropagation: Basics and New Developments. In The Handbook of Brain Theory and Neural Networks (Ed. Arbib M), pp. 134–139. Cambridge, MIT Press

A Trend on Regularization and Model Selection in Statistical Learning: *A Bayesian Ying Yang Learning Perspective*

Lei Xu

Department of Computer Science and Engineering, Chinese University of Hong Kong, Shatin, NT, Hong Kong, P.R. China

Summary. In this chapter, advances on regularization and model selection in statistical learning have been summarized, and a trend has been discussed from a Bayesian Ying Yang learning perspective. After briefly introducing Bayesian Ying-Yang system and best harmony learning, not only its advantages of automatic model selection and of integrating regularization and model selection have been addressed, but also its differences and relations to several existing typical learning methods have been discussed and elaborated. Taking the tasks of Gaussian mixture, local subspaces, local factor analysis as examples, not only detailed model selection criteria are given, but also a general learning procedure is provided, which unifies those automatic model selection featured adaptive algorithms for these tasks. Finally, a trend of studies on model selection (i.e., automatic model selection during parametric learning), has been further elaborated. Moreover, several theoretical issues in a large sample size and a number of challenges in a small sample size have been presented. The contents consist of

Lei Xu: *A Trend on Regularization and Model Selection in Statistical Learning: A Bayesian Ying Yang Learning Perspective*, Studies in Computational Intelligence (SCI) **63**, 365–406 (2007)
www.springerlink.com

5. A Trend and Challenges
   - *A Trend for model selection*
   - *Theoretical issues in a large sample size*
   - *Challenges in a small sample size*
6. Concluding Remarks

Key words: Statistical learning, Model selection, Regularization, Bayesian Ying-Yang system, Best harmony learning, Best matching, Best fitting, AIC, BIC, Automatic model selection, Gaussian mixture, Local factor analysis, theoretical issues, challenges.

1 Best Fitting vs Over-fitting

Statistical learning is usually referred to a process that a learner discovers certain dependence relation underlying a set of samples $\mathcal{X}_N = \{x_t\}_{t=1}^N$. The learner is equipped with a device or model M to accommodate this dependence relation. Such a relation is featured by a specific structure S^o and a specific setting θ^o taken by a set of parameters θ. Keeping the same structure S^o, we can get a family of specific relations $S^o(\theta)$ by varying θ within a given domain Θ that includes θ^o. Provided that every x_t comes from $S^o(\theta^o)$ without noise or disturbance, if we know S^o but do not directly know θ^o, we can get $\theta = \theta^o$ via the principle of searching one $\theta \in \Theta$ such that $S^o(\theta)$ best fits the samples in $\mathcal{X}_N$, as long as N is large enough. Usually, samples come from $S^o(\theta^o)$ subject to certain uncertainties, e.g., noises, disturbances, random sampling, etc. When N is large enough, we may still get a unique estimate value θ^* to approximate θ^o via this best fitting principle. Such a task of determining θ is usually called *parameter learning.*

The task becomes not so simple in the cases that either S^o is unknown or N is not large enough even when S^o is known. If we do not know S^o, we have to assign an appropriate structure to M. More specifically, a structure is featured by its structure type and its complexity or scale. E.g., considering relations described by $y(x) = a_3x^3 + a_2x^2 + a_1x + a_0$, its structure type is a polynomial and its scale is simply an integer that is equal to 3. For two structures S_a and S_b of a same type, S_a is actually a sub-structure (or S_a is included in S_b, shortly denoted by $S_a \prec S_b$) if S_a has a scale smaller than that of S_b. E.g., a polynomial of the order 2 is a sub-structure in a polynomial of the order 3. For two structures S_a and S_b of different types, if one is not a sub-structure of the other, we can always enlarge the scale of one structure to a large enough one such that it includes the other as a sub-structure. For this reason, we let M to consider a family of structures $S(\theta_{\mathbf{k}}, \mathbf{k})$, where S may not be same as the unknown one of S^o, but is pre-specified by one of typical structures, depending on a specific learning task encountered. Readers are referred to [41] for a number of typical structure types. $\mathbf{k}$ is a tuple that consists of one or

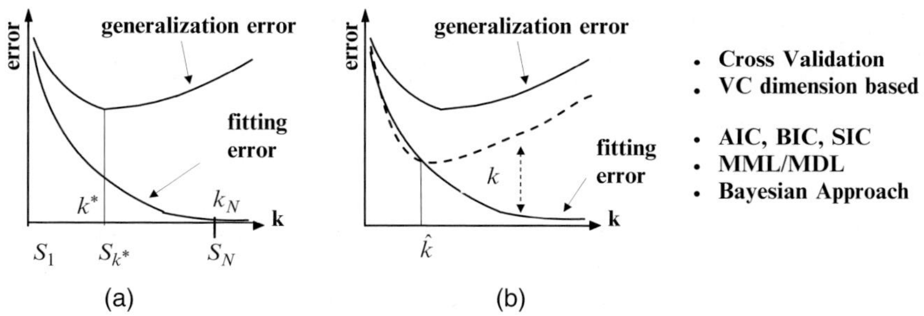

Fig. 1. Fitting error vs generalization error

several integers. By enumerating in a certain manner the values that **k** takes, we can get a series of embedded structures $S_1 \prec S_2 \prec \cdots \prec S_k \cdots$ such that $S_{k^*-1} \prec S^o \prec S_{k^*}$.

It is not difficult to find k^* if $\mathcal{X}_N$ comes from $S^o(\theta^o)$ without noise or disturbance. Searching a best value of θ_1 such that $S_1(\theta_1)$ best fits the samples in $\mathcal{X}_N$, there will be a big fitting error. This fitting error will monotonically decrease as k increases. Also, it reaches zero when $k = k^*$ and remains to be zero as k further increases. That is, k^* can be found by the smallest k where a zero fitting error is reached. However, the best fitting error by $S_{k^*}(\theta_{k^*})$ will still be nonzero, if the samples in $\mathcal{X}_N$ have been infected by noises while N is a finite size. As shown in Fig. 1(a), the fitting error will keep to decrease monotonically as k further increases, until it reaches zero at k_N that relates to N and but is usually much larger than k^*. In other words, a large part of structure with a much larger scale has actually been used to fit noises or disturbances. As a result, we can not get k^* by the principle of finding the smallest scale at which the best-fitting error is zero. This is usually called *over-fitting* problem.

We also encounter the same problem even when we known that the samples $\mathcal{X}_N = \{x_t\}_{t=1}^N$ come from $S^o(\theta^o)$ without noise but the size N is not large enough. In such a case, we are unable to determine a unique θ^* via the best fitting principle, because there will be infinite many choices of θ by which the best fitting error is zero. In other words, the samples in $\mathcal{X}_N$ actually come from a unknown sub-structure inside S^o. That is, we are lead to the same situation as the above ones with S^o unknown.

2 Trends on Regularization and Model Selection

2.1 Regularization

In the literatures of statistics, neural networks, and machine learning, many efforts in two directions have been made on tackling the *over-fitting* problem in the past 30 or 40 years. One direction consists of those made under the name of *regularization* that is imposed during parameter learning [35, 24].

Though we do not know the original structure underlying the samples in $\mathcal{X}_N$, we may consider a structure $S_k(\theta_k)$ with its scale large enough to include the original structure, which is also equivalent to the cases that the samples $\mathcal{X}_N = \{x_t\}_{t=1}^N$ come from a known structure $S^o(\theta^o)$ but with N being not large enough. Instead of searching an appropriate substructure, we impose certain constraint on θ or certain regularity on the structure $S(\theta) = S_k(\theta_k)$ with a scale k such that we can find a unique θ^* by best fitting to get a specific structure $S(\theta^*)$ but still with a scale k as an effective approximation to a substructure in a lower scale.

The existing typical regularization techniques can be roughly classified into two types. One is featured by a corrected best fitting criterion in a format of *best-fitting plus correction*, as summarized below:

- *Tikhonov regularization* One popular approach relates to the well known Tikhonov regularization [24, 11, 35], featured by the following format

$$\theta^* = arg \min_\theta [F(S(\theta), \mathcal{X}_N) + \lambda P(S(x, \theta))], \tag{1}$$

  where $F(S(\theta), \mathcal{X}_N)$ denotes the fitting error for implementing the best fitting principle, and $P(S(x,\theta))$ is usually called a stabilizer that describes the irregularity or non-smoothness of the manifold $S(x,\theta)$ specified by the structure $S(\theta)$. Moreover, λ is called a regularization strength that controls how strong the stabilizer is in action.
- *Ridge regression* It has been widely used in the literature of neural networks (see a summary by Sec. 2.3 in [67]), featured by the following format

$$\theta^* = arg \min_\theta [F(S(\theta), \mathcal{X}_N) + \lambda \Omega(\theta)], \tag{2}$$

  where $\Omega(\theta)$ denotes a regularized term that attempts to shrink the dynamic range that θ varies. One typical example is $\Omega(\theta) = \|\theta\|^2$.
- *Bayesian approach* Another widely studied apporach is called maximum a posteriori probability (MAP), featured by maximizing the posteriori probability

$$p(\theta|\mathcal{X}_N) = p(\mathcal{X}_N|\theta)p(\theta)/p(\mathcal{X}_N), \tag{3}$$

  or equivalently minimizing $-[\ln p(\mathcal{X}_N|\theta) + \ln p(\theta)]$ with $\ln p(\mathcal{X}_N|\theta)$ taking the role of $F(S(\theta), \mathcal{X}_N)$, while $\ln p(\theta)$ takes the role of $\Pi(S(x,\theta))$ and $\Omega(\theta)$.

This type has both one crucial weakness and one key difficulty. The crucial weakness comes from the choices of $P(S(x,\theta))$ in eq.(1), $\Omega(\theta)$ in eq.(2), and $p(\theta)$ in eq.(3), which usually have to be imposed in an isotropic manner.

Thus, regularization works only to a single mode data set in a symmetrical, isotropic, or uniform structure. However, such a situation is usually not encountered for two main reasons. First, a data set may include multiple disconnected substructures. Second, even for a single mode data set in a symmetrical, isotropic, or uniform structure, we need to use a structure with an appropriate scale k^* to fit it. If a structure with a larger sale $k > k^*$ is used, it is desired that the part corresponding to those extra scales can be ignored or discarded in fitting. For an example, given a set of samples (x_t, y_t) that come from a curve $y = x^2 + 3x + 2$, if we use a polynomial of an order $k > 2$ to fit the data set, i.e. $y = \sum_{i=0}^{k} a_i x^i$, we desire to force all the parameters $\{a_i, i \geq 3\}$ to be zero. In other words, we have to treat the parameters $\{a_i, i \geq 3\}$ differently from the parameters $\{a_i, i \leq 2\}$, instead of being in an isotropic or uniform way.

In addition to the above problem, we also encounter a key difficulty on how to appropriately control the strength λ of regularization, which is usually able to be roughly estimated only for a rather simple structure via either handling the integral of marginal density or in help of cross validation, but with very extensive computing costs [31, 32, 33].

The second type of regularization techniques consists of those not directly guided by a corrected best fitting criterion, but rather heuristically based. While some are quite specific problem dependent, some are generally applicable to many problems, for examples we can

- add noises to the samples in $\mathcal{X}_N$;
- add noises to a solution θ^* obtained by certain approach;
- terminate a learning process before it converges.

Though the adding noise approaches can be qualitatively related to the Tikhonov regularization [5], we do not know what type of noises to add and thus usually add a Gaussian noise with a variance λ, which is still an isotropic type regularization. In other words, we still can not avoid being suffering from the previously discussed crucial weakness. Also, it is very difficult to determine an appropriate variance λ.

As to the early termination approach, it has also been qualitatively related to Tikhonov regularization in some simple structure. However, it is very difficult to guide when to terminate.

Actually, all the regularization efforts suffer the previously discussed crucial weakness and difficulty, which come from its nature of searching a structure $S(\theta^*)$ with a scale k to approximate a substructure in a lower scale, while the part related to those extra scales has not been discarded but still in action to blur those useful ones. To tackle the problems, we turn to consider the other direction that consists of those efforts made under the name of *model selection*, featured by searching a structure with an appropriate scale k^*.

2.2 Model Selection

As discussed previously in Fig. 1, we can not find an appropriate k^* according to the best fitting principle. Several theories or principles have been proposed to guide the selection of k^*, which can be roughly classified into two categories.

One directly bases on the generalization error, i.e., the fitting error of an estimated $S_k(\theta_k^*)$ not only on the samples in $\mathcal{X}_N$ but also on all the other samples from $S^o(\theta^o)$ subject to noises and certain uncertainty. Using $p_o(x) = p(x|S^o(\theta^o))$ to describe that x comes from $S^o(\theta^o)$ subject to certain noises or uncertainty, and letting $\varepsilon(x,k)$ to denote the fitting error of x by $S_k(\theta_k^*)$, if we measure the following expected error (also called generalization error):

$$E_g(k) = \int \varepsilon(x,k) p_o(x) dx, \tag{4}$$

which takes in consideration not only the best fitting error on the samples in $\mathcal{X}_N$ but also all the other samples from $S^o(\theta^o)$.

However, it is impossible to estimate the generalization error by knowing all the samples from $S^o(\theta^o)$. We have to face the difficulty of estimating the generalization error merely based on the samples in $\mathcal{X}_N$. Two representative theories for this purpose are as follows:

- *Estimating generalization error by experiments* Studies of this type are mostly made under the name of cross-validation (CV) [31, 32, 33, 28], by which generalization error is estimated in help of experiments of making training and testing via repeatedly dividing a same set of samples into a different training set and a different testing set.
- *Estimating bounds of general error via theoretical analysis* Though it is theoretically impossible to get an analytical expression for the generalization error merely based on the samples in $\mathcal{X}_N$, we may estimate some rough bound $\Delta(k)$ on the difference between the best fitting error and the generalization error. As shown in Fig. 1(b), we consider

$$\hat{k} = arg \min_k J(k, \theta_k^{fit}), \; J(k, \theta_k^{fit}) = F(\mathcal{X}_N, \theta_k^{fit}) + \Delta(k), \tag{5}$$

where $\theta_k^{fit} = arg \min_{\theta_k} F(\mathcal{X}_N, \theta_k)$, and $F(\mathcal{X}_N, \theta_k)$ is a cost measure for implementing a best fitting, e.g., it is usually the negative likelihood function $-\ln p(\mathcal{X}_N|\theta_k)$. One popular example for this $\Delta(k)$ is the VC dimension based learning theory [39]. A bound $\Delta(k)$ relates to not only N and the fitting error $F(\mathcal{X}_N, \theta_k^{fit})$, but also a key index called VC dimension. Qualitatively, such a bound $\Delta(k)$ is useful for theoretical understanding. However, it is usually difficult to implement because the VC dimension is very difficult to estimate except for some simple cases.

The other category summarizes all the other efforts not directly based on estimating the generalization error. They are closely related but proposed from different aspects, as summarized below:

- *Minimizing information divergence* The discrepancy between the true model and the estimated model is minimized via minimizing $KL(p_o\|p_{\theta_k})$, where $p_{\theta_k} = p(x|\theta_k) = p(x|S_k(\theta_k))$ and $KL(p\|q)$ is the well known Kullback-Leiber information, i.e.,

$$KL(p\|q) = \int p(x) \ln \frac{p(x)}{q(x)} dx. \tag{6}$$

It further follows that $KL(p_o\|p_{\theta_k}) = H(p_o\|p_o) - H(p_o\|p_{\theta_k})$, where

$$H(p\|q) = \int p(x)\ln q(x)dx. \tag{7}$$

Its negation $-H(p\|q)$ is also called the cross entropy, and $-H(p\|p)$ is the entropy of $p(x)$. Since $H(p_o\|p_o)$ is constant, minimizing $KL(p_o\|p_{\theta_k})$ is equivalent to maximizing $H(p_o\|p_{\theta_k})$. If the unknown true density $p_o(x)$ is simply replaced by the empirical density from the sample set $\mathcal{X}_N$, i.e.,

$$p_{\mathcal{X}_N}(x) = \frac{1}{N}\sum_{t=1}^{N}\delta(x - x_t), \tag{8}$$

maximizing $H(p_{\mathcal{X}_N}\|p_{\theta_k})$ becomes the maximum likelihood function, which gives $\theta_k^{ML} = arg\max_{\theta_k} H(p_{\mathcal{X}_N}\|p_{\theta_k})$. However, as discussed before, the best fitting measure $H(p_{\mathcal{X}_N}\|p_{\theta_k^{ML}})$ will monotonically decrease as k and thus is not good to be used for selecting k. A better choice is to use the unknown $E_{\mathcal{X}_N}[H(p_{\mathcal{X}_N}\|p_{\theta_k})]_{\theta_k=\theta_k^{ML}}$. To estimate it, we consider the bias

$$b(k,N) = E_{\mathcal{X}_N}\{E_{\mathcal{X}_N}[H(p_{\mathcal{X}_N}\|p_{\theta_k})]_{\theta_k=\theta_k^{ML}}\} - E_{\mathcal{X}_N}H(p_o\|p_{\theta_k^{ML}}), \tag{9}$$

which relates to both k and N. With this bias $b(k,N)$ estimated, we are lead to eq.(5) with

$$F(\mathcal{X}_N, \theta_k^{fit}) = H(p_{\mathcal{X}_N}\|p_{\theta_k^{ML}}),\ \Delta(k) = b(k,N) \tag{10}$$

for selecting k. Along this line, we are further lead to the well known Akaike information criterion (AIC) with $b(k,N) = 0.5d_k/N$ [1, 2, 3] and a number of its extensions AICB, CAIC, etc, [34, 6, 7, 14, 8].

- *Optimizing marginal likelihood* Introducing a prior $p(\theta_k)$, we consider to maximize the likelihood of the following marginal distribution:

$$p(x|S_k) = \int p(x|S_k(\theta_k))p(\theta_k)d\theta_k. \tag{11}$$

Instead of solving this integration, we consider $h(\theta_k) = \ln p(x|S_k(\theta_k))$ that is expanded into a second order Taylor series with respect to θ_k around θ_k^{ML}. Let $p(\theta_k) = 1$ noninformatively inside the intergal, we get

$$\begin{aligned} p(x|S_k) &= p(x|S_k(\theta_k^{ML}))\int p(\theta_k)e^{-0.5N(\theta_k-\theta_k^{ML})^T I(\theta_k^{ML})(\theta_k-\theta_k^{ML})}d\theta_k, \\ &= p(x|S_k(\theta_k^{ML}))(2\pi)^{0.5d_k}|I(\theta_k^{ML})N|^{-0.5} \end{aligned} \tag{12}$$

where d_k is the dimension of θ_k, and $I(\theta_k)$ is the observed Fisher information matrix. After ignoring some terms approximately, we are lead to eq.(5) again with

$$F(\mathcal{X}_N, \theta_k^{fit}) = H(p_{\mathcal{X}_N}\|p_{\theta_k^{ML}}),\ \Delta(k,N) = 0.5\frac{d_k \ln N}{N}, \tag{13}$$

which is usually referred as Bayesian inference criterion (BIC) [29, 15, 23].

- *Ockham Razor (minimizing two parts of coding)* The idea is to minimize the sum of the description length for the model and the description length for residuals that the model fails to fit. Typical examples include those studies under the name of Minimum Message Length (MML) theory [36, 37, 38] and the name of Minimum Description Length (MDL) [26, 27, 22, 9, 13]. Though being different from the above BIC type approach conceptually, the implementation of either MML or MDL actually crash back to be exactly equivalent to the above BIC type approach.

Though each of the above approaches can provide a criterion $J(k, \theta_k^*)$ in a format of eq.(5), we still have to face two problems for selecting an appropriate scale k^*. First, in the cases that the size N of samples is small or not large enough, each criterion actually provides a rough estimate that can not guarantee to give k^*, and even results in a wrong result especially when k consists of several integers to enumerate. Moreover, one criterion works better in this case and the other criterion may work better in that case, none can be said to be better than the others. Second, in addition to the performance problem, another difficulty is its feasibility in implementation, because it has to be made in the following two phases:

Phase 1: Enumerate k within a range from k_d to k_u, that is assumed to contain the optimal k^*. For each specific k, we make *parameter learning* to get $\theta_k^{fit} = arg\min_{\theta_k} F(\mathcal{X}_N, \theta_k)$ according to the best fitting principle (e.g., minimizing a square error or maximizing a likelihood function).

Phase 2: Select $k^* = arg\min_k J(k, \theta_k^{fit})$ for every k within $[k_d, k_u]$.

This two-stage implementation is very expensive in computing, which makes it infeasible in many real applications.

2.3 Model Selection: from Incremental to Automatic

There are also efforts made in literatures towards the difficulty of the above two-stage implementation. One type of efforts is featured by an incremental implementation. Parameter learning is made incrementally in a sense that it attempts to incorporate as much as possible what learned at k into the learning procedure at k+1. Also, the calculation of $J(k, \theta_k^*)$ is made incrementally. Such an incremental implementation can indeed save computing costs in certain extent. However, parameter learning has to be made still by enumerating the values of k, and computing costs are still very high. As k increases to $k + 1$, an incremental implementation of parameter learning may also lead to suboptimal performance because not only those newly added parameters but also the old parameter set θ_k have to be re-learned.

Another type of efforts has been made on a widely encountered category of structures, with each consisting or composing of k individual substructures, e.g., a Gaussian mixture structure that consists of k Gaussian components. A local error criterion is used to check whether a new sample x belongs to each substructure. If x is regarded as not belonging to any of the k substructures, the $k+1$-th substructure is added to accommodate this new x. This incremental implementation for determining k is much faster. However, the local evaluating nature makes it very easy to be trapped into a poor performance, except for some special cases that $\mathcal{X}_N = \{x_t\}_{t=1}^N$ come from substructures that are well separated from each other.

Another new road has also been explored for more than ten years, with a feature that model selection can be implemented automatically during parameter learning, in a sense that making parameter learning on a structure $S_k(\theta_k)$ with its scale large enough to include the correct structure, will not only determine parameters but also automatically shrink its scale to an appropriate one, while those extra substructures are discarded during learning. It combines the good feature of *regularization* and the good feature of *model selection*. On one hand, it takes the good feature of *regularization* (i.e., parameter learning is only implemented on a structure $S_k(\theta_k)$ with a larger scale), but discards the crucial problem of regularization (i.e., those extra substructures are still kept to blur the useful ones). On the other hand, it takes the good feature of *model selection*, i.e., only a structure with an appropriate scale is in action without any extra substructures to deteriorate its performance. Moreover, it not only keeps the model selection performance as good as that by a two-stage implementation, but only performs parameter learning only once with a drastic reduction in computing costs.

One early effort along such a new road started from Rival Penalized Competitive Learning (RPCL) for clustering analysis and detecting curves in an image [64, 66]. A structure in a scale k consists of k individual substructures, with each being simply one point as one cluster's center. Initially, k is given a value larger than the appropriate number of clusters. A coming sample x is allocated to one of the k centers via competition, and the winning center moves a little bit to adapt the information carried by this sample. Moreover, the rival (i.e., the second winner) is repelled a little bit away from the sample to reduce a duplicated information allocation. Driving those extra centers away, this rival penalized mechanism will keep an appropriate number of centers. In other words, RPCL makes the number of clusters determined automatically during learning. This is a favorable feature that the conventional competitive learning or clustering algorithm (e.g., k-means) does not have. RPCL has been further extended from spheral clusters to elliptic clusters via Gaussian mixture [55, 52, 49]. Readers are referred to [40, 47] for a recent elaboration and to [17, 18, 16] for successful applications.

RPCL learning was heuristically proposed on a bottom level (i.e., a level of learning dynamics or updating rule), which is quite different from our previous discussions on a global level of using one learning principle or theory to guide parameter learning and model selection in a top-down manner. Proposed firstly in [59] and systematically developed in past years [53, 51, 52, 49, 47, 42, 43, 41], the Bayesian Ying-Yang (BYY) harmony learning is such a global level theory that guides various statistical learning tasks with model selection achieved automatically during parameter learning.

3 BYY Harmony Learning: A Trend on Regularization and Model Selection

3.1 Bayesian Ying-Yang System

Instead of letting M to consider a parametric structure for directly best fitting the observed samples $\mathcal{X}_N$, M is considered or designed as a system that jointly describes the observed samples and their inner representations $\mathcal{Y}$ via two different but complementary parts. As shown in Fig. 2(a), one is named as *Yang* that consists of one component for representing $\mathcal{X}_N$ and one component for describing a path from $\mathcal{X}_N$ to $\mathcal{Y}$. The other part is named as *Ying* that consists of one component for representing $\mathcal{Y}$ and one component for describing a path from $\mathcal{Y}$ to $\mathcal{X}_N$. Each of the four components may have several choices for its corresponding structure. The four components can be integrated into a system in more than one choices under the name of architecture. In such a Ying-Yang system, a principle of using a structure in a specific type to best fit $\mathcal{X}$ can be generalized into a principle for a Ying-Yang system to best match each other, which includes using a structure to directly best fit $\mathcal{X}$ as a subtask.

The Ying-Yang system can be further formulated in a statistical framework by considering the joint distribution of $\mathcal{X}$ and $\mathcal{Y}$, which can be described via the following two types of Bayesian decomposition:

$$p(\mathcal{X},\mathcal{Y}) = p(\mathcal{Y}|\mathcal{X})p(\mathcal{X}), \ q(\mathcal{X},\mathcal{Y}) = q(\mathcal{X}|\mathcal{Y})q(\mathcal{Y}). \tag{14}$$

In a compliment to the famous Chinese ancient Ying-Yang philosophy, the decomposition of $p(\mathcal{X},\mathcal{Y})$ coincides the Yang concept with $p(\mathcal{X})$ describing samples from an observable domain (called the Yang space) and $p(\mathcal{Y}|\mathcal{X})$ describing the forward path from $\mathcal{X}_N$ to $\mathcal{Y}$ (called the Yang pathway). Thus, $p(\mathcal{X},\mathcal{Y})$ is called the Yang machine. Similarly, $q(\mathcal{X},\mathcal{Y})$ is called the Ying machine with $q(\mathcal{Y})$ describing representations in an invisible domain (thus regarded as a Ying space) and $q(\mathcal{X}|\mathcal{Y})$ describing the backward path (called the

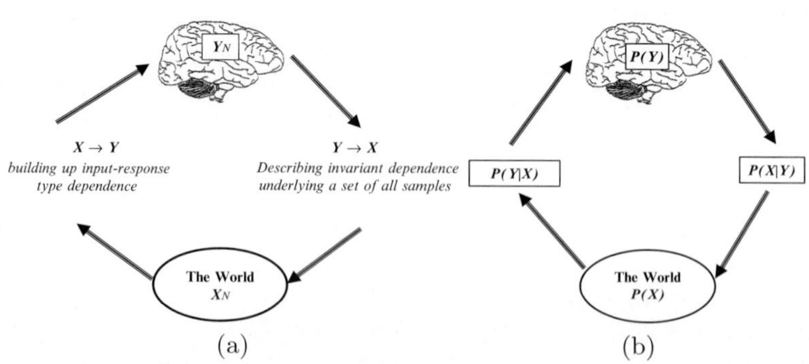

Fig. 2. (a) Ying-Yang system, (b) Bayesian Ying-Yang system

Ying pathway). As shown in Fig. 2(b), such a pair of Ying-Yang machines is called *Bayesian Ying-Yang (BYY) system.*

Each of the above four components may have more than one choices for its structure, as summarized below:

- $p(\mathcal{X})$ is either a nonparametric estimation via data smoothing, e.g.,
$$p_h(\mathcal{X}_N) = \prod_{t=1}^{N} G(x_t|\bar{x}_t, h^2 I), \tag{15}$$
or a direct use of the samples in $\mathcal{X}_N$, i.e., by $p_0(\mathcal{X}_N) = p_h(\mathcal{X}_N)|_{h=0}$. That is, each specific sample $\bar{x}_t$ is blurred or smoothed by a Gaussian noise with a variance h^2, resulting in a Gaussian random variable x_t.
- The structure of $q(\mathcal{Y})$ takes at least three crucial roles. One is bridging the Ying and Yang two parts. The other is specifying the nature of learning tasks via a number of choices summarized by a general structure [41]. The third role is that the scale of its structure actually dominates the scale of the system architecture, as to be introduced in the next subsection.
- With the settlements of $q(\mathcal{Y})$ and $p(\mathcal{X})$, each of $p(\mathcal{Y}|\mathcal{X})$ and $q(\mathcal{X}|\mathcal{Y})$ has also more than one different choices.

Moreover, there are different ways for integrating the four components together, which result in different system architectures.

As a result, $p(\mathcal{X}, \mathcal{Y})$ and $q(\mathcal{X}, \mathcal{Y})$ in eq.(14) are actually not the same though both represent the same joint distribution of $\mathcal{X}$ and $\mathcal{Y}$. In such a formulation, the best fitting principle is generalized into a principle that $p(\mathcal{X}, \mathcal{Y})$ and $q(\mathcal{X}, \mathcal{Y})$ best match each other. This match can be measured by the Kullback divergence [59] and several non-Kullback divergences [57, 45], summarized in the following general expression:

$$D(p\|q, \theta_k) = \int p(\mathcal{Y}|\mathcal{X})p(\mathcal{X}) f\left(\frac{p(\mathcal{Y}|\mathcal{X})p(\mathcal{X})}{q(\mathcal{X}|\mathcal{Y})q(\mathcal{Y})}\right) d\mathcal{X} d\mathcal{Y}, \tag{16}$$

where $f(r)$ is a convex function. Particularly, we have the Kullback divergence when $f(r) = \ln r$. In this setting, our task becomes to specify each of the four components via determining all the unknowns subject to all the known parts, i.e., $\mathcal{X}_N$ and the pre-specified structure of each component. In a summary, we have

$$\min_{\substack{p(\mathcal{X}),\ p(\mathcal{Y}|\mathcal{X}),\ q(\mathcal{X}|\mathcal{Y}), q(\mathcal{Y}), \textit{and } \theta_k \textit{ subject to} \\ \textit{their pre-specified structures and } \mathcal{X}_N}} D(p\|q, \theta_k). \tag{17}$$

In the special case that $p(\mathcal{Y}|\mathcal{X})$ is free of any pre-structure, minimizing $D(p\|q, \theta_k)$ with respect to $p(\mathcal{Y}|\mathcal{X})$ will lead to a special case that is equivalent to using $q(\mathcal{X}) = \int q(\mathcal{X}|\mathcal{Y})q(\mathcal{Y})d\mathcal{Y}$ to best fit $\mathcal{X}_N$ in a sense of the maximum likelihood principle. This nature, together with the feature that the special settings of $q(\mathcal{Y})$ as well as of other three components $q(\mathcal{X}|\mathcal{Y})$, $p(\mathcal{Y}|\mathcal{X})$ and $p(\mathcal{X})$ lead to specific structures of a number of existing typical learning tasks, makes a number of existing statistical learning approaches summarized in a unified perspective with new variants and extensions [44, 45, 41].

3.2 BYY Harmony Learning

Still, we encounter an over-fitting problem for such a best Ying-Yang match by eq.(17). Similar to what discussed in Sec. 1, we consider the BYY system with a family of system architectures of the same type but in different scales.

A system architecture type is an integration of the four components with each in its own specific structure type. As introduced in the previous subsection, the structure type of $q(\mathcal{Y})$ takes a leading role, and the scale of the entire system architecture is dominated by the scale of the structure for $q(\mathcal{Y})$. We can denote this scale by an integer k that usually represents an enumeration of a set of integers $\mathbf{k}$ embedded within the structure of $q(\mathcal{Y})$. Also we can use θ_k to denote all the unknowns in an architecture of a scale k. Suffering an expensive computing cost, we may learn a best value $\theta^*_{k^*}$ at a best scale k^* via the following two-phase implementation:

Phase 1: Enumerate a series of architectures of a same type but in different scales. For each one at k, we make *parameter learning* to get $\theta^*_k = arg \min_{\theta_k} D(p\|q, \theta_k)$.

Phase 2: Select $k^* = arg \min_k J(k, \theta^*_k)$ according to one of the existing model selection criteria.

Much more importantly, the Ying-Yang system is motivated together with the ancient Chinese philosophy that the Ying and the Yang should be best harmony in a sense that two parts should not only best match but also are matched in a compact way. Applying this philosophy into a BYY system, we have a best harmony principle in the following twofold sense:

- *Best matching* the difference between the two Bayesian representations in eq.(14) should be minimized.
- *Least complexity* the resulted architecture for the BYY system should be in a least complexity, i.e., its inner representation has a least complexity.

The above can be further mathematically measured by the following functional

$$H(p\|q, \theta_k) = \int p(\mathcal{Y}|\mathcal{X})p(\mathcal{X})f(q(\mathcal{X}|\mathcal{Y})q(\mathcal{Y}))\mu(d\mathcal{X})\mu(d\mathcal{Y}) - \ln Z, \qquad (18)$$

where $f(r)$ is again a convex function as in eq.(16), and a most useful case is $f(r) = \ln r$. Instead of eq.(17), we specify the four components via determining all the unknowns subject to all the known parts as follows [59, 51, 49]:

$$\max_{\substack{p(\mathcal{X}),\ p(\mathcal{Y}|\mathcal{X}),\ q(\mathcal{X}|\mathcal{Y}), q(\mathcal{Y}),\ and\ \theta_k\ subject\ to \\ their\ pre\text{-}specified\ structures\ and\ \mathcal{X}_N}} H(p\|q, \theta_k), \qquad (19)$$

which guides not only learning on θ^*_k but also model selection on k^*. This is a significant difference from the conventional approaches, by which θ^*_k is learned under a best fitting principle but k^* is selected in help of another learning theory.

The term $\ln Z$ imposes certain regularization into learning on $\mathcal{X}_N$ with a small size N, which will be discussed in Sec. 3.5. Here, we give a further insight on $H(p\|q)$ via the following decomposition

$$\begin{aligned}
&H(p\|q,\theta_k) = H_{x|y} + H_y, \\
&H_{x|y} = \int p(\mathcal{Y}|\mathcal{X})p(\mathcal{X})\ln q(\mathcal{X}|\mathcal{Y})\mu(d\mathcal{X})\mu(d\mathcal{Y}), \\
&H_y = \int p(\mathcal{Y})\ln q(\mathcal{Y})\mu(d\mathcal{Y}),\ p(\mathcal{Y}) = \int p(\mathcal{Y}|\mathcal{X})p(\mathcal{X})\mu(d\mathcal{X}).
\end{aligned} \tag{20}$$

On one hand, the term $H_{x|y}$ accounts for a best fitting of the samples in $\mathcal{X}_N$ by $q(\mathcal{X}|\mathcal{Y})$ in help of the corresponding inner representation $\mathcal{Y}$. If $p(\mathcal{X})$ is given by eq.(8) and a set $\mathcal{Y}_N$ is given for pairing $\mathcal{X}_N$, $H_{x|y}$ degenerates into the likelihood function of $q(\mathcal{X}|\mathcal{Y})$. On the other hand, the term H_y accounts for two purposes. When the structure of $q(\mathcal{Y})$ is not pre-imposed with too much constraint, maximizing H_y results in $q(\mathcal{Y}) = p(\mathcal{Y})$ such that $-H_y$ becomes exactly the entropy of the inner representation. Thus, maximizing H_y leads to an inner representation in a least complexity. Usually, $q(\mathcal{Y})$ is pre-imposed with different structures for different learning tasks, maximizing H_y forces the resulted $p(\mathcal{Y})$ to satisfy these constraints correspondingly. It can be observed that $H_{x|y}$ increases while H_y decreases as the scale k increases, which trades off for an appropriate k^*. In other words, $H(p\|q,\theta_k)$ can be used at Phase 2 of a two-phase implementation given at the beginning of this subsection, in place of a model selection criterion. That is, we get $\theta_k^* = arg\min_{\theta_k} D(p\|q,\theta_k)$ at Phase 1, and then, as shown in Fig. 3(a), we select a best k^* at Phase 2 by

$$k^* = arg\min_k\ J(k),\ \ J(k) = -H(p\|q,\theta_k)|_{\theta_k=\theta_k^*}. \tag{21}$$

With this two-phase implementation as a link, we can compare the performances of this new criterion with those typical model selection criteria discussed in Sec. 2.2.

3.3 BYY Harmony Learning and Automatic Model Selection

The BYY harmony learning by eq.(19) has a salient advantage that an appropriate scale k^* can be obtained by implementing parameter learning only once on an architecture in a larger scale.

Considering $q(\mathcal{Y})$ in a scale reducible structure that its scale k can be effectively reduced into a smaller one by forcing a critical subset of parameters within the structure becoming zero. That is, we consider a distribution $q(y|\theta_k^y), \theta_k^y \in \Theta_k^y$ that demonstrates its maximum scale k^y when θ_k^y takes values within some specific domain $\hat{\Theta}_k^{\ y}$ of Θ_k^y while it effectively reduces into a smaller scale when a critical subset ϕ_k of θ_k^y becomes zero. For an example, a mixture distribution $q(y|\theta_k^y) = \sum_{\ell=1}^k \alpha_\ell q(y|\phi_\ell)$ has a scale k when $\alpha_\ell > 0$ for all ℓ, but will reduce into a scale $k-1$ if one $\alpha_\ell = 0$. For another example, an independent product $q(y|\theta_k^y) = \prod_{j=1}^m q(y^{(j)}|\theta_k^{y^{(j)}})$ has a scale m in general

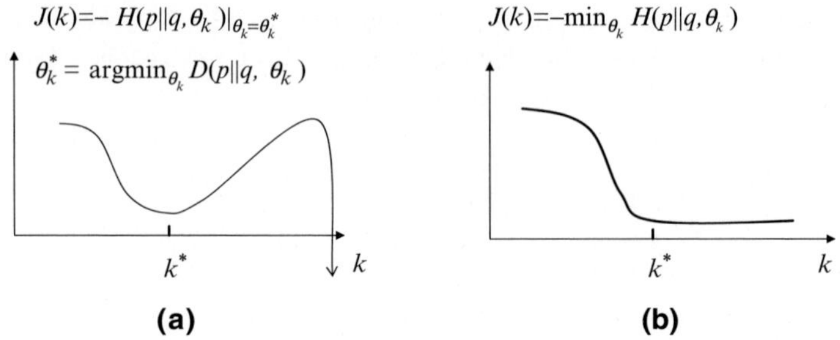

Fig. 3. Selection of an appropriate k^*

but a reduced scale $m-1$ when there is one j with $var(y^{(j)})=0$, i.e., the variance parameter of $y^{(j)}$ becomes zero. Readers are referred to Sec. 22.5 in [44], Sec. 23.3.2 in [45], Sec. II(B) in [42], and Sec. III(C) in [41].

For $q(\mathcal{Y})$ in a scale reducible structure, we have two different types of choices. First, let

$$J(k) = -\max_{\theta_k,\ subject\ to\ \theta_k^y \in \hat{\Theta}_k{}^y} H(p\|q,\theta_k^*), \tag{22}$$

we are lead to a case as shown in Fig. 3(a) for a two-phase implementation. E.g., we get such a case when $\alpha_\ell = 1/k$ for all ℓ or $var(y^{(j)})=1$ for all j. Second, we let $J(k)=-\max_{\theta_k} H(p\|q,\theta_k)$ without any constraint, maximizing H_y will push the part $\theta_k^{y\ (2)}$ of those extra substructures to zeros such that $q(\mathcal{Y})$ effective shrinks to a scale smaller than k. As a result, the curve shown in Fig. 3(a) becomes the curve shown in Fig. 3(b).

In other words, considering $q(\mathcal{Y})$ in a scale reducible structure with an initial scale k that is larger than an appropriate one, we can implement the following parameter learning

$$\max_{\theta_k} H(p\|q,\theta_k), \tag{23}$$

which results in not only a best θ_k^* but also an appropriate k^* determined automatically during this learning. Readers are referred to [41, 43, 44, 49].

3.4 BYY Model Selection Criteria on a Small Size of Samples

In a situation that the sample size N is too small, the performance of automatic model selection by the BYY harmony learning will deteriorate, and we have to turn back to a two phase implementation. For this purpose, we seek to find improved model selection criteria from eq.(21).

We consider a more general BYY system with an augmented inner-system representation $\mathcal{R}$ that consists of not only $\mathcal{Y}$ featured by a per sample pairing relation between $\mathcal{X}_N$ and $\mathcal{Y}_N$ (i.e., each x_t gets its inner representation y_t), but also all the unknown parameters θ_k (including h in eq.(15)). With such an extension, eq.(14) becomes

$$p(\mathcal{X},\mathcal{R}) = p(\mathcal{R}|\mathcal{X})p(\mathcal{X}), \; q(\mathcal{X},\mathcal{R}) = q(\mathcal{X}|\mathcal{R})q(\mathcal{R}). \tag{24}$$

Specifically, $q(\mathcal{R}) = q(\mathcal{Y},\theta_k) = q(\mathcal{Y}|\theta_k)q(\theta_k)$ that consists of two parts. One is $q(\theta_k)$ that describes a priori distribution for the values that θ_k may take. The other is actually the previous one under the notation $q(\mathcal{Y})$, which is featured by a family of parametric functions that vary as a set of parameters θ_k^y that is a subset of θ_k. That is, $q(\mathcal{Y}) = q(\mathcal{Y}|\theta_k) = q(\mathcal{Y}|\theta_k^y)$. Coupling with the representation of $\mathcal{Y}$, $q(\mathcal{X}|\mathcal{R}) = q(\mathcal{X}|\mathcal{Y},\theta_k) = q(\mathcal{X}|\mathcal{Y},\theta_k^{xy})$ is actually the previous one under the notation $q(\mathcal{X}|\mathcal{Y})$, defining a family of parametric functions with a set of parameters θ_k^{xy} that is also a subset of θ_k. Moreover, $p(\mathcal{R}|\mathcal{X}) = p(\mathcal{Y},\theta_k|\mathcal{X}) = p(\mathcal{Y}|\mathcal{X},\theta_k)p(\theta_k|\mathcal{X})$ that comprises of two parts. $p(\mathcal{Y}|\mathcal{X},\theta_k^{yx})$ is actually the previous one under the notation $p(\mathcal{Y}|\mathcal{X})$, associated with a set of parameters θ_k^{yx} that is another subset of θ_k too. The other part $p(\theta_k|\mathcal{X})$ describes the uncertainty of estimating θ_k from $\mathcal{X}$, which is actually the posteriori counterpart of the a priori $q(\theta_k)$.

Correspondingly, we can observe that the harmony functional by eq.(18) actually comes from

$$H(p\|q) = \int p(\mathcal{R}|\mathcal{X})p(\mathcal{X})f(q(\mathcal{X}|\mathcal{R})q(\mathcal{R}))\mu(d\mathcal{X})\mu(d\mathcal{R}). \tag{25}$$

In the case $f(r) = \ln r$, it can be further rewritten into

$$\begin{aligned}
&H(p\|q) = \int p(\theta_k|\mathcal{X})H(p\|q,\theta_k)d\theta_k \\
&H(p\|q,\theta_k) = \int p(\mathcal{Y}|\mathcal{X},\theta_k^{yx})p_h(\mathcal{X})\ln\left[q(\mathcal{X}|\mathcal{Y},\theta_k^{xy})q(\mathcal{Y}|\theta_k^y)\right]d\mathcal{X}d\mathcal{Y} - Z(\theta_k), \\
&Z(\theta_k) = -\ln q(\theta_k),
\end{aligned} \tag{26}$$

where $H(p\|q,\theta_k)$ is actually the one given in eq.(18) at the case $f(r) = \ln r$.

Given the structures of $q(\mathcal{Y}|\theta_k^y)$, $q(\mathcal{X}|\mathcal{Y},\theta_k^{xy})$, and $p(\mathcal{Y}|\mathcal{X},\theta_k^{yx})$, the task of learning is featured by $\max_{\{p(\theta_k|\mathcal{X}),\ \mathbf{k}\}} H(p\|q)$. By expanding $H(p\|q,\theta_k)$ with respect to θ_k around the following θ_k^* up to the second order and ignoring its first order term since $\nabla_{\theta_k}H(p\|q,\theta_k) = 0$ at $\theta_k = \theta_k^*$, the task can be approximately decomposed into the following two parts:

$$\begin{aligned}
&\theta_k^* = arg\max_{\theta_k} H(p\|q,\theta_k), \; \mathbf{k}^* = arg\min_{\mathbf{k}} J(\mathbf{k}), \; J(\mathbf{k}) = -H(p\|q), \\
&H(p\|q) = H(p\|q,\theta_k^*) - 0.5d(\theta_k^*), \\
&d(\theta_k) = -Tr[\Sigma_{\theta_k}\frac{\partial^2 H(p\|q,\theta_k)}{\partial\theta_k\partial{\theta_k}^T}]_{\theta_k=\theta_k^*}, \\
&\Sigma_{\theta_k} = \int(\theta_k-\theta_k^*)(\theta_k-\theta_k^*)^T p(\theta_k|\mathcal{X})d\theta_k.
\end{aligned} \tag{27}$$

That is, to get rid of the difficulty of estimating $p(\theta_k|\mathcal{X})$ and the related computing cost, we can implement learning in two phases as follows:

Phase 1: Enumerate **k** for a series of architectures of a same type but in different scales. For each candidate, we estimate $\theta_k^* = arg\max_{\theta_k} H(p\|q,\theta_k)$.

Phase 2: Select a best architecture by $\mathbf{k}^* = arg\min_{\mathbf{k}} J(\mathbf{k})$, where $d(\theta_k^*)$ can be further approximately simplified into an integer as follows:

$$d(\theta_k) = \begin{cases} d_k, & (a) an\ under - constraint\ choice, \\ 2d_k, & (b) an\ over - constraint\ choice, \end{cases} \tag{28}$$

where d_k is the number of free parameters in θ_k.

Eq. (28) comes from a reason related to the celebrated Cramer-Rao inequality. We roughly regard that the obtained θ_k^* suffers a uncertainty under $p(\theta_k|\mathcal{X})$ with a covariance matrix Σ_{θ_k} such that $\Sigma_{\theta_k} \frac{\partial^2 H(p\|q,\theta_k)}{\partial \theta_k \partial \theta_k{}^T} \approx I$ at $\theta_k = \theta_k^*$, especially when we consider a noninformative priori $q(\theta_k) = 1$ or $\ln q(\theta_k) = 0$, which leads to eq.(28)(a). Generally, it may be too crude to simply count the number of parameters in θ_k. Instead, $d(\theta_k^*)$ is an effective number closely related to how $p(\theta_k|\bar{\mathcal{X}}_N)$ is estimated. For an estimator $\theta_k^* = T(\bar{\mathcal{X}}_N)$ basing on a sample set $\bar{\mathcal{X}}_N$, if this estimator is unbiased to its true value θ^o, it follows from the celebrated Cramer-Rao inequality that $p(\theta_k|\bar{\mathcal{X}}_N)$ can be asymptomatically regarded as $p(\theta_k|\bar{\mathcal{X}}_N) = G(\theta_k|\theta^o, [NF(\theta^o)]^{-1})$ with $F(\theta) = -\frac{1}{N}\frac{\partial^2 \ln q(\bar{\mathcal{X}}_N|\theta)}{\partial\theta\partial\theta^T}$, and thus we have $\Sigma_{\theta_k} = \int(\theta_k - \theta^o + \theta^o - \hat{\theta}_k)(\theta_k - \theta^o + \theta^o - \hat{\theta}_k)^T G(\theta_k|\theta^o, [NF(\theta^o)]^{-1}) d\theta_k \approx 2[NF(\theta^o)]^{-1}$. Moreover, if we roughly regard that $\frac{\partial^2 H(p\|q,\theta_k)}{\partial \theta_k \partial \theta_k{}^T}|_{\theta_k = \theta^o} = [NF(\theta^o)]^{-1}$ as N becomes large enough, we are alternatively lead to eq.(28)(b).

3.5 BYY Learning Integrates Regularization and Model Selection

Recall Sec. 2.1 and Sec. 2.2, the conventional regularization approaches have only a limited help on those learning problems due to a small sample size. Also, these regularization approaches suffer one crucial weakness caused by an isotropic regularization and a key difficulty on controlling a regularization strength. The conventional model selection approaches aim at tackling the weaknesses, but it suffers a huge cost to enumerate a number of candidate models with different values of k. Associated with a BYY system under the best harmony principle, the roles of regularization and model selection can be integrated in a sense that the crucial weakness caused by an isotropic regularization can be avoided by the model selection ability of the best harmony principle, while types of companion regularization can still be imposed to improve the weakness caused by the model selection mechanism of the BYY harmony learning. Moreover, some of these companion regularization approaches can also be decoupled from a BYY system and become directly applicable to the conventional maximum likelihood learning on a parametric model $p(x|\theta)$.

Considering $H(p\|q,\theta_k) = \int p_h(\mathcal{X})H_{\mathcal{X}}(p\|q,\theta_k)d\mathcal{X} - Z(\theta_k)$ in eq.(19), we can also expand $H_{\mathcal{X}}(p\|q,\theta_k)$ with respect to $\mathcal{X}$ around $\bar{\mathcal{X}}_N$ up to the second order, resulting in

$$H(p\|q,\theta_k) = H_{\bar{\mathcal{X}}_N}(p\|q,\theta_k) + 0.5h^2 Tr[\frac{\partial^2 H_{\mathcal{X}}(p\|q,\theta_k)}{\partial\mathcal{X}\partial\mathcal{X}^T}]_{\mathcal{X}=\bar{\mathcal{X}}_N} - Z(\theta_k),$$
$$H_{\mathcal{X}}(p\|q,\theta_k) = \int p(\mathcal{Y}|\mathcal{X},\theta_k^{yx})\ln[q(\mathcal{X}|\mathcal{Y},\theta_k^{xy})q(\mathcal{Y}|\theta_k^y)]d\mathcal{Y}. \quad (29)$$

The term $0.5h^2 Tr[\cdot]$ is usually negative and thus increases as $h^2 \to 0$. What a value h will take depends on what type of the priori term $Z(\theta_k)$, for which there are three typical situations.

The simplest and also most usual case is $q(\theta_k) = 1$ or $Z(\theta_k) = 0$. In this case, $\max_h H(p\|q,\theta_k)$ will force $h = 0$, and thus we simply have $H(p\|q,\theta_k) = H_{\bar{\mathcal{X}}_N}(p\|q,\theta_k)$. When thc function forms of $q(\bar{\mathcal{X}}_N|\mathcal{Y},\theta_k^{xy})$ and $q(\mathcal{Y}|\theta_k^y)$ are given while there is no priori constraint on the function form $p(\mathcal{Y}|\mathcal{X},\theta_k^{yx})$, we can consider the learning task in a sequential maximization, i.e., $\max_{\theta_k}\{\max_{p(\mathcal{Y}|\mathcal{X})} H(p\|q)\}$. It follows from $\max_{p(\mathcal{Y}|\mathcal{X})} H_{\mathcal{X}}(p\|q,\theta_k)$ that

$$p(\mathcal{Y}|\bar{\mathcal{X}}_N) = \delta(\mathcal{Y} - \bar{\mathcal{Y}}_N),\ \bar{\mathcal{Y}}_N = arg\max_{\mathcal{Y}}[q(\bar{\mathcal{X}}_N|\mathcal{Y},\theta_k^{xy})q(\mathcal{Y}|\theta_k^y)],$$
$$H_{\bar{\mathcal{X}}_N}(p\|q,\theta_k) = H_{\bar{\mathcal{X}}_N,\bar{\mathcal{Y}}_N}(p\|q,\theta_k),$$
$$H_{\bar{\mathcal{X}}_N,\mathcal{Y}}(p\|q,\theta_k) = \ln[q(\bar{\mathcal{X}}_N|\mathcal{Y},\theta_k^{xy})q(\mathcal{Y}|\theta_k^y)]. \quad (30)$$

That is, $\max_{\theta_k}\{\max_{p(\mathcal{Y}|\mathcal{X})} H(p\|q)\}$ becomes $\max_{\theta_k} H_{\bar{\mathcal{X}}_N,\bar{\mathcal{Y}}_N}(p\|q,\theta_k)$.

On one hand, the winner-take-all (WTA) maximization in eq.(30) indirectly implements a mechanism of selecting an appropriate value **k** that enables the automatic model selection discussed in Sec. 3.3. Readers are referred to Sec. 22.5 in [44], Sec. 23.3.2 in [45], Sec. II(B) in [42], and Sec. III(C) in [41]. However, there is also an other hand. If we subsequently implement $\max_{\theta_k} H_{\bar{\mathcal{X}}_N,\bar{\mathcal{Y}}_N}(p\|q,\theta_k)$ by ignoring the relation of $\bar{\mathcal{Y}}_N = arg\max_{\mathcal{Y}}[q(\bar{\mathcal{X}}_N|\mathcal{Y},\theta_k^{xy}) q(\mathcal{Y}|\theta_k^y)]$ to θ_k, we have to re-update $\max_{p(\mathcal{Y}|\mathcal{X})} H_{\mathcal{X}}(p\|q,\theta_k)$ by eq.(30). Though such an alternative maximization will gradually increase $H_{\mathcal{X}}(p\|q,\theta_k)$, it cannot avoid to get stuck in a local maximum or perhaps even a saddle point. Moreover, such an iterative maximization has to be made on the whole batch $\bar{\mathcal{X}}_N$ per step and thus is computationally very expensive. In an actual implementation [59, 53, 51, 52, 49, 47, 42, 43, 41], such an iteration is made per sample per step in a form $\bar{y}_t = arg\max_{y_t}[q(\bar{x}_t|y_t,\theta_k^{xy})q(y_t|Y_{t-1},\theta_k^y)]$, where Y_{t-1} is either an empty set or a set that consists of a number of past samples of $y_{t-1},\cdots,y_{t-p}$.

When $\bar{\mathcal{Y}}_N = arg\max_{\mathcal{Y}}[q(\bar{\mathcal{X}}_N|\mathcal{Y},\theta_k^{xy})q(\mathcal{Y}|\theta_k^y)]$ becomes an explicit expression with respect to $\bar{\mathcal{X}}_N$ and θ_k or $\bar{y}_t = arg\max_{y_t}[q(\bar{x}_t|y_t,\theta_k^{xy})q(y_t|Y_{t-1},\theta_k^y)]$ becomes an explicit expression with respect to x_t, Y_{t-1} and θ_k, we can take these explicit expressions into $\max_{\theta_k} H_{\bar{\mathcal{X}}_N,\bar{\mathcal{Y}}_N}(p\|q,\theta_k)$ by considering a compound dependence on θ_k, which will be helpful to improve the problem of local maximum. However, except for certain particular cases, it is usually difficult to get such explicit expressions. Therefore, we have to ignore the dependence of $\bar{\mathcal{Y}}_N$ with respect to θ_k.

The local maximum problem can be remedied by a unique regularization type coped with a BYY system, *structural regularization* or shortly BI-regularization. Details are referred to Sec. 3 in [46].

The key idea of the BI-regularization is to let $p(\mathcal{Y}|\mathcal{X})$ in an appropriately designed structure, three examples are given as follows:

(a) Let the optimization procedure for $\bar{\mathcal{Y}}_N$ in eq.(30) to be approximated by a parametric function:

$$\bar{\mathcal{Y}}_N = F(\bar{\mathcal{X}}_N, \theta_k^{yx}),\ p(\mathcal{Y}|\bar{\mathcal{X}}_N) = \delta(\mathcal{Y} - \bar{\mathcal{Y}}_N),$$
$$H_{\bar{\mathcal{X}}_N}(p\|q, \theta_k) = H_{\bar{\mathcal{X}}_N, \mathcal{Y}}(p\|q, \theta_k)_{\mathcal{Y}=F(\bar{\mathcal{X}}_N, \theta_k^{yx})}, \tag{31}$$

where two examples of $F(\bar{\mathcal{X}}_N, \theta_k^{yx})$ are as follows,
(1) $\bar{y}_t = f(x_t, \theta_k^{yx})$ for i.i.d. samples,
(2) $\bar{y}_t = f(x_t, \Xi_t, \theta_k^{yx})$ for temporal samples,
Ξ_t consists of past samples from either or both of $\bar{\mathcal{X}}_N$ and $\bar{\mathcal{Y}}_N$.

(b) We can also consider jointly a number of functions $\mathcal{Y} = F_\ell(\bar{\mathcal{X}}_N, \theta_{\ell,k}^{yx})$, $\ell = 1, \cdots, n_{yx}$ as follows

$$\bar{\mathcal{Y}}_N = F_{\ell^*}(\bar{\mathcal{X}}_N, \theta_k^{yx}),\ \ell^* = arg \max_\ell [q(\bar{\mathcal{X}}_N|\mathcal{Y}, \theta_k^{xy})q(\mathcal{Y}|\theta_k^y)]_{\mathcal{Y}=F_\ell(\bar{\mathcal{X}}_N, \theta_{\ell,k}^{yx})},$$
$$H_{\bar{\mathcal{X}}_N}(p\|q, \theta_k) = H_{\bar{\mathcal{X}}_N, \mathcal{Y}}(p\|q, \theta_k)_{\mathcal{Y}=F_{\ell^*}(\bar{\mathcal{X}}_N, \theta_{\ell^*,k}^{yx})}. \tag{32}$$

(c) In the special cases that $\mathcal{Y}$ is represented in discrete values or both $q(\bar{\mathcal{X}}_N|\mathcal{Y}, \theta_k^{xy})$ and $q(\mathcal{Y}|\theta_k^y)$ are Gaussian, it is also possible to let

$$p(\mathcal{Y}|\mathcal{X}) = \frac{q(\bar{\mathcal{X}}|\mathcal{Y}, \theta_k^{xy})q(\mathcal{Y}|\theta_k^y)}{\int q(\mathcal{X}|\mathcal{Y}, \theta_k^{xy})q(\mathcal{Y}|\theta_k^y)d\mathcal{Y}}. \tag{33}$$

In these two cases, the integral over $\mathcal{Y}$ either becomes a summation or is analytically solvable. Readers are referred to Sec. 3.4.2 in [40] some discussions on the summation cases.

Another type of regularity lost comes from that $H_{\bar{\mathcal{X}}_N, \bar{\mathcal{Y}}_N}(p\|q, \theta_k)$ is computed only based on the samples in $\mathcal{X}_N$ via $p_0(\mathcal{X}_N) = p_h(\mathcal{X}_N)|_{h=0}$ by eq.(15). Though the above discussed structural regularization is helpful to remedy this problem indirectly, another regularization type coped with a BYY system is the Z-regularization featured by the term $Z(\theta_k) \neq 0$ in eq.(29), with two typical examples as follows:

- When $q(\theta_k)$ is irrelevant to h but relates to a subset of θ_k, $\max_h H(p\|q, \theta_k)$ will still force $h = 0$, and thus force the second term $0.5h^2Tr[\cdot]$ disappeared, while $Z(\theta_k)$ will affect the learning on θ_k. A typical example is

$$q(\theta_k^{xy}, \theta_k^y) \propto [\sum_{t=1}^N \int q(\bar{x}_t|y_t, \theta_k^{xy})q(y_t|\theta_k^y)dy_t]^{-1}, \tag{34}$$

$$or\ q(\theta_k) = q(\theta_k^{xy}, \theta_k^y) \propto [\sum_{t=1}^N q(\bar{x}_t|\bar{y}_t, \theta_k^{xy})q(\bar{y}_t|\theta_k^y)]^{-1},\ \text{if we get } \bar{y}_t \text{ per } \bar{x}_t,$$

which normalizes a finite size samples of $q(\bar{x}_t|y_t, \theta_k^{xy})q(y_t|\theta_k^y)$. Thus, it is called *normalization learning*. Readers are referred to Sec. 2.2 in [52], Sec. 22.6.1 in [44], and [47].

- Another typical case is $q(\theta_k) = q(h) \propto [\sum_{t=1}^N \sum_{\tau=1}^N G(x_t|x_\tau, h^2 I)/N]^{-1}$ that merely relates to h but is irrelevant to any other parameters in θ_k. In this case, the term $Z(\theta_k)$ together with the term $0.5h^2 Tr[\cdot]$ will trade off to give an appropriate h, which then affects the learning on other parameters in θ_k via $0.5h^2 Tr[\cdot]$. This type of regularization is equivalent to smoothing the samples in $\mathcal{X}_N$ via adding Gaussian noises with a noise variance h^2. Thus, it is called *data smoothing*. Details are referred to Sec. II(B) in [51], Sec. 23.4.1 in [45], and Sec. III(E) in [41].
Furthermore, *data smoothing* can also be imposed on $\bar{\mathcal{Y}}_N$ given by eq.(31) or eq.(32) by considering

$$p_{h_y}(\mathcal{Y}|\bar{\mathcal{X}}_N) = G(\mathcal{Y}|\bar{\mathcal{Y}}_N, h_y^2 I),$$
$$H_{\bar{\mathcal{X}}_N}(p\|q, \theta_k) = H_{\bar{\mathcal{X}}_N, \bar{\mathcal{Y}}_N}(p\|q, \theta_k) + 0.5h_y^2 Tr[\frac{\partial^2 H_{\bar{\mathcal{X}}_N, \mathcal{Y}}(p\|q, \theta_k)}{\partial \mathcal{Y} \partial \mathcal{Y}^T}]_{\mathcal{Y}=\bar{\mathcal{Y}}_N},$$
$$q(\theta_k) = q(h, h_y) \propto [\sum_{t=1}^N \sum_{\tau=1}^N G(x_t|x_\tau, h^2 I) G(y_t|y_\tau, h_y^2 I)/N]^{-1}. \tag{35}$$

In this case, the term $Z(\theta_k)$ together with the term $0.5h^2 Tr[\cdot] + 0.5h_y^2 Tr[\cdot]$ will trade off to give both h and h_y, which will affect the learning on other parameters in θ_k via $0.5h^2 Tr[\cdot] + 0.5h_y^2 Tr[\cdot]$.

Both the above Z-regularization approaches can be decoupled from a BYY system and become directly applicable to the conventional maximum likelihood learning on a parametric model $p(x|\theta)$, featured by their implementable ways for controlling regularization strength. For *data smoothing*, the regularization strength h^2 (equivalently the noise variance) is estimated in an easy implementing way. For *normalization learning*, the regularization strength is controlled by the term Z, with a conscience de-learning behavior. Readers are also referred to [71] for a rationale for the priors by eq. (34) and eq. (35).

In addition to all the above discussed, regularization can also be implemented by appropriately combining the best match principle and the best harmony principle. Readers are referred to Sec. 23.4.2 in [45] for a summary of several methods under the name KL-λ-HL spectrum, and also to [43] for the relations and differences of the best harmony principle from not only the best match principle but also several existing typical learning theories. In addition, the $\ln(r)$ function in eq.(19) can also be extended to a convex function $f(r)$, and a detailed discussion can be found in Sec. 22.6.3 of [44].

3.6 Best Harmony, Best Match, Best Fitting: BYY Learning and Related Approaches

The differences and relations between the BYY learning and several typical approaches have been discussed in [50, 43]. Here we further elaborate this issue via more clear illustrations in Fig. 4 and Fig. 5.

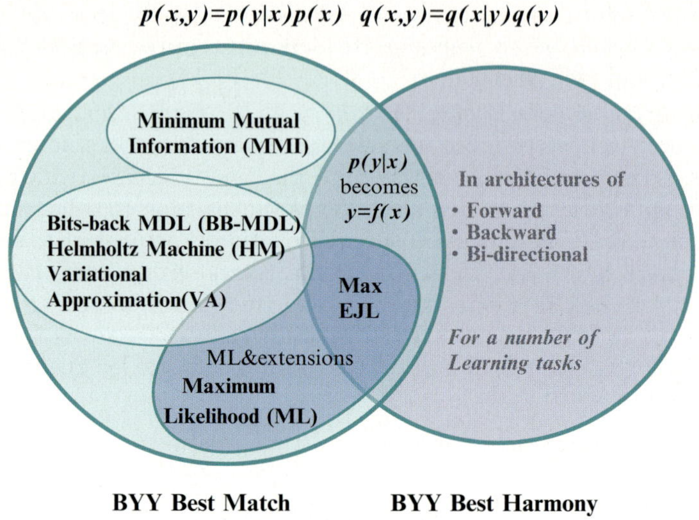

Fig. 4. BYY learning and related approaches (I)

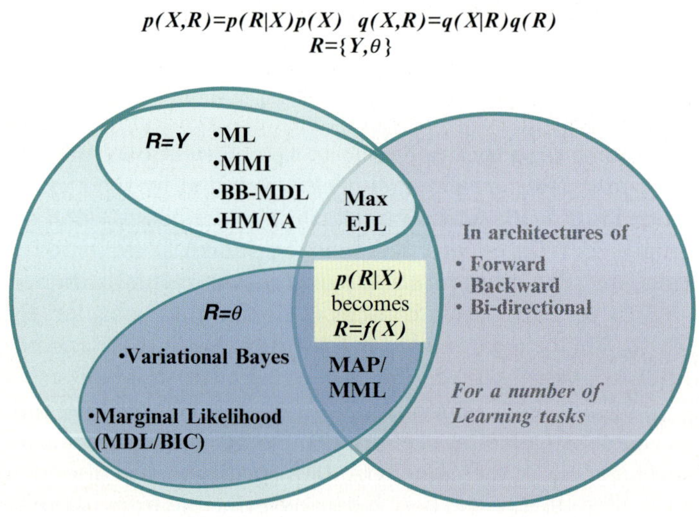

Fig. 5. BYY learning and related approaches (II)

As introduced in Sec. 3.1 and Sec. 3.2, learning in a BYY system can be implemented via either the best match principle by eq.(16) or the best harmony principle by eq.(18). The key difference of the best harmony principle by eq.(18) from the best match principle by eq.(16) is its model selection ability such that it can guide both parameter learning and model selection under a same principle, while the best match principle by eq.(16) can only guide parameter learning while model selection needs another different criterion.

This difference can be further understood in depth from a *Projection Geometry* Perspective (See Sec. III in [43]). The two principles correspond to two different types of *Geometry*, which become equivalent to each other only in a special case that $H(p\|p)$ is fixed at a constant H_0 that is irrelevant to both k and θ_k (see eqn(45) in [43]). For a BYY system, this case means that $p(\mathcal{Y}|\mathcal{X})$ in a format $p(\mathcal{Y}|\mathcal{X}) = \delta(\mathcal{Y} - f(\mathcal{X}))$ or $\mathcal{Y} = f(\mathcal{X})^{\dagger}$, as shown by the overlap part of two big circles in Fig. 4.

This common part has not been studied in the literature before it is studied under the best harmony principle in the BYY harmony learning on $H(p\|q, \theta_k)$ by eq.(18) with $Z = 1$. We say that this case has a backward architecture since only the backward direction $\mathcal{Y} \to \mathcal{X}$ has a specific structure while $p(\mathcal{Y}|\mathcal{X})$ is free from any structural constraint. In these cases, we get eq.(30) from eq.(19). Ignoring $-\ln Z$, it equivalently considers

$$\max_{\{\theta_k, k, \ \mathcal{Y}\}} \ln [q(\mathcal{X}_N|\mathcal{Y}, \theta_k^{xy}) q(\mathcal{Y}|\theta_k^{y})]. \tag{36}$$

That is, we implement a best fitting of samples in $\mathcal{X}_N$ by a joint distribution via maximizing the extreme of the joint likelihood of $\mathcal{X}, \mathcal{Y}$ with respect to the unknown inner representation $\mathcal{Y}$. So, we call eq.(36) the maximum extremal joint likelihood or shortly Max-EJL, which will be further discussed in Sec. 5.2, especially after eq.(62).

In the BYY Best Match domain shown in Fig. 4, the other part of learning on a BYY system includes the widely studied Maximum Likelihood (ML), Minimum Mutual Information (MMI), the bits-back based MDL [13], Helmholtz Machine and Variational Approximation [10, 12]. The detailed discussions are referred to Sec. II in [43]. There are also other cases that have not been studied previously, e.g., *data smoothing* and *normalization learning*, which have already been introduced in Sec. 3.5.

Furthermore, we consider the most general case with $\mathcal{Y}$ replaced by $\mathcal{R} = \{\mathcal{Y}, \theta_k\}$ such that the best harmony principle is turned from eq.(18) into eq.(25), by which we got eq.(26), eq.(27), and eq.(29), as well as those studied in Sec. II(B) of [43].

Similarly, the best match principle is turned from eq.(16) into

$$D(p\|q) = \int p(\mathcal{R}|\mathcal{X})p(\mathcal{X}) f(\frac{p(\mathcal{R}|\mathcal{X})p(\mathcal{X})}{q(\mathcal{X}|\mathcal{R})q(\mathcal{R})}) d\mathcal{X} d\mathcal{R}, \tag{37}$$

which shares with eq.(25) a common part that $p(\mathcal{R}|\mathcal{X})$ in a format $p(\mathcal{R}|\mathcal{X}) = \delta(\mathcal{R} - f(\mathcal{X}))$ or $\mathcal{R} = f(\mathcal{X})^{\dagger}$, as shown by the overlap part of two big circles in Fig. 5. This common part includes not only *Max EJL* at $R = \mathcal{Y}$ but also what

† Strictly speaking, $H(p\|p)$ is still relevant to k even when $p(\mathcal{Y}|\mathcal{X})$ or $P(R|x)$ becomes a δ density, since it can be regarded as a limit of a type $-(c + \ln h)k$ as h $\to 0$. In such a sense we should regard the above discussed common part in Fig. 4 and Fig. 5 is actually not common but only belongs to BYY Best Harmony domain, e.g., Max-EJL is a special case of BYY harmony learning only but not of BYY matching learning.

is usually called *MAximum Posterior (MAP)* at $R = \theta_k$, as well as closely relates to Minimum Message Length (MML) [36, 38].

In the *BYY best match* domain shown in Fig. 5, the other parts includes not only those listed in Fig. 4 at its special case $R = \mathcal{Y}$, but also the existing marginal likelihood related Bayesian approaches at $R = \theta_k$, such as MDL, BIC, and Variational Bayes.

4 Typical Examples

4.1 Gaussian Mixture

The first example that has been studied in [59] under the BYY harmony learning principle is the following Gaussian mixture, that is,

$$p(x|\theta_k) = \sum_{j=1}^{k} \alpha_j G(x|\mu_j, \Sigma_j). \tag{38}$$

In the literature, its parameters θ is learned via the maximum likelihood by the EM algorithm [25, 58], which has been widely studied in the literature of machine learning and neural networks. Readers are referred to Sec. 22.9.1(a) and Sec. 22.9.2(1) in [44] for a summary on a number of further results on the EM algorithm.

To determine k, we need one of typical model selection criteria such as AIC, BIC/MDL in help of a two-phase implementation. In [59], studies on the BYY learning for eq.(38) have been made in the common part shown in Fig. 4. On one hand, a criterion

$$J(k) = 0.5 \sum_{j=1}^{k} \alpha_j \ln|\Sigma_j| - \sum_{j=1}^{k} \alpha_j \ln \alpha_j, \tag{39}$$

and its special cases have been obtained [56]. On the other hand, an adaptive algorithm for implementing eq.(23) is firstly proposed in [59] under the name of the hard-cut EM algorithm for learning with automatic selection on k.

In eq.(38), x_t is a d-dimensional vector and y takes only discrete values $j = 1, \cdots, k$. When $p(\mathcal{X})$ is given by eq.(15), from eq.(18) we have

$$H(p\|q) = \sum_{t=1}^{N} \sum_{j=1}^{k} \int p(j|x) G(x|x_t, h^2 I) \rho_j(x|\theta_j) dx,$$
$$\rho_j(x|\theta_j) = \ln[\alpha_j G(x|\mu_j, \Sigma_j)]. \tag{40}$$

In its simplest case that $h = 0$ and $Z = 1$, from eq.(21) we can get $J(k)$ in eq.(39) after discarding a constant $0.5d \ln(2\pi)$ and ignoring a term

$$\frac{0.5}{N} \sum_{j=1}^{k} Tr[\sum_{t=1}^{N} p(j|x_t)(x_t - \mu_j)(x_t - \mu_j)^T \Sigma_j^{-1}] = 0.5 \frac{d(N-k)}{N}. \tag{41}$$

When N is small, we can no longer regard $0.5d(N-k)/N$ as constant but need to consider $-0.5dk/N$. For eq.(27) and eq.(28), we have $d_k = dk+k-1+\sum_{j=1}^{k} d_{\Sigma_j}$, where d_A denotes the number of free parameters in the matrix A. It further follows that

$$J_C(k) = 0.5\sum_{j=1}^{k}\alpha_j \ln|\Sigma_j| - \sum_{j=1}^{k}\alpha_j \ln\alpha_j - 0.5\frac{kd}{N},$$

$$J_R(k) = J_C(k) + c_k\frac{d_k}{N},\ d_k = dk+k-1+\sum_{j=1}^{k} d_{\Sigma_j}, \tag{42}$$

$$d_{\Sigma_j} = \begin{cases} 1, & \text{for } \Sigma_j = \sigma_j^2 I, \\ d, & \text{for } \Sigma_j \text{ is diagonal}, \\ 0.5d(d+1), & \text{for } \Sigma_j \text{ in general}, \end{cases}$$

where it follows from eq.(28) that

$$c_k = \begin{cases} 0.5, & \text{(a) corresponding to the case (a) in eq.(28)}, \\ 1, & \text{(b) corresponding to the case (b) in eq.(28)}. \end{cases} \tag{43}$$

Moreover, it follows from eq.(23) and eq.(40) that the updating formulae on $\alpha_j, \mu_j, \Sigma_j$ are same as their counterparts in the EM algorithm, while it follows from eq.(30) that

$$p(j|x_t) = \bar{\delta}_{jj_t^*},\ j_t^* = arg\max_j \rho_j(x_t|\theta_j),\ where\ \bar{\delta}_{ji} = \begin{cases} 1, & \text{if } j=i, \\ 0, & \text{otherwise}, \end{cases} \tag{44}$$

which is a hard-cut version of the posteriori probability of j upon x_t

$$p(j|x_t,\theta_j) = \frac{e^{-\rho_j(x_t|\theta_j)}}{\sum_{j=1}^{k} e^{-\rho_j(x_t|\theta_j)}}, \tag{45}$$

in the conventional EM algorithm. Thus, an algorithm that uses $p(j|x_t)$ in eq.(44) to replace the above one is named the hard-cut EM algorithm [59].

Generally, we consider to regularize learning on a small size N by

$$Z = \begin{cases} \frac{1}{N}\sum_{t=1}^{N}\sum_{\tau=1}^{N} G(x_t|x_\tau, h^2 I), & \text{(a) data smoothing with } h\neq 0, \\ \frac{1}{\sum_{t=1}^{N}\sum_{j=1}^{k} e^{\rho_j(x_t|\theta_j)}}, & \text{(b) normalization with } h=0. \end{cases} \tag{46}$$

Interestingly, as shown in [52, 49], it follows from the above case (b) that we can get an algorithm from eq.(23) to demonstrate a mechanism similar to RPCL learning previously introduced in Sec. 2.3.

The type of bi-directional regularization by eq.(33) can also be imposed. E.g., as suggested by Eqn.(40) in [52], it follows from eq.(7) that we also get an algorithm that demonstrates another RPCL-like mechanism [19]. Actually, different types of bi-directional regularization demonstrate different versions of RPCL-like mechanism [46]. Readers are referred to Sec. 23.7 in [45] for a historic remark on RPCL-like mechanisms versus the BYY harmony learning, and to Eqn.(28) in [47] for a unified procedure to implement RPCL and adaptive EM as well as the hard-cut EM algorithm.

4.2 Local Subspaces and Local Factor Analysis

We further consider Σ_j in the following decomposition

$$\Sigma_j = \sigma_j^2 I + \sum_{i=1}^{m_j} \lambda_j^{(i)\ 2} a_j^{(i)} a_j^{(i)\ T}, \quad a_j^{(i)\ T} a_j^{(i)} = 1, \quad a_j^{(i)\ T} a_j^{(\iota)} = 0, \ i \neq \iota, \tag{47}$$

where $\lambda_j^{(1)\ 2} \geq \lambda_j^{(2)\ 2} \cdots \geq \lambda_j^{(m_j)\ 2}$ with each $\lambda_j^{(i)\ 2}$ being the variance of the projection $a_j^{(i)\ T} x$ on the direction of the i-th principal vector $a_j^{(i)}$. This expression actually represents a subspace located at μ_j, shown in Fig. 6(a). Here, our task becomes finding several subspaces at different locations, which is thus called local subspace analysis. Readers are referred to Secs. 3.2 & 3.3 of [49] for not only other variants of elliptic RPCL but also RPCL based local subspaces.

When only the first principal component is considered, we can use this local PCA for collectively detecting multiple lines in an image, as shown in Fig. 6(b). We can also detect multiple planes in an image as shown in Fig. 6(c), as well as multiple curves and surfaces. Some applications are referred to [17, 18].

As illustrated in Fig. 6(a), the subspace obtained via the decomposition by eq.(47) is equivalent to orthogonally projecting each sample x onto a subspace that is located at μ and spanned by vectors a_1, a_2, a_3, such that the average square error $\|e\|^2$ between x and its projection $\hat{x}$ is minimized. It follows from $e = x - \hat{x}, \hat{x} = Ay + \mu$ that this subspace analysis is equivalent to the special case $\Sigma_j = \sigma_j^2 I$ of the following Factor Analysis (FA) [4]:

$$x = A_j y + \mu_j + e_j, \ e_j \sim G(e_j|0, \Sigma_j), \ y \sim G(y|0, I), \ E(e_j y^T) = 0, \tag{48}$$

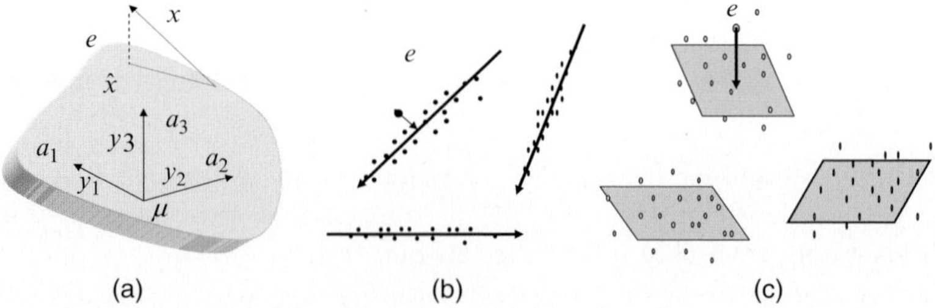

Fig. 6. Subspaces

where $u \sim p(u)$ means that u comes from the distribution $p(u)$. In a general case $\Sigma_j \neq \sigma_j^2 I$, the project $x \to \hat{x}$ is still linear but its direction is no longer orthogonal to the subspace. Also, the average square error $\|e\|^2$ is no longer minimized.

Since a rotation transform on $y \sim G(y|0,I)$ results in $y' \sim G(y|0,I)$ still, it has no difference to $p(x)$ by eq.(48) whether A_j is a general matrix or subject to the following constraint

$$A_j = U_j \Lambda_j,\ U_j U_j^T = I,\ \Lambda_j^2 = diag[\lambda_j^{(1)\ 2}, \cdots, \lambda_j^{(m_j)\ 2}] \tag{49}$$

When $p(\mathcal{X})$ is given by eq.(15), putting eq.(48) into eq.(18) we have

$$\begin{aligned} H(p\|q) &= \sum_{t=1}^{N}\sum_{j=1}^{k} \int p(y|x,j)p(j|x)G(x|x_t,h^2 I)\rho_j(x,y|\theta_j)dxdy - \ln Z, \\ \rho_j(x,y|\theta_j) &= \ln\left[\alpha_j G(x|U_j\Lambda_j y + \mu_j, \Sigma_j)G(y|0,I)\right]. \end{aligned} \tag{50}$$

Similar to eq.(44), it follows from eq.(30) that

$$\begin{aligned} &p(y|x_t,j) = \delta(y - \hat{y}_j(x_t)),\ \hat{y}_j(x_t) = arg\max_y \rho_j(x_t,y|\theta_j), \\ &\hat{y}_j(x_t) = W_j(x_t - \mu_j),\ W_j = \Lambda_j U_j^T (U_j \Lambda_j^2 U_j^T + \Sigma_j)^{-1}. \end{aligned} \tag{51}$$

Also, we can get $p(j|x_t)$ by inserting $\rho_j(x_t|\theta_j) = \rho_j(x_t, \hat{y}_j(x_t)|\theta_j)$ into eq.(44). Again, $p(y|x_t,j)$ in eq.(51) can be regarded as a hard-cut version of the following posteriori probability of y upon getting x_t and j

$$p(y|x_t,j) = G(y|\hat{y}_j(x_t), I - \Pi_j),\ \Pi_j = W_j(U_j\Lambda_j^2 U_j^T + \Sigma_j)W_j^T. \tag{52}$$

In a way similar to eq.(39) and eq.(41), it follows from eq.(21) in the case $h = 0$ and $Z = 1$ that

$$J(k, \{m_j\}_{j=1}^k) + c_{N,k} = 0.5\sum_{j=1}^{k} \alpha_j\{\ln|\Sigma_j| + J_j^y + m_j \ln(2\pi)\} - \sum_{j=1}^{k} \alpha_j \ln \alpha_j,$$

$$J_j^y = \begin{cases} Tr[I - \Pi_j], & \text{(a) for } p(y|x_t,j) = G(y|\hat{y}_j(x_t), I - \Pi_j), \\ Tr[\Gamma_j], & \text{(b) for } p(y|x_t,j) = \delta(y - \hat{y}_j(x_t)), \end{cases}$$

$$\begin{aligned} &\text{where}\quad c_{N,k} = \frac{0.5}{N}(kd + \sum_{j=1}^{k} m_j), \\ &\Gamma_j = \frac{1}{\alpha_j N - 1}\sum_{t=1}^{N} p(j|x_t)\hat{y}_j(x_t)\hat{y}_j^T(x_t) = W_j S_j W_j^T, \\ &S_j = \frac{1}{\alpha_j N - 1}\sum_{t=1}^{N} p(j|x_t)(x_t - \mu_j)(x_t - \mu_j)^T. \end{aligned} \tag{53}$$

Specifically, the case (a) of J_j^y comes from $\int yy^T p(y|x_t,j)dy$, and the case (b) of J_j^y comes from $\sum_{j=1}^k \sum_{t=1}^N p(j|x_t)\hat{y}_j(x_t)\hat{y}_j^T(x_t)$, while $c_{N,k}$ comes in a way similar to eq.(41).

Moreover, similar to eq.(42) we have

$$J_R(k, \{m_j\}_{j=1}^k) = J(k, \{m_j\}_{j=1}^k) + c_k \frac{d_k}{N},$$
$$d_k = dk + k - 1 + \sum_{j=1}^k (m_j + d_{U_j} + d_{\Sigma_j}). \tag{54}$$

where c_k is same as in eq.(43), and $d_{U_j} = dm_j - 0.5m_j(m_j+1)$.

Next, it follows from eq.(48) that the distribution $p(x)$ still remains unchanged when

$$A_j = U_j, \; y \sim G(y|0, \Lambda_j^2) \tag{55}$$

where the components of y remain uncorrelated. Correspondingly, eq.(50), eq.(51), eq.(52), and eq.(53) are modified with the following replacements

$$\rho_j(x, y|\theta_j) = \ln[\alpha_j G(x|U_j y + \mu_j, \Sigma_j) G(y|0, \Lambda_j^2)], \tag{56}$$
$$W_j = \Lambda_j^2 U_j^T (U_j \Lambda_j^2 U_j^T + \Sigma_j)^{-1},$$
$$p(y|x_t, j) = G(y|\hat{y}_j(x_t), \Lambda_j^2 - \Pi_j),$$
$$J_j^y = \begin{cases} \ln|\Lambda_j^2 - \Pi_j| + m_j, & \text{(a) for } p(y|x_t, j) = G(y|\hat{y}_j(x_t), \Lambda_j^2 - \Pi_j), \\ \ln|\Gamma_j| + m_j, & \text{(b) for } p(y|x_t, j) = \delta(y - \hat{y}_j(x_t)). \end{cases}$$

Still, we can get $p(j|x_t)$ by inserting the above $\rho_j(x_t|\theta_j) = \rho_j(x_t, \hat{y}_j(x_t)|\theta_j)$ into eq.(44), and also get $J_R(k, \{m_j\}_{j=1}^k)$ by eq.(54).

There are two specifical cases that deserve a particular mention. One is the special case $\Sigma_j = \sigma_j^2 I$ of eq.(48), which becomes local subspace analysis. The other is the special case $k = 1$, which becomes factor analysis. In the latter case, $J(k, \{m_j\}_{j=1}^k)$ and $J_R(k, \{m_j\}_{j=1}^k)$ are degenerated into $J(m_1)$ and $J_R(m_1)$ for determining the number of factors. Moreover, if we jointly have $k = 1$ and $\Sigma_1 = \sigma_1^2 I$, the problem becomes equivalent to *Principal Component Analysis (PCA)*, $J(m_1)$ and $J_R(m_1)$ can be used for determining the subspace dimension.

4.3 A Unified Learning Algorithm

On one hand, learning on the local factor analysis model by eq.(48) can be implemented in a two-phase implementation. That is, the first phase considers a set of possible candidate models by enumerating $k, \{m_j\}_{j=1}^k$ and then learns parameters in every candidate model in help of the EM algorithm under the maximum likelihood principle. The second phase selects a best candidate $k^*, \{m_j^*\}_{j=1}^{k^*}$ by either $J(k, \{m_j\}_{j=1}^k)$ or $J_R(k, \{m_j\}_{j=1}^k)$ given in the previous subsection. However, not only the computing cost will be impractically huge, especially for selecting $\{m_j\}_{j=1}^k$, but also this criterion type of multiple discrete variables becomes unable to provide a correct minimum point due to a finite sample size.

On the other hand, we can implement learning by eq.(23) in the cases of eq.(49) or eq.(55), during which $k^*, \{m_j^*\}_{j=1}^{k^*}$ is able to be automatically determined. For eq.(49), learning is made via eq.(50) during which m_j is determined via minimizing $-\ln G(y|0, I) = c + 0.5\|y\|^2$ that pushes $y^{(j)\ 2}$ of an extra dimension to 0. For eq.(55), learning is made via eq.(50) modified with eq.(56), during which m_j is determined via minimizing $-\ln G(y|0, \Lambda_j^2)$ that directly pushes each extra $\lambda_j^{(r)\ 2}$ to 0. From eq.(23), eq.(50), and eq.(56), we can obtain the detailed implementing algorithms. One example is the one given by Eqn.(72) plus Table 2 in [42] with $B_j = 0, \forall j$.

A general adaptive learning procedure is introduced in the sequel, which includes not only the one for implementing BYY harmony learning by eq.(23) in the form of eq.(50) and eq.(56) with $Z = 1$ and $h = 0$, but also an adaptive EM algorithm for the maximum likelihood learning, as well as a RPCL learning algorithm via the following rival penalized competition:

$$p_{j,t} = \begin{cases} 1, & \text{if } j = c, \quad c = arg\max_j \rho_j(x_t|\theta_j), \\ -\gamma, & \text{if } j = r, \quad r = arg\max_{j\neq c} \rho_j(x_t|\theta_j), \\ 0, & \text{otherwise}, \end{cases} \tag{57}$$

where $\rho_j(x_t|\theta_j)$ is either the one in eq.(40) or $\rho_j(x_t|\theta_j) = \rho_j(x_t, \hat{y}_j(x_t)|\theta_j)$ by eq.(50) or eq.(56).

The general procedure consists of iterating the following two steps:

Yang step: take a sample x_t, get $y_j(x_t) = W_j(x_t - \mu_j)$ with W_j by eq.(51) for eq.(49) or by eq.(56) for eq.(55). Then, with $\rho_j(x_t|\theta_j) = \rho_j(x_t, \hat{y}_j(x_t)|\theta_j)$ by eq.(50) or eq.(56), further get $p_{j,t}$ as follows

$$p_{j,t} = \begin{cases} p(j|x_t) \text{ by eq.(45)}, & \text{(a) ML} - \text{Learning}, \\ p(j|x_t, \theta_j) \text{ by eq.(44)}, & \text{(b) BYY harmony}, \\ \text{by eq.(57)}, & \text{(c) RPCL} - \text{Learning}, \\ p(j|x_t, \theta_j) - \gamma q(j|x_t), & \text{(d) RPCL} - \text{BYY harmony}, \end{cases} \tag{58}$$

where $\gamma > 0$ is a small number, and $q(j|x_t) \geq 0, \sum_{j=1}^k q(j|x_t) = 1$ are either pre-specified or estimated by some means, e.g., via a normalizing term Z.

Ying step: adaptively update all the parameters, check and discard extra dimensions. The details consist of

$$(a)\ \hat{x}_{j,t} = U_j^{old} y_{j,t} + \mu_j^{old},\ e_{j,t} = x_t - \hat{x}_{j,t},\ \mu_j^{new} = \mu_j^{old} + \eta p_{j,t} e_{j,t},$$
$$g_{U_j} = y_t e_{j,t}^T \Sigma_j^{old\ -1},\ U_j^{new} = U_j^{old} + \eta(g_{U_j}^T - U_j^{old} g_{U_j} U_j^{old}),$$

$$(b)\ \lambda_j^{(i)\ new} = \lambda_j^{(i)\ old} + \eta p_{j,t} \frac{y_{j,t}^{(i)\ 2} - (\lambda_j^{(i)\ old})^2}{\lambda_j^{(i)\ old}},$$

if $\lambda_j^{(i)} \to 0$ is detected, remove the i-th coordinate in the j-th subspace, $m_j \leftarrow m_j - 1$;

$$(c)\ \alpha_j = \frac{\beta_j^{2\ new}}{\sum_{j=1}^k \beta_j^{2\ new}},\ \beta_j^{new} = \beta_j^{old} + \eta \frac{p_{j,t} - \alpha_j^{old} \sum_{j=1}^k p_{j,t}}{\beta_j^{old}}, \tag{59}$$

if $\alpha_j \to 0$ is detected, discard the j-th subspace, $k \leftarrow k - 1$.

$$(d)\ \Sigma_j = S_j S_j^T,\ S_j^{new} = S_j^{old} + \eta p_{j,t} G_{\Sigma_j}^{old} S_j^{old},$$

$$G_{\Sigma_j} = \Sigma_j^{-1} e_{j,t} e_{j,t}^T \Sigma_j^{-1} - \Sigma_j^{-1}.$$

$$For\ \Sigma_j = \sigma_j^2 I,\ simply\ \sigma_j^{new} = \sigma_j^{old} + \eta p_{j,t} \frac{\|e_{j,t}\|^2/d - \sigma_j^{old\ 2}}{\sigma_j^{old}}.$$

The updating on U_j^{new} guarantees the satisfaction of $U_j U_j^T = I$. Moreover, even when $p_{j,t} < 0$, the updating rules (d)&(c) guarantee the satisfaction of $\alpha_j \geq 0, \sum_{j=1}^k \alpha_j = 1$ and that Σ_j remains non-negative definite.

4.4 Other Examples

The above general procedure degenerates back to the unified learning procedure by Eqn.(28) in [47] for Gaussian mixture by eq.(38) at $U_j = 0, \Lambda_j = I$. The above procedure also directly applies to the following two special cases:

- *Local subspace analysis* at the special case $\Sigma_j = \sigma_j^2 I$, which actually provides a unified scheme as well as improvements on the previous studies under the name of local PCA, local subspaces, and multi-sets mixture learning (MML) [62, 60]. Also, it can be applied to detecting lines, planes, curves, and surfaces in pattern recognition tasks [17, 18, 16].
- *Factor analysis* at the special case $k = 1$ that also includes principal components analysis (PCA). Readers are referred to Secs.2.2-2.4 in [48] and Sec. IV in [41] for recent summaries.

Advances on the BYY harmony learning have also been made along the following directions:

- *Independence subspace analysis* Extensions have been made from the specific case $G(y|0, \Lambda_j^2)$ to a general case $\prod_{j=1}^m q(y^{(j)})$, including independence components dependence (ICA), binary factor analysis (BFA), nonGaussian factor analysis (NFA), and LMSER, as well as three layer net. Readers are referred to Secs.4 & 5 in [48] and Sec. IV in [41].
- *Independence state space analysis* Extensions have further been made from $G(y|0, \Lambda_j^2)$ and $\prod_{j=1}^m q(y^{(j)})$ to temporal state spaces by taking temporal relations in consideration, including temporal factor analysis (TFA), independent hidden Markov model (HMM), temporal LMSER, and variants. Readers are referred to [42] and Sec. 6 in [48].
- *Mixture of shape-structures* In the computer vision field, finding multiple shapes is an important task called object detection. In [69], a new approach was proposed under the name of randomized Hough transform (RHT)

[65]. In [62, 60], a multi-set modelling method has been proposed, under situations of strong noise, partially observable objects, and a large amount of objects. In [40], a unified problem solving paradigm has been developed.

- *Combination of multiple inference* In [68], an early systematic study has been made on multiple classifier combination. In [63], a number of results have been obtained on statistical consistency and convergence rates for RBF nets. An alternative model of mixture of experts has been proposed and easily implemented by the EM algorithm [61], which is further applied to replace the existing suboptimal two stage algorithm for RBF nets [54]. Also, the number of basis functions are determined via either RPCL or BYY harmony learning. Readers are referred to Sec. 22.9.1(d) in [44].

5 A Trend and Challenges

5.1 A Trend for Model Selection

Summarizing the discussions on model selection in Sec. 2.2, Sec. 2.3, and Sec. 3, we roughly have two categories of studies. One is local cost based, usually for a learning task on a model that consists of several individual units or components. There is a local cost measure for each individual, e.g., $-\ln[\alpha_j G(x|\mu_j, \Sigma_j)]$ can be such a local cost for the j-th component in the Gaussian mixture by eq.(38). A sample x is excluded from one individual if its corresponding local cost is higher than a pre-specified threshold. If this x is excluded by all the current individual components, a new component is created to accommodate this x. As a result, a number of components are allocated to a set of samples. However, it is difficult to appropriately assign such a pre-specified threshold.

The other category is a global cost based, which is applicable to any model selection tasks. That is, after all its unknown parameters have been learned, a model with a scale k is globally evaluated by a cost $J(k)$ that is computed based on the learned parameters. Studies of this category can be further classified according to the configuration of $J(k)$.

When we have a large sample size N, as discussed in Sec. 1, a negative likelihood and a best fitting error, as well as a best matching error, will vary in a way illustrated by the top part of Fig. 7(a). That is, a correct scale k^* can be determined at the smallest k that either makes $J(k)$ reach its minimum or equivalently makes $\Delta J(k) = J(k+1) - J(k) = 0$. However, as the sample size N drops below a limit, this negative likelihood type $J(k)$ will degenerate into those shown by the top cases of Fig. 7(b) and Fig. 7(c). As a result, a correct scale k^* can not be obtained via searching a minimum or detecting a zero. One heuristic remedy is to check whether $\Delta J(k)$ is smaller than a pre-specified threshold. Alternatively, the conventional learning theories aim at modifying the top cases of Fig. 7(b) and Fig. 7(c) into the bottom cases

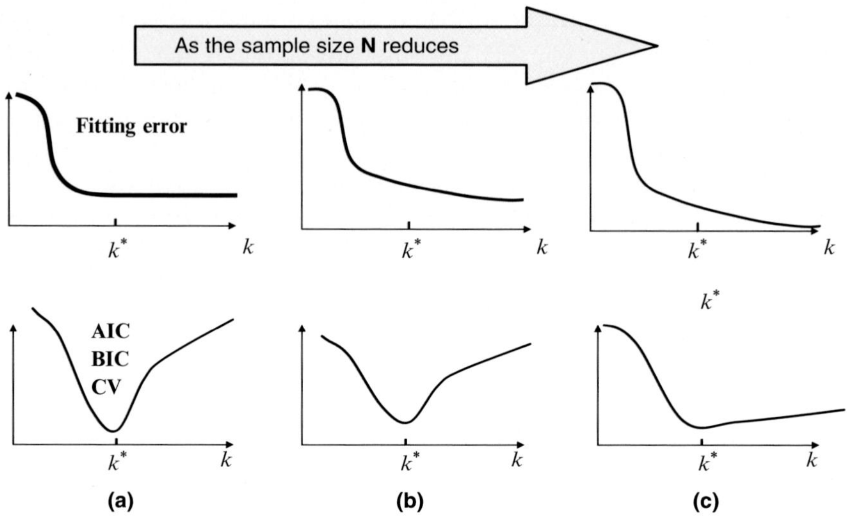

Fig. 7. Best matching error vs model selection criteria as the sample size N reduces

of Fig. 7(b) and Fig. 7(c) such that k^* can still be obtained via searching a minimum.

At each k, a cost $J(k)$ is computed once all the unknown parameters in the corresponding model have been estimated. Except for certain special task (e.g., determining the dimension k of orthogonal subspace), the estimated unknown parameters at one value of k usually can not be carried over to another value of k. More specifically, neither the estimated parameters at a lower value k' can be directly used as a part of the parameters at a higher value k'', nor the estimated parameters at a higher value k'' can be directly adopted for a use at a lower value k'. Therefore, all the unknown parameters have to be estimated completely at each different value k. That is, the model selection has to be implemented expensively.

On one hand, the BYY harmony learning provides us new criteria $J(k)$ by eq.(21), eq.(22), and eq.(27). Illustrated in the middle of Fig. 8 are those $J(k)$ curves by eq.(21) at different sample sizes of N, which are usually same or slightly worse than the popular model selection criterion BIC or equivalently MDL. The curve $J(k)$ by eq.(21) or eq.(22) are used as a bridge to illustrate the equivalent performance of automatic model selection by eq.(23) in a comparison with those existing conventional criteria, instead of being actually used in a two phase implementation for model selection. In a two phase implementation, it is suggested to use $J(k)$ by eq.(27), as illustrated in the bottom of Fig. 8, which are usually better than several typical criteria such as AIC, CAIC, BIC/MDL, CV, etc.

On the other hand, more important is that the BYY harmony learning provides a new trend that integrates model selection and parameter learning into one single process with a considerably reduced computing cost, that is,

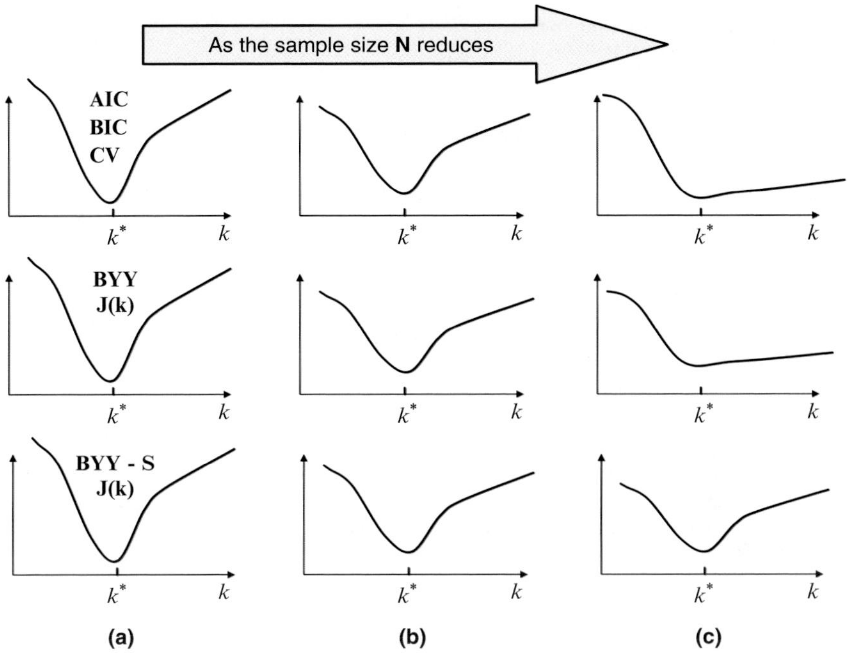

Fig. 8. BYY harmony learning criteria vs model selection criteria as the sample size N reduces

a trend of seeking automatic model selection during parameter learning. Further efforts are deserved along this trend. Generally speaking, in additional to the BYY harmony learning, an approach that follows this trend should be either explicitly or implicitly featured by a cost measure that varies with both the scale integer k and the parameters θ_k, in a format $f(\theta_k, k)$ or shortly $f(\theta_k)$ with the following nature:

$$f(\theta_k) \begin{cases} = f^*, & \text{when } k \geq k^*, \theta_{k^*} = \theta^*_{k^*} \text{ and } \phi_k = 0, \\ > f^*, & \text{otherwise}, \end{cases} \tag{60}$$

where $f^* = f(\theta^*_{k^*})$ is reached at the correct k^* and the correct value $\theta^*_{k^*}$. Moreover, $\theta_k = \{\theta_{k^*}, \theta^r_{k-k^*}\}$ with $\theta^r_{k-k^*}$ denoting the remaining part of θ_k after removing the subset θ_{k^*}, and ϕ_k is a critical subset $\phi_k \subseteq \theta^r_{k-k^*}$. For an example, we have $\phi_k = \{\alpha_j\}^k_{j=k^*+1}$ in eq.(38). When $\phi_k = 0$, a Gaussian mixture with k components actually becomes one with only k^* components.

Given a $k \geq k^*$ initially, minimizing $f(\theta_k)$ with respect to θ_k will force $\theta_{k^*} = \theta^*_{k^*}$ and $\phi_k = 0$, such that a model with a higher scale k actually becomes one with the correct k^* effectively. That is, model selection is made automatically during parameter learning. Moreover, it follows from eq.(60) that $J(k) = \min_{\theta_k} f(\theta_k)$ will illustrate as shown by Fig. 3(b) and that $J(k) = \min_{\theta_k,\ s.t.\ \phi'_k = c} f(\theta_k)$ will illustrate as shown by Fig. 3(a), where c is certain constant and $\phi'_k \supset \phi_k$ is a subset in θ_k, e.g., we have $c = 1/k$ and $\phi_k = \{\alpha_j\}_{j=1}^k$ in eq.(38).

5.2 Theoretical Issues in a Large Sample Size

Corresponding to the studies on asymptotic natures (i.e., the behaviors as the sample size $N \to \infty$) of the maximum likelihood approach or a best fitting type approach (e.g., the Kullback divergence based one by eq.(16)), it is also an interesting problem to study asymptotic natures of the BYY harmony learning. Such studies can be made in two stages.

First, we consider maximizing $H(p\|q, \theta_k)$ in eq.(18). Considering $f(r) = \ln r$ and that samples of x_t are i.i.d., we have $Z \to 1$ as $N \to \infty$ and thus eq.(18) becomes equivalent to the following form

$$H_o(p\|q, \theta_k) = \int p(y|x)p_o(x) \ln [q(x|y, \theta_k^{x|y})q(y, \theta_k^y)]\mu(dx)\mu(dy), \qquad (61)$$

where $p_o(x)$ is the original density that samples of x come from, k is one or a set of integers that represent the scale of y, and $\theta_k = \{\theta_k^{x|y}, \theta_k^y\}$ consists of the unknown parameters sets in the distribution functions $q(x|y, \theta_k^{x|y})$ and $q(y, \theta_k^y)$ respectively, with their structures pre-designed.

When $p(y|x)$ is free of any constraint, $\max_{p(y|x)} H_o(p\|q, \theta_k)$ results in

$$p(y|x) = \delta(y - y(x, \theta_k)),\ y(x, \theta_k) = arg\max_y [q(x|y, \theta_k^{x|y})q(y, \theta_k^y)], \qquad (62)$$
$$H_o(p\|q, \theta_k) = \int p_o(x) \ln Q(x, \theta_k)\mu(dx),\ Q(x, \theta_k) = q(x|y(x, \theta_k))q(y(x, \theta_k)),$$

which was previously discussed after eq.(36), as well as in Fig. 4 and Fig. 5, under the name Max-EJL.

In a comparison of the corresponding maximum likelihood counterpart, i.e.,

$$L_o(\theta_k) = \int p_o(x) \ln q(x, \theta_k)\mu(dx),$$
$$q(x, \theta_k) = \int q(x|y, \theta_k^{x|y})q(y, \theta_k^y)\mu(dy), \qquad (63)$$

we can observe that the above marginal integral $q(x, \theta_k)$ (i.e., a projected sum of $q(x|y, \theta_k^{x|y})q(y, \theta_k^y)$ to the domain of x) is replaced by $Q(x, \theta_k)$ in eq.(62) that is the peak point of $q(x|y, \theta_k^{x|y})q(y, \theta_k^y)$ in the domain of y per each fixed x. Illustratively, we can regard $q(x|y, \theta_k^{x|y})q(y, \theta_k^y)$ as a mountain in a $x - y$ coordinate system, $q(x, \theta_k)$ lumps the total sum of all the masses along the y-axis perpendicularly to the x-axis, while $Q(x, \theta_k)$ only places the mass on the highest ridge of the mountain perpendicularly to the x-axis. Noticing that the mountain is constrained to have a unit total mass, i.e., $\int q(x|y, \theta_k^{x|y})q(y, \theta_k^y)\mu(dx)\mu(dy) = 1$, maximizing $H_o(p\|q, \theta_k)$ will force the mountain shrink to concentrate swiftly to its highest ridge, while maximizing L_o can be achieved by those mountains with a unit mass that stretches along the y-axis in infinite many ways. This provides another perspective that explains why the BYY harmony learning has a model selection ability while the maximum likelihood learning has not.

Moreover, it follows from eq.(62) that

$$\begin{aligned} &H_o(p\|q,\theta_k) = \int p_o(x) \ln \tilde{Q}(x,\theta_k)\mu(dx) + C(\theta_k), \\ &\tilde{Q}(x,\theta_k) = Q(x,\theta_k)/C(\theta_k),\ C(\theta_k) = \int Q(x,\theta_k)\mu(dx), \end{aligned} \tag{64}$$

Maximizing $H_o(p\|q,\theta_k)$ not only forces the configuration of $q(x|y,\theta_k^{x|y})q(y,\theta_k^y)$ to shrink into its highest ridge for a largest $C(\theta_k)$, but also forces $\tilde{Q}(x,\theta_k)$ to match $p_o(x)$ as close as possible.

Comparing eq.(64) with eq.(63), we observe that the asymptotic nature of the BYY harmony learning can be investigated in two typical situations. First, when $C(\theta_k)$ is only relevant to k but irrelevant to θ_k. The asymptotic nature of the BYY harmony learning is similar to the asymptotic nature of the maximum likelihood learning. That is, the key point is to investigate the discrepancy between $p_o(x)$ and $Q(x,\theta_k)$ versus the discrepancy between $p_o(x)$ and $q(x,\theta_k)$. We consider the notations:

$$\begin{aligned} &\mathcal{P}_q(k) = \{q(x,\theta_k):\ for\ all\ \theta_k \in \Xi_k\}, \\ &\mathcal{P}_Q(k) = \{Q(x,\theta_k):\ for\ all\ \theta_k \in \Xi_k\}, \end{aligned} \tag{65}$$

which denote the distribution families that can be represented by $q(x,\theta_k)$ and $Q(x,\theta_k)$, respectively, where Ξ_k is the domain that θ_k takes values.

One typical asymptotic nature is the so called statistical consistency. For the maximum likelihood learning, having statistical consistency means that $q(x,\hat{\theta}_k) \to p_o(x)$ for a maximum likelihood estimator $\hat{\theta}_k$ as $N \to \infty$, or equivalently $p_o(x) \in \mathcal{P}_q(k)$, which is always possible when k becomes large enough. For the BYY harmony learning, when $C(\theta_k)$ is only relevant to k but irrelevant to θ_k, statistical consistency means $p_o(x) \in \mathcal{P}_Q(k)$, which is also possible when k become large enough.

It is interesting to further study the cases where a statistical consistency is not satisfied. Such cases are encountered either when k is not large enough or when $C(\theta_k)$ is relevant to θ_k. The following are several theoretical issues to be explored:

(a) When $C(\theta_k)$ is only relevant to k but irrelevant to θ_k, it deserves to investigate how the bias between $p_o(x)$ and $Q(x,\theta_k)$ and the bias between $p_o(x)$ and $q(x,\theta_k)$ vary as k in the cases with $N \to \infty$.
(b) With the unknown original density $p_o(x)$ replaced by the empirical density by eq.(8) in the above case (a), it deserves to further investigate how the bias between $p_o(x)$ and $q(x,\hat{\theta}_k)$ and the bias between $p_o(x)$ and $Q(x,\hat{\theta}_k)$ vary as k and N vary, where $\hat{\theta}_k$ is obtained by the maximum likelihood learning for $q(x,\hat{\theta}_k)$, and by the BYY harmony learning via eq.(19) or eq.(23) for $Q(x,\hat{\theta}_k)$.

(c) When $C(\theta_k)$ is relevant to θ_k, it follows from eq.(64) that the BYY harmony learning is somewhat similar to a Bayesian learning, with $C(\theta_k)$ taking a role similar to a priori, which usually introduces certain bias. It is interesting to investigate how the bias between $p_o(x)$ and $Q(x, \hat{\theta}_k)$ changes as k (or as both k and N, when $p_o(x)$ is replaced by eq.(8)).

(d) In the above cases, there are no regularization taking its role. As introduced in Sec. 3.5, several ways can be used for imposing certain regularization, and it deserves to study how the asymptotic nature of the BYY harmony learning is affected by regularization. Particularly, it deserves to investigate how the bias between $p_o(x)$ and $Q(x, \hat{\theta}_k)$ changes as k, N, and h, with $p_o(x)$ replaced by eq.(15). It also deserves to study how the bias between $p_o(x)$ and $Q(x, \hat{\theta}_k)$ changes as k and N, when $p(y|x)$ is free but in a structure as discussed in Sec. 3.5 (e.g., given by eq.(31) and eq.(32)).

(e) As discussed in [43], the maximum likelihood learning and the BYY harmony learning can be interpreted from two different views of geometry. It is interesting to further investigate the nature of manifold of $H(p\|q, \theta_k)$, as well as its relations to k, N, and h.

5.3 Challenges in a Small Sample Size

In the cases of a small sample size, we encounter more challenges on a number of theoretic and algorithmic aspects for not only the BYY harmony learning but also other model selection approaches. In the following, we discuss a number of typical challenges:

(a) One is to estimate $H(p\|q) = \int p(\theta_k|\mathcal{X})H(p\|q, \theta_k)d\theta_k$ by eq.(26) more accurately. It is approximated by $J(\mathbf{k}) = -H(p\|q, \theta_k^*) + 0.5d(\theta_k^*)$ in eq.(27) via considering a noninformative priori $q(\theta_k) = 1$ and a rough estimator $\theta_k^* = T(\bar{\mathcal{X}}_N)$ in help of the celebrated Cramer-Rao inequality. Interestingly, this $J(\mathbf{k})$ with $d(\theta_k^*)$ in the case (a) of eq.(28) can also be reached in help of using the idea of eq.(9) to estimate the bias

$$b(k, N) = E_{\mathcal{X}_N}\{E_{\mathcal{X}_N}[H(p\|q, \theta_k)]_{\theta_k = \theta_k^*}\} - E_{\mathcal{X}_N} H(p\|q, \theta_k^*). \qquad (66)$$

Considering the case $H(p\|q, \theta_k) = H_{\bar{\mathcal{X}}_N, \bar{\mathcal{Y}}_N}(p\|q, \theta_k)$ given by eq.(30), i.e., $H(p\|q, \theta_k) = \ln [q(\bar{\mathcal{X}}_N|\bar{\mathcal{Y}}_N, \theta_k^{xy})q(\bar{\mathcal{Y}}_N|\theta_k^y)]$ that can be approximately regarded as the likelihood function jointly on $\bar{\mathcal{X}}_N, \bar{\mathcal{Y}}_N$ if we ignore the dependence of $\bar{\mathcal{Y}}_N$ on θ_k, we can directly get $b(k, N) = 0.5d_k/N$ from AIC [1, 2, 3]. That is, this road also leads to $J(\mathbf{k})$ by eq.(27) with $d(\theta_k^*)$ in the case (a) of eq.(28). Intuitively, the performance changing trend from the case (a) to the case (b) in eq.(28) will be somewhat similar to the changing trend from AIC to BIC [29, 15, 23]. A further improvement may be obtained via mathematical analysis on one $d(\theta_k^*)$ somewhere between the case (a) and the case (b) in eq.(28). It may also deserve to consider $q(\theta_k)$ in a priori distribution instead of simply setting $q(\theta_k) = 1$.

(b) In addition to making empirical comparisons with those typical model selection criteria discussed in Sec. 2.2, it remains to be challenges to make mathematical analysis on the chances and the magnitudes that k^* deviates from the correct one of the underlying distribution, with k^* given by eq.(21) and eq.(27) versus by those typical model selection criteria. Moreover, the studies on k^* by eq.(21) illustrate the performances of automatic model selection during parameter learning, while the studies on k^* by eq.(27) illustrate the performances in a two phase implementation. The performance gain by eq.(27) should also been evaluated together with its computation cost in a quantative way.

(c) More importantly, challenges lay on building up a mathematical link from the BYY harmony measure by eq.(27) or eq.(23) to the generalization error by eq.(4) either in a general sense or specifically for different learning tasks with different structures [41]. In other words, how to mathematically analyzes the performances of BYY harmony learning in the term of the generalization error with respect to the sample size N. A possible direction is to rewrite the BYY harmony measure in a format $\int p(\mathcal{X})R(\mathcal{X},k)\mu(d\mathcal{X})$, and then investigate it in an analogy of those studies on eq.(4). Similar challenges apply to those typical model selection criteria in Sec. 2.2 too.

(d) As introduced in Sec. 3.5, the BYY harmony learning integrates the roles of regularization and model selection. It is interesting to examine this feature in term of generalization error, i.e., how the generalization error is affected by regularization, especially how it varies as k, N, and h with $p_o(x)$ replaced by eq.(15), how it relates to $p(y|x)$ in a structure given by eq.(31) and eq.(32), as well as how accurate the scale k^* obtained by automatic model selection is, with respect to the sample size N and k^*.

(e) As discussed after eq.(30), alternatively making $\max_{\theta_k} H_{\bar{\mathcal{X}}_N,\bar{\mathcal{Y}}_N}(p\|q,\theta_k)$ and $\bar{y}_t = arg\max_{y_t}[q(\bar{x}_t|y_t,\theta_k^{xy})q(y_t|Y_{t-1},\theta_k^y)]$ lead to the problem of local maximum or saddle point. It needs further investigation on how this problem affects the accuracy of the obtained scale k^* and the generalization error with its relation to k, N, and h, via either an implementation with automatic model selection or a two phase implementation.

(f) Also discussed after eq.(30), if we have an explicit expression for $\bar{y}_t = arg\max_{y_t}[q(\bar{x}_t|y_t,\theta_k^{xy})q(y_t|Y_{t-1},\theta_k^y)]$, we can use it to improve the problem of local maximum. It deserves to study such an improvement in term of the accuracy of k^* and the generalization error. As introduced in Sec. 3.5, in the cases without such explicit expressions, one way to remedy is to approximate the desired explicit expression by a parametric structure, e.g., by eq.(31) and eq.(32). In this way, the relation of $\bar{\mathcal{Y}}_N$ or of $\bar{y}_t$ to θ_k can be approximately taken in consideration during updating θ_k. On the other hand, this pre-designed parametric structure may constrain $\max_{p(\mathcal{Y}|\mathcal{X})} H_{\mathcal{X}}(p\|q,\theta_k)$ to reach its optimal solution by eq.(30). Thus, it needs to examine the two-fold role of imposing such a pre-designed parametric structure.

(g) The difficulty of getting the above discussed explicit expression also brings an implementation difficulty. I.e., $\bar{\mathcal{Y}}_N = arg\max_{\mathcal{Y}}[q(\mathcal{X}_N|\mathcal{Y},\theta_k^{xy})q(\mathcal{Y}|\theta_k^y)]$ or $\bar{y}_t = arg\max_{y_t}[q(\bar{x}_t|y_t,\theta_k^{xy})q(y_t|Y_{t-1},\theta_k^y)]$ has to be iteratively solved as an inner loop within a parameter learning process. It can be very expensive to wait for this iterative inner loop to converge. In practice, this convergence is approximately replaced by running the iteration only for a few steps, which can be far before its convergence. It would be also a challenge on appropriately developing such an approximation and on examining how it affects the performance.

(h) Related closely to the above case (f) and case (g), it would be helpful to investigate the manifold of $H_{\mathcal{X}}(p\|q,\theta_k)$, especially on the distribution of local maxima. Since the gradient based technique takes a major role in implementing the BYY harmony learning, it is likely to be stuck at a local maximum. Thus, it deserves to study how the global versus local maximum issue will affect the accuracy of k^* and the generalization error.

(i) As introduced in Sec. 2.3, Rival Penalized Competitive Learning (RPCL) can also perform automatic model selection [64, 66]. Its relation to the BYY harmony learning has been discussed in [52, 49, 47, 19]. Though the convergence behavior of RPCL has already been qualitatively interpreted in a top-down way via the BYY harmony learning (e.g., as discussed in Sec. 4.1), it is interesting to investigate in a bottom up way on converging behaviors of both RPCL learning and adaptive algorithm for BYY harmony learning. Some preliminary studies have been made in [20, 21]. However, challenges still remain on getting the conditions (especially the penalizing strength) for guaranteeing a RPCL learning to correctly converge with a correct scale k^*.

(j) As discussed in Sec. 5.1, the key point of automatic model selection comes from the nature by eq.(60). That is, a critical subset ϕ_k of parameters, e.g., the proportional parameter α_j in eq.(38) and one component variance $\lambda_j^{(r)\ 2}$ of Λ_j^2 in eq.(55), will be driven towards 0 during the maximization process. However, waiting for ϕ_k converging to zero will waste a large computing cost, which is usually unnecessary. It deserves to develop effective techniques to detect the evidences for $\phi_k \to 0$ (e.g., $\alpha_j \to 0$, $\lambda_j^{(r)\ 2} \to 0$). One possible direction is to develop some statistical test for this purpose.

(k) The BYY harmony learning has already been extended to model temporal relations among samples [51, 48]. Many of the above discussed challenges should also be investigated on these temporal extensions.

(l) In a two-stage implementation given at the end of Sec. 2.2, we have to enumerate every k within $[k_d, k_u]$ in a general case without considering the internal structure of $J(k,\theta_k^{fit})$. Such an enumeration can be made in either a forward way (i.e., increasing k from a small initial value) or a backward way (i.e., decreasing k from a large initial value). In a forward implementation, as k increases to $k+1$, not only those newly appeared parameters but also the existing parameter set θ_k have to be re-learned.

In a backward implementation, as $k+1$ decreases to k, we are unable to directly take a subset θ_k from the set θ_{k+1}. However, $J(k, \theta_k^{fit})$ may have a specific structure for certain learning tasks. As k increases to $k+1$, only those newly appeared parameters need to be re-learned while the existing parameter set θ_k remain unchanged. In other words, the learning can be made in an incremental way. Furthermore, we may also consider $J(k, \theta_k^{fit})$ with a more complicated structure in help of certain enumerating or searching technique that was developed for feature selection tasks in the pattern recognition field [70].

6 Concluding Remarks

Advances, trends, and challenges on regularization and model selection in statistical learning have been discussed from a Bayesian Ying Yang learning perspective. After briefly introducing the Bayesian Ying-Yang system and the best harmony learning principle, its advantage of automatic model selection and of integrating regularization and model selection have been addressed, and its differences and relations to typical existing learning methods have been elaborated. Taking the tasks of Gaussian mixture, local subspaces, local factor analysis as examples, not only detailed model selection criteria are given, but also a general learning procedure is provided to unify adaptive algorithms for these learning tasks. Finally, a new trend for model selection has been elaborated; theoretical issues in a large sample size and challenges in a small sample size have been further presented.

Acknowledgement

The work described in this paper was fully supported by a grant from the Research Grant Council of the Hong Kong SAR (Project No:CUHK4173/06E).

References

[1] Akaike, H (1974), "A new look at the statistical model identification", *IEEE Tr. Automatic Control*, 19, 714-723.
[2] Akaike, H (1981), "Likelihood of a model and information criteria", *Journal of Econometrics*, 16, 3-14.
[3] Akaike, H (1987), "Factor analysis and AIC", *Psychometrika*, 52, 317-332.
[4] Anderson, TW, & Rubin, H (1956), "Statistical inference in factor analysis", *Proc. Berkeley Symp. Math. Statist. Prob. 3rd 5*, UC Berkeley, 111-150.
[5] Bishop, C.M., (1995), "Training with noise is equivalent to Tikhonov regularization", *Neural Computation 7*, 108-116.

[6] Bozdogan, H (1987) "Model Selection and Akaike's Information Criterion: The general theory and its analytical extension", *Psychometrika*, **52**, 345-370.
[7] Bozdogan, H & Ramirez, DE (1988), "FACAIC: Model selection algorithm for the orthogonal factor model using AIC and FACAIC", *Psychometrika*, 53 (3), 407-415.
[8] Cavanaugh, JE (1997), "Unifying the derivations for the Akaike and corrected Akaike information criteria", *Statistics & Probability Letters*, 33, 201-208.
[9] Cooper, G & Herskovitz, E (1992), "A Bayesian method for the induction of probabilistic networks from data", *Machine Learning*, 9, 309-347.
[10] Dayan, P. & Hinton, GE (1995), "The Helmholtz machine", *Neural Computation 7*, No.5, 889-904.
[11] Girosi, F, et al, (1995) "Regularization theory and neural architectures", *Neural Computation*, 7, 219-269.
[12] Hinton, GE, Dayan, P, Frey, BJ, & Neal, RN (1995), "The wake-sleep algorithm for unsupervised learning neural networks", *Science 268*, 1158-1160.
[13] Hinton, GE & Zemel, RS (1994), "Autoencoders, minimum description length and Helmholtz free energy", *Advances in NIPS*, 6, 3-10.
[14] Hurvich, CM, & Tsai, CL (1989), "Regression and time series model in samll samples", *Biometrika*, 76, 297-307.
[15] Kashyap, RL (1982), "Optimal choice of AR and MA parts in autoregressive and moving-average models", *IEEE Trans. PAMI*, 4, 99-104.
[16] Z.Y. Liu, H. Qiao, & L. Xu, "Multisets Mixture learning based Ellipse Detection", *Pattern Recognition 39*, pp 731-735, 2006.
[17] Z.Y. Liu, K.C. Chiu, & L. Xu, "Strip Line Detection and Thinning by RPCL-Based Local PCA", *Pattern Recognition Letters* **24**, 2335-2344, 2003.
[18] Liu, ZY, Chiu, KC, & Xu, L (2003), " Improved system for object detection and star/galaxy classification via local subspace analysis", *Neural Networks 16*, 437-451.
[19] Ma, J, Wang, T, & Xu, L (2004), "A gradient BYY harmony learning rule on Gaussian mixture with automated model selection", *Neurocomputing 56*, 481-487.
[20] Ma, J & Xu, L (2002), "Convergence Analysis of Rival Penalized Competitive Learning (RPCL) Algorithm", *Proc. of Intl. Joint Conf. on Neural Networks (IJCNN '02)*, Hawaii, USA, May 12-17, 2002, pp 1596-1602.
[21] Ma, J & Xu, L "The Correct Convergence of the Rival Penalized Competitive Learning (RPCL) Algorithm", *Proc. of Intl. Conf. on Neural Information Processing (ICONIP'98)*, October 21-23, 1998, Kitakyushu, Japan, Vo.1, pp239-242.
[22] Mackey, D (1992) "A practical Bayesian framework for backpropagation", *Neural Computation*, 4, 448-472.

[23] Neath, AA & Cavanaugh, JE (1997), "Regression and Time Series model selection using variants of the Schwarz information criterion", *Communications in Statistics A*, 26, 559-580.

[24] T.Poggio & F.Girosi, "Networks for approximation and learning", *Proc. of IEEE*, **78**, 1481-1497 (1990).

[25] Redner, RA & Walker, HF (1984), "Mixture densities, maximum likelihood, and the EM algorithm", *SIAM Review*, 26, 195-239.

[26] Rissanen, J (1986), "Stochastic complexity and modeling", *Annals of Statistics*, 14(3), 1080-1100.

[27] Rissanen, J (1989), *Stochastic Complexity in Statistical Inquiry*, World Scientific: Singapore.

[28] Rivals, I & Personnaz, L (1999) "On Cross Validation for Model Selection", *Neural Computation*, 11, 863-870.

[29] Schwarz, G (1978), "Estimating the dimension of a model", *Annals of Statistics*, 6, 461-464.

[30] Stone, M (1974), "Cross-validatory choice and assessment of statistical prediction", *J. Royal Statistical Society B*, **36**, 111-147.

[31] Stone, M (1977), "An asymptotic equivalence of choice of model by cross-validation and Akaike's criterion", *J. Royal Statistical Society B*, 39 (1), 44-47.

[32] Stone, M (1978), "Cross-validation: A review", *Math. Operat. Statist.*, 9, 127-140.

[33] Stone, M (1979), "Comments on model selection criteria of Akaike and Schwartz. *J. Royal Statistical Society B*, **41** (2), 276-278.

[34] Sugiura, N (1978), "Further analysis of data by Akaike's information criterion and the finite corrections", *Communications in Statistics A*, **7**, 12-26.

[35] Tikhonov, AN & Arsenin, VY (1977), *Solutions of Ill-posed Problems*, Winston and Sons.

[36] Wallace, CS & Boulton, DM (1968), "An information measure for classification", *Computer Journal*, 11, 185-194.

[37] Wallace, CS & Freeman, PR (1987), "Estimation and inference by compact coding", *J. of the Royal Statistical Society*, 49(3), 240-265.

[38] Wallace, CS & Dowe, DR (1999), "Minimum message length and Kolmogorov complexity", *Computer Journal*, 42 (4), 270-280.

[39] Vapnik, VN (1995), *The Nature Of Statistical Learning Theory*, Springer.

[40] Xu, L., (2007), "A Unified Perspective and New Results on RHT Computing, Mixture Based Learning, and Multi-learner Based Problem Solving", *Pattern Recognition*, Vol. 40, pp. 2129–2153, 2007.

[41] Xu, L., (2005), "Fundamentals, Challenges, and Advances of Statistical Learning for Knowledge Discovery and Problem Solving: A BYY Harmony Perspective", Keynote talk, *Proc. of Intl. Conf. on Neural Networks and Brain*, Oct. 13-15, 2005, Beijing, China, Vol. 1, pp. 24-55.

[42] Xu, L. (2004), "Temporal BYY Encoding, Markovian State Spaces, and Space Dimension Determination", *IEEE Tr. Neural Networks, V15, N5*, pp. 1276-1295.

[43] Xu, L (2004), "Advances on BYY harmony learning: information theoretic perspective, generalized projection geometry, and independent factor auto-determination", *IEEE Tr. Neural Networks, V15, N4*, pp. 885-902.

[44] Xu, L. (2004), "Bayesian Ying Yang Learning (I): A Unified Perspective for Statistical Modeling", *Intelligent Technologies for Information Analysis*, N. Zhong and J. Liu (eds), Springer, pp. 615-659.

[45] Xu, L. (2004), "Bayesian Ying Yang Learning (II): A New Mechanism for Model Selection and Regularization", *Intelligent Technologies for Information Analysis*, N. Zhong and J. Liu (eds), Springer, pp. 661-706.

[46] Xu, L. (2004), "BI-directional BYY Learning for Mining Structures with Projected Polyhedra and Topological Map", Invited talk, in *Proc. of FDM 2004: Foundations of Data Mining*, eds., T.Y.Lin, S.Smale, T. Poggio, and C.J. Liau, Brighton, UK, Nov. 01, 2004, pp. 5-18.

[47] Xu, L. (2003), "Data smoothing regularization, multi-sets-learning, and problem solving strategies", *Neural Networks, V. 15, No. 5-6*, 817-825.

[48] Xu, L. (2003), "Independent Component Analysis and Extensions with Noise and Time: A Bayesian Ying-Yang Learning Perspective", *Neural Information Processing Letters and Reviews*, Vol.1, No.1, 1-52.

[49] Xu, L (2002), "BYY Harmony Learning, Structural RPCL, and Topological Self-Organizing on Mixture Models ", *Neural Networks, V15, N8-9*, 1125-1151.

[50] Xu, L, (2002), "Bayesian Ying Yang Harmony Learning", *The Handbook of Brain Theory and Neural Networks*, Second edition, (MA Arbib, Ed.), Cambridge, MA: The MIT Press, pp. 1231-1237.

[51] Xu, L (2001), "BYY Harmony Learning, Independent State Space and Generalized APT Financial Analyses ", *IEEE Tr. Neural Networks*, **12** (4), 822-849.

[52] Xu, L (2001), "Best Harmony, Unified RPCL and Automated Model Selection for Unsupervised and Supervised Learning on Gaussian Mixtures, Three-Layer Nets and ME-RBF-SVM Models", *Intl J of Neural Systems* **11** (1), 43-69.

[53] Xu, L (2000), "Temporal BYY Learning for State Space Approach, Hidden Markov Model and Blind Source Separation", *IEEE Tr. Signal Processing 48*, 2132-2144.

[54] Xu, L (1998), "RBF Nets, Mixture Experts, and Bayesian Ying-Yang Learning", *Neurocomputing*, Vol. 19, No.1-3, 223-257.

[55] Xu, L, (1998), "Rival Penalized Competitive Learning, Finite Mixture, and Multisets Clustering", *Proc. of IJCNN98*, Anchorage, Vol.II, pp. 2525-2530.

[56] Xu, L (1997), "Bayesian Ying-Yang Machine, Clustering and Number of Clusters", *Pattern Recognition Letters 18*, No.11-13, 1167-1178.

[57] Xu, L, (1997), "New Advances on Bayesian Ying-Yang Learning System with Kullback and Non-Kullback Separation Functionals", *Proc. IEEE-INNS Intl. Joint Conf. on Neural Networks (IJCNN97)*, Houston, Vol. III, pp. 1942-1947.

[58] Xu, L, & Jordan, MI (1996), "On convergence properties of the EM algorithm for Gaussian mixtures", *Neural Computation, 8*, No.1, 1996, 129-151.

[59] Xu, L, (1995), "Bayesian-Kullback Coupled YING-YANG Machines: Unified Learnings and New Results on Vector Quantization", *Proc. Intl. Conf. on Neural Information Processing*, Oct 30-Nov.3, 1995, Beijing, pp. 977-988.

[60] L. Xu, "A Unified Learning Framework: Multisets Modeling Learning", Invited Talk, *Proc. of World Congress on Neural Networks (WCNN95)*, Washington, DC, July 17-21, 1995, Vol.I, pp. 35-42.

[61] Xu, L, Jordan, MI, & Hinton, GE (1995), "An Alternative Model for Mixtures of Experts", *Advances in Neural Information Processing Systems 7*, eds, Cowan, JD, et al, MIT Press, 633-640, 1995.

[62] L. Xu, "Multisets Modeling Learning: An Unified Theory for Supervised and Unsupervised Learning", Invited Talk, *Proc. of IEEE ICNN94*, Orlando, Florida, June 26-July 2, 1994, Vol.I, 315-320.

[63] Xu, L, Krzyzak, A, & Yuille, AL (1994), "On Radial Basis Function Nets and Kernel Regression: Statistical Consistency, Convergence Rates and Receptive Field Size", *Neural Networks*, **7**, 609-628.

[64] Xu, L, Krzyzak, A & Oja, E (1993), "Rival Penalized Competitive Learning for Clustering Analysis, RBF net and Curve Detection", *IEEE Tr. on Neural Networks 4*, 636-649.

[65] Xu, L & Oja, E. (1993), "Randomized Hough Transform (RHT): Basic Mechanisms, Algorithms and Complexities", *Computer Vision, Graphics, and Image Processing : Image Understanding*, Vol.57, No.2, pp. 131-154.

[66] Xu, L, Krzyzak, A & Oja, E (1992), "Unsupervised and Supervised Classifications by Rival Penalized Competitive Learning", *Proc. of 11th Intl Conf. on Pattern Recognition (ICPR92)*, Hauge, Netherlands, Vol.I, pp. 672-675.

[67] Xu, L, Klasa, A, & Yuille, A.L. (1992), "Recent Advances on Techniques Static Feedforward Networks with Supervised Learning", *International Journal of Neural Systems*, Vol.3, No.3, pp. 253-290.

[68] Xu, L., Krzyzak, A., & Suen, C.Y. (1992), "Several Methods for Combining Multiple Classifiers and Their Applications in Handwritten Character Recognition", *IEEE Tr. System, Man and Cybernetics*, Vol. 22, No.3, pp. 418-435.

[69] Xu, L, Oja, E., & Kultanen, P. (1990), "A New Curve Detection Method: Randomized Hough Transform (RHT)", *Pattern Recognition Letters*, Vol.11, pp. 331-338.

[70] Xu, L, P.F. Yan, & T. Chang (1988), "Best First Strategy for Feature Selection", *Proc. of 9th Intl Conf. on Pattern Recognition (ICPR98)*, Nov. 14-17, 1988, Rome, Italy, Vol.II, pp. 706-709.

[71] Xu, L, (2007), "Bayesian Ying Yang Learning", *Scholarpedia,* p. 10469, http://scholarpedia.org/article/Bayesian_Ying_Yang_Learning.

Computational Intelligence in Mind Games

Jacek Mańdziuk

Faculty of Mathematics and Information Science
Warsaw University of Technology, Poland.
mandziuk@mini.pw.edu.pl

Summary. The chapter considers recent achievements and perspectives of Computational Intelligence (CI) applied to mind games. Several notable examples of unguided, autonomous CI learning systems are presented and discussed. Based on advantages and limitations of existing approaches a list of challenging issues and open problems in the area of intelligent game playing is proposed and motivated.

It is generally concluded in the paper that the ultimate goal of CI in mind game research is the ability to mimic human approach to game playing in all its major aspects including learning methods (learning from scratch, multitask learning, unsupervised learning, pattern-based knowledge acquisition) as well as reasoning and decision making (efficient position estimation, abstraction and generalization of game features, autonomous development of evaluation functions, effective preordering of moves and selective, contextual search).

Key words: challenges, CI in games, game playing, soft-computing methods, Chess, Checkers, Go, Othello, Give-Away Checkers, Backgammon, Bridge, Poker, Scrabble.

1 Introduction

Playing games has always been an important part of human activities and the oldest mind games still played in their original form (Go and Backgammon) date back to 1,000 - 2,000 BC.

Games also became a fascinating topic for Artificial Intelligence (AI). The first widely known "AI approach" to mind games was noted as early as 1769 when Baron Wolfgang von Kempelen's automaton Chess player named *The Turk* was presented at the court of Empress Maria Theresa. *The Turk* appeared to be a very clever, actually unbeatable, Chess player who defeated among others Napoleon and the Empress Catherine of All the Russias. It took a few decades to uncover a very smart deception: a grandmaster human player was hidden inside the Turk's machinery and through a complicated construction of levers and straddle-mounted gears was able to perceive opponent's

Jacek Mańdziuk: *Computational Intelligence in Mind Games*, Studies in Computational Intelligence (SCI) **63**, 407–442 (2007)
www.springerlink.com

moves and make its own ones. The history of *The Turk* was described independently by several people, including the famous American novelist Edgar Allan Poe[78]. Although *The Turk* had apparently nothing in common with AI, the automaton is a good illustration of humans' perennial aspiration for creating intelligent machines able to defeat the strongest human players in popular mind games.

Serious, scientific attempts to invent "thinking machines" able to play mind games began in the middle of the previous century. Thanks to seminal papers devoted to programming Chess [93, 110, 74] and Checkers [82] in the 1950s., games remained through decades an interesting topic for both classical AI and CI based approaches.

One of the main reasons for games' popularity in AI/CI community is the possibility to obtain cheap, reproducible environments suitable for testing new search algorithms, pattern-based evaluation methods or learning concepts.

On the other hand the "human aspect" of game playing should not be underestimated. This is why from the very beginning of AI "involvement" in games, it was Chess - the queen of mind games - that attracted special attention and in 1965 was even announced "*the Drosophila of Artificial Intelligence*" by the Russian mathematician Alexander Kronrod.

The focus of this chapter is on the most popular mind games, such as Chess, Checkers, Go, Othello, Backgammon, Bridge, Poker and Scrabble. The reason for choosing these particular games is two-fold: (1) they are all very popular and played all over the world, (2) for decades they have been a target for AI/CI research aiming at surpassing human supremacy. Certainly, there are many other interesting and highly competitive mind games (e.g. Shogi, Chinese Chess, Hex, Amazons, Octi, Lines of Actions, Sokoban), which do not appear in this chapter, mainly due to their lesser popularity - although, some of them are becoming more and more prominent. Also other types of computer games different from mind games, such as skill, adventure, strategic, war, sport, negotiating, racing and others are not considered in this chapter.

In order to make the notation clear and concise, henceforth any reference to CI systems (approaches) will address soft-computing-based systems (*neural networks, genetic or evolutionary algorithms, fuzzy systems, reinforcement learning, Bayesian methods, probabilistic reasoning, rough sets)* capable of learning and autonomous improvement of behavior[1].

It seems worth noting that the aim of this chapter is by no means to criticize the achievements of traditional AI methods in the game playing domain. On the contrary the hitherto accomplishments of AI approaches are undisputable and speak for themselves. Our goal is rather to express the belief that other, alternative ways of developing "thinking machines" are possible and urgently needed. These methods include cognitive, knowledge-free approaches capable of learning from scratch based merely on an unguided training process,

[1] Certainly, a distinction between AI and CI is to some extent a matter of convention. The above proposal is consistent with the author's point of view.

e.g. evolutionary or reinforcement-type, or based on an agent's experience obtained gradually through (self-)playing or in a supervised training process, e.g. with neural nets, but again without explicit implementation of human experts' knowledge.

In our opinion the need for further development of these, knowledge-free methods is unquestionable, and the ultimate goal that can be defined is building a truly autonomous, human-like multi-game playing agent. In order to achieve this goal several challenging problems have to be addressed and solved on the way.

The chapter is organized as follows. In the next section a brief description of state-of-the-art accomplishments in the most popular mind games are presented. Section 3 starts with a general discussion on the challenging issues in the game playing domain and presents further motivation for pursuing this research topic. Next, in several subsections particular challenges are considered one-by-one in more detail. Conclusions are presented in the last section.

2 State-of-the-Art Playing Programs

In this section some of the best playing programs in the most popular games are briefly introduced. In some games (Scrabble, Chess, Checkers, Othello, Backgammon) the supremacy of the presented systems over humans and other artificial agents was officially acclaimed either by gaining the respective World Champion title or by defeating the best human players. In the remaining games considered here (Poker, Bridge, Go) humans are still far ahead of machines.

Scrabble. One of the first programs that achieved a world-class human level in a non-trivial game was *Maven* - a Scrabble playing program written by Brian Sheppard [94]. In the 1990s. Maven successfully challenged several world top Scrabble players including (then) North America Champion Adam Logan and world champion Joel Sherman (both matches took place in 1998). Strictly speaking the supremacy of Maven was demonstrated only in North America - i.e. for US English, but adaptation of Maven to another dictionary is straightforward. Actually, it was later on adapted to UK English, International English, French, Dutch and German.

Scrabble is a game of imperfect information with a large branching factor, and as such is very demanding for AI research. The key to Maven's success lies in efficient, selective move generation and perfect endgame play supported by B^* search algorithm[2]. Maven, similarly to TD-Gammon described in sect. 3.1, uses game scenarios simulation or "rollouts", which proved to be a strong evaluation technique. The implementation details are presented in [94].

[2] As soon as the bag is empty, Scrabble becomes a perfect information game, since one can deduce the opponent's rack by subtracting the tiles one can see from the initial distribution.

In [94] Sheppard stated: "*There is no doubt in my mind that Maven is superhuman in every language. No human can compete with this level of consistency. Anyone who does not agree should contact me directly to arrange a challenge match*". So far, no-one tried ...

Chess. Presumably the most striking achievement of AI in games was *Deep Blue II's* victory over Garry Kasparov - the World Chess Champion (at the time the match was held) and one of the strongest Chess players in the history. This event ended a nearly 50-year era of Chess programming efforts which started in Shannon's paper [93]. The evaluation function of Deep Blue II was composed of over 8,000 features implemented in a single chess chip. 480 such chips formed an extremely fast, massively parallel search system based on 30-node cluster allowing for total search speed between 100 million and 330 million positions per second depending on their tactical complexity [20] or 50 billion positions in three minutes - the average time allotted for each move [48]. Certainly, except for tuning thousands of weights in the evaluation function, a lot of other problems concerning massively-parallel, non-uniform, highly-selective search or creation and analysis of the extended opening book and endgame database had to be solved in order to achieve the final result. There in no doubt that Deep Blue II is a milestone achievement from an engineering and programming point of view [45, 20]. From a CI viewpoint much less can be said since the system did not take advantage of any learning or self-improvement mechanisms.

The victory of Deep Blue II attracted tremendous interest among game playing researches and also had an undisputed social and philosophical impact on other people, not professionally related to science (New York's Kasparov vs. Deep Blue II match was followed on the Internet by thousands of people all over the world). On the other hand the result of the match should not lessen further research efforts aiming at developing an "intelligent" Chess playing program equipped with cognitive skills similar to those used by human Chess grandmasters.

Since Deep Blue's era several other Chess programs, e.g. *Shredder, Fritz, Deep Junior* or the recent Chess supercomputer - *Hydra* played successfully against human grandmasters, but none of these matches gained comparable public attention and esteem.

Checkers. The first computer program that won a human world championship was *Chinook* - the World Man-Machine Champion developed by Jonathan Schaeffer and his collaborators from the University of Alberta [87, 89, 84]. Chinook's opponent in both 1992 and 1994 was Dr Marion Tinsley - the ultimate Checkers genius, who was leading the scene of Checkers competitions for over 40 years losing during that period as few as only 7 games (including the 2 lost to Chinook)! As Schaffer stated: Tinsley was "*as close to perfection as was possible in a human*" [89].

Similarly to Deep Blue, Chinook can be regarded as a large scale AI engineering project including all vital aspects of AI design: efficient search, well-

tuned evaluation function, opening book and endgame database. Special care was taken over specific tactical combinations (e.g. exchanging one of our own pawns for two of an opponent's) and these situations were carefully analyzed and coded in special tables. The evaluation function was linear and composed of over 20 major components, each of which having several heuristic parameters. All evaluation function weights were hand-tuned. During the re-match in 1994 Chinook was equipped with a complete 7-piece endgame database (i.e. exact solutions of all endings of 7 pieces or less) and with a 4×4 subset of an 8-piece database (i.e. all endings in which each side was left with exactly 4 pieces) [89]. At the time of writing this chapter the 9-piece database is completed and the 10-piece one is on the way [85].

The ultimate goal of Schaeffer and his group is to build the perfect Checkers player by solving the game of Checkers. Recently it was announced that another opening (already the second one) has been solved - proven to be a draw [85].

Othello. Another game in which computers outperformed over humans is Othello. In 1997, just a few months after Deep Blue's victory, Michael Buro's program *Logistello* [19], running on a single PC machine, decisively defeated the then Othello World Champion Takeshi Murakami with the score $6-0$. Taking into account that Logistello was not implemented in a special hardware and considering the convincing result of the match as well as post-mortem analysis of the games which showed that program was not in trouble in any of the six games played, further advances Buro's achievement.

The main factors contributing to this strong victory were: (1) new, efficient way of feature selection for the evaluation function [17]. Starting from a set of predefined, atomic features, various Boolean conjunctions of these simple features were considered and their weights calculated by the linear regression based on several million training positions labelled either by their true mini-max value or an approximation of it. (2) Forward pruning method ProbCut (and Multi-ProbCut) capable of cutting out the most probably irrelevant subtrees with predefined confidence [16]. In short, the method generalizes from shallow search results to deeper search levels by statistically estimating the coefficients in the linear model approximating the relationship between shallow and deep mini-max search results. (3) Automatic opening book development, which takes advantage of the search results along the promising lines not played so far and consequently allows potentially interesting opening alternatives in the future [18].

All three above mentioned aspects of game playing are game independent and possibly applicable to other two-player board games. On a more general note these methods are in line with a human way of playing which includes building up the evaluation function, performing selective search or looking for new variants in the known game openings.

Backgammon. The state-of-the-art Backgammon playing program is *TD-Gammon* [103, 104, 105] written by Gerald Tesauro. The program implements

Temporal Difference learning and is one of the archetypal examples of successful CI approaches in games. The main features of TD-Gammon are discussed in more detail in sect. 3.1.

In the above mentioned games the human supremacy has already been successfully challenged by AI/CI programs. Among the most popular games there are only three left in which humans have not been conquered - (yet!). These are: Poker, Bridge and Go.

Poker. According to the results of the World Poker Robot Championship that took place in July 2005 the world's best playing Poker program is *PokerProbot*^{TM} [40] - the Amateur Robot Champion and, at the same event, the winner of the match with *Poki-X* [86] written by Jonathan Schaeffer and his group from the University of Alberta.

Since PokerProbot, written by Hilton Givens, is commercial software, very little is known about its internal characteristics and the history of its development. On the contrary the knowledge behind its main opponent Poki-X has been revealed [13]. Both programs were also confronted with one of the game's most accomplished professionals and World Series of Poker champion, Phil Laak - and both lost by a large margin.

The reasons why Poker is so difficult for AI/CI is related to the fact that it is an *imperfect information* game since the other player's cards are hidden. Additionally Poker players often use various kinds of deception or bluffing. Non-accessibility to the whole information (as opposed to Chess, Checkers, and other board games not involving elements of chance) requires (1) modeling of the opponents, (2) applying risk management techniques and (3) using a dynamical, context sensitive and in most cases probabilistic evaluation function rather than a static one. These issues are expanded in [13] with regard to the Poki system. Poki uses a sophisticated, multi-stage betting strategy which includes the evaluation of effective and potential hand strength, opponent's modeling and probabilistic simulations based on selective sampling[3]. All the above issues are essential for successful machine Poker playing. For the scope of this chapter the problem of *opponent modeling* (discussed further in sect. 3.6) is of particular interest.

Bridge. Since 1997 the World Computer-Bridge Championship has been organized each year by The American Contract Bridge League. The regular participants in this annual event are *Jack*, *Bridge Baron*, *WBridge5*, *Micro Bridge*, *Q-Plus Bridge* and *Blue Chip Bridge*. Each of these programs enjoyed some success in previous contests but the most renowned one is the Dutch program named Jack by Hans Kuijf and his team [53]. Jack won the title in 2001, 2002, 2003, 2004 and was placed second in 2005 after WBridge5, though was in the first place in the so-called Round Robin - the aggregated result of direct pairwise comparison.

---

[3] The idea of selective sampling in a world-class playing programs was also applied to Backgammon [105], Scrabble [94] and Bridge[38].

An interesting phenomenon among top bridge programs is GIB (Ginsberg's Intelligent BridgePlayer) written by Matthew L. Ginsberg - historically the first strong bridge playing program. GIB uses partition search (the cutting tree technique defined by Ginsberg) and Monte Carlo sampling techniques for both the bidding and cardplay phases [39].

The level of play of the best computer programs is gradually improving and currently they are able to play on equal terms against intermediate human players.

Go. The game of Go is widely considered as the most demanding, grand AI/CI challenge in the mind games domain. Despite simple rules and no pieces differential, playing the game well is yet a non-achievable task for machines. The most advanced Go programs can still be easily beaten by intermediate amateur human players [4]. There are several reasons for this situation. First of all, Go has a very high branching factor, which effectively eliminates brute-force-type exhaustive search methods. But the huge search space is not the only impediment in efficient play. The very distinctive feature that separates Go and other popular board games is the fact that static positional analysis of the board is orders of magnitude slower in Go than in other games [73]. Additionally, proper positional board judgement requires performing several auxiliary tactical searches oriented on particular tactical issues [73]. Due to the variety of positional features and tactical threats it is highly probable that, as stated in [73], "*no simple yet reasonable evaluation function will ever be found for Go*". Another difficult problem for machine play is the "pattern nature" of Go. On the contrary to humans, who posses strong pattern analysis abilities, machine players are very inefficient in this task, mainly due to the lack of mechanisms (either predefined or autonomously developed) allowing flexible subtask separation. The solutions for these subtasks need then to be aggregated - considering complex mutual relations - at a higher level and provide the ultimate estimation of the board position. Instead, only relatively simple pattern matching techniques are implemented in the current playing programs [72, 73].

Due to the still preliminary stage of Go playing programs' development it is hard to point out the stable leader among them. Instead, there exists a group of about ten programs playing on a more or less comparable level. These include: *Many Faces of Go*, *Go4++*, *Handtalk*, *GoIntellect*, *Explorer*, *Indigo* and a few more. A detailed discussion on the development of Go playing agents can be found in [14, 73]. An interesting proposition for researchers aiming to write their own Go program is the open source application GnuGo [15].

---

[4] Unless otherwise stated in the whole paper we will refer to the game played on a regular 19 x 19 board.

3 The Challenges

Recent advances of AI in the most popular mind games, which led to spectacular challenging the human supremacy in Chess, Checkers, Othello or Backgammon provoke the question: "*Quo vadis mind games research?*". Do we still need to pursue mind game research or maybe defeating human world champions is (was) the ultimate, satisfying goal?

Naturally, when considering the quality of machine playing in particular game as the sole reference point the only remaining target might be the further extension of the machines' leading margin in the man-machine competition (since it is doubtful that the improvement of human players would be adequate to the one of computer players). But improvement of efficiency is not the only and not even a sufficient motivation for further research.

I would argue that good reasons for game research concern ***the way*** in which high playing competency is accomplished by machines. On one side there are extremely powerful AI approaches in which playing agents are equipped with carefully designed evaluation functions, look-up tables, perfect endgame databases, opening databases, grandmaster game repositories, sophisticated search methods (e.g. B*[11], SSS*[99], NegaScout [80], MTD(f) [76, 77], conspiracy numbers [65]) or search enhancements (e.g. singular extensions [2], null moves [9, 44], ProbCut [16], and other [83]) and a lot of other predefined, knowledge-based tools and techniques that allow making high quality moves with enormous search speed. On the other side there are soft, CI-based methods relying mainly on knowledge-free approaches, extensive training methods including reinforcement learning, neural networks, self-playing and even learning from scratch based merely on the final outcomes of the games played. Application and development of these soft techniques pose several challenging questions which are discussed in the remainder of this section. First, in section 3.1 some well-known successful examples of **autonomous learning** in game playing and challenging, open problems related to this type of learning are discussed. Section 3.2 addresses the issue of **creativity** understood as *ad hoc* **knowledge discovery** which may emerge as a result of deliberately designed training process. Section 3.3 is devoted to **intuition**, implementation of which in artificially built systems seems to be one of the grand challenges not only in game playing domain. Section 3.4 considers the problem of **abstraction and generalization** of knowledge possessed during the learning process. In particular the problem of how to generalize from shallow-depth search is still unsolved and considered a challenge. Another challenging problem is efficient **pre-ordering of moves** in the search algorithms (section 3.5). Although a lot of results have been published in this area, the problem - addressed generally - still remains open. Section 3.6 touches on the problem of **opponent modeling** which is one of the central issues in games with imperfect information, especially those in which deception and bluffing are inherent elements, such as Poker or Perudo. Finally, section 3.7 concerns **universality of approaches** and tools applied within CI. The

development of game independent, universal training processes applicable to a wide range of games is one of the relevant current research problems. Possible approaches include **multitask learning** and **lifelong learning**.

3.1 Autonomous Learning

One of the most distinctive features of CI-based systems is the ability to improve themselves through a (self)-learning process. Unlike classical AI approaches which rely on carefully designed, hand-crafted evaluation functions reflecting expert knowledge about various game aspects, the CI systems, given some initial knowledge, are able to improve their performance through learning or evolution.

Construction of game playing agents capable of learning based on experience is one of the challenging issues. There are several notable examples of such systems, e.g. Tesauro's Neurogammon and TD-Gammon, Baxter's KnightCap, Schaeffer's TDL-Chinook, Thrun's NeuroChess, or Fogel's Anaconda - to mention only a few of them. A brief description of the above seminal achievements is presented in the remainder of this section, followed by a general discussion on their strengths and weaknesses as well as related open problems.

Certainly the systems described below by no means pretend to be a complete catalogue of CI achievements in games. They are rather a partial collection of milestone accomplishments subjectively chosen by the author. Other CI approaches to most popular games include for example [112, 109, 36, 50, 41, 34] in Chess, [1, 61] in Checkers, [63, 75, 54] in Give-Away Checkers, [69, 113] in Othello, [52] in Rummy, [26, 22, 92] in Iterated Prisoner's Dilemma, [79] in Backgammon, [4, 5] in Poker, [90, 91, 29, 81] in Go.

Neurogammon and **TD-Gammon**. The first world-class accomplishment in the field of CI in games was TD-Gammon program [103, 104, 105] and its predecessor - Neurogammon [102], both written by Gerald Tesauro for playing Backgammon - an ancient two-player board game. In short, the goal of the game is to move one's checkers from their initial position on the one-dimensional track to the final position (players make moves in the opposite directions). The total distance that pieces belonging to one player can move at a given turn depends on the score of the two dices which are thrown by a player at the beginning of a move. Rolling dices introduces randomness into the game. Based on the dices' score the player makes a decision regarding which pieces to move forward and of how many fields. The game has a high branching factor (due to dices' throwing) and when played by masters becomes a highly complex battle, full of tactical and positional threats including sophisticated blocking strategies.

The evaluation function in Neurogammon was implemented by Multilayer Perceptron (MLP) neural network trained with backpropagation, having as the input the location of pieces on the board and a set of game features carefully designed by human experts. Board positions were extracted from the corpus of master-level games. Neurogammon achieved a steady human intermediate level of play which allowed it to convincingly win the computer olympic competition [102].

Quite a different approach was adopted in TD-Gammon - the successor of Neurogammon. TD-Gammon was also utilizing the MLP network, but it differed from Neurogammon in three key aspects: (1) instead of backpropagation training the temporal difference learning introduced by Sutton [100, 101, 49] was used; (2) the input to the network was a raw board state without any expert features[5]; (3) training was essentially based on self-playing as opposed to training based on board positions that occurred in games played by experts.

Initially, i.e. in a knowledge-free approach, TD-Gammon reached an intermediate human level of play roughly equivalent to Neurogammon's level. In subsequent experiments - still in a self-playing regime, but with the input layer extended by adding expert board features (the ones used in Neurogammon) to the raw board data - the level of play eventually became equivalent to the best world-class human players.

Following Tesauro's work, various attempts to repeat his successful TD approach in other game domains were undertaken, but none of the subsequent trials reached as high level of playing competency in any other game as TD-Gammon did in Backgammon. One of the possible reasons of TD-Gammon's striking efficiency is the stochastic nature of the game which allows broad search of the entire state space and a real-valued, smooth, continuous target evaluation function, as opposed to discrete and discontinuous functions in most of the popular perfect information, deterministic games. Another reason is ascribed to the impact of TD learning strategy on the course of neural net's training: first simple linear associations were learnt, and only then a representation of nonlinear, context-sensitive concepts and exceptional cases was built [105].

KnightCap. Another well known example of TD-type learning in games is Chess playing program KnightCap written by Baxter, Tridgell and Weaver [6, 7, 8]. The authors applied TDLeaf(λ) method - a variant of TD(λ) introduced in [10][6]. As opposed to Samuel [82], Tesauro [103], Thrun [106], Beal and Smith [10] and later on Schaeffer et al. [88] KnightCap's designers found self-playing to be a very poor way of learning and preferred the use of external trainers instead. Hence, the TDLeaf(λ) learning was carried out by playing on the Internet Chess site. The program started from the blitz rating of 1650 and required only three days of playing (308 games) to reach the blitz rating of 2150, which is roughly equivalent to master candidate player. Afterwards the rating curve entered a plateau.

---

[5] Some experiments with adding expert features to the input vector were carried out in subsequent studies.

[6] Although the idea of TDLeaf(λ) was first presented in [10], the algorithm's name was coined in Baxter et al.'s papers.

The success of KnightCap laid, according to the authors, in appropriate choice of TD learning parameters, and first of all in "intelligent material parameters" initialization, which reflected the common knowledge of the pieces' values in Chess. Additional contribution to rapid rating increase was attributed to the fact that the weights of all other (i.e. non-material) parameters were initially set to zero, and therefore even small changes in their values potentially caused a relatively significant increase in the quality of play.

The main lesson from KnighCap's experiment was that the choice of initial weights in the evaluation function is crucial for the speed and quality of training. Another conclusion concerned the choice of training opponents who, according to authors' suggestions, should be comparable in playing strength to the learning program. This observation is in line with common human intuition that too strong or too weak opponents are not as valuable as the ones playing on approximately the same level.

The weakest feature of KnightCap was playing in the opening phase[7]. One possible remedy to this problem is the idea of "permanent brain" introduced in Crafty [46] - the strongest publicly available freeware Chess program and a direct descendant of a former Computer Chess Champion - Cray Blitz [47]. "Permanent brain" stores a number of losing positions and their evaluations in a hash table, which is used in every search. Thus the program avoids playing into these unfavorable lines.

TDL-Chinook. Jonathan Schaeffer, the author of Chinook (the Man-Machine Checkers Champion described in sect. 2), together with Markian Hlynka and Vili Jussila applied TD learning to Checkers [88] in order to verify its efficacy in another (after Backgammon and Chess) demanding game. The authors used the TDLeaf(λ) learning scheme. Their direct goal was a comparison between Chinook's evaluation function and the TD-based learnt one. In order to make this comparison the TD learning player was initially equipped with Chinook's evaluation function but with a different set of weights assigned to its components. Two main approaches were considered: in the first one, Chinook served as the training opponent for the TD player whereas the second approach relied on self-playing. In the first case training was performed in a predefined regime involving the use of some number of standard Checkers openings (afterwards the game was continued and finally completed by the players).

Surprisingly enough it turned out that by applying the TDLeaf(λ) learning scheme the program was capable of reaching the level of play comparable to the teacher's even though Chinook evaluation function's weights had been carefully tuned for more than five years. More surprisingly, the other approach (self-playing) also led to the Chinook caliber program, which implies that

[7] Due to the way TD learning is performed the relatively poorer play in the openings is common to practically all TD implementations regardless of the choice of the game.

external teacher is not indispensable for achieving the human championship level of play in a complex game as Checkers is!

It is interesting that the weighting of features in the evaluation function of the learning program was very different from that of Chinook. Closer examination of weights developed during training revealed several interesting insights into how some "human-type" features (i.e. the ones which very rarely occur in machine vs machine play) are compensated by other components in the evaluation function.

NeuroChess. Another interesting application of CI methods in games is NeuroChess program written by Sebastian Thrun [106], which combines TD learning with Explanation-Based Neural Network learning (EBNN) described in [68, 107]. The evaluation function in NeuroChess is represented by a neural network which inputs and outputs are board features (defined by a human expert) of the current position and the one expected after the next two half-moves, respectively. The challenge of Thrun's approach is to learn the evaluation function (i.e. weights of a network) with TD algorithm based solely on the final outcomes of the training games. Training is also supported by self-playing. The role of EBNN is to speed up the training process by allowing better generalization. This goal is accomplished by defining a separate neural network called the *Chess model* which represents the domain knowledge, obtained based on a large number of grandmaster games.

Although NeuroChess never reached the level of play of GNU-Chess (being its test opponent) defeating it in about 13% of times, the experiment pointed out some important advantages and weaknesses of TD learning based on the final games' outcomes. First of all NeuroChess's ability of playing openings was very poor, which was the consequence of increasing inaccuracy of position estimation from the final position backwards to the opening one. Another characteristic feature of NeuroChess play was mixing very strong moves with schoolboy mistakes, which according to Thrun happened quite frequently.

The main conclusion from Thrun's work is that learning based solely on observation of grandmaster play (TD learning in here) is not efficient enough and may lead to several artifacts in agent's evaluation function. An example of such inefficiency it the tendency of NeuroChess (when trained without self-playing) to move its queen into the center of the board in the early stage of the game. This behavior was learnt from grandmasters' games, but the program was unable to observe that grandmasters make such moves only when the queen is safe from being harassed by the opponent. In other words the basic idea of EBNN, i.e. using domain knowledge for finding *explanations* for a given set of examples in order to generalize based on them is not sufficient in the game of chess since some moves cannot be fully explained based exclusively on the accessible domain theory, *ergo* cannot be properly learnt.

Anaconda (vel Blondie24). Kumar Chellapilla and David Fogel carried out an experiment in which an ensemble of feed-forward neural networks, each representing an evaluation function for the game of Checkers, was evolved in appropriately designed evolutionary process. The input data for each network consisted of locations of pieces on a game board. This data was further decomposed in the first hidden layer into all possible subsets of size 3×3, 4×4, ..., 8×8 of the entire board. The two subsequent hidden layers operated on features originated in the first hidden layer. A single output neuron represented the evaluation of a Checkers' position presented in the input.

In each generation offspring networks were created and then each network (being either parent or offspring) played against five randomly selected opponents from that population. The best networks constituted the population for the next generation. After 250 generations the top network was tested against human competitors on the Internet site where it achieved the rating of an A-class player (immediately below the expert level) [22, 23]. After another 590 evolutionary generations the best network achieved the rating of an expert player (just below the master level) according to the U.S. Chess Federation rating system on the same Internet gaming site [25, 33]. This network was also tested against three characters (Beatrice, Natasha, and Leopold) from the *Hoyle's Classis Games* - commercially available software - winning a six game match with the score 6 : 0 [24]. Chellapilla and Fogel used two names for their network: Anaconda and Blondie24. The former one was related to the system's style of playing (this issue is further discussed in the next section) while the latter - most probably - to attract other player's attention on the Internet gaming zone.

It should be underlined that except for the sum of all board inputs (reflecting difference in material), which was presented as an additional input value directly to the output neuron, no expert knowledge about the game of Checkers was incorporated into the neural networks or the evolutionary process. The only "knowledge" available during the process was the location of pieces on the board, the rules of making (generating) all legal moves in a given position and the minimax heuristic for selecting the most favorable move at a given search depth. In majority of the games the search depth was defined to be equal to 6 or 8. The search was extended further for non-quiescent positions.

The common feature of all playing agents described above is the ability to autonomously improve their playing strength basing on experience (games played). This improvement is achieved either in the self-playing regime or in the course of playing against external opponents. Interestingly, the above mentioned experiments and other works presented in the literature are inconclusive with regard to whether it is more profitably to favor self-playing or rather to train with external opponents. Hence, one of the interesting and challenging issues is further investigation and formalization of the strengths and weaknesses of both training approaches.

In the case of training with external opponents additional key issue is the choice of the opponent players and the scheme of training [6, 63]. According to intuition, too strong or too weak opponents may not lead to expected improvement since weak opponents play badly and the strong ones are too good to be followed by the learner. Also the training scheme, when playing against external opponents, may have a great impact on the speed and quality of the learning process. In particular, in TD learning one may consider updating weights of the evaluation function after each game or only after the games lost or drawn. Another possibility is to update the weights regardless of the game's outcome, but with elimination of weak moves which most probably may be misleading for the training process [6, 7]. One may also consider playing against stronger opponent a few times in a raw if only the learner keeps losing against that opponent. The results for the game of Give-Away Checkers presented in [75] suggest the superiority of such approach over classical TD learning based on either all games played or only the ones not won by the learner. Other constructions of the learning scheme, e.g. the tournament choice of the opponents, can also be considered. The issue of how to define the optimal training scheme deserves further investigation and hopefully new conclusions across various game domains will come into light.

Another relevant issue is the choice of initial weights in the evaluation function. Regardless of the training method (being either TD, neural nets or evolutionary approach) the choice of the starting point is in most cases crucial for the final outcome of the learning process. Usually these initial settings are based on human expert knowledge. Another possibility would be to define a universal, game-independent procedure allowing the development of "reasonable" initial settings that approximate the relative importance of particular features or their combinations.

Naturally, the problem of how to define the optimal set of features that compose the evaluation function for a particular game is also a challenge. A more demanding question would be how to define the human-guided, but semi-autonomous and *game-independent* process that would have led to the construction of suboptimal set of board (game) features. This issue was discussed by Paul Utgoff who stated in [111]: "*Constructing good features is a major development bottleneck in building a system that will make high quality decisions. We must continue to study how to enhance our ability to automate this process*". Utgoff suggested that game features should be overlapping and form a layered, hierarchical system in which more complex features are built based on simpler ones.

Another challenge concerns autonomous learning with zero initial knowledge (except for the rules of the game). The majority of game playing programs rely on carefully designed expert features reflecting positional and tactical nuances of the game. A good counterexample is Anaconda which does not rely on built-in human knowledge at all, and as such is an apparent, successful example of learning from scratch using Computational Intelligence techniques. Chellapilla and Fogel's success contradicts Allen Newell's opinion (supported also by Marvin Minsky): "*It is extremely doubtful whether there is enough information in 'win, lose, or draw' when referred to the whole play of the game to permit any learning at all over available time scales*" [67].

3.2 Creativity - Knowledge Discovery

One of the long-term goals of CI in game playing is development of **creativity** mechanisms which implemented in the playing program might lead to spontaneous **knowledge discovery**.

Some successful examples of such "emerging intelligent behavior" have already been presented in the literature however, according to the author's knowledge, all of them were merely "the side effects" of the training process. The most famous example is probably Tesauro's TD-Gammon described in the previous section, which according to former Backgammon world champion Robertie, came up with genuinely novel strategies that no one had used before. TD-Gammon's play caused revision in human positional judgement in this game leading, for example, to invention of new opening moves - proposed by TD-Gammon and subsequently proved (in exhaustive, statistical analysis as well as tournament play) to be successful. Another interesting observation concerning TD-Gammon is the development of spatial weight patterns in the MLP, responsible for representation of particular game concepts, which *were not explicitly presented in the course of training* [103].

Similar observations about *ad-hoc* feature discovery and feature representation in neural network weights were reported in [70, 62, 71] concerning the game of Bridge. The authors considered the so-called Double-Dummy Bridge Problem, which consists in answering the question about the number of tricks to be taken by a pair of players assuming perfect play of all four hands with all cards being revealed.

Several MLP networks with $0, 1$ or 2 hidden layers were trained in a supervised manner and tested based on the data from the GIB Library [37], created by Ginsberg using his GIB program [39][8]. The input layer was composed of 52 neurons and each of them was assigned to a particular card from a deal. The value of this neuron denoted the hand containing this card (e.g. $N : 1.0$, $S : 0.8$, $W : -1.0$, $E : -0.8$). A single output neuron yielded the predicted number of tricks to be taken (the output range - $[0.1, 0.9]$ was divided into 14 intervals of equal length). Besides deal assignment, no additional information e.g. the rules of the game or the strength of particular cards was provided to the network. Except for achieving satisfying numerical results the other main goal of that research was exploration of networks' knowledge representation and search for patterns in the weight space that possibly represented particular "Bridge features" (e.g. the relative strength of cards). Examining the weights in the trained networks revealed several interesting observations.

Firstly, weights of outgoing connections from input neurons representing *aces* and *kings* always had the biggest absolute values. This feature was simple to explain (for humans) - these cards are the most important in the game of Bridge, especially in no trump contracts.

Secondly, in each trained network there were exactly four connections from input to hidden neurons with weights' absolute values noticeably bigger than

[8] GIB is considered one of the top machine Bridge players.

all the others (about 25.0 vs less than 7.0). Not surprisingly these favored connections started from four input neurons assigned to *aces*.

Thirdly, in all networks it was possible to point out four hidden neurons focused on particular suits (one neuron per suit). Absolute values of connection weights from inputs representing the respective suit to such hidden neuron were much bigger than absolute weight values from the remaining inputs.

Finally, a very interesting feature which appeared in all trained networks with sufficient number of hidden neurons, was the presence of four hidden neurons, each of which focused on five top cards from one particular suit: *ten*, *jack*, *queen*, *king* and *ace*. In each of these five-card groups the most important connections were from *queens* and *kings*, *jacks* were less important, but still much more relevant than *aces* and *tens*. The hypothesis is that these hidden neurons were responsible for a very important aspect of the game - *the finesses*.

All the above observations are in line with human knowledge about the game of Bridge. Estimation of the strength of individual cards as well as entire suits is the basic information considered in the process of a hand's evaluation. Even though these game features are trivial to understand for human players they are not necessarily easy to discover by a neural network. Moreover, quite surprisingly, in the simple training process, the networks were also able to independently discover the notion of *the finesses*, which is a subtle mechanism - not rarely deciding about the final number of tricks taken by a playing pair.

Interesting observations concerning knowledge discovery in the game of Chess were reported in the MORPH experiment [56, 42] which implemented pattern based learning with the weights of patterns being modified through the TD(λ) method combined with *simulated annealing*. Although the strength of MORPH was far inferior to GNU Chess, the patterns learned by the system were consistent with human Chess knowledge. In particular MORPH was able to play openings on a reasonable level, despite the fact that no information about the significance of development or controlling the center of the board in the opening phase had been added to the system. On the other hand, one of the weaknesses of MORPH was poor scalability with respect to the number of patterns, due to the lack of efficient selection mechanisms. Nevertheless, the system was able to defeat human novices while searching only 1-ply.

A general approach to automatic feature generation was presented by Fawcett and Utgoff [32]. Given only domain theory and the ability to solve problems in this domain the system called Zenith was able to automatically generate a set of relevant domain features. The system started from a single feature created automatically from the problem's goal (e.g. "win of white") and by using four predefined types of transformations: decomposition, abstraction, regression and specialization, gradually extended the set of features in an iterative manner. Zenith was applied to Othello with promising results. For example, the system autonomously discovered the importance of stable pieces, i.e. the ones which cannot be reversed [32].

Another well-known example of independent feature discovery in games is Anaconda [25, 33] described in sect. 3.1, which received its name due to the "snake-like" way of playing - in most of the games won by the program its opponent was blocked and therefore forced to make a weak move. However, neither in the input data nor in the evolutionary process of Anaconda's development the concept of mobility was ever explicitly considered. Hence the importance of mobility must have been "invented" by the system or more precisely by the evolutionary process, which guided Anaconda's development.

The potential strength of neuro-evolutionary approach was also reported in Othello [69]. The evolved networks "discovered" positional features and advanced mobility issues indispensable for high-profile tournament play.

All the above examples led to discovering new features, previously unknown to the system, induced from the training data. The ultimate goal that can be put forward in this context is *autonomous* discovering of *all* relevant components of the evaluation function in a way allowing their *separation* and *explanation*. Such a requirement goes beyond Anaconda experiment and other neural or neuro-evolutionary type approaches that resulted in efficient *numerical approximation* of the board state, but lack the *feature-based formulation* of the evaluation function.

3.3 Intuition

Implementation of the concept of intuition is definitely one of the greatest challenges in computer games and also in computer science in general. Nowadays, despite the major breakthroughs made in several disciplines and despite increasingly deeper, scientific understanding of the nature, intuition - paradoxically - becomes more important than ever.

One of the most salient research studies focused on understanding (and implementation of) intuition was performed by Herbert Simon - a Nobel Prize Winner in Economics. According to Simon intuition is nothing mysterious or extraordinary and simply relates to a subconscious pattern recognition process able to immediately provide appropriate pattern(s) among those stored in the memory, based on our knowledge and experience. According to Simon, this does not mean that intuition is an irrational process - he considered it to be a rational but neither conscious nor analytical one [95].

Simon was optimistic about the potential abilities of "thinking machines" and predicted that any "intelligent" human activity (thinking, creativity, decision making, intuition and other) will ultimately be implemented in artificial systems.

In most of mind board games intuition plays a leading role at master level of play. Consider for example Chess. With a branching factor of about 30, in a 50 move (100 ply) game there are about 10^{147} contingencies, which is an enormous number for any human being (grandmasters are believed to search no more than a few hundred contingencies during the entire game). How then it is possible that Chess champions are able to play at such a high level? One of the factors is intuition which allows them to perform highly selective search in this huge space, *although in many cases they are not able to explain why they have chosen to search a particular contingency and skipped the others.* Moreover, when playing simultaneous games, Chess grandmasters usually need only a few seconds to make a move, which generally proves to be very strong (often optimal). This means that they have the ability to immediately find the most relevant information characterizing board position and recognize the most promising continuation (move), usually *without deep, precise calculation of its contingencies.*

Another aspect of intuition in board games is the ability to almost instantaneous recognition of strengths and weaknesses of a given position. A grandmaster usually needs only a few seconds of board analysis in order to tell which side is in the winning or favorable position. One of the possible psychological explanations of this phenomenon is the ability of advanced players to link the new position with previously explored familiar ones and consequently to focus on moves and plans associated with these, already known, positions [27] (this topic is further discussed in sect. 3.4).

Based on the above described results of applying intuition in games, one can provide the operational definition of intuition as an instantaneous, subconscious recognition/reasoning process which does not rely on precise, deep calculations, but instead rather refers to past experiences and previously acquired general knowledge. Consequently, in most mind games, intuition is one of the main factors contributing to the beauty and attraction of the game. Its application often leads, for example, to long term material sacrifices without apparent possibility of its recovery. A well known example in Chess is *the immortal game* played in London in 1851 by Adolf Anderssen and Lionel Kieseritzky in which white sacrificed bishop (on move 11 - see Fig. 1(a)) and subsequently two rooks and a queen (starting on move 18 - see Fig. 1(b)) in order to checkmate on move 23 - (Fig. 1(c)). Certainly, the last three sacrifices were tactical ones, i.e. their consequences could have been precisely calculated by Anderssen, but the introductory sacrifice (bishop on move 11) is an example of an intuitive type of move based on players experience and his "feeling" of the board position.

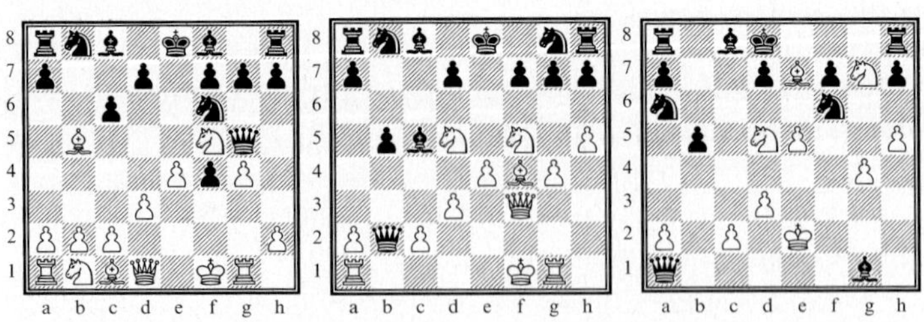

(a) After 11. Rg1 ... (b) After 17. ... Qxb2 (c) 23. Be7++. White won

Fig. 1. Anderssen vs Kieseritzky, *Immortal game*, London, 1851

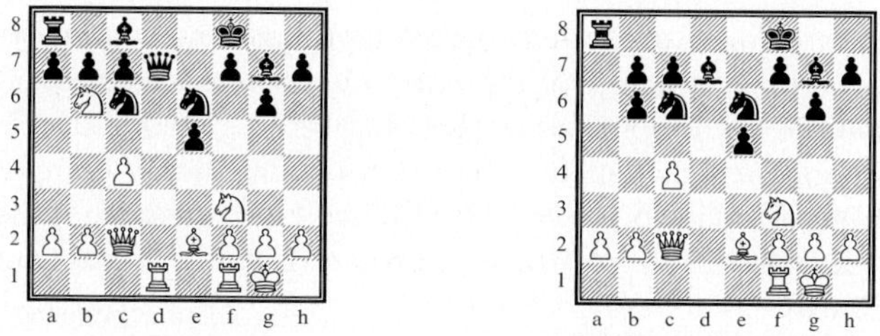

(a) After 16. Nb6 ...

(b) After 16. ... axb6, 17. Rxd7 Bxd7

Fig. 2. Karpov vs Kasparov, New York, 1990

Another interesting example of intuitive sacrifice occurred in the game played between two great archenemies: Anatoly Karpov and Garry Kasparov in the New York match in 1990. In the middle-game position Kasparov sacrificed queen for a rook and knight on moves 16 – 17 (see Fig. 2) and this sacrifice was clearly positional with no immediate tactical or material threats. The game continued up to 53th move, when players agreed for a draw.

Theoretically, human-type intuition in machine playing may possibly emerge as a "side effect" of using a close to optimal evaluation function (on condition that such a function could be practically specified and implemented). Examples of "intuition" of such origin have been observed in the famous Kasparov vs Deep Blue re-match, in which some of the machine's moves were described by grandmasters commentating on the match as *phenomenal* and *extremely human*.

One of very few published attempts focusing on formalization of intuitive concepts in Chess was recently described by Arbiser [3]. The author proposes the way of formalizing such concepts as capture, attack, threat, sacrifice, etc. as well as the notion of *style of opponent's play*, i.e. aggressive, defensive, conservative, tactical or positional. The underlying idea is based on generalization of the null-move heuristic in such a way that instead of hypothetical opponent's moving twice in a row, the opponent is allowed to virtually change one of his or our pieces or add/delete a piece and then make a move. For example the notion of aggressive play will be implemented by exchanging one of the opponent's or our pieces into a strong opponent's piece before deciding a move. Such an exchange would most probably cause immediate threats to us thus forcing the choice of an appropriate response. In short, the following scheme is proposed: modify the board in an adequate manner before calling a regular search algorithm and ensure that the chosen move would be valid and sound in the original board position i.e. the one without initial, fictitious modification. Although the description of the method raises several questions concerning its time complexity as well as the omitted implementation details, overall the algorithm seems to be a step in the right direction.

Understanding and furthermore implementation of the mechanism of intuition in artificial players is one of the main challenges for CI in games. Several issues described in the remainder of this chapter, e.g. geometrical trajectories, positional generalization, feature abstraction may partly compliment to the implementation of intuition, but the efficient and general approach to this wonderful human ability is yet to be specified. I would argue that unless programs (machines) capable of making intuitive moves (in the above described sense) in Chess and other mind board games are created, we should be very cautious about announcing the end of the human era in these games.

In 1931 Albert Einstein wrote [28]: "*The intuitive mind is a sacred gift and the rational mind is a faithful servant. We have created a society that honors the servant and has forgotten the gift*". This aphorism, originally related to religion, can also be referred to other human activities including the domain of machine game playing development.

3.4 Abstraction and Generalization

As discussed in the previous section, one of the facets of human game playing is the ability to abstract particularly relevant game features from a given board position. This skill allows experienced players almost immediate estimation of positional and tactical strengths and weaknesses on both sides as well as to point out future possibilities and potentially promising moves. For example in Chess these crucial features include *pawn structure*, *cooperation of figures* (e.g. two rooks on the 2nd (resp. 7th) line or multiple attack on point F2 (F7 resp.)), *mobility*, *tempo* and many more.

In practically all popular mind board games vital positional and tactical features are context-sensitive. Due to the presence of other pieces on the board their appropriate classification is not a straightforward task for machine players and requires both abstraction and generalization capabilities.

Another generalization task is an attempt to reason on the quality of a move based on shallow search. On one hand it is hard not to agree with Schaeffer, Hlynka and Jussila who stated in [88]: "*There is no free lunch; you can't use shallow search results to approximate deep results*" and therefore advised: "*the weights [of an evaluation function] must be trained using depths of search expected to be seen in practice*".

On the other hand the above claims are not necessarily valid when estimating *the relative* strength of the moves without focusing of their true numerical evaluation. A crude relative estimation of possible moves is crucial for the efficacy of several search algorithms (e.g. the alpha-beta based ones). This issue is further discussed in the next section.

A challenging test of generalization skills applicable to machines is solving game problems defined on arbitrarily large game boards. Intelligent approach to such problems requires efficient generalization from shallow search results. John McCarthy in 1998 in his comments to intelligent Chess problem solving, referring to the famous *Reti problem* (Fig. 3), stated: "*Note that Reti's idea can*

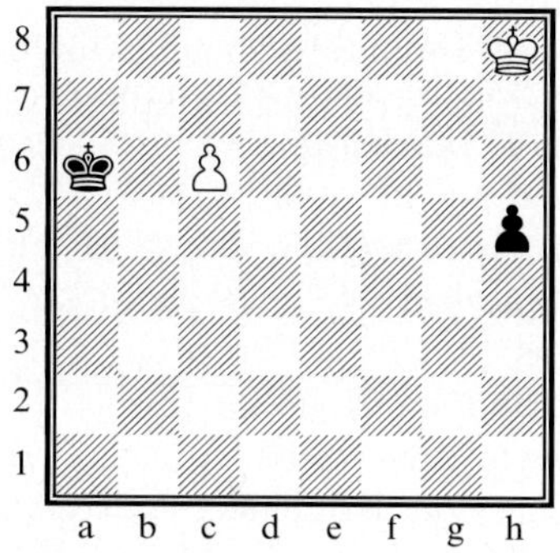

Fig. 3. Reti ending. White to begin and draw.

be implemented on a 100×100 *board, and humans will still solve the problem, but present programs will not ... AI will not advance to human level if AI researchers remain satisfied with brute force as a substitute for intelligence ... Would anyone seriously argue that it is impossible for a computer to solve the Reti problem by other than brute force?*" [66]. An interesting approach to this type of generalization is expressed by the Linguistic Geometry (LG) which focuses on evaluation of *trajectories of possible solutions* rather than on exact exploration of the game tree [98]. Instead of traditional search-based approach, LG proposes methods for construction of problem solving strategies. These strategies can be represented as trajectories on the board (e.g. in the endgame problems) and to some extent allow formalization of expert knowledge and intuition (see [98] for details).

Another challenge related to abstraction and generalization is the quest for learning methods capable of generalizing knowledge across game boards of different sizes. One particularly interesting question is: how to apply the outcomes of learning on small boards to the learning process performed on larger boards? One possible approach is to use *incremental training* methods implemented in neural networks according to the following procedure. Learning starts off in the environment (game board) smaller than the target one. During the training process the environment is gradually increased - up to desired size - and after each change of size the limited number of training examples is modified/added in order to capture new features that arose in this larger, more complicated environment. The claim is that after some training time, the system should be able to recognize features, which are *invariant to the size (degree of complication) of the environment.* In such cases these features will be shared among several instances of the environment and during the training process used to make generalizations about the learning task.

Consider, for example, a game which is played on a board of size n. In the proposed approach the training process begins on a game board of smaller size $k, k < n$ and after the agent learns how to play or solve problems on this board, the board size is increased to $k + t$, where t depends on particular game ($t \in \{1, 2\}$ for the majority of popular games). Then the agent is re-trained *in a limited manner* based on the new set of problems presented on the increased board. The re-training procedure is significantly shorter than the regular training performed on the board of size k. Once again the board size is increased and the agent is re-trained, etc. The whole procedure is stopped after the re-training on board of size n is completed.

The underlying idea is that after the preliminary phase, learning should become relatively easier, and solutions for problems defined on larger boards would be developed by the system based on already defined solutions for problems stated on smaller-size boards. Hence, *subsequent learning would mostly involve efficient use of previously acquired knowledge.*

The above described learning scheme is problem independent, but can be applied only to a certain type of games such as Checkers, Othello or Go, which can be easily defined on boards of different sizes. Such training scheme was used in Othello on 6×6, 8×8 and 10×10 boards (the board of size 10×10 being the target one) [59, 60]. The MLP neural network was fully trained on examples from 6×6 board and subsequently re-trained, in a limited manner, on 8×8 and 10×10 boards' examples. The training goal was to point out the best move in a given position. The results of the above-described incremental training procedure were compared with the full backpropagation training carried out exclusively on 10×10 board examples. The amount of time required for incremental training was considerably lower than in the opposite case. Also numerical results showed a slight improvement over a one-shot training procedure: after incremental training the network responded with the best move (selected according to applied heuristic) in 40% of the cases, compared to 34% achieved in the full backpropagation training on 10×10 board [60].

3.5 Pre-Ordering of Moves

In practical applications, efficient moves pre-ordering should rely on shallow search or no search at all, otherwise, the remaining time devoted to deeper, selective search may be insufficient. Efficacious pre-ordering of moves is again a "very human" skill. Human Chess players, for example, can estimate roughly 2 positions per second - compared to 200 billion ones checked in a second by Deep Blue - and therefore must be extremely effective in preliminary selection of moves.

There exist a few popular, search-free strategies of moves pre-ordering basing on historical goodness of the move in previous games played. These include the *history heuristics*, *transposition tables* or the *killer move* heuristics. In most cases these methods are highly effective since the assumption that a move which often appeared to be efficient in the past is more likely to be suitable for the current game position than other "less popular" moves is generally correct (see e.g. [83] for further discussion and experimental results in Checkers).

An interesting approach to moves pre-ordering in Chess was presented by Greer [43]. The method relies on pattern-oriented classification of moves based on heuristically defined *influence* of a particular move on certain board regions. At first, each square is assigned a label that represents heuristical belief in which of the two players controls this square. Combining this information for all squares leads to the chessmap which represents the regions of the board that are in favor for each side as well as the neutral areas, where none of the players has a visible advantage[9]. Additionally, for each square (or more generally each sector composed of some number of squares) the so-called valueboard is defined based on the relative strength of the control that one side has over that square (sector). The influence of a move on a given square (or sector) is defined as the sign of a difference between the valueboard after that move would have been made and before (i.e. in a current position). This allows to detect the squares that would be strengthened by that move as well as the ones that would be weakened.

In order to learn the influence relationship for Chess positions a neural network was trained based on $10,000$ positions extracted from master and grandmaster games. The input layer represented influence labels of 64 squares and the kings' locations. The desired output values in the 64 element output layer (one neuron per square) were the influence labels after a move had been made in the actual game. After training, the outputs of the network were used to order board squares according to their predicted influence values. Consequently moves that influenced the highest ranked sector(s) of the board were considered as the most promising ones.

The above procedure was further enhanced by giving priority to forced and capture moves. Several tests of this interesting pattern-based heuristic were carried out including the ones on the set of 24 Bratko-Kopec positions [51], for which the quality of the method - calculated as the number of searched nodes - was comparable to the result of applying the history heuristic.

Pattern-based pre-ordering of moves is in line with psychological observations of how human grandmasters make decisions about which moves to consider first. As Johannes Fürnkranz stated in his excellent review of a decade of research in AI and computer chess [35] referring to the work of deGroot [27] "*the differences in playing strengths between experts and novices are not so much due to differences in the ability to calculate long moves sequences, but to which moves they start to calculate. For this preselection of moves chess players make use of chess patterns and accompanying promising moves and plans*".

Pattern-based approaches seem to be perfectly suited for Go, which is a territory game. Since today's Go programs are far from being a threat to human players and rely only on simple pattern matching [73] application

9 The idea of calculating the *influence* of white and black pieces in order to divide the board into sections controlled by the respective players was initially introduced by Zorbist [114] in Go.

of pattern-oriented methods of moves pre-selection in this game may be a promising research direction.

3.6 Opponent Modeling

Modeling the opponent is another fundamental issue in current AI/CI research. The problem actually extends far beyond the game playing domain and is considered as a crucial aspect of any competitive multi-agent environment (e.g. decision support systems, stock markets, trading systems, etc.). In game domain the relevance of opponent modeling strongly depends on the choice of a game. Relatively lesser impact on the quality of playing programs concerns perfect information games, such as Chess, Checkers, Go, Othello, etc. However, also in these games the problem is not negligible. The style of play (tactical vs positional, aggressive vs conservative, etc.), if properly modeled, can provide an important indication for a game playing program. For example, in a disadvantageous position a program could use a specific style of play in order to hinder the potential victory of the opponent and strive to achieve a draw. Another example is seeking the chance to win an even game by steering it to inconvenient (for the opponent) positions and thus provoking an opponent's mistake.

A similar situation is observed among human players. Nearly each of the top players in any popular board game has opponents who are "less convenient for him", i.e. achieve relatively better results against that player than is indicated by their ranking (e.g. ELO in Chess). In other words the "winning relation" is non-transitive and the ranking points provide only general, statistical information about player's strength.

Modeling the opponent is far more important in imperfect information games, especially the ones, in which deception (bluffing) is an inherent part of the rules. A simple, though instructive, example is the kids' game Rock-Paper-Scissors, also known as RoShamBo [12]. In this game, each of the players independently and simultaneously chooses among Rock, Paper and Scissors. In the simplest case of two players the winner is the one whose choice "beats" the opponent's choice under the following rules: Rock beats Scissors, Scissors beat Paper and Paper beats Rock. If both players point out the same object the turn ends with a draw. Even though the rules of the game are trivial, the game itself is more demanding that one might expect at first glance. A simple solution is of course choosing actions randomly according to uniform distribution. Such approach would statistically lead to a draw, however it does not take into account the opponent's playing policy which may possibly be inferred from his/her previous play. Moreover, except for trying to predict the opponent's next move, a skilful player (program) should avoid to be predictable itself. Hence, simple rule-based approaches are not sufficient and more sophisticated methods are to be employed.

Another example of the game in which opponent modeling is crucial for efficient playing is a dice game Perudo [57] also known as Liar's Dice. The rules of Perudo are not very complicated [58], though not as simple as those of RoShamBo. Playing the game well requires quite sophisticated analysis of opponents' past actions in order to detect possible bluffing, since the game significantly consists in bluffing and straightforward playing most probably would not lead to success against experienced opponents.

Certainly the most popular "game of deception" is Poker. Here the notion of (objectively) optimal playing is hard to define due to the huge amount of uncertainty regarding hidden opponents' cards (the *hole cards*) and the latent *community cards*[10]. Hence the optimal behavior can only be estimated with some probability and its calculation strongly depends on the opponents' actions. A simple example given in [13] concerns the frequency of bluffing by the opponent. The one who bluffs more frequently should be called more often compared to the one who bluffs relatively rarely. One possibility of modeling opponent's behavior is to construct a statistical model for the next move prediction based on sufficiently large number of games already played against that opponent. Another possibility is to train a neural network to predict the opponent's next action. Poki-X team (cf. sect. 2) employed a standard, one-hidden-layer MLP network with 19 inputs representing particular aspects of the game and three outputs corresponding to three possible opponent's decisions (*fold*, *raise* or *call*). The outcome of the network provided a probability distribution (after output normalization) of the opponent's action in a given context defined by the input features. After training on deals played by a given opponent, magnitudes of network's weights reflected the relative impact of particular input features on the predicted outcome, which allowed further exploration of the input data, e.g. finding new features and defining a relatively small number of context equivalence classes in the input space.

Additional advantage of using neural nets for the opponent modeling task is their ability to adapt to changes observed in the training patterns (representing the opponent's behavior) by adequate weights' tuning[11].

Computational Intelligence methods are very well suited to the problem of opponent modeling. Probabilistic methods allow building a generic opponent's model for a given game, which can be further optimized, e.g. with the use of genetic algorithms. An alternative approach, especially in the case when the opponent's patterns of activity vary in time, is to use neural networks, due to their capability of adapting internal parameters to the gradually changing input training data. Another promising avenue is to use TD learning and adapt internal parameters of the playing system according to achieved results.

Definitely, on-line, adaptable and close to reality modeling of the opponent is one of the fundamental challenging problems in any non-trivial game,

[10] We refer to the Texas Hold'em variant of Poker, which is now the most popular version of this game. The rules of the game can be found for example in the excellent books by David Sklansky [96, 97]

[11] Certainly such network weights' adjustment requires the use of appropriate training scheme and is possible on condition that changes in the opponent's behavior are relatively smooth.

in particular in imperfect information games, in which data hidden by the opponent can only be inferred by analysis of his/her past actions in similar game situations. Proper prediction of this hidden information increases the expected outcome by decreasing the amount of uncertainty.

It is worth to note that the problem of opponent modeling becomes much more demanding when multi-player situation is considered, in which players can form ad-hoc coalitions (formal or informal) against the current leader or in order to gain some benefits. In such a case players' decisions are highly contextual and strongly depend on short-term and long-term goals of the coalitions they belong to.

3.7 Universality of Tools

One of the grand challenges in game playing is associated with designing general-purpose methods and algorithms that abstract from particular games. CI is well located in this stream and several CI-based attempts to create game-independent methods were presented in the literature.

Game-Independent Learning

Research concerning game-independent learning aims at developing systems capable of learning any game belonging to a certain class of games. This topic was very popular in the mid 1990's when some renowned methods and systems originated.

One of the well known universal learning systems is Michael Gherrity's SAL program [36] capable of learning any two-player, perfect information, deterministic game. SAL consists of a kernel that uses TD learning combined with neural network's backprop learning in order to learn the evaluation function for a given game. The kernel is game-independent and remains unchanged for different games. The rules of making valid moves for any particular game are represented by the game-specific module. SAL learns by trial and error from the games it has played hitherto. It generates two evaluation functions, one for each playing side which allows learning non-symmetric games or imposing asymmetry in symmetric games, if necessary. The system uses only 2-ply search of the game tree. SAL's success is strongly hindered by slow learning. For example, it took the program $20,000$ games to learn to play the Tic-Tac-Toe game.

Another interesting approach to game-independent learning is represented by Susan Epstein's HOYLE system [30, 31], able to learn two-person, deterministic, perfect information games. The underlying idea of HOYLE is to use a set of game independent advisors each specializing in a narrow, specific aspect of game-playing (e.g. one advisor may focus on material advantage while another one on finding the winning moves or sequences of moves, etc.). Each of the advisors may recommend some moves and all of them can comment on these proposals from their specialized viewpoint. Finally the advisors vote using a simple arithmetic voting system. Similarly to SAL, HOYLE uses only shallow search (2-ply ahead at most).

The diversity of advisors plays a crucial role in learning a new game. Each of the advisors learns patterns from played games chosen according to its individual priorities. One advisor may be focused on patterns related to the opening moves while another, for example, on those related to strong, winning moves, etc. Besides game-specific knowledge that can be acquired by the advisors based on analysis of the played games, HOYLE is *a priori* equipped with some general knowledge about the domain of two-person, deterministic games.

The efficacy of HOYLE has been demonstrated by playing Tic-Tac-Toe and Nine-Men's Morris. The potential of Epstein's approach in more complicated games (Chess, Checkers, ...) has not been experimentally proven.

MORPH II developed by Robert Levinson [55] is another example of game-independent learning system and also a problem solver. MORPH II is a direct extension of MORPH - a Chess learning program mentioned in sect. 3.2. MORPH II uses several CI learning techniques which include neural network-like weights propagation and genetic algorithm-type pattern evolution. Additionally, MORPH II implements symbolic learning. The system is capable of autonomous abstraction of new features and patterns and development of its own learning modules. Like its predecessor, MORPH II relies on shallow search equal to only 2 plies on average. The system has successfully learnt to play Chess on a novice level. Its strength against more demanding opponents hasn't been demonstrated.

All the above-mentioned general learning systems are potentially capable of picking up any game within a certain class of games. They use CI techniques combined with AI symbolic learning and multi-agent approach. They all rely on shallow search, just 1 or 2 plies. All of them have demonstrated the ability to efficiently learn how to play very simple games or more complicated ones, but at a novice level only. Definitely this direction deserves further exploration and the issue of how to design universal, game-independent learning systems is one of the grand challenges for CI community in the area of intelligent game playing.

Multitask Learning

The idea of designing game-independent learning systems can also be realized within the multitask or incremental learning schemes. Multitask learning utilizes simultaneous learning of a few tasks and sharing the representation issues, experience and knowledge among all of them in order to make the overall learning process faster and more effective. Incremental, lifelong learning is usually implemented as a sequential learning process, i.e. tasks are being learnt one after another, but again representation of problems as well as knowledge acquired in previous learning are widely shared and involved in subsequent learning, consequently making the latter easier.

Several approaches to multitask and lifelong learning (not only within game playing domain) has been developed in the last 10 years (e.g. [108, 107, 21, 64]), but there is still a strong demand for new concepts in this area. In case of game playing one may think of using the experience gained in learning one or more games in order to alleviate the effort of learning another, similar game. This type of learning is typical for humans. For example, previous experience in playing cards is one of the fundamental factors in efficient learning of a new card game.

Generally speaking, in case of similar games some representation issues and high level game rules are either common or very similar. Therefore, when learning a new game there is no need to start from the very beginning. The learning system may exploit already possessed knowledge - albeit usually its appropriate tuning will be required.

4 Conclusions

Mind games provide cheap, replicable environments, perfectly suited for testing new Computational Intelligence learning methods and search algorithms. They also serve as an excellent framework for testing and validating various implementations of human-type cognitive skills, e.g. intuitive behavior, creativity, knowledge discovery, or unguided, autonomous and context-sensitive learning.

All the above skills are crucial in achieving the long-term goal of CI in mind game playing research, which is the ability to mimic human methods of learning, reasoning and decision making. Several challenging problems have to be addressed on this path. Some of them are proposed and motivated in this chapter.

One of the most interesting issues is implementation of mechanisms of autonomous knowledge discovery that would lead to creation of new game features and new playing strategies. In particular a very challenging task is autonomous choice of board features that compose efficient (close to optimal), descriptive game representation allowing adequate evaluation of board positions. At the moment the development of a world-class playing program requires that the set of features be predefined by human experts. Even though there exist a few notable examples of learning how to play certain games without human expertise, there is still a lot of work ahead.

Another fundamental issue is the ability to improve artificial player's behavior through the learning process resting solely on the experience-based knowledge acquired from previously played games. An interesting, open problem in this area is analysis of pros and cons of two main learning schemes used so-far, i.e. playing against external opponents vs self-playing. In the former case, additional open problems concern the optimal choice of the training opponents and the training schedule. Both types of learning were successfully applied to various board games especially with TD learning algorithms. Further exploration of these two directions may ultimtely lead to the development of a general purpose learning engine (system) capable of learning any board mind game.

Another formidable challenge is implementation of intuition in the game-playing systems or, more precisely, implementation of mechanisms that would efficiently pretend human-type intuitive behavior. Such achievement would straightforwardly lead to the efficacious search-free pre-selection of moves and instantaneous estimation of position strength as well as the ability to play strong positional moves relying on shallow search only. All three above mentioned skills are typical for experienced human players, but still generally non-attainable for machines.

Yet another challenging issue concerns game independent learning, in particular incremental learning methods allowing for sequential or simultaneous learning of a few games. Sequential incremental game learning may rely on appropriate tuning of already possessed knowledge and generation of new features only when necessary, i.e. when they are "sufficiently different" from already discovered ones. Simultaneous learning requires that representational and computational issues be shared on-line among various learning tasks (games) and each learning task benefits from this synergy.

Achieving the above challenging goals does not necessarily mean construction of an omnipotent, unbeatable program/machine capable to play "in God's way". On the contrary, as humans make mistakes and are not infallible, also the CI-based playing systems may possibly suffer from "human" weaknesses, though to a much lesser extent.

In summary, CI-based methods focus on the way the game playing issues are implemented and solved rather than on the quality of play, which is regarded as relevant, yet subservient, supplementary goal. This distinguishes CI-based game-learning systems from traditional AI-based approaches focused mainly on maximization of the level of play. Consequently, in the near future AI playing systems are with high probability hardly to be defeated by CI-based ones.

The above conclusion, on the other hand, does not mean that application of CI methods to mind game playing is not interesting or not advantageous. On the contrary, I would argue that the mind game playing research will more and more be tied with Computational Intelligence methods and its future will be closely related to the development of psychologically motivated learning processes attempting to follow higher-level human competencies.

Acknowledgement

The author is grateful for the support from the Warsaw University of Technology, grant no. 504G 1120 0008 000.

References

[1] I. Aleksander. Neural networks - evolutionary checkers. *Nature*, 402(6764):857, 1999.

[2] T. Anantharaman and M. Campbell. Singular extensions: Adding selectivity to brute-force searching. *Artificial Intelligence*, 43:99–109, 1990.

[3] A. Arbiser. Towards the unification of intuitive and formal game concepts with applications to computer chess. In *Proceedings of the Digital Games Research Conference 2005 (DIGRA'2005)*, Vancouver, B.C., Canada, 2005.

[4] L. Barone and L. While. An adaptive learning model for simplified poker using evolutionary algorithms. In *Proceedings of the Congress of Evolutionary Computation (GECCO-1999)*, pages 153–160, 1999.

[5] L. Barone and L. While. Adaptive learning for poker. In *Proceedings of the Genetic and Evolutionary Computation Conference*, pages 566–573, 2000.

[6] J. Baxter, A. Tridgell, and L. Weaver. Experiments in parameter learning using temporal differences. *ICCA Journal*, 21(2):84–99, 1998.

[7] J. Baxter, A. Tridgell, and L. Weaver. Knightcap: A chess program that learns by combining td(λ) with game-tree search. In *Machine Learning, Proceedings of the Fifteenth International Conference (ICML '98)*, pages 28–36, Madison Wisconsin, July 1998.

[8] J. Baxter, A. Tridgell, and L. Weaver. Learning to play chess using temporal differences. *Machine Learning*, 40(3):243–263, 2000.

[9] D. Beal. A generalised quiescence search algorithm. *Artificial Intelligence*, 43:85–98, 1990.

[10] D. F. Beal and M. C. Smith. Learning piece values using temporal differences. *ICCA Journal*, 20(3):147–151, 1997.

[11] H. Berliner. The B* tree search algorithm: A best-first proof procedure. *Artificial Intelligence*, 12(1):23–40, 1979.

[12] D. Billings. Thoughts on RoShamBo. *ICGA Journal*, 23(1):3–8, 2000.

[13] D. Billings, A. Davidson, J. Schaeffer, and D. Szafron. The challenge of poker. *Artificial Intelligence*, 134:201–240, 2002.

[14] B. Bouzy and T. Cazenave. Computer Go: an AI oriented survey. *Artificial Intelligence*, 132(1):39–103, 2001.

[15] D. Bump. GNU Go. http://www.gnu.org/software/gnugo/gnugo.html, 1999.

[16] M. Buro. Probcut: An effective selective extension of the alpha-beta algorithm. *ICCA Journal*, 18(2):71–76, 1995.

[17] M. Buro. From simple features to sophisticated evaluation functions. In H. J. van den Herik and H. Iida, editors, *Proceedings of Computers and Games Conference (CG98)*, volume 1558 of *Lecture Notes in Computer Science*, pages 126–145, Springer, Berlin, 1999.

[18] M. Buro. Toward opening book learning. *ICCA Journal*, 22(2):98–102, 1999.

[19] M. Buro. Improving heuristic mini-max search by supervised learning. *Artificial Intelligence*, 134:85–99, 2002.
[20] M. Campbell, A. J. Hoane Jr., and F.-h. Hsu. Deep Blue. *Artificial Intelligence*, 134:57–83, 2002.
[21] R. Caruana. Multitask learning. *Machine Learning*, 28:41–75, 1997.
[22] K. Chellapilla and D. B. Fogel. Evolution, neural networks, games, and intelligence. *Proceedings of the IEEE*, 87(9):1471–1496, 1999.
[23] K. Chellapilla and D. B. Fogel. Evolving neural networks to play checkers without relying on expert knowledge. *IEEE Transactions on Neural Networks*, 10(6):1382–1391, 1999.
[24] K. Chellapilla and D. B. Fogel. Anaconda defeats Hoyle 6-0: A case study competing an evolved checkers program against commercially available software. In *Congress on Evolutionary Computation, La Jolla, CA, USA*, pages 857–863, 2000.
[25] K. Chellapilla and D. B. Fogel. Evolving a neural network to play checkers without human expertise. In N. Baba and L. C. Jain, editors, *Computational Intelligence in Games*, volume 62, pages 39–56. Springer Verlag, Berlin, 2001.
[26] P. Darwen and X. Yao. On evolving robust strategies for iterated prisoner's dilemma. volume 956 of *LNCS*, pages 276–292. Springer, 1995.
[27] A. D. de Groot. *Thought and Choice in Chess.* Mouton Publishers, The Hague, 1965.
[28] A. Einstein. *Cosmic Religion, with Other Opinions and Aphorisms.* 1931.
[29] M. Enzenberger. Evaluation in Go by a neural network using soft segmentation. In *Advances in Computer Games: Many Games, Many Challenges: Proceedings of the International Conference on Advances in Computer Games (ACG-10)*, pages 97–108, Graz, Austria, 2003.
[30] S. Epstein. Identifying the right reasons: Learning to filter decision makers. In R. Greiner and D. Subramanian, editors, *Proceedings of the AAAI 1994 Fall Symposium on Relevance*, pages 68–71, New Orleans, 1994. AAAI Press.
[31] S. L. Epstein, J. Gelfand, and J. Lesniak. Pattern-based learning and spatially-oriented concept formation in a multi-agent, decision-making expert. *Computational Intelligence*, 12(1):199–221, 1996.
[32] T. E. Fawcett and P. E. Utgoff. Automatic feature generation for problem solving systems. In D. Sleeman and P. Edwards, editors, *Proceedings of the 9th International Conference on Machine Learning*, pages 144–153. Morgan Kaufmann, 1992.
[33] D. B. Fogel. *Blondie24: Playing at the Edge of Artificial Intelligence.* Morgan Kaufmann, 2001.
[34] D. B. Fogel, T. J. Hays, S. L. Hahn, and J. Quon. A self-learning evolutionary chess program. *Proceedings of the IEEE*, 92(12):1947–1954, 2004.

[35] J. Fürnkranz. Machine learning in computer chess: the next generation. *ICGA Journal*, 19(3):147–161, 1996.
[36] M. Gherrity. A game-learning machine. PhD Thesis, University of California, San Diego, CA, 1993.
[37] M. L. Ginsberg. GIB Library. http://www.cirl.uoregon.edu/ginsberg/gibresearch.html.
[38] M. L. Ginsberg. GIB: Steps toward an expert-level bridge-playing program. In *International Joint Conference on Artificial Intelligence (IJCAI'99)*, pages 584–589, Stockholm, SWEDEN, 1999.
[39] M. L. Ginsberg. GIB: Imperfect information in a computationally challenging game. *Journal of Artificial Intelligence Research*, 14:303–358, 2001.
[40] H. Givens. PokerProbot. http://www.pokerprobot.com/, 2006.
[41] D. Gomboc, T. A. Marsland, and M. Buro. Evaluation function tuning via ordinal correlation. In *Advances in Computer Games: Many Games, Many Challenges: Proceedings of the International Conference on Advances in Computer Games (ACG-10)*, pages 1–18, Graz, Austria, 2003.
[42] J. Gould and R. Levinson. Experience-based adaptive search. In R. Michalski and G. Tecuci, editors, *Machine Learning: A Multi-Strategy Approach*, pages 579–604. Morgan Kaufmann, 1994.
[43] K. Greer. Computer chess move-ordering schemes using move influence. *Artificial Intelligence*, 120:235–250, 2000.
[44] E. A. Heinz. Adaptive null-move pruning. *ICCA Journal*, 22(3):123–132, 1999.
[45] F.-h. Hsu. *Behind Deep Blue.* Princeton University Press, Princeton, NJ, 2002.
[46] R. M. Hyatt. Crafty. ftp.cis.uab.edu/pub/hyatt, 2006.
[47] R. M. Hyatt, H. L. Nelson, and A. E. Gower. Cray Blitz. In T. A. Marsland and J. Schaeffer, editors, *Computers, Chess, and Cognition*, pages 111–130. Springer Verlag, New York, 1990.
[48] IBM Corporation. Deep Blue technology. http://www.research.ibm.com/know/blue.html, 2006.
[49] L. Kaelbling, M. L. Littman, and A. W. Moore. Reinforcement learning: A survey. *Journal of Artificial Intelligence Research*, 4:237–285, 1996.
[50] G. Kendall and G. Whitwell. An evolutionary approach for the tuning of a chess evaluation function using population dynamics. In *Proceedings of the 2001 Congress on Evolutionary Computation CEC2001*, pages 995–1002. IEEE Press, 2001.
[51] D. Kopec and I. Bratko. The Bratko-Kopec experiment: A comparison of human and computer performance in chess. In M. R. B. Clarke, editor, *Advances on Computer Chess 3*, pages 57–72. Pergamon Press, Oxford, 1982.
[52] C. Kotnik and J. K. Kalita. The significance of temporal-difference learning in self-play training td-rummy versus evo-rummy. In T. Fawcett

and N. Mishra, editors, *Machine Learning, Proceedings of the Twentieth International Conference (ICML 2003)*, pages 369–375, Washington, DC, USA, August 2003. AAAI Press.

[53] H. Kuijf. Jack - computer bridge playing program. http://www.jackbridge.com, 2006.

[54] M. Kusiak, K. Walędzik, and J. Mańdziuk. Evolution of heuristics for give-away checkers. In W. Duch et al., editors, *Artificial Neural Networks: Formal Models and Their Applications - Proc. ICANN 2005, Part 2, Warszawa, Poland*, volume 3697 of *LNCS*, pages 981–987. Springer, 2005.

[55] R. Levinson. MORPH II: A universal agent: Progress report and proposal. Technical Report UCSC-CRL-94-22, Jack Baskin School of Engineering, Department of Computer Science, University of California, Santa Cruz, 1994.

[56] R. A. Levinson and R. Snyder. Adaptive pattern-oriented chess. In L. Birnbaum and G. Collins, editors, *Proceedings of the 8th International Workshop on Machine Learning*, pages 85–89. Morgan Kaufmann, 1991.

[57] A. Macleod. Perudo as a development platform for Artificial Intelligence. In *13th Game-On International Conference (CGAIDE'04)*, pages 268–272, Reading, UK, 2004.

[58] A. Macleod. Perudo game. http://www.playperudo.com/, 2006.

[59] J. Mańdziuk. Incremental learning approach for board game playing agents. In *Proceedings of the 2000 International Conference on Artificial Intelligence (IC-AI2000)*, volume 2, pages 705–711, Las Vegas, USA, 2000.

[60] J. Mańdziuk. Incremental training in game playing domain. In *Proceedings of the International ICSC Congress on Intelligent Systems & Applications (ISA2000)*, volume 2, pages 18–23, Wollongong, Australia, 2000.

[61] J. Mańdziuk, M. Kusiak, and K. Walędzik. Evolutionary-based heuristic generators for checkers and give-away checkers. *Expert Systems*, 2007, (*accepted*).

[62] J. Mańdziuk and K. Mossakowski. Looking inside neural networks trained to solve double-dummy bridge problems. In *5th Game-On International Conference on Computer Games: Artificial Intelligence, Design and Education (CGAIDE04)*, pages 182–186, Reading, UK, 2004.

[63] J. Mańdziuk and D. Osman. Temporal difference approach to playing give-away checkers. In L. Rutkowski et al., editors, *7th Int. Conf. on Art. Intell. and Soft Comp. (ICAISC 2004), Zakopane, Poland*, volume 3070 of *LNAI*, pages 909–914. Springer, 2004.

[64] J. Mańdziuk and L. Shastri. Incremental Class Learning approach and its application to handwritten digit recognition. *Information Sciences*, 141(3–4):193–217, 2002.

[65] D. McAllester. Conspiracy numbers for min-max search. *Artificial Intelligence*, 35:287–310, 1988.

[66] J. McCarthy. Homepage of John McCarthy. http://www-formal.stanford.edu/jmc/reti.html, 1998.
[67] M. L. Minsky. Steps towards artificial intelligence. In *Proceedings of IRE*, volume 49, pages 8–30, 1961.
[68] T. M. Mitchell and S. Thrun. Explanation based learning: A comparison of symbolic and neural network approaches. In P. E. Utgoff, editor, *Proceedings of the 10th International Conference on Machine Learning*, pages 197–204, San Mateo, CA, 1993. Morgan Kaufmann.
[69] D. E. Moriarty and R. Miikkulainen. Discovering complex othello strategies through evolutionary neural systems. *Connection Science*, 7(3):195–209, 1995.
[70] K. Mossakowski and J. Mańdziuk. Artificial neural networks for solving double dummy bridge problems. *Lecture Notes in Artificial Intelligence*, 3070:915–921, 2004.
[71] K. Mossakowski and J. Mańdziuk. Neural networks and the estimation of hands' strength in contract bridge. In L. Rutkowski et al., editors, *8th International Conference on Artificial Intelligence and Soft Computing (ICAISC06)*, Lecture Notes in Artificial Intelligence, pages 1189–1198, Zakopane, POLAND, 2006.
[72] M. Müller. Computer Go as a sum of local games: An application of combinatorial game theory. PhD Thesis, ETH Zürich, Switzerland, 1995.
[73] M. Müller. Computer Go. *Artificial Intelligence*, 134:145–179, 2002.
[74] A. Newell, J. C. Shaw, and H. A. Simon. Chess-playing programs and the problem of complexity. *IBM Journal of Research and Development*, 2(4):320–335, 1958.
[75] D. Osman and J. Mańdziuk. Comparison of tdleaf(λ) and td(λ) learning in game playing domain. In N. R. Pal et al., editors, *11th Int. Conf. on Neural Inf. Proc. (ICONIP 2004), Calcutta, India*, volume 3316 of *LNCS*, pages 549–554. Springer, 2004.
[76] A. Plaat, J. Schaeffer, W. Pijls, and A. de Bruin. Best-first fixed-depth minimax algorithms. *Artificial Intelligence*, 87(1–2):255–293, 1996.
[77] A. Plaat, J. Schaeffer, W. Pijls, and A. de Bruin. Exploiting graph properties of game trees. In *13th National Conference on Artificial Intelligence (AAAI-96)*, volume 1, pages 234–239, Menlo Park, CA, 1996.
[78] E. A. Poe. Maelzel's chess player. *Southern Literary Messenger*, (April), 1936.
[79] J. B. Pollack, A. D. Blair, and M. Land. Coevolution of a backgammon player. In C. G. Langton and K. Shimokara, editors, *Proceedings of the Fifth Artificial Life Conference*, pages 92–98. MIT Press, 1997.
[80] A. Reinefeld. An improvement to the scout tree-search algorithm. *ICCA Journal*, 6(4):4–14, 1983.
[81] T. P. Runarsson and S. M. Lucas. Coevolution versus self-play temporal difference learning for acquiring position evaluation on small-board Go. *IEEE Transactions on Evolutionary Computation*, 9(6):628–640, 2005.

[82] A. L. Samuel. Some studies in machine learning using the game of checkers. *IBM Journal of Research and Development*, 3(3):210–229, 1959.

[83] J. Schaeffer. The history heuristic and alpha-beta search enhancements in practice. *IEEE PAMI*, 11(11):1203–1212, 1989.

[84] J. Schaeffer. *One Jump Ahead: Challenging Human Supremacy in Checkers.* New York: Springer-Verlag, 1997.

[85] J. Schaeffer. Chinook. http://www.cs.ualberta.ca/˜ chinook/, 2006.

[86] J. Schaeffer. Poki-X. http://www.cs.ualberta.ca/˜ games/poker/, 2006.

[87] J. Schaeffer, J. C. Culberson, N. Treloar, B. Knight, P. Lu, and D. Szafron. A world championship caliber checkers program. *Artificial Intelligence*, 53(2-3):273–289, 1992.

[88] J. Schaeffer, M. Hlynka, and V. Jussila. Temporal difference learning applied to a high-performance game-playing program. In *International Joint Conference on Artificial Intelligence (IJCAI)*, pages 529–534, 2001.

[89] J. Schaeffer, R. Lake, P. Lu, and M. Bryant. Chinook: The world man-machine checkers champion. *AI Magazine*, 17(1):21–29, 1996.

[90] N. N. Schraudolph, P. Dayan, and T. J. Sejnowski. Temporal difference learning of position evaluation in the game of go. In J. D. Cowan, G. Tesauro, and J. Alspector, editors, *Advances in Neural Information Processing 6*, pages 817–824. Morgan Kaufmann, San Francisco, 1994.

[91] N. N. Schraudolph, P. Dayan, and T. J. Sejnowski. Learning to evaluate go positions via temporal difference methods. In N. Baba and L. C. Jain, editors, *Computational Intelligence in Games*, volume 62, pages 77–98. Springer Verlag, Berlin, 2001.

[92] Y. G. Seo, S. B. Cho, and X. Yao. Exploiting coalition in co-evolutionary learning. In *Proceedings of the 2000 Congress on Evolutionary Computation*, volume 2, pages 1268–1275. IEEE Press, 2000.

[93] C. E. Shannon. Programming a computer for playing chess. *Philosophical Magazine*, 41 (7th series)(314):256–275, 1950.

[94] B. Sheppard. World-championship-caliber scrabble. *Artificial Intelligence*, 134:241–275, 2002.

[95] H. Simon. Making managenet decisions: The role of intuition and emotion. In Weston Agor, editor, *Intuition in Organizations*, pages 23–39. Sage Pubs., London, 1987.

[96] D. Sklansky. *Hold'Em Poker.* Two Plus Two Publishing, Nevada, USA, 1996.

[97] D. Sklansky and M. Malmuth. *Hold'Em Poker for Advanced Players, 21st Century Edition.* Two Plus Two Publishing, Nevada, USA, 2001.

[98] B. Stilman. *Liguistic Geometry. From search to construction.* Kluwer Academic Publishers, Boston, Dordrecht, London, 2000.

[99] G. Stockman. A minimax algorithm better than alfa-beta? *Artificial Intelligence*, 12(2):179–196, 1979.

[100] R. Sutton. Learning to predict by the method of temporal differences. *Machine Learning*, 3:9–44, 1988.

[101] R. S. Sutton and A. G. Barto. *Reinforcement Learning: An Introduction.* MIT Press, Cambridge, MA, 1998.

[102] G. Tesauro. Neurogammon wins computer olympiad. *Neural Computation*, 1:321–323, 1989.

[103] G. Tesauro. Practical issues in Temporal Difference Learning. *Machine Learning*, 8:257–277, 1992.

[104] G. Tesauro. TD-Gammon, a self-teaching backgammon program, achieves master-level play. *Neural Computation*, 6(2):215–219, 1994.

[105] G. Tesauro. Temporal Difference Learning and TD-Gammon. *Communications of the ACM*, 38(3):58–68, March 1995.

[106] S. Thrun. Learning to play the game of chess. In G. Tesauro, D. Touretzky, and T. Leen, editors, *Advances in Neural Information Processing Systems 7*, pages 1069–1076. The MIT Press, Cambridge, MA, 1995.

[107] S. Thrun. *Explanation-Based Neural Network Learning: A Lifelong Learning Approach.* Kluwer Academic Publishers, Boston, MA, 1996.

[108] S. Thrun and T. M. Mitchell. Learning one more thing. Technical report, Carnegie Mellon University, USA, CMU-CS-94-184, 1994.

[109] W. Tunstall-Pedoe. Genetic algorithms optimizing evaluation functions. *ICCA Journal*, 14(3):119–128, 1991.

[110] A. M. Turing. Digital computers applied to games. In B. V. Bowden, editor, *Faster than thought: a symposium on digital computing machines*, chapter 25. Pitman, London, UK, 1953.

[111] P. E. Utgoff. Feature construction for game playing. In J. Fürnkranz and M. Kubat, editors, *Machines that Learn to Play Games*, pages 131–152. Nova Science Publishers, Huntington, NY, 2001.

[112] A. van Tiggelen. Neural networks as a guide to opitimization. The chess middle game explored. *ICCA Journal*, 14(3):115–118, 1991.

[113] T. Yoshioka, S. Ishii, and M. Ito. Strategy acquisition for the game "othello" based on reinforcement learning. *IEICE Transactions on Information and Systems*, E82-D(12):1618–1626, 1999.

[114] A. Zorbist. Feature extractions and representation for pattern recognition and the game of go. PhD Thesis, University of Wisconsin, 1970.

Computer Go: A Grand Challenge to AI

Xindi Cai and Donald C. Wunsch II

University of Missouri – Rolla

Summary. The oriental game of Go is among the most tantalizing unconquered challenges in artificial intelligence after IBM's DEEP BLUE beat the world Chess champion in 1997. Its high branching factor prevents the conventional tree search approach, and long-range spatiotemporal interactions make position evaluation extremely difficult. Thus, Go attracts researchers from diverse fields who are attempting to understand how computers can represent human playing and win the game against humans. Numerous publications already exist on this topic with different motivations and a variety of application contexts. This chapter surveys methods and some related works used in computer Go published from 1970 until now, and offers a basic overview for future study. We also present our attempts and simulation results in building a non-knowledge game engine, using a novel hybrid evolutionary computation algorithm, for the Capture Go game.

1 Introduction

Games have served as one of the best test benches in artificial intelligence fields since shortly after computers were invented. Human-designed computer game engines have beaten their designers in varieties of games, from those as simple as Tic-Tac-Toe to as complex as Chess. The brute-force search algorithm, combined with an expert database, achieved noteworthy success, given the computational power of current machines, when IBM DEEP BLUE beat the World Chess Champion Garry Kasparov in 1997 [21]. Unfortunately, this game's tree search approach is hampered by the traditional Chinese game Go.

Unlike most other games of strategy, Go has remained an elusive skill for computers to acquire, and it is increasingly recognized as a "grand challenge" of artificial intelligence, which attracts researchers from a diversity of domains, such as game theory [2], pattern recognition, reinforcement learning [47], and even cognitive psychology [10].

Even though computers still play Go at an amateur level, numerous publications demonstrate some quite meaningful exploration that helps us to understand the nature of this computer game better. In our review of

Xindi Cai and Donald C. Wunsch II: *Computer Go: A Grand Challenge to AI*, Studies in Computational Intelligence (SCI) **63**, 443–465 (2007)
www.springerlink.com

computer Go literature on what has been achieved, where the barriers lie, and how to make a breakthrough, we emphasize that developing an efficient self-learning mechanism in Go may best reveal the nature of goal-driven decision making and interacting in a range of environments, where strategies are acquired to allocate available resources in achieving maximum payoffs, either short-term or long-term. Even though other AI approaches are promising in computer Go, we particularly favor applying reinforcement learning and neural network techniques to build a zero-knowledge board evaluation function. Such neural network evaluators have been successfully trained by evolutionary algorithms in Checkers and Chess [6, 7, 16]. We utilize a hybrid of an evolutionary algorithm and particle swarm optimization to train our neural network evaluator and obtain encouraging results on Capture Go, a simplified version of Go. The success in learning one of the key strategies of Go, i.e., capture and defense, through Capture Go will demonstrate that other strategies of similar complexity are solvable with the same technique. With more strategy boxes accumulated, this divide-conquer approach may result in an overall game engine incrementally built in a hierarchy of high levels, employing those boxes as the building blocks and eventually becoming competitive against human beings.

The chapter is organized as follows: Section 2 provides background information about Go and some comparison between computer Go and a Chess program. In Section 3, we track the development of computer Go and focus on several successful programs. The theme for Section 4 is the current promising approaches in self-learning with little built-in knowledge where conventional architecture and algorithms have failed in computer Go. In Section 5, we present our simulation results in training a zero knowledge game engine to play Capture Go using an innovative hybrid population computation. We round off the article with an overview for future directions of study and conclusions in Section 6.

2 Background

This section furnishes some background knowledge for both the traditional game Go and the computer version so that we can discuss the methods presented in this chapter. First, we introduce the game Go with its terms and rules. Then, we briefly discuss the techniques used for computer games. Finally, we compare computer Go and computer Chess to show why the conventional methods failed.

2.1 The Game Go

Go is a deterministic, perfect information, zero-sum game of strategy between two players. Players take turns placing black and white pieces (called stones) on the intersections of the lines in a 19x19 grid called the Go board. Once played, a stone cannot be removed unless captured by the other player. To win the game, each player seeks to surround more territory (empty grids) with one's own stones than the opponent.

Adjacent stones of the same color form strings, and hence groups; an empty intersection adjacent to a stone, a string, etc. is called its liberty. A group is captured when its last liberty is occupied by the opponent's stone. A player cannot make a suicidal move by placing a stone on an intersection with no liberty. An eye is a formation of stones of special significance in Go. When an empty intersection is completely surrounded by stones of the same color, it is known as an eye. An opponent cannot place a stone on that intersection unless it is a capturing move, i.e. unless placing the stone causes one or more of the stones surrounding the intersection to be captured. A string with two eyes cannot be captured because filling one of the eyes would be a suicidal, and therefore illegal, move. Having formed two eyes, or having the potential to do so, is the line between "alive" strings and "dead" ones. Those strings that are incapable of forming two eyes will be considered as captured and hence are removed when calculating the territories at the end of the game. Evaluating whether stones can be formed into strings, furthermore into groups, and whether strings are capable of forming two eyes is the fundamental skill in Go as it represents the game strategies of "attack," making ourselves strong and being aggressive, and "defense," avoiding vulnerabilities and surviving.

To prevent loop, it is illegal to make moves that recreate prior board positions (rule of Ko). The rule for Go is simple: one can place his/her stone on any empty intersection unless it is a suicidal or Ko move. A player can pass his/her turn at any time. The game ends when both players pass in consecutive turns (see Fig. 1). There are excellent books available on the game of Go [43].

2.2 Techniques Used in Computer Games

Claude Shannon [42] proposed that a mechanical algorithm could play a game if that algorithm contained two ingredients: an evaluation function—a mathematical formula that assigns credits according to different board positions—and a rationale, which he called "minimax," that seeks to minimize the maximum damage that the opponent can do in any circumstance.

The evaluation function quantifies how good or bad each legal move on the board was, while the minimax procedure provides a way to evaluate the possible alternative positions, given the credits from the evaluation function, by favoring the position that had the least advantage for the rival. This mechanism is the bar code of almost every computer game product.

Numerous algorithms, such as minimax, Alpha-Beta search, MFD, and different parallel versions, are proposed in the minimax procedure for the purpose of achieving a deeper and wider search with the same computational power. On the other hand, the evaluation function is evolved from static

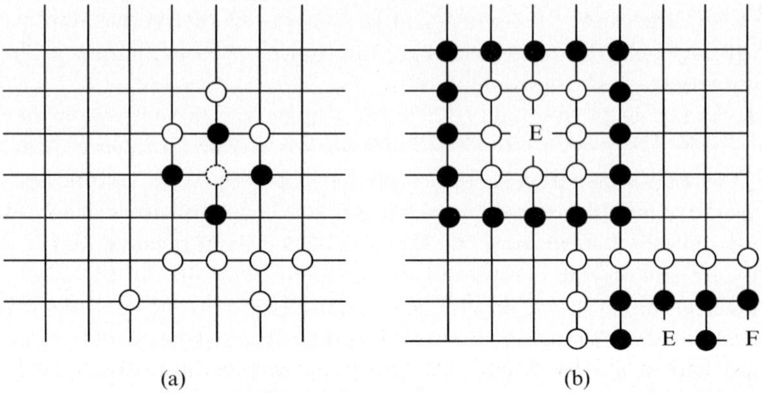

Fig. 1. Go board, terms, and rules. Only a portion of the 19x19 Go board is shown. In (a), the lower 5 T-like white stones form a string, and hence a group with the white stone to their left. In the middle, black captures a white stone, the dashed one, and removes it from the board by placing the top black stone on the board. However, white cannot immediately put a stone onto the dashed-stone intersection to capture the top black stone because such a move would repeat the previous board position, thus violating the rule of "Ko." In the middle of (b), white has an eye at "E." White cannot put his stone at "E," which results in no liberty of the whole string. That is a suicide move and forbidden by the rules. Black can put his stone at "E" even though there is no liberty of that stone either, but it is allowed because it captures the white string. So the white string with only one eye is dead and will be removed from the board at the end of the game. The black string at the corner has two eyes, i.e., "E" and "F," and hence is alive because white cannot seize both of the eyes simultaneously

knowledge-based pattern recognition to dynamic neural networks trained by Heuristic Dynamic Programming (HDP) or Evolutionary Algorithm (EA).

2.3 Computer Go and Computer Chess

Given the same programming techniques that were so successful in Chess, computer Go has so far significantly bypassed what Chess programs have achieved in a high level of game playing. We separate these two games into parts, based on Shannon's plot, to illustrate the gap we need to overcome.

In the game tree aspect, good Chess programs look seven plys, a move by each player, ahead or deeper, and there are about 35–40 moves available, on average, to a player. Exploration on such a search tree results in an evaluation of about 60 billion scenarios with good prune skills. The IBM Deep Blue is capable of evaluating 200 million positions a second, which allows Chess programs to execute a massive search and evaluation with the help of expert knowledge.

Unlike Chess, Go starts with an empty board and fills with stones as the game progresses. Theoretically speaking, there are 361! leaves at the bottom of the game tree with no capture occurring during the game. The branching factor of the search tree, or the number of legal moves on the board, is around 200 on average, more at the beginning and less at the end of the game. The game length varies from 150 to 300 moves. All these factors result in a game tree varying from 10^{360} to 10^{690}. In general, a game tree of approximately 10^{575} leaves [1] is accepted by most researchers. A 7-ply look-ahead search needs to handle ten thousand trillion positions, which makes a brute force search approach infeasible, even with the most powerful computer.

The evaluation function is much easier in Chess because the pieces have their own rank values. Assessing the strength of either side can be simply reduced to comparing the piece value, plus some measure of strength position and threat to the King, since normally a piece advantage will lead to a victory. The idea of rank in Go is very vague. A single stone has no rank at all. Moreover, the importance of a stone, a string, and a group changes based on a number of factors, such as connectivity, liberty, position on the board, correlation with neighboring friend and/or foe, and even the player's preferences, as the game progresses. Sometimes, the same stones can be either protected or sacrificed with the same aim to gain the maximum territory. Additionally, tactical skills play a more important role in Chess. A tactical evaluation based on piece quality of the board correctly yields the likely winner. In Go, winning a tactical struggle over a group may not clearly lead to winning the game. In other words, it is more important to know how to apply those tactical skills properly at different board areas, under different game conditions, or even in different sequences than how to play each tactic correctly. Unfortunately, the former is very game/situation dependent, and even human masters do not agree with each other on which tactical method to pick (sometimes totally opposite methods are selected just because of masters' personal tastes). Therefore, installation of an expert knowledge database, which usually contains the procedures of tactical skills, as the evaluation function, as used in Chess, is implausible for Go because defining and formulating such knowledge into Go moves is difficult.

Besides the above two facets, computer Go also lacks the capability to perform deep, narrow look-ahead. "Ladder situation" is a typical example. Another problem in Go is to determine the finish of the game. In Chess, the game ends when one of the players resigns, when a checkmate is achieved, or when stalemate/draw positions by rule (e.g. king + knight vs. king, position repetition, 50-move rule) are reached, which all are immediate, definite, and easily recognizable states. In Go, the game ends when both players choose to pass consecutively. They agree to pass when they feel that it would not improve their territory by placing more stones on the board. Beginners frequently play beyond the optimal point at which an expert would stop. Current Go programs display similar behavior, especially in self-playing.

3 Development of Computer Go

The history of computer Go is rather short. The first paper discussing computer Go was published in 1970 by Albert Zobrist [49], who wrote the first computer program to play a complete Go game. Since then, many programs have been developed, including, but not limited to: *Go Nemesis* [46], *Wally* [29], *Many Faces of Go* [17], *Handtalk* [11] and *Go4++* [33], some of which will be discussed in full later in this section. Also, quite a few international computer Go tournaments have been held since the 1st international computer Go congress took place in 1986.

3.1 Early Attempts

Computer Go pioneers inherited most of their techniques from computer Chess. A small game tree was searched due to limited computational power, and certain board patterns, based on fixed, pre-built expert knowledge working as feature evaluation, were checked in order to generate candidate moves.

Zobrist introduced the idea of using influence functions to quantify the impact of black and white stones, and hence segment the board into black and white territories. Each stone radiates its influence, measured by a numeric value with positive for black stones and negative for white ones on the board. The influence attenuates as the hamming distance increases. Also, a black stone itself is given a value of $+50$ and a white stone, -50. For every intersection on the board, the influence function accumulates the influence credits spread by all black and white stones. The black and white territories can be recognized as the areas of contiguous positive and negative values. On the other hand, Zobrist quantified the board in various 19x19 arrays that contain information such as the occupation of an intersection (black, white, or empty); the number of white and black neighbors; the number of stones and liberties for each string; size (including empty points) and number of stones in each segment. This internal representation was constructed for future feature evaluation and pattern recognition.

The move generator in Zobrist's program was built by integrated feature evaluations. With the help of internal representation, the program performed pattern recognition over the entire board, hoping to find some patterns matching those stored in the expert knowledge database. In this database, each pattern was associated with a candidate move and a numeric value to reflect the priority of the move. At the end of the process, the pattern with the highest value was picked, and its associated move became the program's next move. In addition, a limited look-ahead, a depth of three moves, or heuristic search was executed on a local area in handling forming eyes, ladders, saving/capturing strings, and connecting/cutting strings.

Zobrist's program performed somewhat weakly. It beat some beginners but was fairly vulnerable when playing against experienced players.

Ryder's program [37] extended Zobrist's work with refined influence functions and larger summed feature evaluations. The contribution Ryder made was that he combined both strategic, long-term objectives and tactical, short-term objectives because he saw that Go requires a balance between fortifying actual and/or potential territories and avoiding loss of key stones to the opponent. In maintaining this balance, Ryder formulated three domains of interest: how well each side controls his/her regions on the board; where the best moves for both sides are located; and what the life-and-death statuses of strings for both sides are.

Move generation in Ryder's program consisted of two phases. At first, the summed feature evaluation, working as a move filter, reduced the candidate moves from all legal ones to about the 15 best moves according to the tactical status of all strings. In the second phase, the best 15 moves were further analyzed with respect to both tactical and naïve strategy theories. The move with the highest combined score after the first and second round was then recommended as the next move.

Besides refining the credits each stone contributed through the influence function, Ryder also used the influence function to describe the strength of connectedness and relation between stones, and hence to form individual stones and neighboring empty points into some local properties, such as walls, strings, groups, and armies, based on some predefined influence thresholds.

Reitman and Wilcox [34, 35, 36] built their Go program, from INTERIM.2 to Nemesis, by replicating human perceptual and cognitive abilities in Go. The three essential ingredients they incorporated into their Go program were perception, knowledge, and coordination. By perception, they meant the representation of different board positions as a skilled Go player would do. Types of knowledge stored in the program were: tactical – including how to save/kill a group and how to make territory; strategy – evaluating the board position as the game proceeded; coordination – controlling the flow of both perceptions and knowledge for the purpose of generating the next move.

The INTERIM.2, and later the Nemesis [46], had tremendous data structures cascading from simple to complex: stringboard, linkboard, gameboard, and gamemap. They were designed to contain the representations mentioned above. The program also employed a tactician, PROBE, to answer specific questions and propose reasonable initial moves. Those moves, instead of all legal candidates, worked as the branches of the game tree for certain depths of plays until a board position could be judged as either a success or a failure for the specific question. Typically, around 60–80 moves were searched for a particular problem based on a hierarchy of experts in PROBE. Therefore, such a look-ahead drive was rather a narrow, goal-driven but faster one than a general, full-board yet time-consuming one with the price of being greedy at local patterns.

Restricted by the computational powers they had at the time, researchers of computer Go in the 70s and 80s implemented quite an amount of Go knowledge, borrowed directly from human experts, in their programs. The

Go knowledge encoded in the form of patterns greatly pruned the game tree, which by no means brute force can handle, by concentrating on the most promising candidate moves. On the other hand, the room for improvement for these knowledge-intensive Go engines was limited due to their hard learning mode. No ability in Go beyond their fixed database had been demonstrated in their performance. Besides rigid playing, these early programs lacked overall game strategy because the search involved was local rather than full-width. The preferred patterns, and hence corresponding moves, on the sub-board may not be the right play for a long-term goal. These were the reasons why early Go programs ranked at low Kyu level.

3.2 Knowledge Representation and Rule-Based Go Engines

Computer Go engines saw some breakthrough in the 1990s. *Many Faces of Go*, *Handtalk*, and *Go4++* were among the strongest Go programs commercially available during that time. All these programs employed complicated data structures to describe board position, sophisticated tactical strategies to generate candidate moves, and expert patterns to modify move values.

Many Faces of Go is a typical representative of this category. It contains dynamic data, such as intersections, eye, string, connection, and group, to illustrate board positions. The dynamic data structure is modified incrementally as stones are added to or removed from the board. The data are also recalculated, ether locally or globally, when regions of the board have been affected by a move or move sequence.

The evaluation function, which maintains the data structure, is a tactical analyzer. It reads strings to evaluate liberties, eye, connection, group strength, and territories. Therefore, it assigns a score to board positions depending on how strongly they are controlled by white or black.

The game engine of *Many Faces of Go* is driven by a strategy function, which scans the dynamic data to pick up important areas on the board to act. It also switches the engine among open game, middle game, and end game, hence utilizing different knowledge/rule databases for move generation. Other functionalities include game judgment ("ahead," "even," "behind," etc.), sente evaluation, and urgent defense/capture.

Knowledge and rules are implemented intensively in *Many Faces of Go* for candidate move generation. A rule-based expert system with more than 200 rules is responsible for suggesting plausible moves for full board level. A Joseki database consisting of over 36,000 moves is designed for corner fights. In the pattern database, each of the 1200 patterns of size 8x8 is associated with a move tree. Patterns match leads to move suggestion, eye/connection clarification, in different game phases (middle or end game), and at different areas of the board (middle, edge, or corner).

Combining a strategy function, a move suggestion expert system, and a pattern database, the *Many Faces of Go* evaluates only a small number of moves and plays the one with the highest score.

Handtalk and *Go*4++ process similar procedures; *Handtalk* implements patterns in assemble code to enhance the matching speed, and *Go*4++ emphasizes the connectivity in its data structure and evaluation function.

The common and distinguishing feature of computer Go programs in this category is that candidate moves are heuristically generated by an expert knowledge database and/or pattern matching. The success of the program depends heavily on the sound design of the expert knowledge database and sufficient patterns to cover every case, which is usually very difficult, especially in middle game.

3.3 Combinatorial Game Theory Approach in Computer Go

Combinatorial game theory [12] treats a game as a sum of local subgames and provides a mathematical basis to analyze the game in a divide-conquer manner.

Combinatorial game theory has been applied to many aspects of computer Go. This approach achieved a major breakthrough in computer Go by beating a human professional master at the end game [3]. Another attempt of combinatory game theory is to employ it in position evaluation by focusing on different zones of the board and then computing a full board evaluation from these local evaluations [27]. It is also utilized to decompose Go game tree search [26].

However, the theory has one drawback for its emphasis on being exact. At the end game, the territories held by each side are very clear; thus, subgames can be precisely divided. In open and middle games, however, such assumption is usually not true. Current research involves using heuristic rules/conditions to replace [9] or relax [28] the theory.

4 Learning in Computer Go

Unlike the knowledge-based approaches, which retrieve the built-in expert solutions via pattern recognition, the learning approaches really focus on teaching the computer to analyze the environment and then play the game from its own experience. Neural networks, reinforcement learning, and evolutionary computation are heavily involved in these approaches. Neural networks usually act as a game engine, such as a board evaluator and/or move filter for game tree search, utilizing reinforcement learning techniques to describe the game environment/goals in sequence and employing evolutionary computation methods to tune the weights in order to minimize the cost function.

Temporal difference [44] methods are incremental learning procedures specialized for prediction problems where the sensory inputs are applied in sequence. They have been successfully employed for the prediction evaluation function at different board positions in backgammon [45]. Temporal difference algorithms minimize the following criterion function:

$$J(w) = \sum_{p=1}^{P} \sum_{k=1}^{N_p} \lambda^{N_p - k} (zN_p - G(x_p(k))^2 \tag{4.1}$$

by adjusting the weights in a neural network as:

$$\Delta w_t = \alpha(P_{t+1} - P_t) \sum_{k=1}^{t} \lambda^{t-k} \Delta_w P_k \tag{4.2}$$

In both equations, P is the number of examples, i.e., the number of games; N_p is the number of steps in the pth example, which is not known until the outcome is determined; zN_p is the actual outcome of the pth example; $G(x_p(k))$ is the output of the network when presented with $x_p(k)$; and $\lambda[0,1]$ is a parameter used to place more emphasis on predictions temporally close to the outcome.

Using the TD approach, a computer Go program can pick up moves that result in better board positions once appropriate weights are learned. In addition, as the weight evolves, the program strategy will also evolve, leading to different performance and triggering, and, in return, another round of weight adjustment. Such a process guides the computer Go engine to play a game without supervision, i.e., explicitly labeled good/bad moves. In fact, any legal move can be used for training, and the game engine gradually learns the strategies itself by playing both sides according to its current evaluation function, with little expert knowledge involved during the whole process.

Schraudolph et al. [39, 40] implemented a Go position evaluator on a neural network system, trained by the TD(0) algorithm. The architecture of the neural system was designed to reflect the spatial board information and eliminate symmetric effects on the board (color and position reflection/rotation) with hidden units and weight sharing. The network predicts the fate of every point on the board rather than just the overall score and then evaluates whole positions accordingly. After extensive self-play training (moves acquired stochastically by Gibbs sampling to avoid duplication and local minima), the system managed to edge past *Wally*, a weak computer Go program, and even beat *Many Faces of Go* at some low level playing.

Zaman et al. [47, 48] used an Heuristic Dynamic Programming (HDP) type adaptive critic design [32] for evaluating a Go board. The main difference between HDP and the above-mentioned TD approaches is that HDP uses an additional utility function (per step cost/reward) in the training signal.

An HDP-type critic estimates the function J (cost-to-go) in the Bellman equation of dynamic programming expressed as:

$$J(t) = \sum_{k=0}^{\infty} \gamma^k U(t+k) \tag{4.3}$$

where γ is a discount factor for finite horizon problems ($0 < \gamma < 1$), and U(.) is a non-negative utility function or local cost/reward. The critic is trained forward in time and tries to minimize the following error measure over time:

$$\|E\| = \sum_t E^2(t) \tag{4.4}$$

where,

$$E(t) = J(t) - [\gamma J(t+1) + U(t)] \tag{4.5}$$

the terms inside the square bracket make the desired signal at time t, if t is not the terminal state. At the end of the game, the desired signal is simply U(t). J(t) is a function of R(t), i.e., the observable states. In terms of Go, R(t) can be the board representations at step t. The function U(t) denotes an incremental area measure from board R(t − 1) to R(t). When the area associated with R(t − 1) is larger than that of R(t) (loss of area between two steps, t − 1 and t), U(t) is set to zero because U(.) is strictly non-negative by the principle of dynamic programming. In [47], the U(t) function is given by:

$$U(t) = \begin{cases} util(t) - util(t-1); & if\ util(t) > util(t-1) \\ 0; & otherwise \end{cases} \tag{4.6}$$

where,

$$util(t) = \eta \frac{N_W}{N_W + N_B} + \nu \frac{A_W}{A_W + A_B} + \rho \frac{P_B}{P_W + P_B} \tag{4.7}$$

N_W and N_B are the number of WHITE and BLACK stones on the board, respectively; A_W and A_B are the areas occupied by WHITE and BLACK stones, respectively; P_B and P_W are BLACK and WHITE prisoners held by the opponent, respectively.

The architecture of Zaman's computer Go engine includes five distinct modules: Critic, Action, Wally, Go, and Utility. The Critic module is a multi-layer perceptron trained by the TD(0) method. It estimates the sum of discounted reward/cost for the network's future actions starting from the current state. The Action network plays as a move filter. Firstly, it lists all legal moves for the current state; secondly, it generates a new board position for each move; after that, it evaluates the critic's evaluation of the resulting board; and finally, it selects the move based on the critic's board evaluation. The utility module is used to measure network cost/reward at the current state of the board. The Go module incorporates rules of the game, and Wally serves as the network's opponent. The game engine surpassed the strength of Wally.

Besides the reinforcement technique discussed above, evolutionary computation, especially genetic algorithm, is also very popular in computer Go engine implementation. Like natural evolution, the genetic algorithm evolves the neural networks to tackle evaluation function problems in Go.

Genetic operators, such as selection, crossover, and mutation, are applied to neural networks for effective architecture and/or weights to solve the credit assignment problem. Instead of punishing or rewarding individual moves, the evolutionary approach evaluates and selects networks, i.e., game strategies, based on their overall performance in the game. SANE [38] evolved both architecture and weights of a three-layer feedforward network simultaneously to evaluate move value for each board position. After a few hundred generations of evolution, SANE defeated Wally 75% of the time and exhibited several aspects of general Go playing, indicating a good scale-up.

Genetic algorithms are also used to direct minimax searches away from poor information [25] and optimize search heuristics for life-and-death problems [31].

5 An Example of Training a Non-Knowledge Game Engine for Capture Go

Evolutionary algorithms have shown to be a promising approach to solving complex constrained optimization problems. Chellapilla and Fogel succeeded in evolving an expert-level neural board position evaluator for Checkers without any domain expertise [6, 7, 15]. Their work concludes that computer game engines can learn, without any expert knowledge, to play a game at an expert level, using a co-evolutionary approach.

The trend of [6, 7] is followed, and PSO is applied in combination with an EA to develop a neural evaluator for the game of Capture Go. As a simplified version of Go, Capture Go has the same rules but a different goal – whoever captures first, wins. The system for this game, with minor modifications, should be a useful subsystem for an overall Go player. Previous work [24] on Capture Go showed that the simplified game is a suitable test bench to analyze typical problems involved in evolution-based algorithms, such as lifetime learning, incremental evolution [20], open ended evolution, and scalability. Growing from zero knowledge, this game engine extends our work [4]. The large-scale game engine, a neural network with more than 6000 parameters, is trained by a particle swarm optimization (PSO)-enhanced evolutionary algorithm through self-playing. The innovative hybrid training algorithm inheriting the advantages, i.e., fast convergence and good diversity, of both PSO and EA is more powerful than its individual parts, and this has been shown in other applications [5]. It is compared with a Hill-Climbing (HC) algorithm as a first-order benchmark. The hybrid, method-based game engine is also played against a hand-coded defensive and a web player to show its competence. The Capture Go games are played on a 9x9 board.

5.1 Particle Swarm Optimization

Particle swarm optimization is a form of evolutionary computation developed by Kennedy and Eberhart [22, 23]. Similar to EAs, PSO is a population-based optimization tool, where the population is initialized with random potential solutions and the algorithm searches for optima, satisfying some performance index over iterations. It is unlike an EA, however, in that each potential solution (called a particle) is also assigned a randomized "velocity" and is then "flown" through an m-dimensional problem space.

Each particle i has a position represented by a position vector $\mathbf{x}_i$ (the possible solution for the given problem). A swarm of particles moves through the problem space, with the velocity of each particle represented by a vector $\mathbf{v}_i$. At each time step, a function f representing a quality measure is calculated by using $\mathbf{x}_i$ as input. Each particle keeps track of its own best position, which is recorded in a vector $\mathbf{p}_i$, and $f(\mathbf{p}_i)$, the best fitness it has achieved so far. Furthermore, the best position among all the particles obtained so far in the population is recorded as $\mathbf{p}_g$, and its corresponding fitness as $f(\mathbf{p}_g)$.

At each time step t, by using the individual's best position, $\mathbf{p}_i(t)$, and the global best position, $\mathbf{p}_g(t)$, a new velocity for particle i is calculated using (5.1) below

$$
\begin{aligned}
v_i(t+1) = w \times v_i(t) + c_1 r_1 (p_i(t) - x_i(t)) \\
+ c_2 r_2 (p_g(t) - x_i(t))
\end{aligned}
\tag{5.1}
$$

where c_1 and c_2 are positive acceleration constants, r_1 and r_2 are uniformly distributed random numbers in the range [0, 1], and w is the inertia weight, with a typical value between 0.4 and 0.9. The term $\mathbf{v}_i$ is limited to the range $\pm v_{\max}$. If the velocity violates this limit, it is set at its proper limit. Changing velocity this way enables the particle i to search around the individual's best position $\mathbf{p}_i$ and global best position $\mathbf{p}_g$. Based on the updated velocities, each particle updates its position according to the following:

$$
x_i(t+1) = x_i(t) + v_i(t+1) \tag{5.2}
$$

Based on the above equations, the population of particles tends to cluster together with each particle initially moving in a random direction. Fig. 2 illustrates the procedure of the PSO algorithm. Computing PSO is easy and adds only a slight computational load when incorporated into an EA.

5.2 Evolutionary Algorithm

The evolutionary algorithm (also called evolution strategy in [41]) begins with a uniformly random population of n neural networks, K_i, $i = 1, \ldots, n$. Each neural network has an associated self-adaptive parameter vector σ_i, $i = 1, \ldots, n$, where each component controls the step size of mutation applied to its corresponding weights or bias.

Each parent generates an offspring strategy by varying all associated weights and biases. Specifically, for each parent K_i, $i = 1, \ldots, n$, an offspring $K_i, i = 1, \ldots, n$, was created by

$$
\sigma_i'(j) = \sigma_i(j) \exp(\tau N_j(0,1)), \quad j = 1, \ldots, N_w \tag{5.3}
$$

$$
w_i'(j) = w_i(j) + \sigma_i' N_j(0,1), \quad j = 1, \ldots, N_w \tag{5.4}
$$

where N_w is the number of weights and biases in the feedforward neural network, $\tau = 1/\sqrt{2\sqrt{N_w}}$, and $N_j(0,1)$ is a standard Gaussian random variable resampled for every j[6, 7].

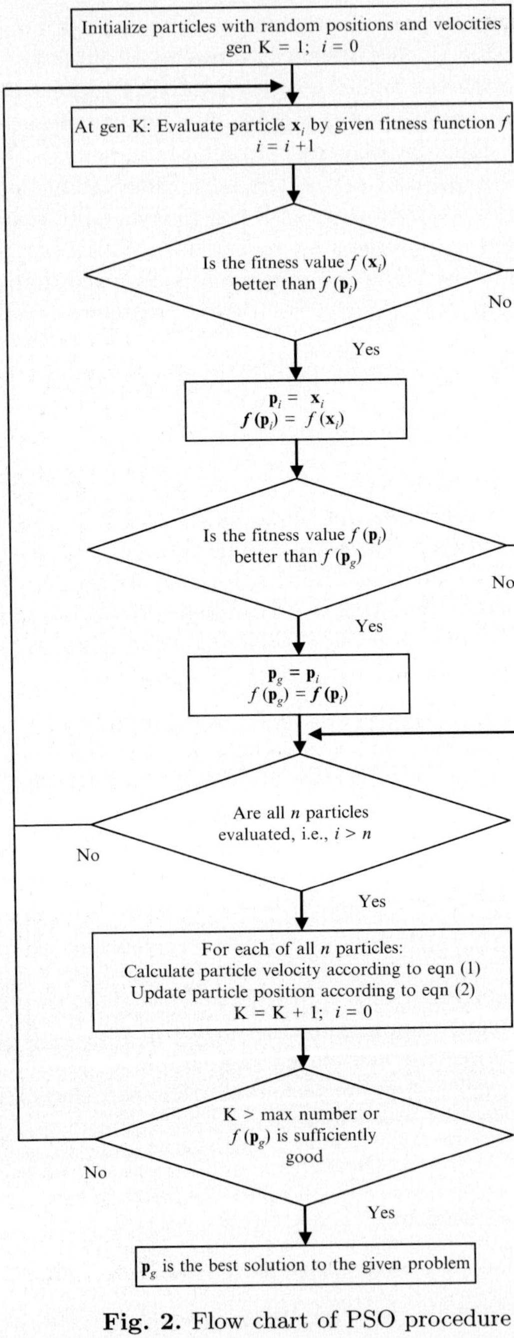

Fig. 2. Flow chart of PSO procedure

5.3 Hybrid of PSO and EA

PSO focuses more on the cooperation among the particles. With memory, each particle tracks the best performance in its own history and its neighborhood throughout the entire evolution when sharing the memory. Such a mechanism guides particles to pursue high fitness values more quickly than mere selection operation in EA. However, particles of PSO are not eliminated even if they are ranked to have the worst fitness in the population, which may waste the limited computational resources. On the other hand, individuals in EA compete for survival. Also, their diversity, maintained by mutation, prevents the population from the premature convergence often found in PSO. Clearly, the advantage of one algorithm can complement the other's shortcoming. Thus, the motivation is to develop a hybrid-based learning algorithm.

Based on the complementary properties of PSO and EA, a hybrid algorithm is used to combine the cooperative and competitive characteristics of both. In other words, PSO is applied to improve the surviving individuals and maintain the properties of competition and diversity in EA. In each generation, the hybrid algorithm selects half of the population as the winners according to their fitness and discards the rest as losers. These elites are enhanced, sharing the information in the community and benefiting from their learning history, by the standard PSO procedure. The enhanced elites then serve as parents for an EA mutation procedure. The offspring also inherit the social and cognitive information from the corresponding parents, in case they become winners in the next generation. Fig. 3 illustrates this hybrid PSO + EA algorithm.

5.4 Simulation Results

A feedforward neural network (multi-layer perceptron MLP) is designed to carry out the board evaluation function, assigning credits for leaves in the game search tree of Capture Go. The best candidate move is then chosen according to the alpha-beta minimax search from the game tree. The board information is represented by a vector of length 81, with each element corresponding to an intersection on the board. Elements in the vector are from $\{-1, 0, +1\}$, where "-1" denotes that the position on the board is occupied by a black stone, "0" denotes an empty position, and "1" denotes a white stone. The feedforward neural evaluator consists of three hidden layers and one output node. The second, third, and output layers are fully connected. Each neuron has a bipolar sigmoid activation function:

$$\tanh(\lambda x) = \frac{e^{\lambda x} - e^{-\lambda x}}{e^{\lambda x} + e^{-\lambda x}} \tag{5.5}$$

with a variable bias term.

The first hidden layer is designed specially, following [6, 7], to process spatial information from the board. In order to grasp the spatial characteristics such as neighborhood and distance of the board, each neuron in the first

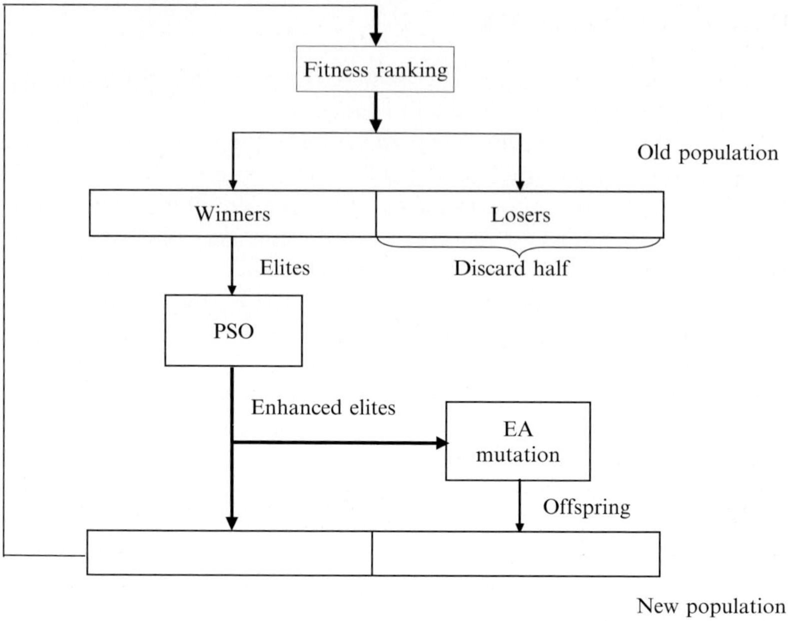

Fig. 3. Flow chart of the hybrid PSO-EA method. The winners, which contain half of the population, are enhanced by PSO and kept in the population for the next generation. Those enhanced winners also work as the parents in EA to produce offspring. The offspring replace the discarded losers to keep a constant number of individuals in the population for the next generation. If the PSO block is removed, the hybrid algorithm is reduced to the conventional EA

hidden layer covers an $n \times n$, $n = 3, \ldots, 9$, square overlapping a subsection of the board. In addition, the connecting weights between the input layer and the first hidden layer are designed specially to reflect the symmetric property of the board. Fig. 4 shows the general structure of the game engine.

In the self-play training, a population of 40 individuals, each representing a game engine, is evolved by playing games of Capture Go. Each individual earns credits based on its game results. Each player, always black, plays one game against each of eight randomly selected opponents, always white, from the population. The game is scored for each player as -2, 0, or $+1$ points depending on the results of loss, draw, or win. In total, there are 320 games per generation, with each engine participating in an average of 16 games. After all games are complete, each individual accumulates the scores it earned as its fitness value and updates according to the algorithms employed, i.e., PSO, EA, or the hybrid.

The weights of each swarm neuro-engine are generated randomly from a uniform distribution over $[-0.2, 0.2]$. The self-adaptive parameters for the EA are initially set to 0.05. The value of v_{max} for PSO is set to 2.0. The whole evolutionary process is iterated for 100 generations. At last, the best

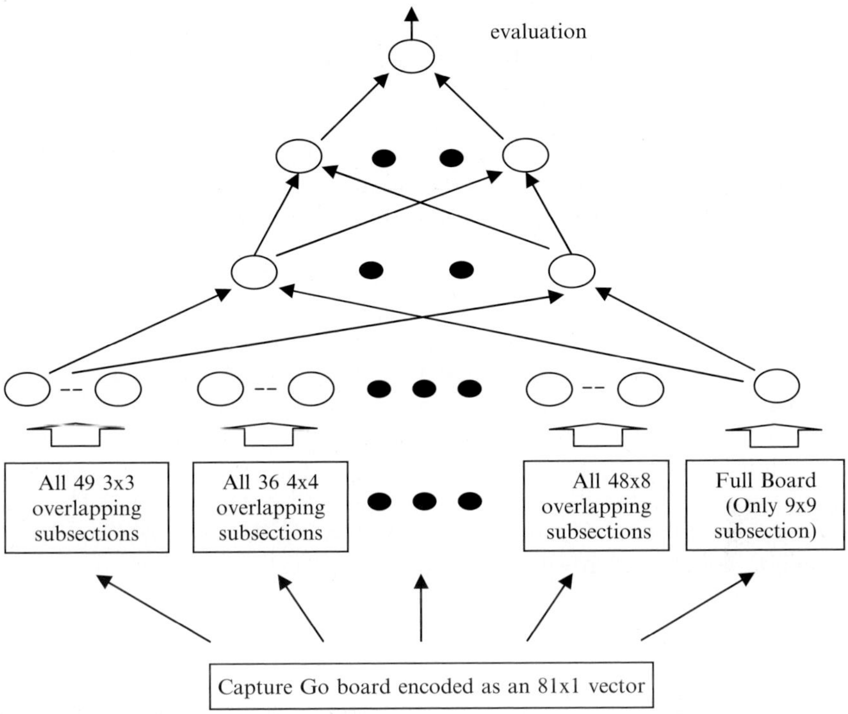

Fig. 4. Architecture of the game engine. This feedforward neural network evaluates the given board. Different sets of parameters of the neuro-game engine lead to different credit assignments for the given board, and hence represent different strategies. The board pattern is interpreted by 140 subsquares. Each of these subsquares is assigned to a node in the first layer of the network for spatial preprocessing purposes. The outputs are then passed through two hidden layers of 40 and 10 nodes, respectively. The output node of the entire network is scaled between [−1, 1] with "−1" in favor for the white and "1" for black

Table 1. Tournament results among hybrid PSO-EA, PSO, EA, HC and random players in 100 games. Black players are listed in rows and white players in columns. For example, the result in row 2, column 4 means that the hybrid PSO-EA player in black wins 90 to 10 against the EA player in white

	Hybrid PSO-EA	PSO	EA	HC	Random
Hybrid PSO-EA	/	79/21	90/10	76/24	100/0
PSO	62/38	/	76/24	70/30	100/0
EA	53/47	68/32	/	70/30	100/0

neuro-engine (at generation 100) of each category, i.e., PSO, EA, and the hybrid, is then used to play against each other and a random player (in each game, roughly 20% of the moves are randomly generated for both sides to avoid duplication). Table 1 summarizes their performance in 100 games in the tournament. All players illustrate success in learning strategies in Capture Go because they overwhelm the random player with only 28 moves per game, on average. The hybrid PSO-EA player in black dominates both PSO and EA players. Considering the advantage that black plays first, the hybrid in white is roughly equivalent to the EA and slightly weaker than the PSO. The PSO and EA players are at the same level. Following the same self-play methodology, another game engine is trained by a simple learning method, hill-climbing (HC), for Capture Go, to verify if the dynamics of the game and the co-evolutionary setup of the training are the key issues of the game engine learning [30]. The tournament results show that PSO, EA, and hybrid players outperform the HC player (with training parameter beta = 0.05 [30]), which indicates that the improvement of game engines comes mainly from the learning algorithms. Finally, a web Capture Go player [19] is brought for illustration, and the best hybrid player wins 23 of 25 games.

In addition to self-play, a defensive player of Capture Go is hand-coded. This player takes defensive strategies with the following priorities: 1) connect all its stones into one string; 2) choose a move that maximizes its liberties (the liberty count saturates when it makes two eyes); 3) surround more empty intersections with a wall; and 4) attack the weak stone(s) of its opponent. The player is hard to capture because it is difficult to seize all its liberties before it makes two eyes or captures an opponent's stone(s) instead (see Fig. 5). The game's result indicates that an opponent is more likely to defeat this defensive player by occupying more territories rather than by capturing its stones. Competing with this player teaches our hybrid engine to manage the balance between seizing its own territories and capturing enemy stones (see Figs. 6 & 7).

6 Conclusion

The game Go remains unconquered for traditional artificial intelligence techniques due to the tremendous size of its game tree, vague rank information of its stones/strings/groups, and dynamic changes of the crucial building blocks.

Early computer Go programs built the game board evaluation function by employing pattern recognition, tactical search, and rule-based reasoning, mainly based on matching the expert knowledge. Even though such techniques are predominant in the top computer Go engines existing now, the rigid or hand-coded look-up table in a knowledge database prevents the game engines from correctly handling the complex and subtle environment beyond the provided expert knowledge.

The emergence of neural networks, reinforcement learning, and evolutionary computation techniques guide the computer Go engines to learn, rather than embed, the game strategies through playing Go games. The game engines without pre-defined knowledge gradually adapt to the nonlinear stimulus-response mappings of the game from their own experience. Multiple at-

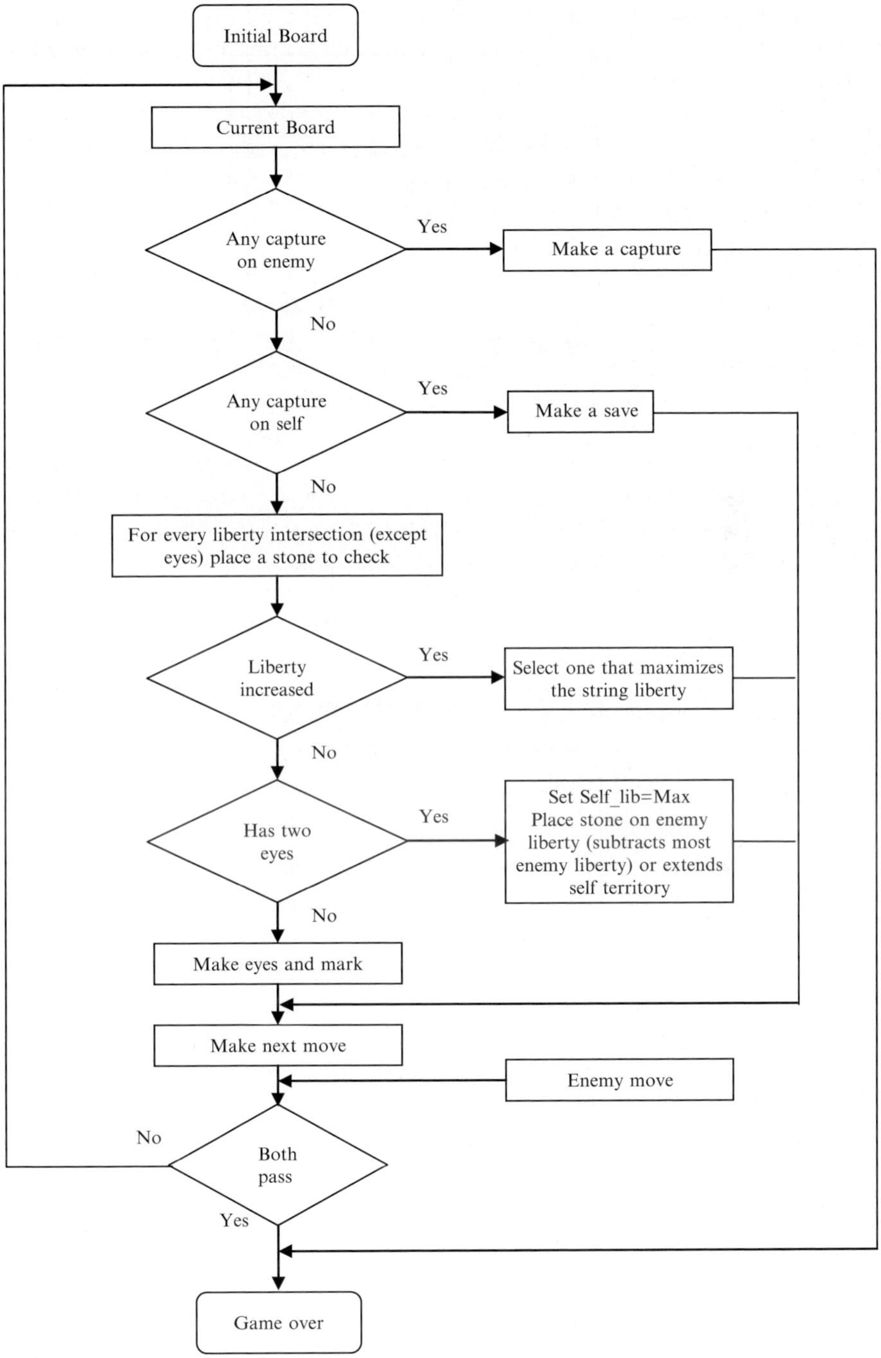

Fig. 5. Flowchart of the defensive player

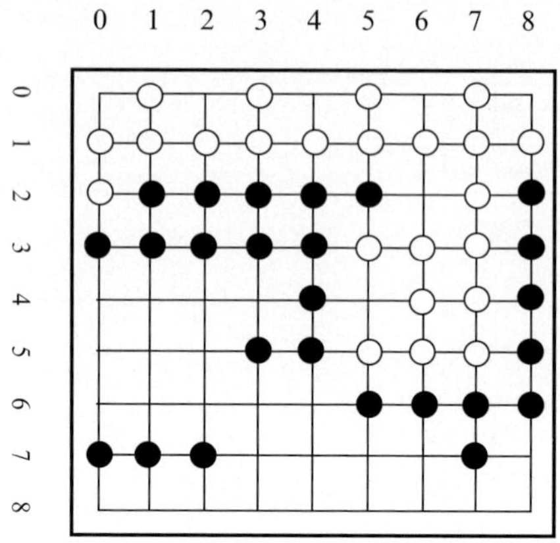

Fig. 6. Final board of game one between the hybrid PSO-EA and the defensive player

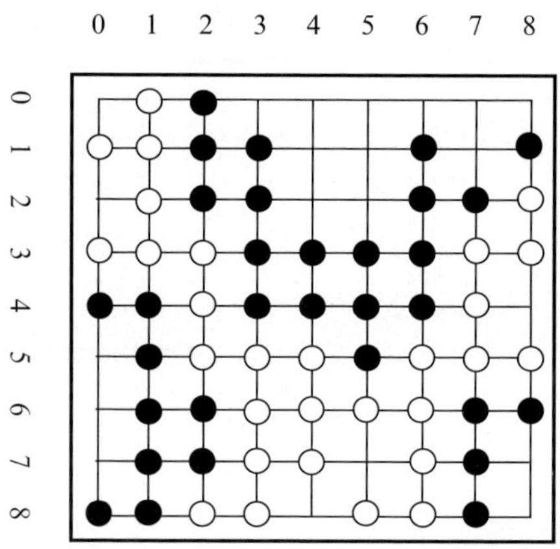

Fig. 7. Final board of game two between the hybrid PSO-EA and the defensive player

tempts in this category have achieved primary success in beating some of the knowledge-based counterparts and have demonstrated the ability to scale up and grasp the general Go strategy.

The acquisition, integration, and use of knowledge are critical to the progress of computer Go programs. Allowing the network to access mature features instead of looking at the raw board may allow it to learn faster and deal with more complex situations. The possibility of merging approaches of network evolution and pattern matching is worthy of further exploration. Evolving a hierarchy of networks where the lower levels, mainly the well-studied features, would provide the inputs for the networks at a high level may be the next breakthrough towards the ultimate goal – defeating human masters in Go.

References

[1] Allis LV, Van den Herik HJ, Herschberg IS (1991) "Which games will survive?" In D. N. L. Levy and D. F. Beal, editors, *Heuristic Programming in Artificial Intelligence 2-The Second Computer Olympiad*, pp. 232–243, Ellis Horwood.

[2] Berlekamp E, Conway J, Guy R (1982) *Winning Ways*, Academic Press, New York.

[3] Berlekamp E, Wolfe D (1994) *Mathematical Go: Chilling Gets the Last Point.* A. K. Peters., MA, USA

[4] Cai X, Wunsch II DC (2004) "Evolutionary computation in playing CaptureGo game," *Proc. of ICCNS'04*, Boston.

[5] Cai X, Zhang N, Venayagamoorthy GK, Wunsch II DC (2005) "Time series prediction with recurrent neural networks using hybrid PSO-EA algorithm," *Neurocomputing, (Accepted).*

[6] Chellapilla K, Fogel D (1999) "Evolution, neural networks, games, and intelligence," *Proceedings of the IEEE*, vol. 87, no. 9, pp. 1471–1496, September.

[7] Chellapilla K, Fogel D (1999) "Evolving neural networks to play Checkers without relying on expert knowledge," *IEEE Trans. on Neural Networks*, vol. 10, no. 6, pp. 1382–1391, November.

[8] Chellapilla K, Fogel D (2001) "Evolving an expert Checkers playing programs without using human expertise," *IEEE Trans. on Evolutionary Computation*, vol. 5, no. 4, pp. 422–428, August.

[9] Chen K, Chen Z (1999) "Static analysis of life and death in the game of Go," *Information Science*, Vol. 121, pp. 113–134.

[10] Chen X, Zhang D, Zhang X, Li Z, Meng X, He S, Hu X (2003) "A functional MRI study of high-level cognition II. The game of GO," *Cognitive Brain Research*, 16(1): 32–37.

[11] Chen Z (1995) "Programming technics in Handtalk," http://www.wulu.com/ht-techn.htm.

[12] Convey J (1976) *On Numbers and Games*, Academic Press, London/New York.

[13] Enderton HD (1991) "The Golem Go program," Tech. Rep. CMU-CS-92-101, Carnegie Mellon University.

[14] Enzenberger M (1996) "The integration of a priori knowledge into a Go playing neural network".
Available: http://www.markus-enzenberger.de/neurogo.ps.gz

[15] Fogel D (2002) *Blondie24: Playing at the Edge of AI.* SF, CA: Morgan Kaufmann.

[16] Fogel D, Hays TJ, Hahn SL and Quon J (2004) "A self-learning evolutionary Chess program," *Proc. of the IEEE*, vol. 92, no. 12, pp. 1947–1954, December.

[17] Fotland D (1999) The 1999 FOST (Fusion of Science and Technology) cup world open computer championship, Tokyo. Available: http://www.britgo.org/results/computer/fost99htm
[18] Fürnkranz J (2001) Machine learning in games: A survey. J. Fürnkranz & M. Kubat (eds.): Machines that Learn to Play Games, Nova Scientific Publishers, Chapter 2, pp. 11–59, Huntington, NY.
[19] Goerlitz S http://www.schachverein-goerlitz.de/Foren/Fun/Go/go.htm
[20] Gomez F, Miikkulainen R (1997) "Incremental evolution of complex general behavior," *Adaptive Behavior*, vol. 5, pp. 317–342.
[21] Hsu F (2002), *Behind Deep Blue.* Princeton, NJ: Princeton Univ. Press.
[22] Kennedy J,. Eberhart R (1995) "Particle Swarm Optimization," *IEEE International Conference on Neural Networks*, vol. 4, pp. 1942–1948, Nov. 27-Dec. 1, Perth, Australia.
[23] Kennedy J,. Eberhart R, Shi Y (2001) *Swarm Intelligence.* San Meteo, CA: Morgan Kaufmann.
[24] Konidaris G, Shell D, Oren N (2002) "Evolving neural networks for the capture game," *Proc. of the SAICSIT postgraduate symposium*, Port Elizabeth, South Africa. Available from: http://www-robotics.usc.edu/~dshell/res/evneurocapt.pdf
[25] Moriarty DE, Miikkulainen R (1994) "Evolving neural networks to focus minimax search," *Proc. of National Conference on Artificial Intelligence (AAAI-94)*, pp. 1371–1377.
[26] Muller M (1999) "Decomposition search: A combinatorial games approach to game tree search, with applications to solving Go endgames," *Proc. of IJCAI*, vol. 1, pp. 578–583.
[27] Muller M (2002) "Position evaluation in computer Go," *ICGA Journal*, Vol. 25, No. 4, pp. 219–228.
[28] Muller M (2003) "Conditional combinatorial games and their application to analyzing capturing race in Go," *Information Science*, Vol. 154, pp. 189–202.
[29] Newman B (1988) "Wally, a simple minded Go-program," ftp://imageek.york.cuny.edu/nngs/Go/comp/.
[30] Pollack JB, Blair AD (1998) "Co-evolution in the successful learning of Backgammon strategy," *Machine Learning*, Vol. 32, pp. 226–240.
[31] Pratola M, Wolfe T (2003) "Optimizing GoTools' search heuristics using genetic algorithms," *ICGA Journal*, vol. 26, no. 1, pp. 28–48.
[32] Prokhorov D and Wunsch II DC (1997) "Adaptive critic designs," *IEEE Trans. on Neural Networks*, vol. 8, no. 5, pp. 997–1007, September.
[33] Reiss M (1995) e-mail sent in January 1995 to the computer Go mailing list, http://www.cs.uoregon.edu/~richard/computer-go/.
[34] Reitman W, Kerwin J, Nado R, Reitman J, Wilcox B (1974) "Goals and plans in a program for playing Go," *Proc. of the 29th National Conference of the ACM*, pp. 123–127.

[35] Reitman W, Wilcox B (1975) "Perception and representation of spatial relations in a program for playing Go," *Proc. of the 30th National Conference of the ACM*, pp. 37–41.
[36] Reitman W, Wilcox B (1978) "Pattern recognition and pattern-directed inference in a program for playing Go," In D. Waterman and F. Hayes-Roth, editors, *Pattern Directed Inference Systems*, pp. 503–523, Academic Press, New York.
[37] Ryder J (1971) "Heuristic analysis of large tree as generated in the game of Go," PhD thesis, Department of Computer Science, Stanford University.
[38] Richards N, Moriarty D, McQuesten P, Miikkulainen R (1998) "Evolving neural networks to play Go," *Applied Intelligence*, vol. 8, pp. 85–96.
[39] Schraudolph N, Dayan P, Sejnowski T (1994) "Temporal difference learning of position evaluation in the game of Go," *Advances in Neural Information Processing*, vol. 6, pp. 817–824.
[40] Schraudolph N, Dayan P, Sejnowski T (2000) "Learning to evaluate Go position via temporal difference methods," In L. Jain and N. Baba Eds, *Soft Computing Techniques in Game Playing*, Springer Verlag, Berlin.
[41] Schwefel (1995) *Evolution and Optimum Seeking.* Wiley, NY.
[42] Shannon CE (1950) "Automatic Chess player," *Scientific American* 182, No. 48.
[43] Smith A (1956) *The Game of Go*, Charles Tuttle Co., Tokyo, Japan.
[44] Sutton R (1988) "Learning to predict by the method of temporal differences," *Machine Learning*, No. 3, pp. 9–44.
[45] Tesauro G (1992) "Practical issue in temporal difference learning,"*Machine Learning*, No. 8, pp. 257–278.
[46] Wilcox B (1985) "Reflections on building two Go programs," *ACM SIGART Newsletter*, pp. 29–43.
[47] Zaman R, Prokhorov DV, Wunsch II DC (1997) "Adaptive critic design in learning to play the game of Go," *Proc. of the International Joint Conference on Neural Networks*, vol. 1, pp. 1–4, Houston.
[48] Zaman R, Wunsch II DC (1999) "TD methods applied to mixture of experts for learning 9x9 Go evaluation function," *Proc. of the International Joint Conference on Neural Networks*, vol. 6, pp. 3734–3739
[49] Zobrist A (1970) "Feature extractions and representation for pattern recognition and the game of Go," PhD thesis, Graduate School of the University of Wisconsin.

This research was supported in part by the National Science Foundation and the M. K. Finley Missouri endowment.

Authors' addresses: Applied Computational Intelligence Laboratory, Dept. of Electrical & Computer Engineering, University of Missouri – Rolla, 1870 Miner Circle, Rolla, MO 65409-0040; email: cai@umr.edu, dwunsch@ece.umr.edu

Noisy Chaotic Neural Networks for Combinatorial Optimization

Lipo Wang and Haixiang Shi

School of Electrical and Electronic Engineering, Nanyang Technological University, Block S1, Nanyang Avenue, Singapore 639798

Summary. In this Chapter, we review the virtues and limitations of the Hopfield neural network for tackling NP-hard combinatorial optimization problems (COPs). Then we discuss two new neural network models based on the noisy chaotic neural network, and applied the two methods to solving two different NP-hard COPs in communication networks. The simulation results show that our methods are superior to previous methods in solution quality. We also point out several future challenges and possible directions in this domain.

1 Introduction

Since Hopfield and Tank's innovative work on solving the traveling sales man problem (TSP) using neural networks, there are numerous research efforts on applying the Hopfield neural network (HNN) and HNN-based neural network techniques to solving combinatorial optimization problems (COPs) [3, 10, 26, 42, 36, 27, 24]. However, Wilson and Pawley [43] raise doubts on the validity of the HNN to solving COPs after they were unable to reproduce the results in Hopfield and Tank's work. They claimed that the original HNN formulation for the TSP is unreliable even for small-sized problems. Many explanations for the poor solution quality of the TSP had been made in terms of energy function formulation [8, 29, 4] and parameter selection [15, 25, 22, 7].

Poor solution quality, dependences on energy function formulations, and the difficulties in parameter selection of the original HNN are due to its gradient descent dynamics leading local minima. This chapter introduces the chaotic neurodynamics which can help to avoid local minima and converge to better solutions in solving NP-hard combinatorial optimizations (COPs).

In 1983, Kirkpatrick *et al* [23] developed simulated annealing, which emulates the annealing processing in metals by first heating the metal to its melting point and then slowly cooling the material. Because of the stochastic nature of the optimization process, simulated annealing can also be called stochastic simulated annealing (SSA) [23]. SSA is known to relax to

Lipo Wang and Haixiang Shi: *Noisy Chaotic Neural Networks for Combinatorial Optimization*, Studies in Computational Intelligence (SCI) **63**, 467–487 (2007)
www.springerlink.com

a global minimum with probability 1 if the annealing takes place sufficiently slowly, i.e., at least inversely proportional to the logarithm of time [13]. In a practical term, this means that SSA is capable of producing good (optimal or near-optimal) solutions for many applications, if the annealing parameter (temperature) is reduced exponentially with a reasonably small exponent [40].

SSA has been widely used in various optimization problems with great success, but it still suffers from several deficiencies:

1) For large problems, the method requires prohibitively long relaxation time in order to find solutions with acceptable quality, i.e., SSA consumes too much iteration time due to its Monte Carlo scheme. To guarantee convergence to an exact solution, SSA will require more iterations than complete enumeration does for some problems [5].
2) SSA often requires subtle adjustments of parameters in the annealing schedule, such as the length of the temperature steps during annealing, the temperature range, the number of re-starts and re-direction during the search [32, 1, 19, 20, 23].

In order to improve the searching ability of the SSA, complex neurodynamics such as chaotic simulated annealing (CSA) was proposed [46]. Compared with the gradient descent dynamics of the HNN models and neural networks with SSA dynamics, neural networks with CSA have a richer spectrum of dynamic behaviors, such as stable fixed points, periodic oscillations, and chaos.

Nozawa demonstrated the search ability of the chaotic neural networks (CNN) [30, 45]. Chen and Aihara [46, 5] proposed the chaotic simulated annealing (CSA) by starting with a sufficiently large negative self-coupling in the Aihara-Takabe-Toyoda [2] network when the dynamics is chaotic, and gradually decreasing the self-coupling so that the network eventually stabilizes, thereby obtaining a transiently chaotic neural network (TCNN). Their computer simulations showed that the CSA leads to good solutions for the TSP much more easily compared to the Hopfield-Tank approach [16, 17] and SSA. Chen and Aihara [6] offered the following theoretical explanation for the global searching ability of the chaotic neural network: its attracting set contains all global and local optima of the optimization problem under certain conditions, and since the chaotic attracting set has a fractal structure and covers only a very small fraction of the entire state space, CSA is more efficient in searching for good solutions for optimization problems compared to other global search algorithms such as SSA.

Other kinds of CSA have also been proposed. Wang and Smith proposed another chaotic annealing by annealing the time-step in the Euler approximation of the continuous Hopfield network [41]. Hayakawa et al. [14] obtained the CSA by adding the chaotic noise into the Hopfield network. Zheng et al. [47] improved the Wang-Smith's chaotic simulated annealing which reaps the benefits of Wang-Smith model and Chen-Aihara model.

There are mainly three significant differences between SSA and CSA [5]:

1) SSA is stochastic on the basis of the Monte Carlo scheme while CSA is deterministic with transiently chaotic dynamics.
2) The convergent processing of SSA is controlled by stochastic "thermal" *fluctuations* while that of CSA is controlled by *bifurcation* structures.
3) SSA essentially searches all possible states while temporally changing probability distributions, whereas CSA restricts to a fractal subspace. Because the searching region in CSA is usually smaller compared with the entire state space, CSA can be expected to perform efficient searching if the restriction is adequate to include a global optimum state or some near-global optimum states.

However, CSA has completely deterministic dynamics and is not guaranteed to settle down at a global optimum no matter how slowly the annealing parameter (the neuronal self-coupling) is reduced [37]. Different from the searching direction of SSA that is probabilistically determined by mutual interactions among neurons, CSA is uniquely determined by mutual interactions among neurons. In practical terms, this means that CSA sometimes may not be able to provide a good solution at the end of annealing for some initial conditions of the network, no matter how slowly annealing takes place, i.e., CSA sometimes may not be able to provide a good solution at the conclusion of annealing even after a long time of searching.

Wang and Tian [40] proposed a new approach to simulated annealing, i.e., stochastic chaotic simulated annealing (SCSA), using a noisy chaotic neural network (NCNN) by adding decaying stochastic noise into the TCNN. Compared with CSA, SCSA performs stochastic searching both before and after chaos disappears and is more likely to find optimal or sub-optimal solutions.

2 Mathematical Formulations of the NCNN

The NCNN model is described as follows [40]:

$$x_{jk}(t) = \frac{1}{1+e^{-y_{jk}(t)/\varepsilon}} , \tag{1}$$

$$\begin{aligned} y_{jk}(t+1) = k y_{jk}(t) + \alpha \left\{ \sum_{i=1, i\neq j}^{N} \sum_{l=1, l\neq k}^{M} w_{jkil} x_{jk}(t) + I_{ij} \right\} \\ - z(t)(x_{jk}(t) - I_0) + n(t) , \end{aligned} \tag{2}$$

$$z(t+1) = (1-\beta_1) z(t) , \tag{3}$$

$$n(t+1) = (1-\beta_2) n(t) , \tag{4}$$

where the notations are:

x_{jk} : output of neuron jk ;

y_{jk} : input of neuron jk ;

w_{jkil}: connection weight from neuron jk to neuron il, with $w_{jkil} = w_{iljk}$ and $w_{jkjk} = 0$;

$$\sum_{i=1,i\neq j}^{N} \sum_{l=1,l\neq k}^{M} w_{jkil}x_{jk} + I_{ij} = -\partial E/\partial x_{jk} \tag{5}$$

E : energy function;

I_{jk} : input bias of neuron jk ;

k : damping factor of nerve membrane ($0 \leq k \leq 1$);

α : positive scaling parameter for inputs ;

β_1 : damping factor for neuronal self-coupling ($0 \leq \beta_1 \leq 1$);

β_2 : damping factor for stochastic noise ($0 \leq \beta_2 \leq 1$);

$z(t)$: self-feedback connection weight or refractory strength ($z(t) \geq 0$), $z(0)$ is a constant;

I_0 : positive parameter, which is used as threshold for each neuron, can be a fixed number or variable one;

ε : steepness parameter of the output function ($\varepsilon > 0$) ;

$n(t)$: random noise injected into the neurons, in $[-A, A]$ with a uniform distribution;

$A[n]$: amplitude of noise n.

This NCNN model is a general form of chaotic neural networks with transient chaos and decaying noise. In the absence of noise, i.e., $n(t) = 0$, for all t, the NCNN as proposed in eqns. (1) - (4) reduces to the TCNN in [5].

3 Gradual Noisy Chaotic Neural Network (G-NCNN)

Compared with the constant number of neurons used in conventional neural networks, the gradual neural network [11] adopts an increasing number of neurons. Usually, the number of neurons that a neural networks needs for solving a COP is determined by the problem, e.g., N-city TSP problem, $N \times N$ neurons are needed to compose a solution space. For the NCNN, N^2 neurons are needed at the start of neural network updating. But if using the gradual neural network, the neural network needs only a fraction of $N \times N$ neurons at the beginning of neuron computation, which are the most likely selected group of neurons in the final results (each neuron stands for a status in the solution matrix based on different problems, e.g., if neuron ij fires, it means the ith element is assigned to jth element in the assignment problem). The rest of all $N \times N$ neurons are gradually added to the current commutating group.

The reason why the gradual scheme is needed instead of a constant number of neurons is that the gradual scheme can be considered as one kind of objective in the optimization. There are objectives in COPs that need to find the minimal "cost", where the cost can have many different meanings in different problems, e.g., interference, delay, or length of a path. In order to achieve the objective that the solution found by the neural network is with the minimal cost, neurons are divided into several groups and activated or added in several stages with each stage only adopt one group of neurons with smallest cost left over. Through the gradual scheme, we do not need to formulate the objective ("minimal cost") in the energy function, because we think that the objective is realized in the gradual expansion stages which let the neurons with smaller cost be included in neuron computation in an earlier stage. When dealing with this kind of objective, we can adopt the gradual scheme naturally through the following steps [11]:

1) Assume there are $N \times M$ neurons needed for the problem. Compute the cost matrix $C[N \times M]$, i.e., the cost of neuron ij if neuron ij is selected in the final solution. Here N and M can have the same value.
2) Sort the neuron in ascending order of the cost.
3) Divide the $N \times M$ neurons into P groups $(G_1, G_2, \ldots, G_P)$, with each group have p neurons, where p is the maximum integer less than or equal to NM/P. Group G_1 is the group of neurons with smallest cost and the other groups G_2 to G_P contain the neurons with larger cost with an ascending order of cost.
4) Add the neurons in the first group G_1 to the neural network and let the neural network update. If a feasible solution is found, exit, the solution found is the final solution. If the pre-defined steps are used up and still no solutions are found, then apply the gradual expansion scheme by adding the neurons in G_2 and let the neural network update again.

We found that although using gradual scheme can help to reduce the objectives in optimizations, it cannot guarantee to achieve the objective. But it can be made up by adding the objective again into the energy function.

4 Noisy Chaotic Neural Network with Variable Threshold (NCNN-VT)

Besides the gradual noisy chaotic neural network, another extension of the noisy chaotic neural network is proposed by using the variable threshold in the neural network, instead of the fixed threshold value. It can be found that the variable threshold can be used to achieve the objective of the NP-hard optimization problem, and again reduces the number of objectives needed to be formulated in the energy function.

4.1 Adaptive Mapping Scheme

Adaptive Mapping Scheme (AMS) aims to map the objective of optimization problem into the probability of firing of each neuron. If the objective of the problem is to find the solution with minimal cost, then the neurons with smaller cost will have larger probability to be selected in the final solution matrix. By investigating the single neuron dynamics of the NCNN model, it can be seen that the positive parameter I_0 in the NCNN model is responsible for the neuron firing probability. The single neuron dynamics of the NCNN by varying the parameter I_0 is shown in Figs. 1 to 4 (the x-axis is the time steps t, the y-axis is the output of neuron $x_{ij}(t)$). We can see clearly that if the value of parameter I_0 is 0.3, the output of neuron is very close to 0.3 after the network passes the last bifurcation-2 in Fig. 2. After the bifurcation point, the neural network slowly converge to a stable point at about 0.5. The same pattern can be observed in other figures. It means that different value of I_0 ($0 < I_0 < 1$) induces different probability of neuron firing. The neuron with large value of I_0 will be selected to be firing (neuron output equals 1) with bigger probability, whereas, the small value of I_0 will result in less chance the neuron to be firing.

4.2 Model Definition

The difference between the NCNN and the NCNN-VT model is that the positive parameter I_0 in eqn. (2) of the NCNN becomes a variant labeled as I_{jk}, where jk stands for the neuron jk, i.e.,

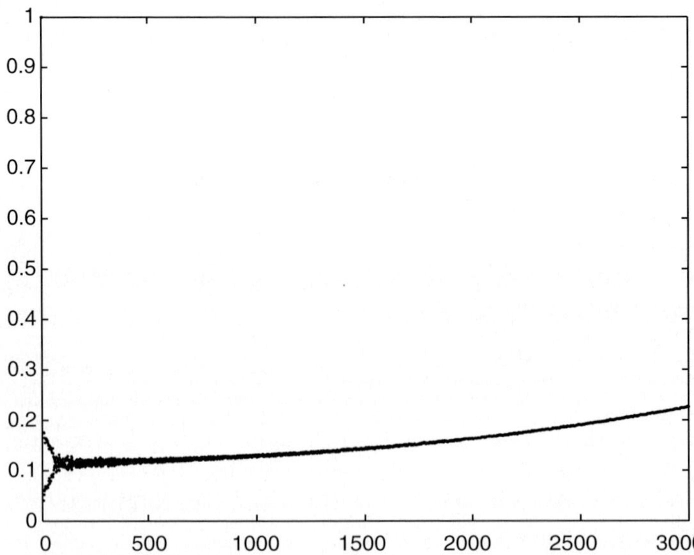

Fig. 1. The single neuron dynamics of the noisy chaotic neural network with variable threshold, $I_0 = 0.1$

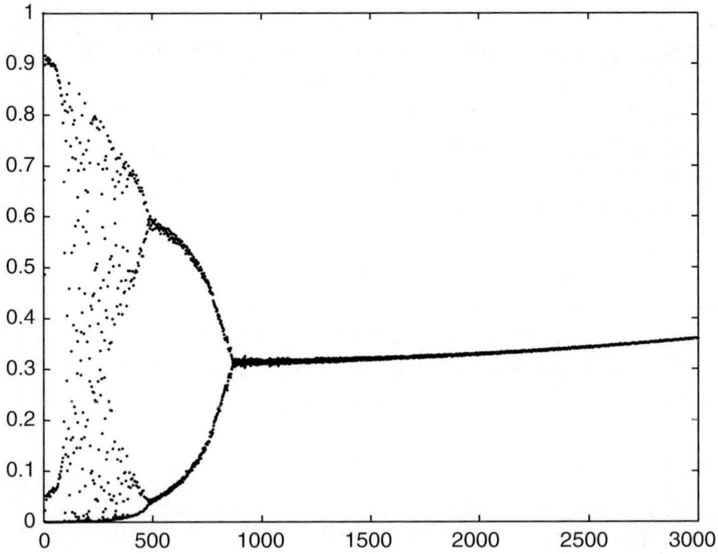

Fig. 2. The single neuron dynamics of the noisy chaotic neural network with variable threshold, $I_0 = 0.3$

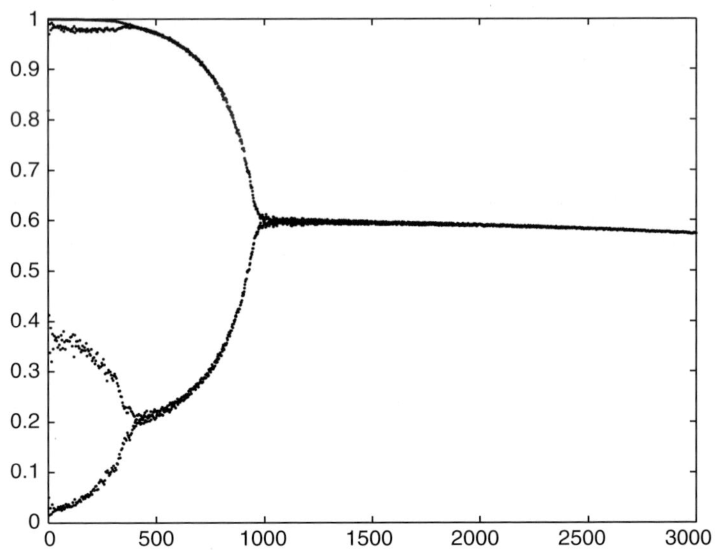

Fig. 3. The single neuron dynamics of the noisy chaotic neural network with variable threshold, $I_0 = 0.6$

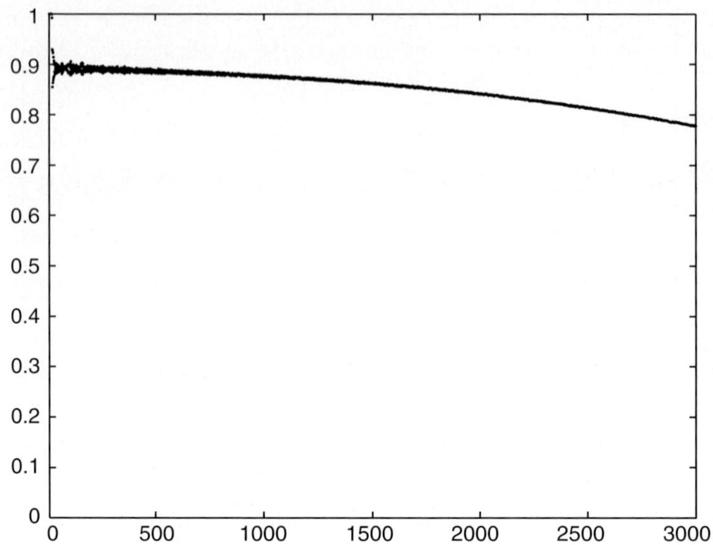

Fig. 4. The single neuron dynamics of the noisy chaotic neural network with variable threshold, $I_0 = 0.9$

$$y_{jk}(t+1) = ky_{jk}(t) + \alpha \left\{ \sum_{i=1,i\neq j}^{N} \sum_{l=1,l\neq k}^{M} w_{jkil} x_{jk}(t) + I_{ij} \right\}$$
$$-z(t) \left[x_{jk}(t) - I_{jk} \right] + n(t) \,. \tag{6}$$

where I_{jk} is not a constant parameter but a variable which determines the selection of firing of each neuron. The value of I_{jk} is related to the problem optimization term.

4.3 Mapping Functions

The mapping function connects the problem's cost with the parameter I_{ij}.

$$I_{jk} = f(c_{jk}), \quad j = 1, 2, \ldots, N; k = 1, 2, \ldots, M \tag{7}$$

The mapping function can be a linear or nonlinear function which transform the cost into the probability value between 0 and 1. Normally, a linear mapping is adopted:

$$f(c_{jk}) = 1 - \frac{c_{jk} - c_{min}}{c_{max} - c_{min}} \,. \tag{8}$$

where c_{jk} is the jk element in the cost matrix C, c_{max} is the maximum value among the cost matrix and c_{min} is the minimum value in the cost matrix.

From the mapping functions in eqn. (8), it can be seen that the neuron ij with smaller cost d_{ij} will have larger or higher probability to fire while the ones with larger cost will be inhibited to fire.

5 Using G-NCNN to Solve the Broadcast Scheduling Problem

5.1 Problem Introduction

In a time-division-multiple-access (TDMA) network, time is divided into frames and each TDMA frame is a collection of time slots. A time slot has a unit time length required for a single packet to be communicated between adjacent nodes. When nodes transmit simultaneously, conflicts will occur if the nodes are in a close range. Therefore, adjacent nodes must be scheduled to transmit in different time slots, while nodes some distance away may be arranged to transmit in the same time slot without causing conflict [39]. The goal of the broadcast scheduling problem (BSP) is to find an optimal TDMA frame structure that fulfills the following two objectives. The first is to schedule transmissions of all nodes in a minimal TDMA frame length without any conflict. The second is to maximize channel utilization or total conflict-free transmissions.

5.2 Energy Function Formulation

For the two objectives, a two-stage methods using the NCNN are used to solve the problem. The first stage aims to find the minimal TDMA frame cycle length (M), whereas the objective in the second stage is to maximize the total node transmissions in order to fulfill the channel utilization.

The G-NCNN consists of $M \times N$ neurons. M is initially set as its lower bound value L_m, which can be easily obtained using graph theory [21]. The network can be formulated to a graph G = (V, E). The graph G can be transformed into $G' = (V, E')$, where E in G stands for one-hop-away edges, and E' in G' stands for one-hop-away and two-hop-away edges. The lower bound is:

$$L_m = \omega(G') . \tag{9}$$

where $\omega(G')$ is the maximal cardinality of a clique in G' [33].

The energy function E_1 for the first stage is given as follows [9]:

$$E_1 = \frac{W_1}{2}\sum_{j=1}^{N}(\sum_{k=1}^{M} x_{jk} - 1)^2 + \frac{W_2}{2}\sum_{j=1}^{N}\sum_{i=1}^{M}\sum_{k=1,k\neq i}^{N} d_{jk}x_{ij}x_{ki} , \tag{10}$$

where W_1 and W_2 are weighting coefficients. The W_1 term represents the constraint that each of the N nodes in the PRN must transmit exactly once during each TDMA cycle. The W_2 term indicates the constraint that any pair of nodes which is one-hop-away or two-hop-away must not transmit simultaneously during each TDMA cycle.

From eqn. (2), eqn. (5), and eqn. (10), we obtain the dynamics of the G-NCNN as follows:

$$y_{jk}(t+1) = ky_{jk}(t) + \alpha \left\{ -W_1(\sum_{k=1}^{M} x_{jk} - 1) - W_2 \sum_{k=1,k\neq j}^{N} d_{jk}x_{ki} \right\} \\ -z(t)\left[x_{jk}(t) - I_0\right] + n(t) \,. \tag{11}$$

The G-NCNN stops when the it finds a feasible assignment and the current number of time slots together with its transmission assignments are the optimal results for phase I of the BSP. In this chapter, different from the GNN in [9], where the neurons are expanded gradually at every P iterations during the iterative computation of the neural network, we implement the GES based on a convergence index $\delta(t)$ of the network energy, which we defined as:

$$\delta(t) = \sum_{q=t-4}^{t} \left|E(q) - E(q-1)\right| / E(0) \,. \tag{12}$$

where $E(q)$ is the value of energy function at time step q. If index $\delta(t)$ is less than a very small value, e.g., $\delta(t) < 10^{-4}$ in our simulation, the neural network is considered as having fully converged. If the network has converged but no feasible solutions are found using the current number of time slots, the number of time slots is increased by 1, i.e., $M \rightarrow M+1$, and the G-NCNN re-starts to search for optimal solutions with the updated number of neurons.

In the second stage, the minimal TDMA frame length M is found and each node is assigned with one and exactly one time slot. In phase II, we aim at maximizing the channel utilization by adding as many conflict-free transmissions as possible to the TDMA frame. Because in phase I one node is assigned with exactly one slot in order to find a minimal frame length, there are many nodes which can use other time slots without violating the no-conflict constraint. Thus, additional transmissions may be found on some nodes but frame length M and the assigned transmissions in phase I are fixed [9].

$$E_2 = \frac{W_3}{2} \sum_{j=1}^{N} \sum_{i=1}^{M} \sum_{k=1,k\neq i}^{N} d_{jk}x_{ij}x_{ki} + \frac{W_4}{2} \sum_{j=1}^{N} \sum_{i=1}^{M} (1 - x_{ij})^2 \,, \tag{13}$$

where W_3 and W_4 are weighting coefficients. W_3 represents the constraint term that any pair of nodes which is one-hop-away or two-hop-away must not transmit simultaneously during each TDMA cycle. W_4 is the optimization term which maximizes the total number of firing neurons.

From eqn. (2), eqn. (5), and eqn. (13), we obtain the dynamics of the NCNN for phase II of the BSP as follows:

$$y_{jk}(t+1) = ky_{jk}(t) + \alpha \left\{ -W_3 \sum_{k=1,k\neq j}^{N} d_{jk}x_{ki} + W_4(1 - x_{ij}) \right\}$$
$$-z(t)\left[x_{jk}(t) - I_0\right] + n(t) \,. \qquad (14)$$

The NCNN is updated cyclically and asynchronously, which means we update the neurons in two loops and the neuron is selected to be computed in a fixed order. The new state information of a updated neuron is immediately available for the other neurons in the computation. The iteration is terminated once a feasible transmission schedule is obtained, i.e., the transmissions of all nodes are conflict-free.

5.3 Results Discussions

The benchmark examples are get from other published papers. Each problem is simulated 50 different times and the best and the average values are displayed in Table 1.

Table 1. Comparisons of average delay time η and numbers of time slots M and computation time T obtained by the G-NCNN and other algorithms for the three benchmark problems in 50 runs, where Best/Avg stands for the best value and average value in multiple runs.

Case		BM 1 Best/Avg	BM 2 Best/Avg	BM 3 Best/Avg
	η	6.8/7.0	9.0/9.5	5.7/6.1
G-NCNN	M	8/8.0	10/10.5	8/8.0
	T	6.0/7.2	16.0/18.3	6.0/6.5
	η	7.0/7.0	9.3/9.	6.3/6.5
HNN-GA	M	8/8.0	10/10.0	8/9.0
	T	4.0/4.7	17/19.0	13.0/14.0
	η	7.2/7.4	10.0/12.0	6.8/7.2
SVC	M	8/8.0	10/11.0	8/10.0
	T	2.5/2.8	15.0/15.4	10.0/12.0
	η	7.1/7.2	9.5/10.0	6.2/6.5
GNN	M	8/8.0	10/10.5	8/8.5
	T	15.0/16.4	18.0/20.0	17.0/19.5
	η	7.2/7.5	10.5/12.5	6.9/8.2
MFA	M	8/9.0	12/13,5	9/10.0
	T	25.0/7.2	32.5/38.5	28.0/29.0

From the results, we can see that our G-NCNN method can find shorter frame length than previous methods. In addition, our proposed method can find the smallest average time delay η among all methods in all three cases. The computation time (T in the table) needed by the G-NCNN is relatively lower than the previous MFA and comparable to other methods.

6 Applying the NCNN-VT to the Frequency Assignment Problem

6.1 Problem Introduction

Due to the economic effect on the average person in everyday life, there is an increasing number of satellites in geostationary orbits. In order to accommodate the crowded satellites in the same orbit, optimal design of satellites are necessary in order to provide high quality transmissions. In satellite communication systems, the major impairments in transmission design include thermal noise, rain attenuation, inter-modulation, and co-channel interference, among which, co-channel interference dominates because it seriously affects system design and operation [28]. Hence, the reduction of the co-channel interference has arisen as a major problem in satellite communications with the dramatic increase of geostationary satellites in orbits.

In order to reduce the interference, re-arrangements of frequency assignments, which take advantage of carrier interleaving, is thought as an effective way in practical situations. Early efforts have focused on various analytical methods for evaluations of co-channel interference [31, 18] and very few systematic methods have been adopted to optimize frequency assignments to reduce co-channel interference. The later work of Muzuike and Ito [28] revealed the importance of mathematical models for reduction of co-channel interference. They proposed a basic mathematical model to formulate the co-channel interference reduction problem as the "assignment problem". The assignment problem aims to minimize the largest interference among carriers. Fig. 6 shows the co-channel interference model for the system in Fig. 5. In the inter-system context, the two sets of carriers share the same frequency band. One set of carriers (C_{11} to C_{13}) is in satellite system 1 and the other set of carriers (C_{21} to C_{24}) corresponds to satellite system 2 in Fig. 5.

In the model shown in Fig. 6, carrier frequencies for one set of carriers are to be rearranged while keeping the other set fixed, i.e., the frequencies for carriers in system 2 are chosen to be re-arranged while the frequencies for system 1 are fixed. Fig. 5 shows the inter-system co-channel interference between two adjacent satellite systems. The communications are assumed to operated between F_a and F_b as showed in Fig. 6, where F_a and F_b are frequency band. The co-channel interference can be evaluated by calculating the each pair of carriers using the same frequency, which varies with different pairs. The objective of this assignment problem is to find the optimal

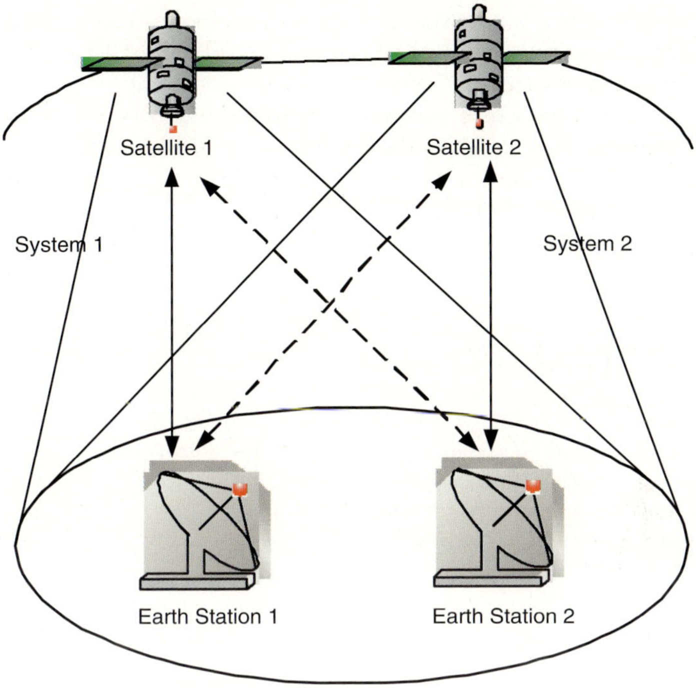

Fig. 5. Inter-system co-channel interference

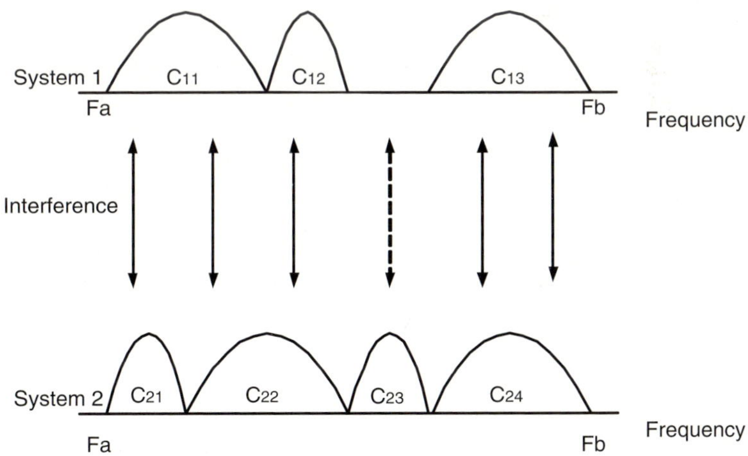

Fig. 6. Co-channel interference model for the system in Fig. 5, where the C_{xy} stands for the carrier y in system x, e.g., C_{23} stands for the carrier 3 in system 2

assignment of frequencies in system 2 in order to reduce the co-channel interference. The largest interference is considered as a limiting factor, and the optimal assignment is the one which can minimize the limiting factor.

In this chapter, we use a two-dimensional neural network which consists of $N \times M$ neurons for the FAP of N carriers and M segments. The output of each neuron V_{ij} will be converted into binary values V_{ij}^d. V_{ij}^d represents whether carrier i is assigned to segment $j-(j+c_i-1), (i=1,\ldots,N; j=1,\ldots,M)$, where c_i indicates the length of carrier i, i.e.:

$$V_{ij}^d = \begin{cases} 1 \text{ carrier } i \text{ is assigned to segment } j-(j+c_i-1), \\ 0 \text{ otherwise.} \end{cases}$$

Fig. 7 shows the neural network formulation for the 4-carrier-6-segment problem. This neural network consists of 24 (= 4 × 6) neurons as shown in Fig. 7 (a). Fig. 7 (b) is the convergence state, with the black squares stand for the neurons with output $V_{ij}^d = 1$. Fig. 7 (c) shows the full assignment for each carrier. And Fig. 7 (d) is the final frequency assignment for the FAP. Note that it can be easy to expand the convergence state in Fig. 7 (c) to the final assignment in Fig. 7 (d) given the carrier length for each carrier. We provide only the solution format in Fig. 7 (c) to represent the final assignment in this Chapter.

Our objective is to minimize the largest element of the interference matrix selected in the assignment and at the same time minimize the sum of interference of all selected elements. Thus we define the choice of the mapping function of I_{ij} as follows:

$$\begin{aligned} I_{ij} &= 1 - \frac{d_{ij} - d_{i,min}}{d_{i,max} - d_{i,min}} \\ &= \frac{d_{i,max} - d_{ij}}{d_{i,max} - d_{i,min}} \,. \end{aligned} \tag{15}$$

where d_{ij} is the ij-th element in cost matrix $D = (d_{ij}, i=1,\ldots,N; j=1,\ldots,M)$, and $d_{i,max}$[1] is the maximum value in line i of matrix D and $d_{i,min}$ is the minimum value line i of matrix D. We will label $d_{i,max}$ as d_{max} and $d_{i,min}$ as d_{min} for simplicity.

Through the mapping in eqn. (15), not only can we achieve the objectives of FAP, but also separate the objective from the energy function, which will make the tuning of weighting coefficients in the energy function easier without the need to balance the optimization term and constraint term in one energy function. Moreover, it will improve the convergence speed of the noisy chaotic neural network as shown in the result discussion section.

[1] Note that the maximum value of cost does not include the infinity value of in the interference matrix. Actually the neurons corresponding to the infinite interference will never fire due to its inhibitive cost.

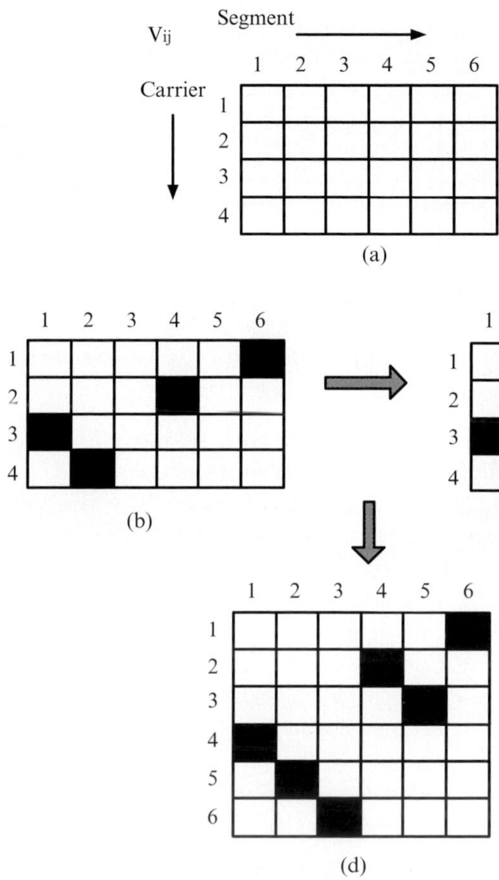

Fig. 7. The neural network formulation for the FAP. (a) The 24 neurons for the 4-carrier-6-segment FAP. (b) the convergence state of the neural network. (c) the full assignment of the neural network formulation. (d) the final assignment of the segments for the FAP through the expansion from the neural network formulation

6.2 Energy Function and Results

Recall that the goal of the FAP has been separated from the constraints. Only the two constraints of the FAP need to be formulated in the energy function. The first constraint of an N-carrier-M-segment problem is that each first segment of the N carriers in system 2 must be assigned to one and only one of the M segments. Hence, one and only one neuron among the M neurons for each carrier has output 1. Then the first constraint can be formulated as [11]:

$$E_1 = \sum_{i=1}^{N} (\sum_{q=1}^{M} V_{iq} - 1)^2 . \quad (16)$$

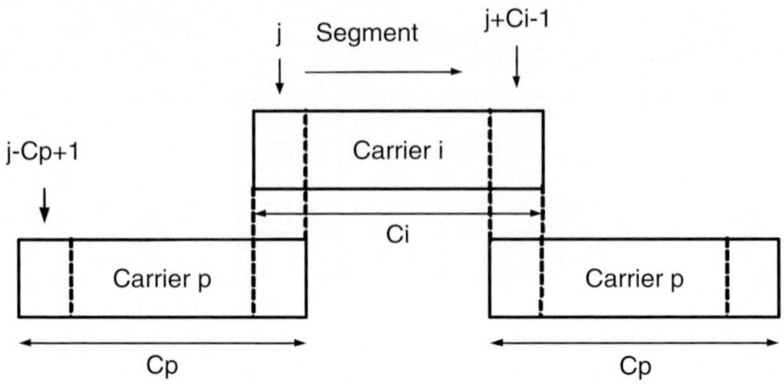

Fig. 8. Violation condition for the second and third constraints of the FAP

The second and third constraints are that each segment in system 1 can be assigned to at most one segment in system 2 and the assignment of carriers in system 2 should be in consecutive segments in system 1 in the same order. If carrier i is assigned to segment $j-(j+c_i-1)$, any other carrier $p(p \neq i)$ must not be assigned to the consecutive segment $(j-c_p+1)-(j+c_i-1))$. The violation condition of the two constraints is shown in Fig. 8, where c_i is the carrier length of carrier i. In other words, if carrier i is assigned with consecutive c_i segments, the segments occupied by any other carrier cannot occupy with the segment $j-(j+c_i-1)$. The first segment of each carrier $p, (p \neq i)$ should be $(j-c_p+1)$ before and $j-(j+c_i-1)$ after the first segment of carrier i.

Hence these two constraints give raise to the second part of the energy function to be minimized, as formulated in [11]:

$$E_2 = \sum_{i=1}^{N} \sum_{j=1}^{M} \sum_{\substack{p=1 \\ p \neq i}}^{N} \sum_{q=j-c_p+1}^{j+c_i-1} V_{ij} V_{pq} \,. \tag{17}$$

Note that because $(j-c_p+1)$ can be negative and $(j+c_i-1)$ can exceed the total number of segments M. The original formulation in eqn. (17) [11] has errors in dealing with the bounds and produces the program bugs when searching the solutions. We formulate the second term in our energy function in a revised version as follows:

$$E_2' = \sum_{i=1}^{N} \sum_{j=1}^{M} \sum_{\substack{p=1 \\ p \neq i}}^{N} \sum_{q=max(j-c_p+1,1)}^{min(j+c_i-1,M)} V_{ij} V_{pq} \,. \tag{18}$$

where function $max(x, y)$ returns the maximum value between (x, y) two numbers and $min(x, y)$ finds the minimum value between (x, y).

We use the following convergence term in our energy function to help the neuron output converge to the corner (0 or 1) of the hypercube:

Table 2. Comparisons of the simulation results (largest interference and total interference) obtained by the NCNN-VT, GNN and HopSA in the benchmark examples.

Instance	GNN[11]		HopSA[34]		NCNN-VT	
	largest	total	largest	total	largest	total
BM 1	30	100	30	100	30	100
BM 2	4	13	4	13	4	13
BM 3	7	85	7	85	7	88
BM 4	64	880	84	886	64	880
BM 5	640	8693	817	6851	640	7246

$$E_3 = \sum_{i=1}^{N}\sum_{j=1}^{M} V_{ij}(1 - V_{ij}) \,, \tag{19}$$

The total energy function of the NCNN-VT is given by the summation of the three parts E_1, E_2', and E_3:

$$E = \frac{W_1}{2}\sum_{i=1}^{N}(\sum_{q=1}^{M} V_{iq} - 1)^2 + \frac{W_2}{2}\sum_{i=1}^{N}\sum_{j=1}^{M}\sum_{\substack{p=1 \\ p\neq i}}^{N}\sum_{q=max(j-c_p+1,1)}^{min(j+c_i-1,M)} V_{ij}V_{pq}$$

$$+\frac{W_3}{2}\sum_{i=1}^{N}\sum_{j=1}^{M} V_{ij}(1 - V_{ij}) \,. \tag{20}$$

where W_1, W_2, and W_3 are weighting coefficients.

6.3 Result Discussions

Table 2 shows the results obtained by the NCNN-VT and a comparison with other previous methods. For the benchmark problems from BM 1 to BM 5, the NCNN-VT algorithm matches or improves the results of other existing algorithms. The results on the benchmark examples show that the NCNN-VT can find better or similar solution compared with the previous methods.

7 Future Challenges

We have reviewed neural-network-based techniques for solving NP-hard COPs, especially neural networks with chaotic neuro-dynamics. Two applications in telecommunication networks demonstrated that chaotic neural networks have effective search abilities compared to other methods. Despite the computational advantages of these methods, there still exist tremendous challenges from both methodology and applications points of view.

The "no free lunch" theorem proposed by Wolpert and Macready [44] showed that all algorithms that search for an extremum of a cost function perform exactly the same, when averaged over all possible cost functions. They claimed that if algorithm A outperforms algorithm B on some cost functions, then loosely speaking there must exist exactly as many other functions where B outperforms A. The no free lunch theorem also applies for the NCNN and the extensions (G-NCNN and NCNN-VT). Actually, the NCNN-based methods are facing difficities on solving problems besides combinatorial optimizations, for example, they are opt for solving combinatorial optimization problems, other than applications like non-linear or multi-dimensional function optimization. We may ask the question: how do chaotic neural networks compare to other computational intelligence methods, such as genetic algorithms and ant colonies, in solving other practical COPs?

Problem modeling can also cause difficulties when solving COPs using neural networks. Every COP, whether simple or hard, needs to be constructed into an energy function formulation before using the NCNN to solve it. And the form of this energy function is critical for neural-network-based methods to search for optimal solutions. Different formulations may lead to different solution quality and search time. It is common to see various formulations of energy funcations made by different researchers on the same problem, e.g. the TSP problem [4, 43, 38]. In order to improve solution quality, the formulation needs to be revised and upgraded from time to time. The state-of-the-art energy function formulation, if not difficult to formulate, is actually time-consuming to find.

Another tough problem when using neural-network-based methods is the selection of the parameters. The various parameters, including system parameters and weighting coefficients in the energy functions, are influential to solution quality and search efficiency. There are basic guidelines for parameter selections [35]; however, it can be challenging to find the optimal parameters, which in turn will lead to optimal solution quality and search efficiency.

For applications in the communications domain, comparisons performed in the research studies have usually been undertaken in simplified scenarios simulated in servers or desktop PCs. Except for some partial implementations on real hardware [12], algorithm testing using hardware is usually not undertaken. It will be necessary to place a greater emphasis on demonstrating advantages of computational intelligence methods in real communications hardware, in order to convince industrialists to adopt computational intelligence methods in telecommunications companies and equipment manufacturers.

References

[1] E. Aarts and J. Korst. *Simulated Annealing and Boltzmann Machines.* John Wiley, Chichester, 1989.

[2] K. Aihara, T. Takabe, and M. Toyoda. Chaotic neural networks. *Physics Letters A*, 144:333–340, 1990.

[3] Mustafa K. Mehmet Ali and F. Kamoun. Neural networks for shortest path computation and routing in computer networks. *IEEE Trans. on Neural Networks*, 4:9, 1993.

[4] R.D. Brandt, Y. Wang, A.J. Laub, and S.K. Mitra. Alternative network for solving the travelling salesman problem and the list-matching problem. In *Proceedings IEEE International Joint Conference on Neural Networks*, volume 2, pages 333–340, 1988.

[5] L. Chen and K. Aihara. Chaotic simulated annealing by a neural network model with transient chaos. *Neural Networks*, 8:915–930, 1995.

[6] L. Chen and K. Aihara. Global searching ability of chaotic neural networks. *IEEE Trans. Circuits and Systems - I: Fundamental Theory and Applications*, 46(8):974–993, 1999.

[7] G.W. Davis. Sensitivity analysis in neural net solutions. *IEEE Trans. on Systems, Man and Cybernetics*, 19:1078–1082, 1989.

[8] D.E. Van den Bout and T.K. Miller. A traveling salesman objective function that works. In *Proceedings IEEE International Joint Conference on Neural Networks*, volume 2, pages 299–303, 1988.

[9] N. Funabiki and J. Kitamichi. A gradual neural network algorithm for broadcast scheduling problems in packet radio networks. *IEICE Trans. Fundamentals*, E82-A:815–824, 1999.

[10] N. Funabiki and S. Nishikawa. A binary hopfield neural-network approach for satellite broadcast scheduling problems. *IEEE Trans. on Neural Networks*, 8:441–445, 1997.

[11] N. Funabiki and S. Nishikawa. A gradual neural-network approach for frequency assignment in satellite communication systems. *IEEE Trans. Neural Networks*, 8:1359–1370, 1997.

[12] V. Catania G. Ficili S. Palazzo G. Ascia and D. Panno. A VLSI fuzzy expert system for real-time tra.c control in atm networks. *IEEE Transactions on Fuzzy Systems*, 5(1):20–31, 1997.

[13] S. Geman and D. Geman. Stochastic relaxation, gibbs distributions, and the bayesian restoration of images. *IEEE Trans. Pattern Analysis and machine Intelligence*, 6:721–741, 1984.

[14] Y. Hayakawa, Marunoto A, and Y. Sawada. Effects of the chaotic noise on the performance of a neural network model for optimization problems. *Physical Review E*, 51:2693–2696, 2002.

[15] S. Hegde, J. Sweet, and W. Levy. Determination of parameters in a hopfield/tank computational network. In *Proceedings IEEE International Conference in Neural Networks*, volume 2, pages 291–298, 1988.

[16] J.J. Hopfield. Neurons with graded response have collective computational properties like those of two-state neurons. In *Proc. Natl. Acad. Sci. USA*, volume 81, pages 3088–3092, 1984.

[17] J.J. Hopfield and D. W. Tank. Neural computation of decisions in optimization problems. *Biological Cybernetics*, 52:141–152, 1985.

[18] M. Jeruchim. A survey of interference problems and applications to geostationary satellite networks. In *Proceedings IEEE*, pages 317–331, 1977.

[19] D.S. Johnson, C.H. Papadimitriou, and M. Yannakakis. Optimzation by simulated annealing: an experimental evalution, part 1, graph partitioning. *Operat. Res.*, 37:865–892, 1989.

[20] D.S. Johnson, C.H. Papadimitriou, and M. Yannakakis. Optimzation by simulated annealing: an experimental evalution, part 2, graph partitioning. *Operat. Res.*, 39:378–406, 1991.

[21] D. Jungnickel. *Graphs, Netwrks and Algorithms.* Springer-Verlag, Berlin, Germany, 1999.

[22] B. Kamgar-Parsi and B. Kamgar-Parsi. Dynamical stability and parameter selection in neural optimization. In *Proceedings IEEE International Joint Conference on Neural Networks*, volume 4, pages 566–571, 1992.

[23] S. Kirkpatrick, C.D. Gelatt, and M. P. Vecchi. Optimization by simulated annealing. *Science*, 220:671–680, 1983.

[24] T. Kwok and K.A. Smith. A noisy self-organizing neural network with bifurcation dynamics for combinatorial optimization. *IEEE Trans. on Neural Networks*, 15:84–98, 2004.

[25] W.K. Lai and G.G. Coghill. Genetic breeding of control parameters for the hopfield/tank neural net. In *Proceedings IEEE International Joint Conference on Neural Networks*, volume 4, pages 618–632, 1992.

[26] O. Lazaro and D. Girma. A hopfield neural-network-based dynamic channel allocation with handoff channel reservation control. *IEEE Trans. on Vehicular Technology*, 49:1578–1687, 2000.

[27] R.S.T. Lee. A transient-chaotic autoassociative network (tcan) based on lee oscillators. *IEEE Trans. on Neural Networks*, 15:1228–1243, 2004.

[28] T. Mizuike and Y. Ito. Optimization of frequency assignment. *IEEE Trans. Communications*, 37:1031–1041, 1989.

[29] H. Nonaka and Y. Kobayashi. Sub-optimal solution screening in optimization by neural networks. In *Proceedings IEEE International Joint Conference on Neural Networks*, volume 4, pages 606–611, 1992.

[30] H. Nozawa. A neural network model as a globally coupled map and applications based on chaos. *Chaos*, 2(3):377–386, 1992.

[31] B. Pontano. Interference into angel-modulated systems carrying multichannel telephony signals. *IEEE Trans. Communications*, 21, 1973.

[32] C.R. Reeves. *Modern Heuristic Techniques for Combinatorial Problems.* Oxford, Blackwell, 1993.

[33] S. Salcedo-Sanz, C. Bouso no Calzón, and A.R. Figueiras-Vidal. A mixed neural-genetic algorithm for the broadcast scheduling problem. *IEEE Trans. on Wireless communications*, 2:277–283, 2003.

[34] S. Salcedo-Sanz, R. Santiago-Mozos, and C. Bouso no Calzón. A hybrid hopfield network-simulated annealing approach for frequency assignment in satellite communications systems. *IEEE Trans. Systems, Man, and Cybernetics-Part B: Cybernetics*, 34:1108–1116, 2004.

[35] H.X. Shi and L.P. Wang. Broadcast scheduling in wireless multihop networks using a neural-network-based hybrid algorithm. *Neural Networks*, 18:765C771, 2005.

[36] H. Tang, K.C. Tan, and Z. Yi. A columnar competitive model for solving combinatorial optimization problems. *IEEE Trans. on Neural Networks*, 15:1568–1573, 2004.

[37] I. Tokuda, K. Aihara, and T. Nagashima. Adaptive annealing for chaotic optimization. *Phys. Rev. E*, 58:5157–5160, 1998.

[38] A. Varma and Jayadeva. A novel digital neural network for the travelling salesman problem. In *Neural Information Processing, 2002. ICONIP '02*, volume 2, pages 1320–1324, 2002.

[39] G. Wang and N. Ansari. Optimal broadcast scheduling in packet radio networks using mean field annealing. *IEEE Journal on Selected Areas in Communications*, 15:250–260, 1997.

[40] L.P. Wang and F. Tian. Noisy chaotic neural networks for solving combinatorial optimization problems. In *Proc. International Joint Conference on Neural Networks*, volume 4, pages 37–40, 2000.

[41] L.P. Wang and K. Smith. On chaotic simulated annealing. *IEEE Transactions on Neural Networks*, 9:716–718, 1998.

[42] R.L. Wang, Z. Tang, and Q.P. Cao. A hopfield network learning method for bipartite subgraph problem. *IEEE Trans. on Neural Networks*, 15:1458–1465, 2004.

[43] G.V. Wilson and G.S. Pawley. On the stalibility of the travelling salesman problem algorithm of hopfield and tank. *Biol. Cybern.*, 58:63–70, 1988.

[44] David H. Wolpert and William G. Macready. No free lunch theorems for optimization. *IEEE Transactions on Evolutionary Computation*, 1:67C82, 1997.

[45] M. Yamaguti, editor. *Solution of the optimization problem using the neural network model as a globally coupled map*, 1994.

[46] M. Yamguti, editor. *Transient chaotic neural networks and chaotic simulated annealing*. Amsterdam: Elsevier Science Publishers, 1994.

[47] L. Zheng, K. Wang, and K. Tian. An approach to improve wang-smith chaotic simulated annealing. *International Journal of Neural Systems*, 12:363–368, 2002.

Printing: Krips bv, Meppel
Binding: Stürtz, Würzburg